SUSTAINABLE STELLENBOSCH

Opening Dialogues

Editors

Mark Swilling

Ben Sebitosi

Ruenda Loots

Sustainable Stellenbosch – Opening dialogues

Copyright © 2012 SUN MeDIA Stellenbosch and the Authors

All rights reserved.

No part of this book may be reproduced or transmitted in any form or by any electronic, photographic or mechanical means, including photocopying and recording on record, tape or laser disk, on microfilm, via the Internet, by e-mail, or by any other information storage and retrieval system, without prior written permission by the publisher.

Views expressed in this publication are those of the authors and do not necessarily reflect those of the publisher.

First edition 2012

ISBN 978-1-920338-55-8 (Print)
ISBN 978-1-920338-86-2 (PDF)

Set in 9/11 Amerigo BT

Cover image by Luke Metelerkamp, taken from a new section of development that was being rolled out in Kayamandi, 12 July 2011.
Section and section introduction images: Luke Metelerkamp
Cover design: Liezel Meintjes
Typesetting: Liezel Meintjes
Language editing: Tania Botha
Proofreading: Natalie Mayer and Davida van Zyl
Indexer: Michel Cozien

SUN PRESS is an imprint of SUN MeDIA Stellenbosch. Academic, professional and reference works are published under this imprint in print and electronic format. This publication may be ordered directly from www.sun-e-shop.co.za.

Printed and bound by SUN MeDIA Stellenbosch, Ryneveld Street, Stellenbosch, 7600.

www.africansunmedia.co.za
www.sun-e-shop.co.za

CONTENTS

ACKNOWLEDGEMENTS ix
FOREWORD – Conrad Sidego xi
PREFACE – Mark Swilling xiii
PREFACE – Russel Botman xv
CONTRIBUTORS xx

INTRODUCTION 1
Sustainable Stellenbosch – From potential to hope in practice 3
Mark Swilling & Ben Sebitosi

Towards an 'ethics of hope'? 12
Johan Hattingh

SECTION 1 – Spaces 21
Section Introduction 23
Chapter 1: **Spatial planning** – Planning a sustainable Stellenbosch 24
   Simon Nicks

Chapter 2: **Land** – Farmland and tenure security 48
   Juanita Pienaar

Chapter 3: **Urban spaces** – Quartering Stellenbosch's urban space 57
   Ronnie Donaldson & Jolanda Morkel

Chapter 4: **Housing** – The challenge of informal settlements 68
   Lauren Tavener-Smith

Chapter 5: **Local economic development** – Promoting a productive Stellenbosch .. 84
   Wolfgang H. Thomas

Response to local economic development – Local economic development – myth or mission? 94
Ann Heyns

SECTION 2 – FLOWS 99
Section Introduction 101
Chapter 6: **Food** – A sustainable system for Stellenbosch 102
   Candice Kelly, Jess Schulschenk, Anri Landman & Gareth Haysom

Chapter 7: **Waste** – The long walk to sustainable waste management 116
   Thys Serfontein & Cobus Kotzé

Chapter 8: **Water services** – Stellenbosch Municipality water services 126
   Eugene Cloete, Marelize Botes & Michéle de Kwaadsteniet

Chapter 9: **Sanitation** – Alternative solutions for sustainable waste water management .. 139
   Ben Sebitosi

Chapter 10: **Energy** – Towards sustainable energy flows for Stellenbosch 151
   Alan Brent, Riaan Meyer & Wikus van Niekerk

Chapter 11: **Transport** – Improving traffic flows in Stellenbosch 160
   Marion Sinclair, Christo Bester & Esbeth van Dyk

Response to Transport 173
Matthew Moody

SECTION 3 – Ecosystem services ... 177
Section Introduction ... 179
Chapter 12: **Soil health** – Sustaining Stellenbosch's roots ... 180
  Johann Lanz

Chapter 13: **Agriculture** – From vulnerability to viability ... 191
  Gareth Haysom & Luke Metelerkamp

Chapter 14: **Water** – Supply and quality ... 204
  Jo Barnes

Chapter 15: **Ecosystem services** – Protecting Stellenbosch's natural systems ... 215
  Blake Robinson

Response to Section on Ecosystem services ... 227
Karen J Esler

SECTION 4 – Social dynamics ... 231
Section Introduction ... 233
Chapter 16: **Heritage** – Contextualising heritage production ... 236
  Albert Grundlingh & Dora Scott

Chapter 17: **Shared space** – Power, heritage, play ... 244
  Marthie Kaden

Chapter 18: **Poverty and inequality** – Stocktaking of the social landscape of Stellenbosch ... 255
  Joachim Ewert

Chapter 19: **Social cohesion** – Pipe dream or possibility? ... 269
  Jerome Slamat, Thumakele Gosa & Chris Spies

Chapter 20: **Business** – Transformation towards sustainability ... 282
  Alan Brent, Pieter van Heyningen & Sumetee Pahwa-Gajjar

Chapter 21: **Health and wellness** – The burden of disease in Stellenbosch ... 291
  Bob Mash

Chapter 22: **Well-being** – Changing human behaviour ... 300
  Leslie Swartz & Kees van der Waal

Chapter 23: **Education** – Socially critical education for a sustainable Stellenbosch ... 310
  Lesley le Grange, Chris Reddy & Peter Beets

Chapter 24: **Sport** – Devising a game plan for a sustainable Stellenbosch ... 322
  Julian F. Smith

Chapter 25: **Children** – Imagining a Stellenbosch where children come first ... 333
  Eve Annecke, Naledi Mabeba, Magdelien Delport & Jess Schulschenk

CONCLUSION – Mark Swilling ... 345

INDEX ... 348
ACRONYMS ... 360

List of Figures

Figure 1	Hierarchy of settlement, nodes and linkages in the Stellenbosch area	5
Figure 1.1	A: Dwelling unit density in Kayamandi. B: Dwelling unit density in the Town Centre South	26
Figure 1.2	The growth of Stellenbosch town: 1710, 1770, 1817, 1905 (all drawings at same scale)	29
Figure 1.3	Stellenbosch town and environs, 1817	30
Figure 1.4	Urban growth of Stellenbosch	32
Figure 1.5	40 year lateral growth comparison: Bellville to Durbanville (A and B), Kuils River to Stellenbosch (C and D)	33
Figure 1.6	Grouping of small farms	35
Figure 1.7	Schematic zoning of a biosphere reserve	35
Figure 1.8	Target sites for urban restructuring and densification	41
Figure 1.9	Stellenbosch Town: Tender sites in 2006	42
Figure 3.1	Spatial growth of the built-up area between 1938 and 2009	58
Figure 3.2	Student population per suburb	59
Figure 3.3	Student numbers according to population group: 1990-2010	62
Figure 4.1	A: Grey water drainage. B: Informal connections in Enkanini. C: Incremental building in Enkanini. D: Urinal in Enkanini	71
Figure 4.2	Projected capital expenditure requirements on infrastructure for Stellenbosch, 2012-2022	79
Figure 4.3	Sources of finance for infrastructure in Stellenbosch, 2012-2022	79
Figure 4.4	A: Interior wall of the iShack, Enkanini. B: Solar-powered light-emitting diode (LED) light in the iShack, Enkanini	81
Figure 6.1	Comparative weekly food consumption in Stellenbosch and associated cost	105
Figure 6.2	Actual current consumption vs. nutritionally optimal consumption by food group for Stellenbosch	107
Figure 7.1	Stellenbosch landfill. A: Before compliance – 2009. B: After compliance – 2011	118
Figure 7.2	Future development plans for the Stellenbosch landfill site	123
Figure 8.1	Sources of microbial risk in a municipal water supply system	128
Figure 8.2a	The river systems and major dams in the Stellenbosch municipal area	132
Figure 8.2b	Critically endangered rivers in the Stellenbosch municipal area	132
Figure 10.2	A: 720 W stand-alone PV installation at Nollie se Kloof in the Ceres district. B: Solar roof tile installation at the Sustainability Institute in the Lynedoch EcoVillage outside Stellenbosch. C: Small 1-kW wind turbine and water filter at the Sustainability Institute	157
Figure 11.1	Age/gender comparison of fatal road accidents in the Western Cape, 2007-2009	164
Figure 11.2	Comparison of fatalities by road user type, 2007-2009	164
Figure 11.3	Stellenbosch collisions reported by month, 2009	165
Figure 12.1	Distribution of high-potential soils in the Stellenbosch area	183
Figure 13.1	Indirect links between the polycrisis and economic pressures on Swartland farms	197
Figure 14.1	The water cycle	205
Figure 15.1	Distribution of vegetation types across the Stellenbosch municipal area	216
Figure 15.2	Ecosystem status across the Stellenbosch municipal area	217
Figure 15.3	Reserves and protected areas within the Stellenbosch municipal area	218
Figure 15.4	The Cape Winelands Biosphere Reserve plan	222
Figure 17.1	Predators and Prey Sculpture Tour by Dylan Lewis	247
Figure 17.2	Cosy for a rhino project on Die Braak, Stellenbosch, 2009	251
Figure 18.1	Population pyramid for Stellenbosch, 2009	258
Figure 18.2	Income distribution in Stellenbosch, 2009	259
Figure 18.3	Inequality in Stellenbosch, 1996 and 2009	260
Figure 18.4	Percentage of population groups living in poverty in Stellenbosch, 1996 to 2009	260
Figure 20.1	Businesses are embedded sub-systems of larger social-ecological systems	288

Figure 21.1a	Age distribution of female deaths in the Winelands District	292
Figure 21.1b	Age distribution of male deaths in the Winelands District	293
Figure 21.2	Causes of post-neonatal infant deaths in the Cape Winelands and Overberg, 2004-2006	293
Figure 21.3	The wellness paradigm	298
Figure 23.1	The four dimensions of environment	311
Appendix 1	A: Estimated age distribution in the Stellenbosch population, 2010	339
	B: Estimated education levels in Stellenbosch Municipality by population group, 2006	339
	C: Estimated employment status in Stellenbosch by population group, 2010	340
	D: Estimated occupational composition in Stellenbosch by population group, 2006	341

List of Tables

Table 1	Changes in political power in Stellenbosch, 1996-2012	9
Table 1.1	Changes in political power in Stellenbosch, 1996-2012	37
Table 3.1	Population of Stellenbosch	57
Table 3.2	Housing needs analysis	63
Table 4.1	Demographic and housing trends in Stellenbosch	69
Table 6.1	Key components of the polycrisis	103
Table 7.1	Major roles and responsibilities of stakeholders in waste management	120
Table 7.2	Waste management issues identified by the Engineering Directorate and submitted to Council	120
Table 7.3	Suburbs identified to participate in Phase 1 of the Recycling at Source project	120
Table 8.1	Main water supply in Stellenbosch	127
Table 8.2	Blue Report Card of Stellenbosch Municipality	131
Table 8.3	The blocks for water tariff charges in Stellenbosch municipal area	134
Table 8.4	The replacement cost of the water infrastructure of Stellenbosch Municipality	134
Table 9.1	The state of WWTWs in Stellenbosch	141
Table 10.1	Bulk electricity purchases, sales and losses for the 2008/2009 financial year	152
Table 10.2	Energy-saving measures and the potential electricity savings they represent	153
Table 10.3	Industrial sectors and processes with the greatest potential for solar thermal uses	154
Table 10.4	Chilling processes using thermal energy	155
Table 11.1	Daily traffic flows around Stellenbosch	161
Table 11.2	Calculation of fuel consumption on major Stellenbosch roads	162
Table 12.1	Generalised summary description of the prominent soils of the Stellenbosch area	182
Table 12.2	Proportions of the different soil potential categories for the cultivation of wine grapes of soils in the Stellenbosch area	183
Table 13.1	Percentage of workers employed in the various economic sectors (2007) with author estimates for indirect contributions added	194
Table 13.2	Swartland case study farms	197
Table 13.3	New Swartland farm management practices	198
Table 13.4	Farms in the Stellenbosch region employing agro-ecological farming methods	200
Table 18.1	2010 matric results for high schools in the Stellenbosch municipal area	264
Table 20.1	Forces and factors that drive the increasing responsibilities of businesses	284
Table 21.1	Top ten causes of premature death in Stellenbosch	292
Table 21.2	Contribution of risk factors to South African disability adjusted life years (DALYs)	294
Table 22.1	The Mindspace checklist of influences on behaviour	302
Table 22.2	Basic building blocks for innovation	306
Table 23.1	National table of targets for school allocations, 2009-2011	314
Table 23.2	Approaches to environmental education	318
Appendix 1	Teacher-learner ratios in Stellenbosch schools	321
Table 25.1	Examples of potential 'barefoot engineering' courses	337

ACKNOWLEDGEMENTS

This work is based on the research supported by the National Research Foundation (NRF). Any opinion, finding and conclusion or recommendation expressed in this material is that of the author(s) and the NRF does not accept any liability in regard thereto.

The editors would like to thank all the authors for the valuable contributions made to this publication. Furthermore, none of this would have been possible without the support of officials employed by Stellenbosch Municipality who have worked with Stellenbosch University academics and researchers within the framework of the Rector-Mayor Forum since 2005.

The publisher and the editors would like to thank the peer reviewers for their valuable contribution to this publication.

Stellenbosch University

Stellenbosch University (SU) has aligned its core functions with five themes derived from national, continental and international development agendas. These are the eradication of poverty and related conditions, the promotion of human dignity and health, democracy and human rights, peace and security, as well as a sustainable environment and a competitive industry.

These themes, as well as the 20-plus academic initiatives in which they find tangible expression, make up the university's HOPE Project (www.thehopeproject.co.za). This is the vehicle through which SU practices its science-for-society approach.

One of the initiatives of SU's HOPE Project is the TsamaHub (www.tsama.org), which is an acronym derived from its full name, the Centre for Transdisciplinarity, Sustainability, Assessment, Modelling and Analysis. (A "tsama" is also a type of wild watermelon that has long sustained life in Southern Africa's driest parts.) The TsamaHub helps coordinate the extensive research programme that supports the joint work of the Rector-Mayor Forum, including the contributions and production of this book.

FOREWORD

CONRAD SIDEGO

This book reflects my commitment to the building of a new spirit of regeneration in Stellenbosch that will foster the following:

- innovation for sustainability;
- new investment and entrepreneurship;
- safer communities;
- dignified living for all; and
- an eco-friendly lifestyle.

A sustainable Stellenbosch, with special emphasis on the building of a green economy, is the long-term goal of Stellenbosch Municipality.

I recognise, however, that this cannot happen without innovations that make it possible to do a lot more with a lot less. The key is finding new technologies in the energy, waste, water and sanitation sectors, in particular, but also in how we manage traffic and mobility. Innovations, however, do not happen just because they are needed. World-wide experience shows that spaces for engagement, dialogue, exploration and creativity need to be opened up and fostered, because it is from these kinds of spaces that innovations tend to emerge. Innovations are usually the outcome of intense interactions between researchers, investors and practitioners who manage to build sufficient trust so that they can jointly tackle complex problems. Without trust and these spaces for innovation, we will not overcome the challenges faced by Stellenbosch.

Many of the chapters of this book are the products of interaction among researchers at Stellenbosch University and officials and councillors from Stellenbosch Municipality. Some were written by officials themselves (like the chapter on waste) and others by people working very closely with officials (like the chapters on housing, sanitation, energy, social development and transport).

What has emerged, is a shared body of knowledge that, for the first time, provides an integrated understanding of the challenges Stellenbosch faces and the possible future solutions at its disposal.

Mayor Conrad Sidego
9 November 2012

PREFACE

MARK SWILLING

It is mid-June 2012. Snow caps the distant mountain peaks as overnight temperatures drop to 5 °C. Most of the 220 000 or so people living in Stellenbosch wake up between 06h00 and 08h00; most wash, dress, eat and head off to work, school or university ... or yet another day's search for a job. As they do, the demand for electricity and water peaks, followed soon after by peak flows into the town's sewage treatment plants. By 07h45, most of the main intersections are logjammed with cars and hundreds of people are disembarking from the trains that arrive quite regularly at peak times. As the number of students walking to class starts to swell in the town centre, hundreds of workers brave the slippery, muddy slopes of the Enkanini informal settlement to walk to work. Meanwhile, most workers on the surrounding farms have already clocked in for work by 08h00, again mostly walking to work from their labourers' cottages on the farm; if not, they walk, take taxis or the train to get there.

Sylvia, Alan and Eric are just three of the people who wake up between 06h00 and 08h00. Sylvia lives in a shack in Enkanini, Alan in a large house in the Jonkershoek Valley, and Eric in a semi-detached clay house in the Lynedoch EcoVillage. Sylvia is unemployed, Alan is a professor at Stellenbosch University, and Eric is a farmer. They all have children of school-going age and all three have participated in one way or another in discussions about the future of their communities and Stellenbosch. When compared, their aspirations are remarkably similar: a decent life for all, a pleasing environment, safety and a promising future for their children.

Perched on a steep slope in Enkanini, Sylvia's shack is one of 16 000 in Stellenbosch. She is fully aware that she had no legal right to build a dwelling there, for Enkanini is not just an informal settlement; unlike most other informal settlements in South Africa, it is an illegal settlement. Her two-roomed shack is cold and draughty and has no services: no water supply, toilet, or energy connection. She collected water the previous afternoon from the communal tap – a 10-minute walk from her house – so she has enough water to drink and wash; as for the toilet, her two children and she wait for daylight before walking to the communal toilet. She prepares breakfast and sees her children off to school. Sylvia is currently unemployed, but is participating in a community project to improve services in Enkanini. Greeting others along the way, she carefully negotiates the slippery pathways to make the meeting at 09h00; all the while wondering whether she will have enough money to buy food for the evening meal.

Alan and his wife are awake by 06h00. Their double-storey house with its neat garden and pool sits comfortably in a quiet suburban lane towards the edge of the built area at the start of the Jonkershoek Valley. They feed and dress their three young children, then bundle them into the car and drop them off at nearby schools. The traffic is bearable and the schools are high quality. Alan starts lecturing at 08h00 and his wife, a psychologist, will see her first patient at 09h00. As they go through their morning routine, the family of five take their services for granted, switching on lights, the stove and heater, opening taps and flushing toilets – without sparing any thought for where it all comes from or where it all goes. Their rubbish will be collected later during the day. Alan and his wife work long hours, including most weekends, to pay for this lifestyle – they need two incomes to cover the bond repayments, property taxes and municipal service charges. Yet, they are thankful not to be in a big city; at least here they can take walks, feel safe and buy local food if they want to.

Eric is a founding member of the Lynedoch EcoVillage. He is what is often called a "small farmer" and benefits from being part of a land reform project, with a share of the cost of his house being subsidised. His wife is a nurse who works the night shift and they survive thanks to their combined incomes. Their semi-detached adobe brick house has all the services they need: drinking water, electricity, hot water from a solar geyser, and a flush toilet inside linked to a local treatment system that recycles the sewage. Their organic waste is collected daily by the Home Owners Association for composting; other separated waste is collected on a weekly basis. Eric's monthly levies and service charges are less than what he would pay for similar services in other parts of Stellenbosch. He is on his farm by 07h00. He has a bakkie and it takes him just five minutes to drive to the farm, where he has three workers and a span of six oxen to help him work the fields. He uses only organic farming methods and cannot keep up with the demand for his produce. It's not lucrative work, but he is passionate about farming in a way that restores the soil.

This collection of essays, written mainly by academics from Stellenbosch University, is about the place that Sylvia, Alan and Eric call "home". As they go through their daily life, with all its struggles, joys and sorrows, they are unwittingly participating in the constant making – and remaking – of this place. But how much do they know about Stellenbosch? Do they know how many people live here or how many live in poverty? Do they know anything about each other as people, as citizens, as parents? Do they know how the town has evolved and why some of its buildings have been preserved? Do they know how it functions as a town? Do they know where their water comes from or where their rubbish goes? These are the kinds of questions that need answering, and it should not require a great effort to find the answers.

As academics, we have a duty to engage with our local context. We need to make sense of how the place works. That is the purpose of this book. The contributors are from many different disciplines, which means that a wide range of approaches and styles will be evident across the chapters. Moreover, it is not a traditional academic book. It is neither structured around a single coherent argument, nor has it been possible to ensure that all contributors strike the same balance between empirical evidence, conceptual rigour and normative commitment. Instead, it should be seen and read as an unfolding conversation.

As the title suggests, the intention of the book is to contribute to opening dialogues about the future of Stellenbosch by applying our craft as academics. We write, and as we write we synthesise many different strands of thought derived from our research and reflections as we interrogate the past and present with particular hopes for the future in mind.

We therefore offer these writings as the opening dialogues in what we believe will be a much longer conversation between the university and the town about the kind of shared future that befits a place of such beauty and such potential.

Mark Swilling
20 June 2012

PREFACE

RUSSEL BOTMAN

I often speak of Stellenbosch and surrounds as our "first laboratory" at the university. What I mean by this is that the people and institutions of our area can be considered a primary source of data for our basic and applied research in the natural and social sciences.

The benefit of such an approach for us as academics is that our work is enriched with input from our immediate environment. At the same time, members of the community stand to benefit from relevant and socially responsive research, teaching and community interaction.

There has long been a town-and-gown tradition in many university towns the world over. But this relationship is not always one in which professors leave the Ivory Tower of academia and literally get their hands dirty at the local dumpsite – a veritable mountain in more ways than one – as is the case in Stellenbosch.

What exactly does the relationship between universities and the rest of society entail? In the late 19th and early 20th century, the Humboldtian notion of the university was that it should "advance knowledge by original and critical investigation, not just to transmit the legacy of the past or to teach skills" (Anderson, online). It could be argued that this laid the foundation for the university to play an important role in civil society.

More recently, Castells (1991) carved out a role for higher education in the current era in a paper titled *The University System: Engine of development in the new world economy* (Cloete, Bailey, Pillay, Bunting & Maassen 2011:5). This he (Castells 2009:3) later reiterated, arguing that "knowledge production" and "technological innovation" have become "the most important productive forces" in the current "global knowledge economy".

However, Divala (2008:199) worries that "globalisation and neo-liberalism push universities to a position where they are more relevant to global demands than local needs" – and Pieterse (2010:13) points out that this is especially true for the developing world and its universities.

At Stellenbosch University (SU) we try to address both these dimensions. For instance, our researchers have experienced that as you tackle the living conditions of shack dwellers on your doorstep (see the chapter on Housing in Section 1 of this book), you are likely to come up with solutions with universal applicability. In fact, the one helps you do the other better.

This links up with Gould's (2004:453) point that the "broadest and most vibrant context for the development of knowledge in higher education is its social mission to empower individuals and to serve the public good."

SU does this through its HOPE Project[1] – a reflection on Freire's[2] notion of a "pedagogy of hope". The HOPE Project is not a stand-alone or add-on initiative, but a broad approach intrinsic to all three core activities of the university – research, learning and teaching, as well as community interaction. In a policy document,[3] Stellenbosch University states that it "is committed to creating hope in and from Africa by means of excellent scholarly practice." This idea is expanded upon as follows (Stellenbosch University 2011:4):

---

1 See http://www.thehopeproject.co.za

2 Paulo Freire, *Pedagogy of the oppressed* (London: Penguin, 1996 [1970]); and Paulo Freire, *Pedagogy of hope: reliving Pedagogy of the oppressed* (London: Continuum, 2004 [1992]).

3 Stellenbosch University. 2011. Hope as guiding concept for Stellenbosch University. Unpublished policy document adopted by the Stellenbosch University Senate on 18 March 2011 and the Stellenbosch University Council on 3 May 2011.

> The University endeavours to create the conditions that will ignite the imagination of scientists to solve problems in creative ways through basic and applied research and through multi-, inter- and transdisciplinary academic activities.

This book, *Sustainable Stellenbosch – Opening Dialogues*, is a tangible expression of what we mean when we say the university is "committed to creating hope" in our own community. It shows exactly how Stellenbosch University is helping to shape a more equitable and sustainable future for our people by throwing the weight of science behind the developmental challenges of the present in our immediate environment as the legacy of the past drags on.

To facilitate this work, we strive to maintain a sound working relationship with local authorities in our region. Since 2007, the university has concluded cooperation agreements with the Stellenbosch, Drakenstein and Hessequa municipalities. And in Stellenbosch, officials of the university and municipality meet at least once a month in the Rector-Mayor Forum to discuss issues of mutual concern and to coordinate our efforts in the promotion of human development.

An exciting initiative that has been rolled out via the Rector-Mayor Forum is the Stellenbosch free Wi-Fi project, which started on a pilot basis in the central business area early in 2012. The idea is to provide free wireless internet access to anyone in Stellenbosch. No registration is needed, and users can surf the net and check their email without incurring any costs, though large downloads are not permitted. The pilot phase is being expanded to other areas, with the eventual aim of covering the whole of Stellenbosch. The motivation is that free access to the internet will pry open the doors of learning a bit wider for all the people of our town and surrounds.

Another issue that enjoys priority within the Rector-Mayor Forum is sustainability. This is in line with a view expressed by the Higher Education Sustainability Initiative,[4] expressed in the run-up to the Rio+20 midway through 2012, that universities should not only teach and research sustainable development, but also green their own campuses and support sustainability efforts in the communities in which they reside. Let me provide some examples of what we are doing in this regard.

SU has experienced unprecedented growth of late. Since the turn of the millennium, our student numbers have shot up from 20 000 to 28 000 – and we are foreseeing the need to take in another 2 000 students in the near future. This is putting enormous pressure on our facilities, which are already overflowing. Over a number of years, we have been developing strategies of making SU more sustainable. This has resulted in our Campus Master Plan for Facilities and Mobility, which can be regarded as two sides of the same coin.

Being situated in the heart of Stellenbosch, we consider meeting the challenge of developing sustainable infrastructure and pioneering green choices a priority.[5] We are developing new learning, living and work spaces for the 21st century, and adapting our existing facilities to meet current norms of energy efficiency, resource usage, recycling and waste management.

At the start of 2012, we launched several mobility innovations. This was motivated as much by necessity as by principle. Neither the university nor the town can handle the increase in motor vehicles in Stellenbosch. Our figures show that 82% of staff members and 43% of students use their own cars. If each of them have to get a dedicated parking space, we would have a shortage of 7 200 bays. We have neither the space nor the money to cater for this. In any case, the more motor vehicles, the larger our carbon footprint, which is the opposite of what we want.

So, we have introduced a shuttle service on campus, called MATIE BUS. Motorists who drive to our main campus are encouraged to park at dedicated sites on the outskirts, where they can board a shuttle to the centre. We have also launched the MATIE BIKE cycling project, making a total of 400 re-purposed commuter bikes from the Netherlands available to students and staff members. It is taking some time to restore a cycling culture to campus, but already more than half of these cycles have been rented.

---

4   The Initiative was formed by the United Nations Department of Economic and Social Affairs to feed into the UN Conference on Sustainable Development in Rio de Janeiro. See http://www.uncsd2012.org/index.php?page=view&type=1006&menu=153&nr=34

5   For more information, visit http://www.sun.ac.za/sustainability

One could say what is good for the town is good for the university, and *vice versa*. Another way of putting this is that the university has a social contract with the town and all of its people. Unlike the more conventional use of the term, which seeks to provide a legitimate basis for political authority, the university's pact with Stellenbosch entails a willingness to be of service to the community.

This can be traced back to an important policy statement back in the year 2000, the university's *Strategic Framework for the Turn of the Century and Beyond*,[6] which signalled a new direction. In it, "[t]he University acknowledges its contribution to the injustices of the past, and therefore commits itself to appropriate redress and development initiatives" (Stellenbosch University 2000:16). This is a reference to South Africa's apartheid history – which affected all universities and other institutions – accompanied by a statement of intent relating to steps required to correct what had gone wrong in the past.

This is further unpacked as follows: "In commitment to a readiness to serve, the university acknowledges the need for development and service in communities and areas previously and currently disadvantaged in the provision of services and infrastructure" (Stellenbosch University 2000:17).

As a university, the primary resources we have at our disposal for this task is knowledge. That is our currency – the knowledge that we have built up over the years and that we are continually re-evaluating and adding to as a normal part of our work as academics. By declaring our "commitment to a readiness to serve", we are saying that we are happy to share our knowledge with those who need it most.

As is apparent from the thoughtful and thorough contributions in this book, Stellenbosch faces many challenges. This presents our researchers with a golden opportunity to help redesign the town for a better future, in cooperation with the people of this area and the local authorities. And that this opportunity is being utilised by the university's researchers and the municipality's officials is also clear from this book.

The revitalisation of Stellenbosch will be particularly high on the agenda the next year or two. Cape Town has been designated World Design Capital (WDC)[7] for 2014 by the International Council of Societies of Industrial Design, and Stellenbosch is an official partner in this process. Awarded to cities based on their commitment to use design as an effective tool for social, cultural and economic development, the WDC has become a global movement and serves to acknowledge that design can, and does, impact the quality of human life.

The Cape Town bid acknowledged the design assets of the region, taking the WDC beyond the city's immediate borders. Stellenbosch is considered a hub of design innovation, sustainable thinking, cutting-edge research and academic excellence within the Cape Town Functional Region.

The WDC 2014 bid pledge of 'Live Design. Transform Life' is our opportunity to work towards the future that we want to create for ourselves in the greater Stellenbosch area. We need design that will help us overcome the divisions of the past and go into our common future together.

One of the enduring legacies of apartheid is the spatial organisation of our towns and cities. The Group Areas Act may be a thing of the past, but different "races" still largely live in geographically separate areas – with class distinctions and discrepancies in terms of the provision of services being very apparent. Stellenbosch is no exception, and also has a history of forced removals – particularly from an area in the centre of town called *Die Vlakte* (The Flats).

The University is committed to making a break with the settlement patterns of apartheid. In 2011, we opened Botmashoogte,[8] a new student residence in Idas Valley, a suburb designated a so-called coloured area in the apartheid era. Idas Valley is separated from the rest of Stellenbosch by a single road, yet many people have remarked that the two areas feel as if they are worlds apart. By acquiring a building that used to be an old hotel and later a block of flats before being converted to a student residence in Idas Valley, the university is in a tangible way establishing a link between itself and an underserved community literally on our doorstep.

---

6   http://www.sun.ac.za/university/stratplan/statengels.doc

7   For more information, visit http://www.capetown2014.co.za

8   See http://www0.sun.ac.za/ssg/index.php?option=com_content&view=article&id=78&Itemid=667&lang=en

An important initiative that is aimed at promoting unity in our town is the Stellenbosch Heritage Project.[9] Launched in July 2012 as a joint initiative between the municipality and the university, the project also involves Stellenbosch 360 (the rebranded Stellenbosch Tourism and Information Authority) and other stakeholders. The aim is to "unite the people of Stellenbosch" by creating a new "mutual culture as a way of life, which could be handed down to future generations."

Community interaction[10] helps us to give expression to our responsibility. It is one of the three core functions of Stellenbosch University, the other two being research, and learning and teaching. Community interaction emphasises the reciprocity between the university and its community. It provides the means whereby both parties can actively and in partnership discover knowledge and teach and learn from each other. Community interaction also contributes to an environment where student learning is enriched and research relevance is enhanced. It supports SU's institutional commitments to reciprocity, redress, development and transformation.

Part of our engagement with our community is our interaction with schools – through our Schools Partnership Programme[11] and various other initiatives. Essentially, the university provides support to three categories of role players in schools – teachers and school management teams, learners, as well as the Education Department and its officials. For teachers and managers we provide curriculum support, management training and the building of professional learning communities. For learners we run programmes that broaden access to higher education, provide mentoring, intervene strategically to improve matric results, build "pipelines" between schools and the university, and provide after-school support. For the Department and its officials we conduct research and provide training and professional development.

An initiative through which we are engaging the people of Stellenbosch and surrounds is Matie Community Service (MCS),[12] a community interaction, non-profit organisation of the university. It runs development programmes such as adult basic education and training, primary health care, entrepreneurial development, the Khanyisa Learning Project, and its One-Stop Service.

That the university is intrinsically part of the fabric of the town is also apparent from the annual Stellenbosch University *Woordfees* (Word Fest),[13] a week-long festival of the arts, featuring a variety of talks by writers and poets, musical productions and book events. Throughout the year, Woordfees also runs a language empowering community project called *Woorde Open Wêrelde* (Words Open Worlds, or WOW).[14]

The bottom line as far as the university's involvement in Stellenbosch is concerned is that we are doing it for future generations. This is what the concept of a "pedagogy of hope" is about. We are here to serve the next generation – the future leaders and business people and professionals getting an education at the university, but also the children of tomorrow who have to grow up in Stellenbosch and become global citizens here, ready to face the challenges of an ever-changing world.

We need to create a more equitable and sustainable town than the Stellenbosch we have at present. This book provides a comprehensive overview of important contributions by researchers of the university and officials of the municipality in this regard. But as the "Opening Dialogues" in the title suggests, it is only the beginning. It is my hope that all role players and stakeholders will take hands and work together towards building a better future for all the people of the greater Stellenbosch area.

H Russel Botman
Rector and Vice-Chancellor of Stellenbosch University
30 October 2012

---

9   http://blogs.sun.ac.za/news/files/2012/07/Verklaring-v-Voorneme-200dpi-001.jpg

10 For more information, visit http://admin.sun.ac.za/ci

11 http://admin.sun.ac.za/ci/schools/vision.html

12 http://www0.sun.ac.za/mgd

13 http://www.woordfees.co.za

14 http://www.woordfees.co.za/wow/2010-2/

REFERENCES

Anderson R. 2010. The 'Idea of a University' today. *History & Policy, Policy Paper, 98*. [Accessed 31 October 2012] http://www.historyandpolicy.org/papers/policy-paper-98.html#S4

Castells M. 1991. *The University System: Engine of development in the new world economy*. Washington DC: The World Bank.

Castells M. 2009. Transcript of a lecture on higher education delivered at the University of the Western Cape, 7 August. [Accessed 31 October 2012] http://www.chet.org.za/files/uploads/events/Seminars/Higher%20Education%20and%20Economic%20Development.pdf

Cloete N, Bailey T, Pillay P, Bunting I & Maassen P. 2011. *Universities and economic development in Africa*. Wynberg: Centre for Higher Education Transformation.

Divala JJK. 2008. Is a liberal conception of university autonomy relevant to higher education in Africa? Unpublished DPhil Ed dissertation. Stellenbosch: Stellenbosch University. [Accessed 31 October 2012] http://hdl.handle.net/10019.1/1168

Gould E. 2004. The university and purposeful ethics. *Higher Education in Europe*, 24(4):451-460.

Pieterse HM. 2010. Democratic citizenship education and the university in a cosmopolitan world. Unpublished MEd thesis. Stellenbosch: Stellenbosch University. [Accessed 31 October 2012] http://hdl.handle.net/10019.1/5434

Stellenbosch University. 2000. A Strategic Framework for the Turn of the Century and Beyond. Unpublished policy document. Stellenbosch: Stellenbosch University.

Stellenbosch University. 2011. Hope as guiding concept for Stellenbosch University. Unpublished policy document. Stellenbosch: Stellenbosch University.

Contributors

EDITORS

Prof **MARK SWILLING** is Programme Director of the Sustainable Development Programme at Stellenbosch University's School of Public Leadership, as well as Academic Director of the Sustainability Institute. He is responsible for the coordination of Master's and Doctoral programmes in sustainable development and has conducted extensive research on ways to make urban systems more sustainable.

Prof **BEN SEBITOSI** is Senior Lecturer at the Centre for Renewable and Sustainable Energy Studies of the Department of Mechanical and Mechatronic Engineering, Stellenbosch University. He teaches two postgraduate modules in renewable and solar energy, and acts as supervisor to Master's and Doctoral students from the Faculty of Engineering and the School of Public Leadership.

RUENDA LOOTS works as a project coordinator for the Sustainability Institute and has coordinated various publications and educational programmes. She is also a member of BiomimicrySA and runs workshops for tertiary institutions. Ruenda is currently completing her PhD in biochemistry at Stellenbosch University.

AUTHORS (IN ALPHABETICAL ORDER)

EVE ANNECKE is the Founding Director of the Sustainability Institute (SI) and co-founder of the Lynedoch EcoVillage. She leads the SI's child-centred focus in building sustainable communities. Her research and teaching at Master's level integrates sustainability, complexity, leadership and environmental ethics.

Dr **JO BARNES** is an epidemiologist and Senior Lecturer at the Division of Community Health of the Faculty of Health Sciences at Stellenbosch University's Tygerberg campus. Her particular research interests are water pollution and waterborne diseases, sanitation and housing, especially in impoverished areas.

PETER BEETS is an Associate Professor in the Department of Curriculum Studies at the Faculty of Education, Stellenbosch University, and Programme Chair of the Postgraduate Certificate in Education. He teaches and conducts research in the fields of geography education, curriculum and assessment, and received Stellenbosch University's Rector's Award for Excellence in Teaching in 2010.

Prof **CHRISTO BESTER** is Head of the Transportation Division of the Department of Civil Engineering, Stellenbosch University. He is responsible for the coordination of the Master's and Doctoral programmes in Transportation Engineering and conducts research on road safety, traffic engineering and intelligent transportation systems.

Dr **MARELIZE BOTES** is co-author of fourteen peer-reviewed scientific articles, co-editor of a book, and co-holder of two patents. She is currently a postdoctoral researcher at the Department of Microbiology, Stellenbosch University, where she is conducting research on the application of nanotechnology in water filtration and the control of biofilm formation in water systems.

ALAN BRENT is Professor of Sustainable Development at the School of Public Leadership of the Faculty of Economic and Management Sciences, and Associate Director of the Centre for Renewable and Sustainable Energy Studies of the Engineering Faculty, Stellenbosch University. His research centres mainly on management and policy issues pertaining to the sustainability of technologies and innovation.

Prof **EUGENE CLOETE** is Vice-Rector (Research and Innovation) and Chairman of the Water Institute Advisory Board of Stellenbosch University. His research focus is on water supply and sanitation, with an emphasis on innovative solutions for sustainable water management.

Dr **MICHÉLE DE KWAADSTENIET** obtained a PhD degree in microbiology from Stellenbosch University in 2009 and is currently a postdoctoral student in the Department of Microbiology at Stellenbosch University. She is the co-author of twelve peer-reviewed scientific articles, co-editor of a book, and co-holder of a patent.

MAGDELIEN DELPORT, a graduate of the MPhil Sustainable Development programme, is currently working in the field of leadership development and organisational change. Her involvement with the Sustainability Institute includes research about the relevance of early childhood development in building sustainable communities and action research highlighting the plight of farm worker women in the Cape Winelands.

RONNIE DONALDSON is an Associate Professor in the Department of Geography and Environmental Studies, Stellenbosch University, where he teaches Urban Studies in the second, third and Honours year programmes. His research interests include urban and tourism development.

Prof **KAREN J ESLER** is Deputy Chairperson of the Department of Conservation Ecology, Stellenbosch University. Her research focus is on understanding how drivers of change (environmental change, overexploitation, habitat fragmentation and alien invasion) influence plant population and vegetation community structure and processes. This knowledge, embedded in social-ecological systems, contributes to best practice management, restoration and conservation actions.

Prof **JOACHIM EWERT** taught Sociology of Development, Economic Sociology and Sociology of Work at Stellenbosch University until 2010. Since the mid-1990s, most of his research has centred on the transformation of the South African wine industry, including comparative studies of the wine sectors in the South of France, Greece and Spain.

THUMAKELE GOSA is currently completing an MPhil in Sustainable Development Planning and Management at Stellenbosch University. He is a board member of IMBADU M-Afrika Development Consortium, an organisation that seeks to promote African cultural heritage and stimulate economic development.

Prof **ALBERT GRUNDLINGH** is Chairperson of the History Department at Stellenbosch University. He is first and foremost a social historian and has a keen interest in heritage construction. He has published several books and numerous academic articles.

JOHAN HATTINGH is Professor of Philosophy and former Vice-Dean of Social Sciences at Stellenbosch University. His graduate and postgraduate teaching and research focus particularly on value disputes and the interface of theory and practice in the field of applied, business, professional, leadership, development, environmental and climate change ethics.

GARETH HAYSOM initiated the Sustainability Institute's Sustainable Agriculture Programme in 2008 and is a research fellow of the programme, as well as extraordinary lecturer at Stellenbosch University's School of Public Leadership. He is currently completing a PhD at the African Centre for Cities, University of Cape Town, where his research centres on food system challenges.

ANN HEYNS is Development Manager of Stellenbosch 360 (NPO), previously known as the Stellenbosch Tourism and Information Association. She founded the Ghoema Route (R44) in 2007 and is responsible for the development of Route 360, a culture tourism route that will include all communities in the greater Stellenbosch municipal region.

MARTHIE KADEN is Senior Lecturer in the Department of Visual Arts, Stellenbosch University. She is responsible for the coordination of the Visual Communication Design study stream and currently conducts PhD research on art and design as a form of social practice. Her research interests include memory, spatiality and the expanded field of drawing.

CANDICE KELLY is a research associate at the Sustainability Institute and extraordinary lecturer at the School of Public Leadership at Stellenbosch University. She coordinates five sustainable agriculture modules that form part of a BPhil in Sustainable Development and supervises MPhil students, and is currently doing her PhD on social learning for sustainable food systems.

COBUS KOTZÉ is the venture leader for the Bill & Melinda Gates Foundation funded BioCycle Project, developing commercially viable and scalable routes for the bio-conversion of human waste into valuable products, using black soldier fly larvae. These larvae can grow on a number of waste streams and be processed into chitin, biodiesel and protein meal for animal feeds.

ANRI LANDMAN holds an MPhil in Sustainable Development Management and Planning from Stellenbosch University. She has worked as a freelance researcher for various sustainable food initiatives in Limpopo and Gauteng, and was also a project coordinator of The Siyakhana Initiative for Ecological Health and Food Security at the University of the Witwatersrand, Johannesburg. She is currently enrolled on a joint PhD at the University of Antwerp, Belgium, and Stellenbosch University.

JOHANN LANZ runs a soil science consulting firm that services both the environmental and agricultural (predominantly wine) industries. Consulting projects involve soil resource evaluations and mapping for agricultural land use planning and management, as well as specialist study inputs into environmental assessments. Past research includes a project pertaining to the concept of soil health.

Prof **LESLEY LE GRANGE** is Vice-Dean (Research) of the Faculty of Education at Stellenbosch University. He teaches and researches in the fields of environmental education, science education, curriculum studies, and research methodology. He has received several research and teaching awards and has authored, co-authored and contributed to more than a hundred and fifty publications.

NALEDI MABEBA chairs all children's projects at the Sustainability Institute and is Programme Coordinator of the Early Childhood Development programme at the Learning for Sustainability FET College. The programme assists teachers working in preschools in disadvantaged communities, enabling them to work in a non-violent space and simultaneously connecting them with nature. She coordinates all activities at the Lynedoch Crèche and Baby Centre.

Prof **BOB MASH** is Head of Family Medicine and Primary Care at Stellenbosch University. He is responsible for training under- and postgraduate students and conducts research on non-communicable chronic diseases, family medicine and district health systems. He also leads the sustainable development initiative at the Faculty of Medicine and Health Sciences.

LUKE METELERKAMP holds an MPhil in Sustainable Development Planning and Management from Stellenbosch University. He is currently a research fellow of the Sustainability Institute and also works as a freelance photographer based in Cape Town.

RIAAN MEYER holds a BEng in Electrical Engineering and an MScEng in Mechanical Engineering. He has been working as a research engineer at the Centre for Renewable and Sustainable Energy Studies of Stellenbosch University since 2006. His main professional interests are renewable energy resource determinations.

MATTHEW MOODY is currently working as a sustainable transport planning assistant for the City of Cape Town. He holds degrees in Political Science, Development Studies and Public Policy, and also completed an MPhil in Sustainable Development Planning and Management from Stellenbosch University in 2012. His thesis focused on sustainable transport, with specific reference to Stellenbosch.

JOLANDA MORKEL is Senior Lecturer in the Department of Architectural Technology at the Cape Peninsula University of Technology (CPUT). Her research interests include computer-supported and online architectural design education, work-integrated learning, sustainable school design, and change management in sensitive heritage environments. She also coordinates a community service programme for sustainable multi-grade rural schools in partnership with the Centre for Multi-Grade Education at CPUT.

SIMON NICKS has qualifications in commerce, architecture, planning and urban design. His recent and current work includes housing and *in situ* upgrading projects in the Western and Eastern Cape, as well as environmental master plans for the Driftsands Nature Reserve, the False Bay Ecology Park, and the Moddergat (Macassar), Kuils and Hout Bay rivers. He has also produced spatial development frameworks in various provinces. He has won awards for planning, urban design and environmental conservation, and an environmental research scholarship tenable at the University of Cape Town is named after him.

SUMETEE PAHWA-GAJJAR is a PhD candidate in the postgraduate Sustainable Development programme at Stellenbosch University. In her thesis, she applies resilience thinking to unravel corporate notions of sustainability. She also conducts research about national policy, city-level initiatives and community projects aimed at a lower carbon future.

Prof **JUANITA PIENAAR** lectures Customary Law and Property Law in the Department of Private Law at Stellenbosch University and has a National Research Foundation B3-rating for the period 2012-2017. She has a particular interest in land reform and is involved with the Land Claims Court on a part-time basis, as well as the Law Faculty's Institutional Strategic Project focussed on Poverty, Eviction and Homelessness.

Prof **CHRIS REDDY** is Chairperson of the Department of Curriculum Studies in the Faculty of Education at Stellenbosch University and programme leader of the Environmental Education Programme University of Stellenbosch. He teaches and researches in the fields of environmental education, science education and teacher education.

BLAKE ROBINSON is a sustainability consultant specialising in sustainable cities and urban resource flows and is involved in a number of local and international sustainability projects as strategist, writer and project manager. He holds a Master's degree in Sustainable Development Planning and Management from Stellenbosch University.

JESS SCHULSCHENK has an MPhil in Sustainable Development Management and Planning from Stellenbosch University. Her thesis focussed on the benefits and limitations of local food economies in terms of promoting sustainability. She has been involved in various sustainability initiatives in both the private and public sector in the Stellenbosch region for the past few years and is currently a senior researcher at Ernst & Young, where her work centres on corporate governance and sustainability.

DORA SCOTT completed an MPhil in Cultural Tourism and Heritage at Stellenbosch University and is the archivist of the Antarctic Legacy Project, a social science project of the Centre of Excellence for Invasion Biology at Stellenbosch University. She is responsible for collecting and archiving historical material pertaining to South Africa's involvement in the Antarctic and Sub-Antarctic regions.

THYS SERFONTEIN was manager of Solid Waste and Area Cleaning in the Engineering Directorate of Stellenbosch Municipality from 2009 to 2012, where his responsibilities included the safe collection, transport and disposal of solid waste in the WCO24 area. During his term as manager, Stellenbosch Municipality resolved the non-compliance issues of the area's landfill site and implemented a long-term Integrated Waste Management solution for Stellenbosch, which resulted in the town achieving third place in the Department of Environmental Affairs and Tourism's 2010 Cleanest Town Competition.

Dr **MARION SINCLAIR** is Coordinator of Road Safety Research in the Department of Civil Engineering at Stellenbosch University. Her work involves developing and undertaking multidisciplinary research initiatives about traffic collisions and traffic injuries. Dr Sinclair's area of specialism is road user behaviour.

Dr **JEROME SLAMAT** is Senior Director: Community Interaction at Stellenbosch University. The division is responsible for policy development; the management and development of community interaction within the university; the management of partnerships on different levels; the promotion of an integrated scholarship of engagement through different forms of experimental learning and community-based research; and the preservation, expansion and development of local cultural histories.

Prof **JULIAN F SMITH** is Vice-Rector (Community Interaction and Personnel) of Stellenbosch University and has been involved in higher education management at an executive level since 1991. He is currently responsible for strategic partnerships at a local, regional, provincial and national level, as well as managing sport as a strategic asset.

CHRIS SPIES is an independent conflict transformation practitioner who works mainly in the field of socio-political conflict. He specialises in the design and facilitation of courses in dialogue and mediation for intergovernmental and international organisations such as the African Union, European Union, United Nations and Folke Bernadotte Academy in Sweden.

LESLIE SWARTZ is a clinical psychologist and Professor of Psychology at Stellenbosch University, with a special interest in the contribution of psychology to development and disability issues. Recent publications include *Disability and international development* and *Able-Bodied*, a memoir dealing with disability issues in Southern Africa.

LAUREN TAVENER-SMITH is currently pursuing a PhD degree with an emphasis on urban infrastructure, particularly alternative sanitation for informal settlements. She is the project leader of a transdisciplinary research group that focuses on informing transitions to liveable and sustainable informal settlements through incremental upgrading.

WOLFGANG H. THOMAS is Professor extraordinaire at the University of Stellenbosch Business School (USB) and coordinator of the Bottom of the Pyramid Learning Lab at the USB. His teaching, research and consulting work centre around local, regional and national economic development, with a strong African emphasis.

KEES VAN DER WAAL is Professor of Social Anthropology at Stellenbosch University. His research interests lie in development, culture and identity, and the use of ethnographic methods for social research. While his past ethnographic fieldwork has focused on rural areas in Limpopo Province, he currently also works in the Cape Winelands.

Dr **ESBETH VAN DYK** is a principal researcher and supply chain analyst in the Logistics and Quantitative Methods Competence Area of the Council for Scientific and Industrial Research. Her main fields of expertise are agricultural supply chains and agro-logistics, and freight transport and logistics operations modelling.

PIETER VAN HEYNINGEN is currently completing his PhD at the School of Public Leadership, Stellenbosch University. His academic focus is on shifting economic systems toward sustainability by means of sustainability-oriented innovation systems.

WIKUS VAN NIEKERK is Professor in the Department of Mechanical and Mechatronic Engineering and Director of the Centre for Renewable and Sustainable Energy Studies of the Faculty of Engineering, Stellenbosch University. He undertakes research on a wide range of renewable energy technology-oriented issues.

Introduction

Introduction

SUSTAINABLE STELLENBOSCH
From potential to hope in practice

MARK SWILLING & BEN SEBITOSI

Introduction

When talking about Stellenbosch, Mayor Conrad Sidego may often be heard repeating, with great passion, the question: "With all that we have in Stellenbosch, if we can't get it right here, where can we get it right?"

What Mayor Sidego is referring to is the well-known fact that there is a remarkable and unique concentration of capabilities, resources and opportunities in Stellenbosch, which, if mobilised by a resolute overarching vision, has the potential to translate into an extraordinary example of hope in practice. Should this be achieved, it would not only attract attention and support from within the South African context, where alternatives to racially and class-divided cities are so desperately needed, but also from within the international arena, with the world becoming increasingly hungry for alternatives as it lurches ever faster into deeper and seemingly intractable economic and ecological crises.

This way of thinking is also what, in part, inspires the HOPE Project[1] initiated by the Rector and Vice-Chancellor of Stellenbosch University (SU), Prof Russel Botman, who argues that research should bring hope for Africa and the world.

This book combines Mayor Sidego's conviction about the potential of Stellenbosch and Prof Botman's call for the kind of knowledge that makes hope a practicable possibility.

The mix of capabilities, resources and opportunities mentioned above include the following:

- a growing and increasingly diversified economy embedded within a wide set of ecosystems that not only sustain it, but are also stunningly beautiful;
- a large, well-resourced research and teaching university that brings together a wide range of expertise and knowledge generation capabilities;
- a cluster of wealthy individuals and profitable businesses for whom Stellenbosch represents more than just a convenient location, and who view it instead as a place where they would like to live, work and support a more sustainable future;
- a network of environmental scientists, government officials, farmers and NGOs working with our natural systems and who have developed the kind of expertise required to inform alternative ways of planning and executing development projects;
- increasingly active and robust social, business, cultural and religious initiatives that connect people across race and class barriers;
- active and well-organised communities, from wealthy ratepayers to impoverished shack dwellers;
- a rural-urban (or "rurban") setting that is not (yet) overwhelmed by large numbers of people whose needs dwarf the size of the local economy;

---

1 Visit http://www.thehopeproject.co.za for more information.

- a long tradition of well-managed municipal infrastructures (water and sanitation, solid waste, energy, mobility), which are, nevertheless, under severe stress at present;
- a pool of knowledge and skills that enables knowledge-intensive businesses and organisations to locate in Stellenbosch rather than a larger metropolitan city; and
- committed and capable leadership in local government, business, civil society, education and the faith sector with the will to work together.

As will become clear from this book, none of the above features are static and the relations between them may be undermined or reversed very quickly. Indeed, as will be shown, poor municipal governance over many years has resulted in infrastructural problems that are already hindering the economy, polluting the environment and reinforcing poverty. Similarly, businesses who disengage from their local contexts effectively exacerbate these problems.

This book has emerged from the so-called Rector-Mayor Forum, which was established in 2005 to facilitate monthly meetings between Stellenbosch University and Stellenbosch Municipality to discuss the challenges facing the town and find ways of combining forces to solve them. By providing these organisations with a platform to engage each other on problems ranging from noisy student houses to macro challenges such as energy shortages, the Forum has gradually fostered a sense of trust and developed a *modus operandi* that connects policy making with knowledge generation. The initiative has also resulted in various spin-offs, including the emergence of the Social Cohesion Movement, a broad community organisation driven by a coalition of well-connected civil society leaders; an initiative for the incremental upgrading of informal settlements (such as Enkanini); a Stellenbosch food security strategy; various collaborations around the 2010 FIFA World Cup; youth development and joint crime control; and, as of 2011, collaboration in respect of infrastructure and spatial planning.

Although the Rector-Mayor Forum led to improved cooperation regarding matters of practical concern, what was lacking was a comprehensive, shared understanding of the challenges faced by Stellenbosch. Researchers and local government officials tend to know their own sectors well, while citizens and businesses have access to complex planning and policy documents relating specifically to the constitutional scope of local government powers and functions. What was needed was a broader picture of Stellenbosch as a place, a community, an economy and an ecosystem – and this book represents an attempt to provide just such a picture.

Researchers from across the various SU faculties (as well as others who had worked with Stellenbosch researchers in the past) were approached to contribute chapters capturing their particular understanding of the problems and trends, the future implications of a business-as-usual approach, and how different ways of thinking about these issues might inform more constructive, equitable and sustainable outcomes. The letter of invitation to the authors invoked the words of Albert Hirschman, one of the most creative economists of the twentieth century, to capture the spirit of this collaborative research endeavour, i.e. to "... set the stage for conceptions of change to which the inventiveness of history and a 'passion for the possible' are admitted as vital actors" (Hirschman 1971:37).

Stellenbosch may mean a number of different things to different people. For most, it refers to the historical town of Stellenbosch, where most of Stellenbosch University's buildings and facilities are located. However, since 1994, it has come to refer to a deracialised and amalgamated town that includes the historically black township of Kayamandi and historically coloured nodes of Cloetesville, Idas Valley and Jamestown. In fact, as of 2000, the municipal area of Stellenbosch (i.e. the area managed by Stellenbosch Municipality), includes a large region with various localised urban nodes (some of which were previously autonomous towns), namely: Stellenbosch town, Lynedoch, Vlottenburg, Koelenhof, Klapmuts, Raithby, Jamestown, Dwars River Valley, Franschhoek, La Motte, Groot Drakenstein and Wemmershoek.

Although there was a workshop in November 2010 for the various chapter contributors (Stellenbosch Municipality was also invited), it must be admitted that this book is more *inter*disciplinary than *trans*disciplinary in nature. Interdisciplinary research is about integrating inputs from different disciplines to generate policy-relevant research and scientific results *for* practitioners. Transdisciplinary research happens when researchers from different disciplines *collaborate with* practitioners from government, business and civil society *to co-produce* knowledge that is not only scientifically robust, but also socially robust and impactful. Due to time constraints, it was impossible to bring busy academics and practitioners together at numerous workshops to

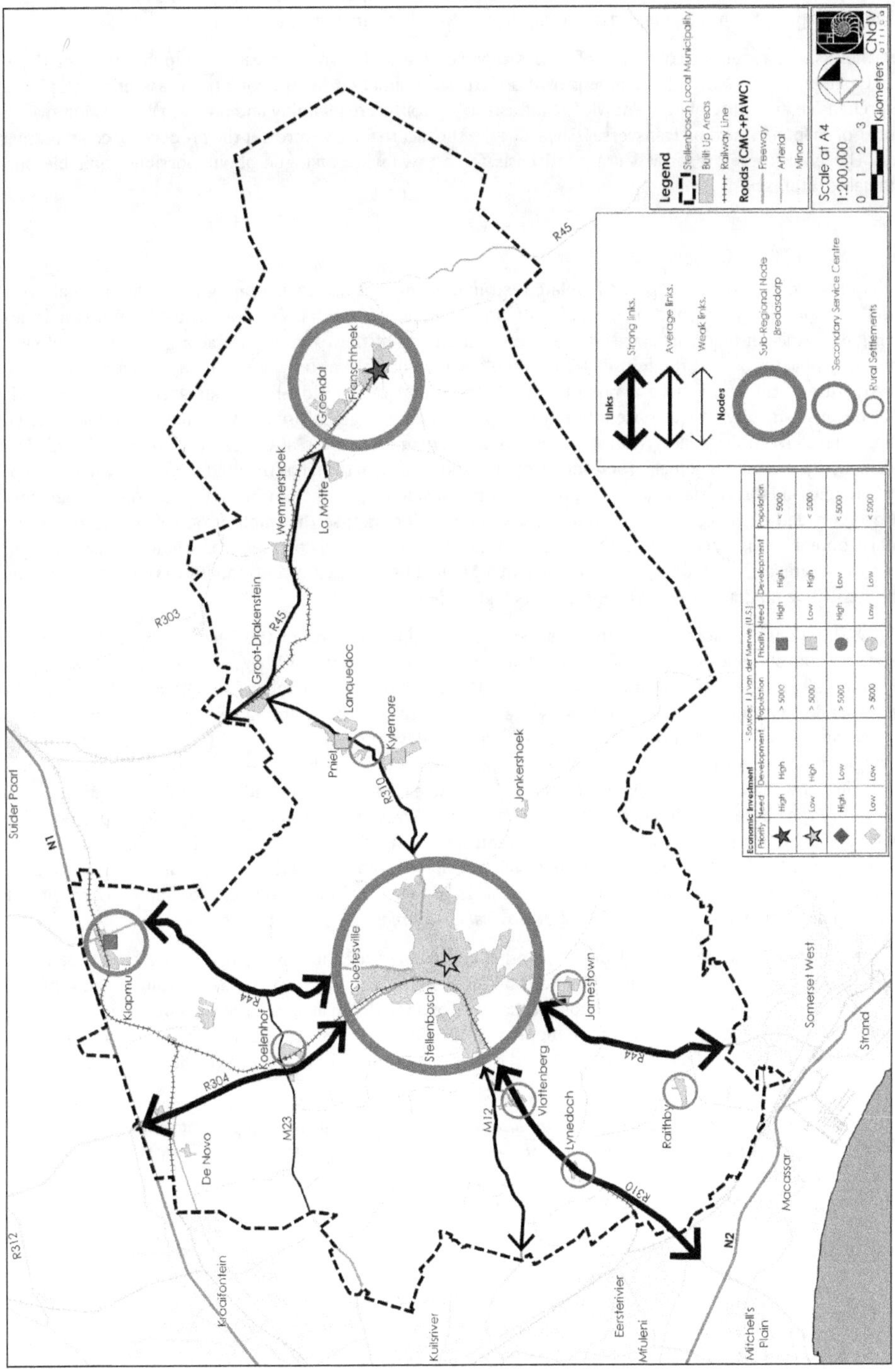

Figure 1   Hierarchy of settlement, nodes and linkages in the Stellenbosch area [CNdV Africa 2009]

tackle a shared set of challenges in a transdisciplinary way. Instead, the various contributors have written their individual chapters from the perspective of their respective disciplinary frameworks.

Nevertheless, the resulting collection of essays represents a major step forward. The publication of these diverse perspectives in a single volume provides a crucial foundation for the long-term task of developing a comprehensive, transdisciplinary knowledge infrastructure for future planning and policy making.[2] Moreover, a number of chapters touch on transverse issues, despite the fact that they were not the products of collaborative work. The book therefore provides a good basis for a town-wide dialogue about possible equitable and sustainable futures.

The global context

Accelerating global crises and (partially related) structural changes are transforming the locational patterns, spatial relationships and political dynamics of urban centres around the world, and Stellenbosch is no exception. A walk through the university campus and streets of the town reveals at a glance the profound impact of globalisation and the information technology revolution. Not only does the high-quality fibre optic infrastructure of the town make it possible for SU to be part of the global, 24/7 knowledge community; it has also transformed what was once a sleepy wine industry into a globally-traded business linked to massive foreign direct investments in agricultural production, tourism and elite real estate development. All this has fundamentally altered the demand for – and price of – land, which, in turn, has resulted in a prolonged period of farm evictions and forced many thousands of the poorest residents into the mushrooming informal settlements and overcrowded backyards of the various urban nodes. Superimpose the implications of the most serious global economic crisis since 1929 and the effects of the global ecological crisis (which includes climate change) onto this picture, and it becomes clear that, whether Stellenbosch likes it or not, thanks to these global forces its future will be fundamentally different from its highly contested past.

In 2009, the global economy contracted for the first time since the end of the Second World War. For this reason, economists tend to view 2009 as marking the end of the post-World War II long-term development cycle (Gore 2010). The first half of this cycle focused on expanding democratisation, liberation from colonial rule, expanded production, and inclusivity via a plethora of welfare states inspired by the economics of John Maynard Keynes. The second half of this long-term development cycle, which started in response to the stagflation crisis of the 1970s, was an era dominated by the information revolution, globalisation, the dismantling of welfare states in favour of free markets, increasing inequality and accelerated urbanisation. In fact, South Africa became a democracy at a time when neo-classical economics was enjoying unprecedented global hegemony. But just over a decade later, all this had changed. During the post-2009 search for the new conceptual coordinates of the next long-term development cycle, not only is Keynesian economics back in play, but we also have in play the intellectual underpinnings of Green Economy policies – the policy outcomes of research generated by the new discipline of *ecological economics*.

While it may take a decade to find the conceptual coordinates for the next long-term development cycle, there is little doubt that they will be influenced by the need to respond to a polycrisis brought about by the convergence of multiple economic and ecological crises (for an in-depth analysis, see Swilling & Annecke 2011). In the introductory chapter of its 626-page report entitled *Towards a Green Economy: Pathways to Sustainable Development and Poverty Eradication*, the United Nations Environment Programme (UNEP) notes:

> *The causes of these crises vary, but at a fundamental level they all share a common feature: the gross misallocation of capital. During the last two decades, much capital was poured into property, fossil fuels and structured financial assets with embedded derivatives. However, relatively little in comparison was invested in renewable energy, energy efficiency, public transportation, sustainable agriculture, ecosystem and biodiversity protection, and land and water conservation.*
>
> [UNEP 2011:14]

---

2   In fact, some transdisciplinary work has already commenced following the National Research Foundation's three-year Community Engagement grant to the TsamaHub. The grant is being used to fund joint research with groups of shack dwellers in Enkanini.

Seven globally significant mainstream documents will, in one way or another, shape the way our generation sees the world which we need to change. They are as follows:

- **Ecosystem degradation**: the United Nations (UN) *Millennium Ecosystem Assessment,* compiled by 1 360 scientists from 95 countries and released in 2005 (with virtually no impact beyond the environmental sciences), has confirmed for the first time that 60% of the ecosystems upon which human systems depend for survival are degraded (UN 2005).
- **Global warming**: the widely accepted reports of the Intergovernmental Panel on Climate Change (IPCC) confirm that global warming is taking place due to release into the atmosphere of greenhouse gases caused by, among other things, the burning of fossil fuels; and that warns that, should average temperatures increase by 2°C or more, this would lead to major ecological and socio-economic changes – most of them for the worse, with the world's poor experiencing the most destructive consequences (IPCC 2007).
- **Oil peak**: the *2008 World Energy Outlook* published by the International Energy Agency (IEA) declared the 'end of cheap oil' (IEA 2008). Although there is still some dispute over whether we have hit peak oil production or not, the fact remains that mainstream perspectives now broadly agree with the once-vilified 'peak oil' view.[3] Even the major oil companies now agree that oil prices are going to rise and that alternatives to oil must be found sooner rather than later. Oil accounts for over 60% of the global economy's energy needs. Our cities and global economy depend on cheap oil, and changing this requires a fundamental rethink of the assumptions underpinning nearly a century of urban planning dogma.
- **Inequality**: according to the UN *Human Development Report* for 1998, 20% of the global population who live in the richest countries account for 86% of total private consumption expenditure, whereas the poorest 20% account for 1.3% (United Nations Development Programme (UNDP) 1998). Only the most callous still ignore the significance of inequality as a driver of many threats to social cohesion and a decent quality of life for all.
- **Urban majority**: according to generally accepted UN reports, the majority (i.e. just over 50%) of the world's population was living in urban areas by 2007 (UN 2006). According to the UN habitat report entitled *The Challenge of Slums*, one billion of the six billion people who live on the planet live in slums; or, put differently, one-third of the world's total urban population (rising to over 75% in the least developed countries) live in slums (United Nations Centre for Human Settlements 2003).
- **Food insecurity**: the International Assessment of Agricultural Science and Technology for Development (IAASTD) is the most thorough global assessment of the state of agricultural science and practice ever conducted. According to this report, modern, industrial, chemical-intensive agriculture has caused significant ecological degradation, which, in turn, will threaten food security in a world where access to food is already highly unequal and demand is fast outstripping supply. Significantly, the report confirmed that "23 per cent of all used land is degraded to some degree" (Watson, Wakhungu & Herren 2008:73).
- **Material flows**: according to a 2011 report by the UNEP International Resource Panel,[4] by 2005 the global economy depended on 60 billion tonnes of primary resources (biomass, fossil fuels, metals, and industrial and construction minerals) and 500 exajoules of energy, an increase in material consumption of 36% since 1980 (Fischer-Kowalski & Swilling 2011).

The above trends combine to conjure up a picture of a highly unequal, urbanised world, dependent on rapidly degrading ecosystem services, with looming threats triggered by climate change, high oil prices, food insecurities and resource depletion. This is what the mainstream literature on unsustainable development is worried about. It marks what is now increasingly referred to as the *Anthropocene* – the era in which humans have become the primary force of historico-geophysical evolution (Crutzen 2002).

Significantly, although they are in the policy domain, these seven documents reflect the outcomes of many years of much deeper research on global change by scientists and researchers working across disciplines and diverse contexts on all continents (Steffen, Sanderson, Tyson et al. 2004). Although this process of scientific inquiry leading to policy change is most dramatic with respect to climate science (Weart 2008), it is also true for the life sciences that fed into the outcomes expressed in the *Millenium Ecosystem Assessment* (MEA) and

---

3   See http://www.peakoil.net

4   See http://www.unep.org/resourcepanel

the resource economics that has slowly established the significance of rising oil prices and, most recently, of the rise of material flow analysis. Our increased ability to "see the planet" has given rise to what Clark and colleagues appropriately called the "second Copernican revolution", which they date to the meeting in 2001 where delegates from over 100 countries signed the Amsterdam Declaration, establishing the Earth System Science Partnership (ESSP) (Clark, Crutzen & Schellnhuber 2005:6-7).[5]

The logical outcome of this profound paradigm shift is an increasingly sophisticated appreciation of what Rockström has called our "planetary boundaries", which define the "safe operating space for humanity" (Rockström 2009:472). The significance of Rockström's article is that it managed to integrate, for the first time, the quantifications of these "planetary boundaries" which had previously been established by various mono-disciplines. Without the 'second Copernican revolution', a new science appropriate to a more sustainable world and associated ethics would be unviable.

South Africa has been fundamentally affected by the global polycrisis.[6] The economic crisis brought an end to the debt-funded consumer boom that drove growth after 1994. The price paid for South Africa's inclusion into the BRIC club[7] is an increased reliance on the export of primary materials, in particular coal, metals and minerals. However, many researchers are asking questions about whether long-term reserves are such that the building of a fossil fuel-based and water-intensive mining and energy industry is geophysically possible and economically viable (Hartnady 2010). Others blame South Africa's 'mineral-energy complex' for the low rate of employment growth, which, in turn, partially explains the high unemployment rate (with poor education being the other key driver) (Mohamed 2010).

In partial recognition of these problems, the South African government adopted, during the period 2008-2011, a wide range of policies aimed at the 'greening' of the South African economy, including the signing of a Green Economy Partnership agreement between government, business and labour in late 2011. These policies included the approval by Cabinet of the Long-Term Mitigation Strategy (LTMS) in 2008 and the National Strategy for Sustainable Development in 2011; adoption of the Industrial Policy Action Plan (Version 2), which made explicit reference to the need to shift away from the current resource- and energy-intensive economic structure, in 2010; and the publication of the White Paper on Climate Change in 2011, which made explicit the need for a transition to a low-carbon economy. Many other related policies in the fields of biodiversity, air pollution, solid waste, water conservation, transport and housing may be referred to. However, all these shifts in favour of a more sustainable future are challenged by a set of powerful vested interests that are locked into resource- and energy-intensive production and consumption systems. The most significant player in this regard is Eskom and its view of the world, as expressed in the Integrated Resource Plan (IRP), which was approved by Cabinet in 2011 and constitutes South Africa's long-term energy strategy. The IRP provides for expenditure of over R1 trillion on new energy-generation capacity, including the building of the third and fourth largest coal-fired power stations in the world. The dollar-denominated loans to pay for this added capacity effectively crowd out investments in alternatives of scale, and thus reduce opportunities for new innovation-driven value chains that are potentially more job-creating than mature technologies.

Stellenbosch in context

Any consideration about the future of Stellenbosch should be shaped by the global and national context outlined above. More so than any other local context in Africa, Stellenbosch could well become the gravitational centre of the knowledge revolution that is required if South Africa is to transcend its dependence on mature resource- and energy-intensive industries, which have proven to have limited capacities for absorbing large numbers of workers. Stellenbosch University clearly has a key role to play, as demonstrated in the HOPE Project. But to become this gravitational centre, the entire town of Stellenbosch will need to become a living expression of a just transition to a future sustainable world. Many of the chapters in this book directly address this challenge, in particular with respect to resource flows such as energy, water, sewage, solid waste and mobility. But the

---

5 The first, of course, goes back to the publication of *De Revolutionibus Orbium Coelestium* by Copernicus in 1530.

6 See Swilling & Annecke 2011, Chapter 8.

7 This refers to the international alliance between Brazil, Russian, India and China. Subsequent to South Africa's inclusion in the alliance, the name changed to BRICS.

wider social-ecological context – biodiversity, public health, urban planning, social development, and the art we produce to express our cultural transformation – is equally relevant. All these aspects are discussed in great detail in the chapters of this book.

However, there is one key condition for future success: stable and democratic governance. Sadly, to date, Stellenbosch has been plagued by highly unstable municipal governance (see Table 1).

Table 1   Changes in political power in Stellenbosch, 1996-2012

Date	Party in power	Event
1996	ANC	Elections
2000	DA	Elections
2002	ANC/NNP	Floor crossing
2006	DA Coalition	Elections
2008	ANC Coalition	By-election
2009	DA Coalition	Motion of no confidence
2011 (ongoing in 2012)	DA	Elections

Although stability seems to have returned after the last municipal elections in early 2011, this record of instability has pushed out capable officials, gutted the institutional memory and capacity of the municipality, compromised long-term strategic planning, undermined town-wide collaborations, incentivised corruption and exacerbated poverty. This, in turn, has freed up space for unscrupulous developers, whose security villages have created a patchwork of disconnected privatised elite enclaves of urban consumption that have contributed nothing to the building of an integrated and sustainable urban culture nurtured by public and street spaces that allow for the celebration of diversity and creativity, or urban markets that reinforce the production and sale of locally produced goods and services. However, since the 2011 elections, political stability has made it possible to rebuild the municipality's institutional capacity through the appointment of a new generation of talented officials. Ideally, the main political parties should agree to maintain this trend regardless of the outcome of future elections.

The chapters that follow should be read against the background of the 'second Copernican revolution' and the tragic consequences of the gross misallocation of capital that caused the current global crisis. The emerging picture is both extremely alarming and surprisingly hopeful. As a collective perspective on Stellenbosch past, present and future by Stellenbosch University's research community, these essays provide vital insights for a wider, all-inclusive public debate about where Stellenbosch is heading, and what kind of world we would like to see emerge globally and in practice in the beautiful valleys so delicately held by the Stellenbosch and Simonsig mountains. The complex trajectories of this story are divided into four sections: spaces, flows, ecosystems and social dynamics. Each section starts with a short overview of each chapter, followed by a brief set of recommendations. These section introductions are highly recommended for readers with little time to read the actual chapters.

Reflections on hope

The introduction by Prof Johan Hattingh of the Philosophy Department of Stellenbosch University provides a thoughtful meditation on the discourse of hope that underpins the university's HOPE Project, which inspired the production of this book. Prof Hattingh invites us to reflect critically on the consequences of grand commitments to creating potentially more sustainable and inclusive futures. Commitments such as those expressed in the HOPE Project and strategic documents of Stellenbosch Municipality have consequences that are rarely captured in the sanitised prose of marketing messages, which are designed to appeal to the broadest possible cross-section of society.

Hattingh explores the possibility of an 'ethics of hope' with reference to three influential texts: Paulo Freire's *Pedagogy of Hope,* Ernst Bloch's *Principle of Hope,* and Jurgen Moltmann's *Theology of Hope.* Moltmann's theology of hope, which speaks to a Christian community, is an invitation to resist injustice and to re-imagine a historical future that establishes "pre-figurations of a life in which humans, society and creation alike flourish according to the divine promise of a better life in the Kingdom of God". Ernst Bloch wrote about hope during the period of 1938 to 1947, when the number of democracies in the world had dwindled to eleven and right- and left-wing authoritarianism was on the rise. Like Moltmann, Bloch believed that emancipation from the injustices of capitalism was inspired by our innate utopian longings for human and social fulfilment. Hattingh quotes Kellner to capture the essence of this perspective:

> *The present moment is thus constituted in part by latency and tendency: the unrealized potentialities that are latent in the present, and the signs and foreshadowings that indicate the tendency of the direction and movement of the present into the future.*

Bloch's notion of "unrealized emancipatory potential" as the key to these "foreshadowings" of the future is clearly relevant for this book; but it is Freire who makes the link to education, and in particular the education of the poor. He argued that emancipation of the poor from capitalist exploitation and ideologies of oppression will only come about when the relationship between teacher and learner is fundamentally transformed. Hattingh argues that, for Freire, "hope is not merely seen as a virtuous characteristic of individuals, or a utopian dream inherent to us all, but rather as a radical act of freedom, i.e. a practice of critical social understanding in the service of transforming society". This will not be achieved as long as the role of the teacher is to present the facts; but rather when teacher and learner jointly explore and discover the *'why'* of facts.

Hattingh concludes his discussion of Freire, Bloch and Moltmann by arguing that –

> *... an ethics of hope is indeed possible as a practice of radicalising the present and the past in order to build a better future. At the same time, in terms of content, an ethics of hope may be an extremely dangerous practice, in that every dream and every desire that we strive to realise in the name of a better future must also be subjected to rigorous ethical critique. On the other hand, an ethics of hope emboldens us in the face of every ideology that distorts our dreams of a humane existence and that may deceive us into believing that they have been, or may be, completely fulfilled within a given socio-political order. Accordingly, working with, or working towards, an ethics of hope may be potentially dangerous to those who strive to practice it effectively in this world.*

Concluding remarks

This book undertakes an extensive audit of nearly the entire spatial, economic, ecological and sociological space of Stellenbosch. It takes a bold step towards – in Hirschman's words as quoted in the letter of invitation to authors – "setting the stage for conceptions of change to which the inventiveness of history and a 'passion for the possible' are admitted as vital actors".

The historical context of this discussion is limited, going back only to 1679, when a well-known former Governor of the Cape, Simon van der Stel, imposed his name on what was then a wild, unconquered space. The very existence of Stellenbosch some 300-odd years later as both a thriving community *and* one faced with seemingly insurmountable challenges is testimony to its deeply contested social and cultural roots.

Contemporary conflicts arising from the need for an economically efficient, yet socially responsive and environmentally sustainable development strategy means that traditional models of societal management and outdated technologies are no longer tenable. As a general template, author after author has documented what Stellenbosch *was*, *is* and (directly or by implication) *will be* in a business-as-usual scenario. However, as the well-known American computer scientist Alan Kay put it, "the best way to predict the future is to invent it". The authors have therefore proceeded to propose alternative futures that may arise if we change our way of thinking about what is desirable, viable and ultimately sustainable, given the global and local challenges we face.

Anatole France, the nineteenth-century French poet-cum-journalist, wrote that "all changes, even the most longed for, have their melancholy; for what we leave behind us is a part of ourselves; we must die to one life before we can enter another". We should not underestimate the implications of the paradigm shift that is required. All modes of implementation – unilateral, collaborative and regulatory – will have to be deployed to ensure that a transition takes place.

The authors of this volume will have to live with the consequences of what they envisage and hope to see achieved. Thus, the solutions put forward do not seek to create a revolution, but rather a purposive transition within severe time constraints. Using empirical as well as analytical methods, the chapters go beyond generic principles and put forward what are in many cases radically pragmatic – rather than extreme – solutions. Consequently, no single *panacea* has been suggested, but rather ways of seeing and thinking about alternative futures, which should, ideally, become the basis for town-wide dialogues about a potentially shared and, for the first time, inclusive future.

REFERENCES

Clark WC, Crutzen PJ & Schellnhuber HJ. 2005. *Science for Global Sustainability: Toward a New Paradigm*. John F. Kennedy School of Government Working Paper Series rwp05-032. Cambridge, MA: Harvard University: John F. Kennedy School of Government.

CNdV Africa. 2009. *Stellenbosch Municipal Spatial Development Framework Draft Strategies Report*. Stellenbosch: Stellenbosch Municipality.

Crutzen PJ. 2002. The Anthropocene: Geology and Mankind. *Nature*, 415:23.

Fischer-Kowalski M & Swilling M. 2011. *Decoupling Natural Resource Use and Environmental Impacts from Economic Growth*. Report prepared for the International Resource Panel. Paris: United Nations Environment Programme.

Gore C. 2010. Global Recession of 2009 from a Long-Term Development Perspective. *Journal of International Development*, 22:714-738.

Hartnady C. 2010. South Africa's Diminishing Coal Reserves. *South African Journal of Science*, 106(9/10):1-5.

Hirschman AO. 1971. *A Bias for Hope: Essays on Development and Latin America*. New Haven: Yale University Press.

IEA (International Energy Agency). 2008. *World Energy Outlook*. Paris: International Energy Agency.

IPCC (Intergovernmental Panel on Climate Change). 2007. *Climate Change 2007*. Geneva: United Nations Environment Programme.

Mohamed S. 2010. The State of the South African Economy. In: J Daniel, P Naidoo, D Pillay & R Southall (eds). *New South African Review 1 – Development or Decline?* Johannesburg: Wits University Press.

Rockström J. 2009. A Safe Operating Space for Humanity. *Nature*, 461(24):472-475.

Steffen WA, Sanderson A, Tyson PD *et al*. 2004. *Global Change and the Earth System: A Planet Under Pressure*. Berlin: Springer.

Swilling M & Annecke E. 2011. *Just Transitions: Explorations of Sustainability in an Unfair World*. Cape Town and Tokyo: United Nations University Press & UCT Press.

UN (United Nations). 2005. *Millennium Ecosystem Assessment*. New York: United Nations. [Accessed 12 August 2006] http://www.maweb.org/en/index.aspx

UN. 2006. *State of the World's Cities 2006/7*. London: Earthscan & UN Habitat.

UNDP (United Nations Development Programme). 1998. *Human Development Report 1998*. New York: United Nations Development Programme. [Accessed 22 October 2006] http://www.hdr.undp.org/reports/global/1998/en/

UNEP (United Nations Environment Programme). 2011. *Towards a Green Economy: Pathways to Sustainable Development and Poverty Eradication*. Nairobi: United Nations Environment Programme. [Accessed 1 November 2011] http://www.unep.org/greeneconomy

United Nations Centre for Human Settlements. 2003. *The Challenge of Slums: Global Report on Human Settlements*. London: Earthscan.

Watson RT, Wakhungu J & Herren HR. 2008. *International Assessment of Agricultural Science and Technology for Development (IAASTD)*. [Accessed 21 April 2008] http://www.agassessment.org

Weart SR. 2008. *The Discovery of Global Warming*. Cambridge, MA: Harvard University Press.

Introduction

TOWARDS AN 'ETHICS OF HOPE'?

JOHAN HATTINGH

Introduction

From examples such as the HOPE Project of Stellenbosch University, it is clear that hope may be easily conceptualised as a project – as something that may be constructed, created, built and provided – or that an institution or a person may be seen as a carrier of hope. But what exactly does it mean to build, provide, or be a carrier of hope, i.e. to turn hope into action? Is it something that we may assume responsibility for? Could it be our duty to build or provide hope? If yes, may we then be blamed and held accountable if we do not, or fail to, build or provide hope?

Viewed from a different angle, such a concept of hope may be highly problematic, since those who build, provide, or carry hope are, at first glance, clearly not the same group of people as those who are in 'need of' or could 'take up' hope. Furthermore, the notion of 'providing' hope seems to presuppose a power differential between those who have the agency to provide hope and those without agency, who are provided with hope. Thus, hope may be cynically viewed as the imposition of a dominant position, or even of a lifestyle, on others who do not have the power to say no to it.

So, besides the obvious questions of how one would go about defining a 'need' for hope or ascertaining success in 'taking up' hope once it has been built or has been provided (while trying to avoid doing violence to others in the process), the conceptualisation of hope as a project confronts us with a much deeper question about the diversity of notions of hope, as well as the human tendency to resist or even oppose some of them, while accepting, embracing and promoting others.

Christians, for instance, have traditionally seen faith, hope and love as core theological virtues, received as gifts from God, and which should characterise every aspect of their lives, but the content that many of them give to these virtues may not be acceptable to other Christians, let alone to many non-Christians. The same applies to the notion of hope as mobility, as found in commercial society today and which, in some of its forms, as Deidre McCloskey (2006:x) once quipped, "makes bureaucrats uneasy". Similarly, what feminists or gays living in relative ease in a city may hope for is not the same as what a labourer on a wine farm, or an elderly couple in a rural shanty town, may hope for.

Accordingly, the question arises whether it is at all possible to develop an ethics of hope; and, if so, how it is possible to conceptualise and practice it without falling into the traps suggested by some of the questions evoked above. Is it at all possible to find a conceptualisation of hope on the basis of which an ethics may be built; and, if so, how may that conceptualisation of hope help us to define duties and responsibilities; to differentiate between good and bad behaviour; and steer us to be good persons in a good society? In short: what would an ethics of hope look like – if not in detail, then in broad outline?

To venture a provisional answer to these questions, I would like to draw on Paulo Freire's *Pedagogy of Hope: Reliving Pedagogy of the Oppressed* (1994), Ernst Bloch's *The Principle of Hope* (1986), and Jürgen Moltmann's *Theology of Hope: On the Ground and Implications of a Christian Eschatology* (1965; 1993a).

While none of these influential writers actually developed a dedicated ethics of hope in its own right, all three of them articulated positions that point in the direction of an ethics of hope. Moltmann, for instance, tried to develop an ethics of hope based on his theology, but (as he conceded in 2009) he gave up on the task, both because it was too overwhelming and because he thought that he did not possess the fundamentals to think it through properly. It took someone else – in the person of Timothy Harvie – to extrapolate a theologically-based ethics of hope from Moltmann's eschatological theology in 2009, years after the publication of Moltmann's seminal works.

In this essay, I would argue that perhaps the task is not one of developing an ethics of hope as a comprehensive thought system – the scope and horizon of which would be so vast as to include every possible theme and issue, which would effectively render the undertaking impracticable. Instead, I would like to explore, in much more limited and modest terms, what the implications of placing hope within an ethical discourse might be, and what would be required to translate it into ethical practice. In order to do so, I will present a small conceptual 'portrait gallery' of hope and discuss how these different approaches may help us to engage each other in a practical way, so that we may make our everyday living spaces and places better and more sustainable.

In presenting this portrait gallery of images of hope, I will follow a thematic rather than a chronological order, starting with Moltmann's *Theology of Hope*, which constitutes one of the strongest Christian positions on hope. Next, I will move to Ernst Bloch's philosophical exposition as an example of a secular conceptualisation of the principle of hope. And lastly, I will touch on the Paulo Freire's *Pedagogy of Hope* in order to discuss some of the practical implications of hope in the educational sphere. Against this background, I will articulate a few thoughts on a radical conception of hope, and how this may contribute towards a practice of hope. Writing from Stellenbosch, this portrait gallery would, however, be incomplete if I were to start without a brief sketch of Stellenbosch University's HOPE Project.

The HOPE Project

One of the best examples of the notion of hope as an implementable project or activity may be found in the HOPE Project of the Stellenbosch University, which was launched in 2009 around five themes selected from the Millennium Development Goals (MDGs) with the aim of focussing Stellenbosch University's mission and vision and creating "synergy between higher education and development and economic growth in a more comprehensive way" (The HOPE Project 2010). The five MDG themes entail:

- the eradication of poverty and related conditions;
- the promotion of human dignity and health;
- the promotion of democracy and human rights;
- the promotion of peace and security; and
- the promotion of a sustainable environment and competitive industry.

Around these strategic goals, the university has formulated a large number of innovative projects, through which its teaching, research and community interaction mobilises scientific knowledge to build hope in Africa. Cutting across the boundaries of disciplines and faculties, many of these initiatives acknowledge the transdisciplinary nature of the challenges that they face and that new, cooperative, approaches to problem formulation, knowledge generation and utilisation will have to be developed for science to become relevant for society by turning hope into action.

While I am convinced that most, if not all, of the projects in the HOPE Project are in one way or another promoting hope, I am not so sure if an overview such as this assists one in grasping exactly what an ethics of hope is, or could be. Some of the things that we need to understand better are how the HOPE Project may (and actually does) transform research, teaching, and community interaction practices into concrete practical terms; and what the actual differences are that start to emerge in these spheres when a 'pedagogy of hope', as Freire would have referred to it, enters into the picture.

Another aspect of the same challenge would be to better understand not only the trends and forces in mainstream university pedagogy that could be seen as stumbling blocks in the way of a pedagogy of hope, but also to ask critical questions about the content that is given to such a pedagogy. Stellenbosch University, for instance, has chosen to use the Millennium Development Goals as a starting point to give content to its HOPE Project. But, is this content the only content that may be used to do so; or are there other goals that could also be pursued to create hope? And, if so, why and how do we choose one set of goals in one place for a hope project, and perhaps another set in another place or at another time? Clearly, it is not possible to answer questions like these in this chapter, but I believe that pointers to a better understanding of the possibility and character of an ethics of hope will emerge if we briefly compare other portraits of hope available to us.

Hope in the theology of Jürgen Moltmann

Moltmann (born in 1926) does not see hope as a quiet, internal expectation of an unseen future good associated with God standing outside of history and this world, bringing a kind of inner peace to the saved that makes no difference to the real world – as hope is often seen, in conjunction with the two other theological virtues of faith and love (see 1 Corinthians 13:13 (Holy Bible)), in traditional (pietist) interpretations of Christian doctrines. Instead, he sees hope as a life-giving, liberating and world-transforming force born from "the eschalogical perspective of the divine promise in Christ for the Kingdom of God" (Moltmann 2009:86).

Formulated in plain language, Moltmann's view entails a future-oriented life informed by the promise of the 'Kingdom of God'. As Moltmann sees it, faith – or the gift of accepting this promise – transforms a person spiritually in such a way that a new identity is formed, in turn creating "an interval of tension" – a "between-space" or "*Zwischenraum*" (Harvie 2009:147) – from which current modernity and social structures may be assessed, resisted and transformed. For Moltmann, hope fills a person with a "spirit for life", a "passion for what is possible", and the energy to work for a "life-giving future"; hope creates a space for liberation from the "creeping acclimatization" to injustice, oppression and man-made catastrophes with a view to attaining freedom, peace and social justice; and hope puts a process of transformation in motion through which human lives and the world are sanctified. The process of "sanctification" consists in resisting and overcoming evil by engaging in acts and structures that affirm life (Moltmann 2009:87-98).

While couched in theological language that not all may find convincing, the radical meaning of these ideas about hope may be traced back to what Moltmann once said about life, war and violence:

> *But anyone who really says 'yes' to life, says 'no' to war. Anyone who really loves life says 'no' to poverty. So the people who truly affirm and love life take up the struggle against violence and injustice. They refuse to get used to it. They do not conform. They resist.*
>
> [Moltmann 1992:xii; 2009:89,91]

This same radical meaning of hope is evident elsewhere in Moltmann's writing as well:

> *Faith, wherever it develops into hope, causes not rest, but unrest; not patience, but impatience. It does not calm the unquiet heart, but is itself this unquiet heart in man.*
>
> [Moltmann 1993a:21; 2009:95]

Building on these theological doctrines, Harvie further develops the idea of a Christian ethics of hope around three central tasks it should discharge: creating a time for hope; creating a space for hope; and translating the vision of a better world into concrete action in the present, first and foremost within the context of economic activity (Harvie 2009:147-208).

Elaborating on the theme of creating a time for hope, Harvie points to the task of reinterpreting the present as the time of promise, the essence of which is futurity, whereby we erect signs ("*vorzeichnen*" or "advance gifts") of the promised future (Harvie 2009:151). Harvie argues that accepting the promise of a better future alters the existential condition of human beings and orientates them towards the future in at least two ways: firstly, by placing and assessing the present in the light of the better future that may be attained; and secondly, preparing for the better future to arrive (as it were) in the present by engaging in creative activities. An ethics of hope thus entails a "qualitative, ethical description of time, rather than a quantitative, linear description of time" (*ibid.*:153).

Formulated differently: a Christian ethics of hope entails the creation of a different, ethical notion of time; it requires the creation of a new history, not narrowly conceptualised in terms of nation state or ideology, but widely in terms of God's Kingdom; it requires the creation of a timeframe ("*Zeitraum*" [*ibid.*:153]) that stands in a relationship of tension with the quantitative, linear conception of time, so that qualitatively different expectations of the future may take shape and be pursued in the present. In the 'time of hope', the future may no longer be imagined as a mere extrapolation of the past; in such a 'time of hope' we make history to enable the realisation of a better future in the present (Harvie 2009:161-162).

Turning to the task of creating a space for hope, Harvie invokes the term "*Zwischenraum*" (between-space) (2009:161), which Moltmann used in earlier articulations of his theology of hope. This was for Moltmann the primary metaphor for the interval of tension created by the divine promise. Harvie explains that what Moltmann points to is the task of "making room" (*ibid.*:163) – or creating a context in which we may move beyond the lives lived by autonomous and isolated individuals, acknowledging that human life and action is unthinkable outside of relationships with other people, things and nature. At the same time, this relational space is seen as a moral space characterised by an awareness that each human decision – and its outcome – affects the whole network of social relations; therefore, the deciding agent is likewise affected by every decision made by others.

Teasing out the implications of these assumptions for an ethics of hope, Harvie (2009) firstly affirms the insight that decision making and action never take place within a vacuum, but always within a network of social and ecological relationships; and secondly, that every decision and action may either promote or undermine the humane life promised to us in the narrative of the Kingdom of God. Thus, Harvie points out, according to Moltmann, human decisions and actions not only affect social relations and the quality of life that they allow us to attain, but also the whole of creation. By acknowledging the fundamental contribution of creation to humane living, Moltmann conceptualises the intersection of social and moral space as ecological space.

Moltmann's "space for hope", which Harvie alludes to, is thus a living space that stands in contrast to the reduction of natural and social environments to their geometrical or quantitative characteristics, and resists their further reduction to utilitarian values (Moltmann 1993b:147; Harvie 2009:164). Instead, hope requires the task of establishing pre-figurations of a life in which humans, society and creation alike flourish according to the divine promise of a better life in the Kingdom of God.

It is not necessary, for the purposes of this chapter, to examine in detail the "third action dimension" of Harvie's sketch of an ethics of hope (2009:169ff), except to mention that it entails a fundamental critique of the dominant mode of economic decision making and action by which the majority of the world's population lives, focusing on guidelines to transform our modes of consumption and trade with a view to restoring human worth and dignity in the world. The important point to note here is that hope, as sketched above, creates the possibility of economic transformation by creating an 'interval of tension' that exposes the manner in which dominant economic activities and decision making place obstacles in the way of achieving a future life of human dignity, worth and the flourishing of creation in the Kingdom of God.

Attractive and radical as this notion of hope is in conceptual terms, it is not clear whether such a Christian eschatology may offer grounds for hope for the masses of people the world over who adhere to a different religion or pursue a form of spirituality that is not religious in the conventional sense of the word. To them, the above may sound like ideas that only speak to the already converted, going around in a big circle that is difficult, if not impossible, to penetrate from the outside. It may therefore be worthwhile to also look at another portrait of hope, painted from a secular point of view and in circumstances where hope for humanity was virtually non-existent.

Ernst Bloch's 'principle of hope'

Ernst Bloch (1885-1977) wrote the three volumes of *The Principle of Hope* from 1938 to 1947, while he was in exile in the USA, seeking safety from German National Socialism. At a time when the world was locked into a state of war against the Nazi regime and feelings of despair and hopelessness reigned supreme in the minds of many people, Bloch wrote his *magnum opus* based on the belief that hope is embedded in human consciousness itself; that we may find traces of hope in the mundane expressions of ordinary life, ranging from daydreams, fairytales and myths, the circus and the fair, the popular novel and the fashion accessory to advertising, literature, theatre, art, music, architecture, political, social, medical, technological and geographical utopias, and also philosophy and religion (see Geoghehan 1996; Kellner n.d.).

Bloch argued that humans, as individuals, were unfinished (Bloch 1986:195); that they are "animated by dreams of a better life and by utopian longings for fulfilment" (Kellner n.d.). He also reasoned that the world is not completely solid, closed and fixed: "Instead of these, there are simply processes" (Bloch 1986:196). In terms remarkably reminiscent of Moltmann, Bloch uses the concept of hope, or the "utopian dimension" of our existence, as a basis to criticise and emancipate us from capitalism, and, at the same time, to build a better future society "… where men and women can at last become like proper human beings, living and working and above all enjoying themselves in a world […] where man is walking upright" (Plaice, Plaice & Knight 1986:xxxiii). For Bloch, this utopian dimension is situated in our temporal existence as humans; it is something concrete and does not refer to a world outside of time. However, it is precisely this temporality of our existence that we have to grasp fully, if, as Bloch argues, we want to unlock the critical power of the utopian dimension.

As Bloch sees it, the present of human existence must be placed in relationship to "latency", which connects the present with the unfulfilled potentialities of the past, and "tendency", which connects the present with possibilities or living options for future action. Kellner (n.d.) reformulates it thus:

> The present moment is thus constituted in part by latency and tendency: the unrealized potentialities that are latent in the present, and the signs and foreshadowings that indicate the tendency of the direction and movement of the present into the future.

Going one step further, Bloch asserts that we must grasp and activate this three-dimensional temporality in "an anticipatory consciousness that at once perceives the unrealized emancipatory potential in the past, the latencies and tendencies of the present, and the realizable hopes of the future" (Kellner n.d.). According to Kellner, by this Bloch means that we have to "dream forward", projecting a vision of a "future kingdom of freedom". The crux of the matter for Bloch, therefore, would lie in developing a critical consciousness that is historical and creative at the same time. In this regard, he offers us a methodology of "ideology critique", which he actively practiced personally.

For Bloch, ideology should not be simplistically seen as the false consciousness used by clever politicians to mislead and manipulate the masses; or, for that matter, as a set of ideas that may be overcome by merely pointing to the truth. This simplistic model, Bloch argues, should be replaced with a conception of ideology as a complex phenomenon, in which the dreams and aspirations of people for a better life are at the same time articulated and deflected, and perhaps even also distorted, by a set of ideas that convince people that a certain *status quo* or a certain promised future has already fulfilled, or will in the near future fulfil, these dreams and aspirations. In this latter format, ideology clearly has the function of embellishing, legitimising and mystifying a certain political order, and by doing so may function to manipulate masses of people (Kellner n.d.).

The worst manifestation of ideology in its distorted and manipulative form is arguably the National Socialism of the previous century. However, as a political ideology, it would not have been able to be "successful" without offering some kind of "fantasy bribe" to the people at whom it was addressed. This "bribe" would be the acknowledgement of a kernel of very real human desires and needs that live in the hearts of people, while its "fantasy" part would be the creation of an illusion that their desires have already been fulfilled by a particular social order (Jameson 1979:144). As such, any ideology contains what Bloch (1986:154) refers to as a "cultural surplus" of human desires, needs and dreams that are actually not fulfilled, and which may be used to criticise the ideologies that dominate the *status quo*.

Bloch furthermore believes that ideologies, as they are typically articulated in the so-called superstructure of political philosophy, religion and art, are also present in everyday phenomena such as "daydreams, popular literature, architecture, department store displays, sports, or clothing". All of these phenomena, Bloch argues, may be analysed or "pressed" to unlock its utopian content, or its "anticipations of a better world", which may serve as a basis to critique the social and political structures of the present, as well as to transform it, foreshadowing in the present what could be the better world of the future (Kellner n.d.).

Bloch's notion of ideology critique thus invites us to turn our eyes towards uncovering and discovering what he refers to as the "not-yet-conscious" (1986:127), which he describes as "revelations of unrealised dreams, lost possibilities, aborted hopes – that may be resurrected and enlivened and realized in our current situation" (Kellner n.d.). In this manner, Bloch emphasises the positive in human nature, culture and ideology, the "utopian-emancipatory possibilities" of the past and the present. What he calls for is a transformative consciousness in which we read the dominant ideologies of today from within and against their grain, as it were, to reach the "not-yet-conscious" – that kernel of unfulfilled human wishes and dreams that has the potential and power to explode any and all deceptions and distortions that have been built around them.

This view is clearly not without its problems, in that it requires a mode of critical interpretation in which the unfulfilled wishes and dreams of humanity, forming the kernel of a utopian vision, may be distinguished from the dreams of the far-right and the far-left, which may be equally destructive and oppressive in their realisation. A case in point is the dream of someone like Anders Behring Breivik, who in 2011 went on a killing spree in Norway to articulate his views of a peaceful and uncontaminated social order. Bloch was aware of this challenge, and spent much of his work on developing the tools for such a critical interpretation, including a thorough critique of Freud and developing his own views on human nature and the human psyche. The details of these reflections do not concern us here, except to mention that Bloch always started his ideology critique with the concrete and very real experiences of the "wishing, hopeful, needy and hungry human being", analysing what prohibits the realisation of their desires and the fulfilment of their needs (Kellner n.d.).

With this radical perspective on ideology critique, Bloch points us in the same direction as contemporary ecofeminist thinking and practice (see for example Warren 1990), in which narratives delivered in the first person singular voice of oppressed, dominated and exploited women serve as the starting point for analysis and transformative action. While Bloch's view of the deceived and the exploited is perhaps somewhat wider than that of ecofeminism, both point us towards focusing on the circumstances and lives of people who clearly deserve a better life, but cannot attain it for many reasons, including being caught up in circumstances created, embellished and justified by oppressive ideologies. It is from this experience that Paulo Freire's 'pedagogy of hope' speaks.

The 'pedagogy of hope' of Paulo Freire

Paulo Freire's (1921-1997) pedagogy of hope may be seen as a sustained initiative to use education to intervene in and transform the circumstances that keep masses of people in a situation of poverty and despair. Published in 1994, Freire's *Pedagogy of Hope: Reliving Pedagogy of the Oppressed* revisits the 'pedagogy of the oppressed', which he originally described in Portuguese in 1968. Freire argues (1970:86) that education must become a "practice of freedom" (which was also the title of his first book, written in Portuguese and published in 1967). In order to make it so, strategies of teaching and learning should be adopted to transform the consciousness of both teacher and learner. When this happens, the resultant transformed consciousness will in turn lead to a practice of transforming society.

Responding to what he refers to as "a culture of silence" (Freire 1970:13,50,129), which allows the dominant ideologies of neo-liberalism and capitalism to be perpetuated in society without critical questioning, Freire maintains that it is a futile exercise to approach the poor as "objects" of teaching who need to learn from their teachers about their situation in the world and how to overcome it. Instead, Freire emphasises that education should rather create conditions that enable the teacher and the learner to cooperate with one another to understand why a situation is oppressive and what could be done to transform that situation so that human dignity may be restored. He thus argues for a radically changed relationship between teacher and learner, one where the teacher becomes learner and the learner teacher. He also shifts the objective of the relationship away from the mastering of facts to understanding the 'why' of facts, which changes the relationship of the learner to those facts. Instead of being the passive and silent victim of facts, the learner's consciousness is altered, so that s/he becomes an agent who wants to and is able to change them.

For Freire, hope enters into the picture when such a critical pedagogy unlocks an expectation of a better life in people and empowers them to take action towards realising that expectation — an attitude that they did not have before. Formulated differently, a critical pedagogy creates an understanding of a situation in which the historical formation of that situation is brought to light, and stimulates discovery and exploration of the possible actions that may be taken to transform that situation for the better. For this reason, Freire always resisted the idea of turning his pedagogy into a methodology. Instead, he argued, each teacher and each learner should discover and appropriate, within a particular situation, what it would entail to adopt a stance of critical inquiry and transformative action. Since this may differ from situation to situation, a prescriptive methodology might undermine precisely that which it claims to strive to achieve.

Some critics may argue that Freire's pedagogy overemphasises the poor. Three responses may be offered to such criticism: firstly, Freire would answer that it is cruel and inhumane to give those who are hungry, homeless, and without security a secondary priority in an educational agenda, leaving their needs pending, as it were, while other priorities were addressed. Indeed, Freire reports that his pedagogy was born out of the rage that he experienced on a personal level, because hunger made it impossible for him to excel at school. He argues that addressing the needs of the poor is a priority issue, because it is a justice issue — and justice should not be put on hold to wait for another day. As long as injustice exists for masses of people, Freire insists, a focus on it must infuse and shape all teaching and all learning in educational institutions.

Secondly, Freire's work demonstrates that a pedagogy of hope that focuses on the poor does not in fact entail a narrowing of scope, but includes whatever it takes to understand the circumstances that create poverty and prevent us from overcoming it. For instance, in 1993 he published *A Pedagogy of the City*, a reflection on his 10 years as Minister of Education in Brazil. In this book, Freire focuses on the urban poor and critically analyses various themes relating to the functioning of the city as a whole, as well as the political and administrative struggles school children were subjected to, and describes how these factors affected the performance of Sao Paulo schools.

Freire himself suggested a third response to the criticism that he narrowly restricts pedagogy to poverty, but did not fully develop it. At the time of his death, he was working on a book on ecopedagogy, parts of which was published posthumously in *Pedagogy of Indignation* (2004). Freire understood ecopedagogy to promote ecoliteracy and a future-oriented ecological and political vision as part of establishing worldwide social justice, using anger as a tool to oppose the globalisation of neo-liberalism, corporate greed and militarism. These ideas were further developed by the ecopedagogy movement, which criticises environmental and sustainable development education as not being effective in addressing the environmental crisis. Instead, ecopedagogy promotes an in-depth critical understanding of the many ways in which modern culture and industry promote unsustainable lifestyles (see Kahn 2007; 2008; 2009; 2010).

In a sense, it may be said that hope indeed becomes a 'project' in Freire's pedagogy: it is something that one must engage in, and failure to do so is also a moral failure. However, hope is not merely seen as a virtuous characteristic possessed by individuals, or a utopian dream inherent to all people, but rather as a radical act of freedom, i.e. a practice of critical social understanding in the service of transforming society. As Freire understood it, the conditions for such an act of freedom must be created through shared reflection and learning. The aim of such reflection is not only to transform the world at large, but also one's knowledge and one's self: what each one of us knows and what we may learn from each other; and how we act in the world as a consequence of these insights (Aronowitz 1998; Freire 2004). However, it is not possible to determine beforehand what the content of this reflection, dialogue and action will be. It will depend on where we are at the time when we start to engage with each other, the world, our knowledge of the world and ourselves, and our actions.

An ethics of hope?

From the three portraits of hope discussed above, it is evident that an ethics of hope is indeed possible as a practice of radicalising the present and the past in order to build a better future. At the same time, in terms of content, an ethics of hope may be an extremely dangerous practice, in that every dream and every desire that we strive to realise in the name of a better future must also be subjected to rigorous ethical critique. On the other hand, an ethics of hope emboldens us in the face of every ideology that distorts our dreams of a humane existence and that may deceive us into believing that they have been, or may be, completely fulfilled within a given socio-political order. Accordingly, working with, or working towards, an ethics of hope may be potentially dangerous to those who strive to practice it effectively in this world.

However, putting these potential dangers aside (and leaving the challenges they pose for others to analyse in other books), it is clearly important that the substantive elements of an ethics of hope are taken on board in any effort to move society, or a part of society, towards a sustainable future. If an ethics of hope creates a time and a space for hope, making it possible for us to transform society, it may help us to create an alternative notion of history, as well as a moral space in which the notion of a sustainable future may be articulated in a manner that may help to transform present society, its structures and processes.

Furthermore, if an ethics of hope may help us to develop a critical consciousness through ideology critique, uncovering how, in our everyday practices, consumer behaviour, financial decisions and administrative rules, we pursue dreams and desires in a manner that closes off the possibility of a sustainable future, then it has a place in any and all efforts to establish a sustainable Stellenbosch. If an ethics of hope entails critical engagement with others about the world, our knowledge of the world, and how we act on the basis of that knowledge, and if this critical engagement helps us to transform ourselves, our knowledge and our world, then such an ethics clearly also has a place in any and all efforts to learn more about a sustainable Stellenbosch and share that knowledge with others.

As such, an ethics of hope is not supplementary to efforts to establish a sustainable future (for Stellenbosch). It forms an integral part of every move we make in this regard, right from the start. The contours of such an ethic cannot, however, be defined in advance as a plan or a programme to execute. An ethics of hope is something that always has to start anew; it is an ethics of a journey rather than a destination. At each and every milepost where we think that we have ended our journey towards a sustainable Stellenbosch, we will have to stop, reflect, and assess whether we are opening up or closing down the time and space for hope and radical social transformation; whether we are, yet again, implementing an ideology in which we deceive ourselves and others into believing that we are moving towards a sustainable future, while in practice we are not. Accordingly, an ethics of hope always requires us to move towards its attainment; never to think that we have actually achieved it.

REFERENCES

Aronowitz S. 1998. Introduction. In: P Freire. *Pedagogy of Freedom*. Boulder, CO: Rowman and Littlefield Publishers.

Bloch E. 1986. *The Principle of Hope*. Translated from German by Plaice N, Plaice S & Knight P. Oxford: Blackwell. (Originally published as *Das Prinzip Hoffnung*. Frankfurt am Main: Suhrkamp Verlag. 1959.)

Bloch E. 2000. *The Spirit of Utopia*. Translated from German by Nasser AA. Stanford, CAL: Stanford University Press. (Translation based on the second edition of *Geist der Utopie: Bearbeitete neuauflage der zweiten Fassung von 1923*. Frankfurt am Main: Suhrkamp Verlag. 1963.)

Freire P. 1970. *Pedagogy of the Oppressed*. Translated from Portuguese by Myra Bergman Ramos. New York: Continuum.

Freire P. 1993. *Pedagogy of the City*. New York: Continuum Press.

Freire P. 1994. *Pedagogy of Hope: Reliving Pedagogy of the Oppressed*. New York: Continuum.

Freire P. 2004. *Pedagogy of Indignation*. Boulder, CO: Paradigm Publishers.

Geoghegan V. 1996. *Ernst Bloch*. London & New York: Routledge.

Harvie T. 2009. *Jürgen Moltmann's Ethics of Hope: Eschatological Possibilities for Moral Action*. Farnham, UK: Ashgate.

Jameson F. 1979. Reification and Utopia in Mass Culture. *Social Text 1*, (Winter 1979):130-148.

Kahn R. 2007. The Ecopedagogy Movement: From Global Ecological Crisis to Cosmological, Technological and Organizational Transformation in Education. PhD thesis. University of California, Los Angeles: Graduate School of Education & Information Studies. Los Angeles: University of California, Los Angeles.

Kahn R. 2008. Towards Ecopedagogy: Weaving a Broad-Based Pedagogy of Liberation for Animals, Nature and the Oppressed Peoples of the Earth. In: A Darder, R Torres & M Baltodano (eds). *The Critical Pedagogy Reader*. 2nd Edition. New York: Routledge.

Kahn R. 2009. Producing Crisis: Green Consumerism as an Ecopedagogical Issue. In: J Sandlin & P McLaren (eds). *Critical Pedagogies of Consumption: Living and Learning Beyond the Shopocalypse*. New York: Routledge.

Kahn R. 2010. *Critical Pedagogy, Ecoliteracy, and Planetary Crisis: The Ecopedagogy Movement*. New York: Peter Lang.

Kellner D. n.d. *Ernst Bloch, Utopia and Ideology Critique*. [Accessed 6 September 2011] http://www.uta.edu/huma/illuminations/kell1.htm

McCloskey DN. 2006. *The Bourgeois Virtues: Ethics for an Age of Commerce*. Chicago: University of Chicago Press.

Moltmann J. 1965. *Theology of Hope: On the Ground and Implications of a Christian Eschatology*. Translated from German by Leitch JW. London: SCM Press.

Moltmann J. 1992. *The Spirit of Life: A Universal Affirmation*. Translated from German by Kohl M. London: SCM Press.

Moltmann J. 1993a. *Theology of Hope: On the Ground and Implications of a Christian Eschatology*. Translated from German by Leitch JW. Minneapolis, MN: Fortress Press.

Moltmann J. 1993b. *God in Creation: A New Theology of Creation and the Spirit of God – The Gifford Lectures 1984-1985*. Translated from German by Kohl M. Minneapolis, MN: Fortress Press.

Moltmann J. 2009. Introduction. In: Harvie T. *Jürgen Moltmann's Ethics of Hope: Eschatological Possibilities for Moral Action*. Farnham, UK: Ashgate.

Plaice N, Plaice S & Knight P. 1986. Translators' Introduction. In: E Bloch. *The Principle of Hope*. Translated from German by Plaice N, Plaice S & Knight P. Oxford: Blackwell.

Stellenbosch University: The HOPE Project. 2010. [Accessed 16 August 2011] http://www.thehopeproject.co.za

Warren K. 1990. The Power and the Promise of Ecological Feminism. *Environmental Ethics*, 12(2) (Summer 1990): 125-146.

Section 1

Spaces

Section Introduction

Driven by ever-changing socio-political planning and reinforced by severe income inequality, space in Stellenbosch is a hotly contested commodity. The South African 'suburban dream', shared by the poor and middle class alike, slips ever further out of reach as insecure land tenure, 'studentification', and a rapidly expanding population exert pressure on the municipality's ability to provide services.

Spatial planning: Spatial planners today face the challenge of protecting vital ecosystem services while attempting to radically restructure land ownership after apartheid; promote economic development; and create housing for the 20 000 people currently on the waiting list. The challenge of achieving these objectives has been complicated by periods of political instability within Stellenbosch Municipality, which has prevented the adoption of the Spatial Development Framework (SDF) for Stellenbosch. Due to the courageous and cohesive work done by senior managers during 2011/12, this problem is being resolved. However, genuine and meaningful public participation processes will be required to ensure that an effective vision for a sustainable Stellenbosch is incorporated into an approved SDF.

Land reform: The vast majority of all farm worker evictions that have happened in the Stellenbosch area after 1994 were unlawful. Many farm workers do not know their rights under the Extension of Security of Tenure Act (ESTA) of 1997 and there is a lack of compliance and enforcement of legislation. While evictions may have resolved certain problems for farmers, these problems have been displaced to the urban nodes, resulting in expanded informal settlements.

Urban spaces: Dramatic population growth since the 1970s has placed significant pressure on urban spaces, with growing 'placelessness' and partitioning indicative of a spatial identity crisis. 'Studentification' is displacing original residents; segregation is on the rise due to gated developments; lifestyle estates are shifting Stellenbosch away from agriculture; and informalisation reinforces historic divisions. The town's challenge is to determine its identity while meeting the demand for more student accommodation, middle-class housing and informal settlements. Solutions are possible if the process of sustainable transformation is skilfully managed: intensifying land use on campus; developing agri-villages; building new mixed settlements in spatially tight nodes; and creatively using abandoned spaces to house the poor.

Housing: The Stellenbosch municipal area includes 10 000 to 12 000 shacks in informal settlements, with a growth rate of 4 600 new shacks per year. This means that nearly a third of all households in Stellenbosch are informal structures. However, the municipality only receives 300 housing subsidies annually. Thousands of people are still on the housing waiting list, with some having waited for well over a decade. Informal settlement upgrading builds on the poor's own work to upgrade their dwellings and is a more realistic solution to the housing backlog than formalised housing.

Local economic development (LED): Stellenbosch is blessed with a convenient location, well-developed infrastructure, and rich agricultural, tourism and human resources. However, compared to a population growth rate of 4.5% to 9.2% per year, the Gross Value Added growth rate of 4% per year suggests negative per capita growth. The challenge now is to reconcile growth with long-term resource constraints. In order to implement a more sustainable LED strategy, partnerships and capacity to implement LED policies must be strengthened and deeper insights into LED processes and challenges are required. Tourism clearly has a special role to play in connecting the need to create employment with the need to build a socially cohesive community with a shared sense of heritage.

Spatial development will be more sustainable if:

- a politically stable municipal council – in partnership with key stakeholders – develops a cohesive and progressive vision for the town's sustainable spatial development;
- the informal settlement upgrading process works with the energies and capacities of the urban poor in an incremental manner that best uses the funds available to meet the needs of the maximum number of households;
- farm workers have better access to information and legal representation, and there is better monitoring to prevent illegal evictions;
- available and appropriate greenfield and brownfield land are developed in ways that facilitate social integration, sustainable resource use, skills development and job creation; and
- growth is reconciled with long-term resource constraints.

Chapter 1

SPATIAL PLANNING
Planning a sustainable Stellenbosch

SIMON NICKS

This chapter examines the effect of spatial planning on the development of the area currently known as Stellenbosch Municipality.[1] It describes the typically contrasting social, economic and physical aspects of Stellenbosch Municipality today. It then tracks how the forces of environmental coexistence, social engineering and economic and financial gain have shaped various planning initiatives, beginning over three centuries ago around this former island in the Eerste River.

It will be seen that the municipality as a construct has been extraordinarily unstable for a local government body, which one might expect to represent the secure bedrock of a community's economic and social affairs. For this reason, the tension that has always existed between political change on the one hand and spatial management on the other, between which the pendulum has been swinging with increasing violence in recent years, constitutes one of the key sub-themes of this chapter. The history of the municipality over this period highlights the challenges faced by spatial management planners who, within this extremely fluid political and public management regime, seek to:

- conserve important ecosystem services;
- achieve a radical restructuring of land ownership and urban morphology patterns; and
- promote economic growth and employment.

Stellenbosch now

The central business district (CBD) of South Africa's 'first town' continues to delight visitors with its oak tree-lined streets fringed with classically proportioned urban Cape Dutch buildings abutting *klompie*-bricked stoeps fronted by *leiwater* furrows often flowing with river water. It offers an urban experience of a quality and intensity unique among South African towns, with the possible exception of some sections of Main Road in Paarl and Church Street in Tulbagh.

In stark contrast most South African town centres present themselves as a melange of *ad hoc* buildings - some old, some new, some poorly renovated – with main streets often devoid of trees and randomly widened in places according to some road engineer's requirements for parking and access. They are challenging to drive through, never mind walk along.

But as one moves away from the town's urban core – either to the richer areas, with their ubiquitous low-density suburban houses (increasingly behind high walls), or into the townships – it becomes clear that Stellenbosch town has not escaped the dead hand of the South African suburban dream of apartheid planning and low-income housing policies.

---

1   Due to the Demarcation Board's insistence on giving the municipality (WC024) the same name as its main town, reference will be made to 'Stellenbosch Municipality' and 'Stellenbosch town' throughout in order to differentiate between the settlement and the municipality.

In the townships, the dwellings reflect each decade's changing housing subsidy policy. Built to accommodate the victims of forced removal from convenient urban streets, such as Van Ryneveld Street in the CBD, they begin with the NE 51/6 and NE 51/9 houses[2] of the 1960s (Calderwood & Connell 1952), followed by the rental flat complexes of the early 1970s. The more generous 60-m^2 freestanding houses and maisonettes of the House of Representatives subsidy scheme came next, followed in turn by the smaller 24-m^2 Administration Board houses of the 1980s and early 1990s. Finally, one comes to the 40-m^2 RDP houses of the early 2000s – the precursors to the houses of the 'Breaking New Ground' housing policy, which seeks to transform housing from an industrialised, mono-functional production process to the creation of 'integrated human settlements'.

The townships are interesting in this regard. There were a number of innovative double-storey row and semi-detached housing projects in Cloetesville (Cloetesville Steps and Orlean Lounge) and in Kayamandi, inspired by the Weltevreden Valley and Missionvale housing projects in Cape Town and Port Elizabeth, respectively. However, for reasons including changes in housing policy, financial affordability and party politics, both at local and national level, none of these housing efforts have been sustained.

The only permanent process that has proved continually capable of delivering basic shelter (but seldom services) has been the construction of shack housing, with informal housing built by residents themselves proving to be the only delivery process able to keep up with the scale of the need for shelter. This is most evident in the vibrant streets of Kayamandi, where there is dense infilling of shacks among formal dwellings, as well as in 'the Zones' on the higher slopes. This has created activity that is similar in intensity to – but very different in character from – that in the southern CBD. Interestingly, the dwelling unit densities in the southern CBD are closer to those of Kayamandi than anywhere else in Stellenbosch, although obviously existing in a very different socio-economic context (see Figure 1.1).

The lesson here is that when urban population densities exceed 100 people per hectare, things start to happen – no matter who the people are. Whether this intensity of activity is negative or positive depends entirely on how it is managed and on the perspective of the onlookers: some people love informal markets; others hate them.

The rest of Stellenbosch town, outside of the CBD, comprises upmarket suburbs, their layout and appearance again reflecting the trends of the time. Suburbs developed during the decades of the 1960s through to the 1980s reflect a ubiquitous low-density American-style 'one house one plot' aesthetic, which for some reason was seized on as the suburban dream by the South African middle classes of the time. Possibly this was because the dominant South African urban populations of the 1930s to the 1950s – like their American counterparts – were also 'incomers' from the rural areas. This trend continued into the 1990s and 2000s, with gated security estates such as De Zalze, a large estate on the southern outskirts of Stellenbosch town, featuring spacious, gabled houses with black, double-pitched roofs clustered among green belts around a golf course. A large neighbourhood shopping centre, Stellenbosch Square, across the road from the estate, is the latest addition to these suburban extensions. The prevailing transport mode of almost all residents and shoppers who frequent it is the private car.

As a result, the urban fabric of Stellenbosch town as a whole does not have a coherent, unifying theme provided by local building materials and inspired by styles and/or colours, with buildings differentiated only by size. Rather, it gives the impression of badly put together pieces from very different jigsaw puzzles. These bits of urban fabric are tenuously held together by tarred access roads, along which minibus taxis and private cars speed when they are not slowed down by peak hour traffic congestion, often endangering pedestrians and small children.

Equally tenuously, the settlement is held together below ground by a network of water, sewerage and storm water pipes whose capillaries and bronchioles generally deteriorate as they serve the poorer parts of town and whose bulk plants are in need of large and expensive upgrades. They often fail to reach the furthest limits of the informal settlements, which are served by communal ablution blocks and stand-alone toilets or, in some cases, the bucket system.

---

2 NE 51/6 and NE 51/9 refer to a set of low-income housing types developed by the National Building Research Institute (NBRI) based on their guidelines completed in 1951.

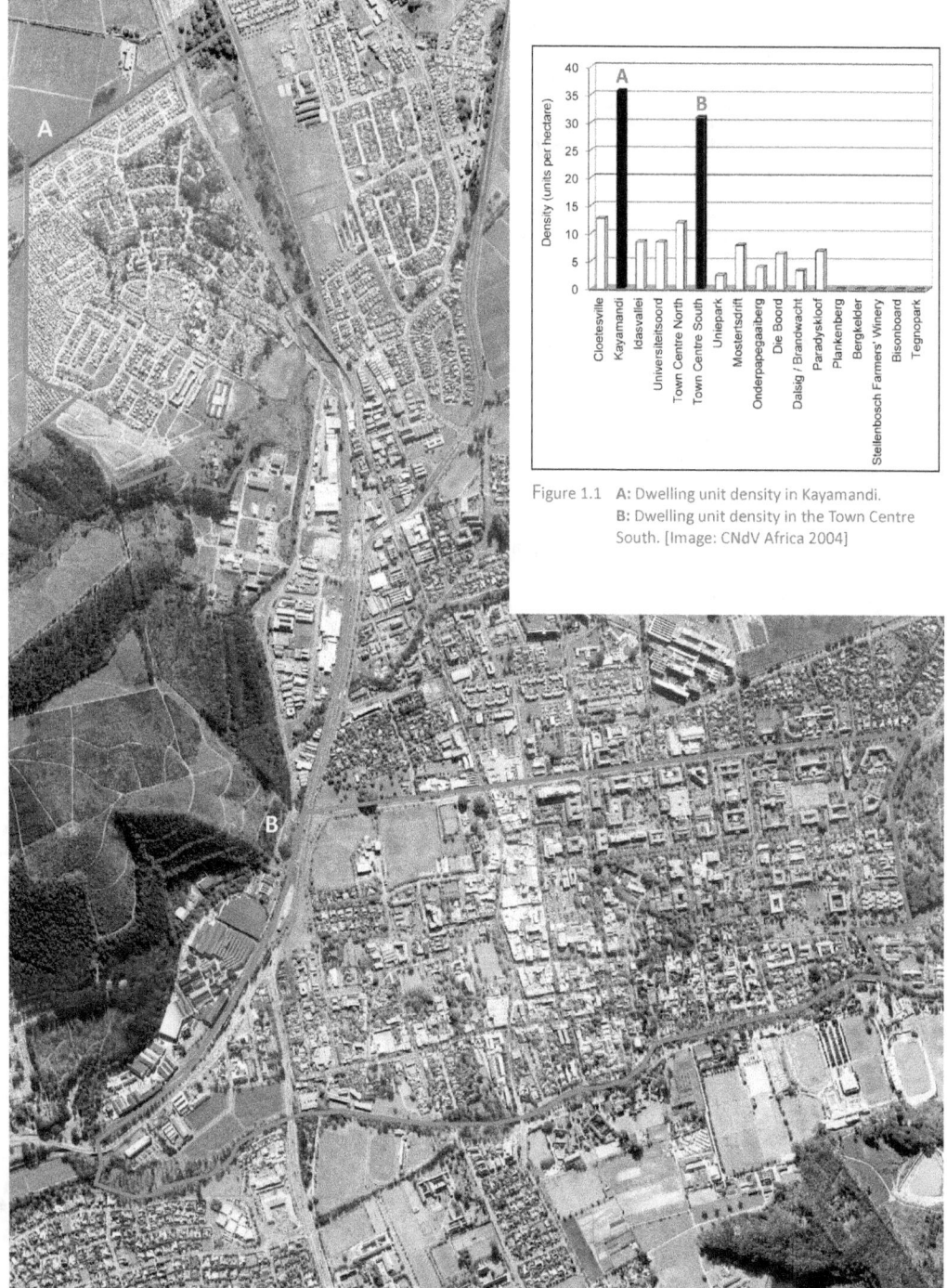

Figure 1.1 **A:** Dwelling unit density in Kayamandi. **B:** Dwelling unit density in the Town Centre South. [Image: CNdV Africa 2004]

There have been innovative efforts to promote a housing form rooted in the town – for example, the double-storey, 40-m^2 row and semi-detached housing schemes of the late 1990s and 2000s in Kayamandi and Cloetesville – but if one took away the vineyards and the mountain backdrops, there would be little to distinguish most of suburban Stellenbosch town from Ladysmith in KwaZulu-Natal, Rustenberg in Gauteng, or Polokwane in Limpopo. In some cases, tall buildings in the town are even obstructing the views of the mountains and vineyards.

The natural systems of the town – fertile land, rivers and indigenous vegetation – are being increasingly obscured – physically, visually and ecologically. The town's carbon footprint is being increased by energy-intensive materials in pipes, cables and road materials, as well as pump stations, water treatment plants and the solid waste transportation system – the result of prevailing civil engineering paradigms for water supply, waste water treatment and storm water drainage. Due to declining urban management capacity, there is continuous pressure to raise the specifications of capital infrastructure, so that, hopefully, less maintenance will be required.

When coupled with formulaic layout and zoning conditions, these trends have resulted in a settlement pattern, form and use of materials that have largely undermined Stellenbosch town's former closely dependent relationship with its ecosystem services, i.e. the water, arable land, and building materials necessary for survival and economic activity.

There are a number of other settlements in the municipality, all very different from one another, with the two secondary ones being Franschhoek and Klapmuts.

Franschhoek is the Western Cape's centre of gourmet cuisine at the scale of an entire village – or at least the upmarket part of it. Surrounded by a belt of wine farms, in turn enclosed by the Franschhoek and Wemmershoek mountains, the village has an extraordinarily beautiful setting at the end of the Berg River Valley. And this part of Franschhoek's aesthetic and ambience is jealously guarded by a variety of citizen institutions.

By contrast, northern Franschhoek (formerly known as Groendal) is similar to Stellenbosch town's townships, and also reflects several decades of different housing subsidy approaches. However, in keeping with the concern to protect the town's aesthetic quality, there have been attempts to ensure that any development that does occur in northern Franschhoek is of a higher quality than the norm. For example, the Mooiwater scheme benefited from certain embellishments as a result of cross-subsidies from the Franschhoek Empowerment and Development Initiative project, which included constructing 'gentlemen's estates' on a former portion of the Franschhoek nature reserve.

Klapmuts is far more strategically located than Franschhoek and Stellenbosch town, because it straddles both the Gauteng-Cape Town rail line and the Old Paarl Road, and has a direct link onto the N1 national route; yet, even though the town was first subdivided over 100 years ago, residents and planners have largely failed to take advantage of this excellent location – perhaps because it was without services for most of this time and, during much of this period, the owners of large, key landholdings in the middle of the settlement were not interested in developing.

Today, Klapmuts consists of a large low-income township to the rear, or south, of the settlement, across the rail line and river; a middle-income suburb around the station; and a slightly more affluent area around the main street, where a filling station has recently been built. Its challenge has been whether, on the one hand, it should develop as a contemporary Boland village forming a gateway to the Winelands, supported by the prevailing spatial development framework; or, on the other, as a massive retail industrial complex, similar to Midrand along the Ben Schoeman highway, as mooted by some developers. The most recent notion is that it could become a model sustainable settlement, hopefully of a form and pattern that reinforces rather than detracts from the sense of place of the Boland and the gateway experience into the Cape Town Metro that it offers.

In respect of the rural areas, Stellenbosch Municipality largely comprises either agriculture – mainly intense belts of vineyards and wine farms, often with wineries; restaurants and tourist accommodation; or wilderness areas encompassing the surrounding mountains. In addition to their role as an impressive scenic backdrop to the municipality, the mountains offer a myriad of mountain biking and hiking opportunities for tourists or those residents able to indulge in these activities.

Stellenbosch Municipality's rural areas provide the inputs – arable land, water and scenic landscapes – for four of its most important, and mutually reinforcing, economic and employment sectors: agriculture, manufacturing, financial services and tourism. Although some of the most scenic wine farming areas in the world are located here, currently largely uncluttered by the kind of massive agro-industrial complexes found in California and in some of the European wine areas, there are nevertheless growing signs of industrialisation as farming becomes more corporate and owners seek to maximise their returns. The quality of this environment is highly sought after by wealthy South African and international business people and retirees as a place to live.

This happy convergence of environmental resources, scenic quality and business opportunities has two other mutually reinforcing spin-offs: the largest number of JSE-listed companies based in any small South African town are headquartered in Stellenbosch, and the town is home to a disproportionately high number of corporate CEOs and executives, which in turn means that it is able to sustain a comparatively high level of economic activity and consumer services for a town of its size. This results in other benefits throughout the value-add chain and for employment, with domestic work, agriculture, and tourism and other service industries constituting the main sources of employment for Stellenbosch's relatively unskilled working poor.

But cracks are appearing in the bucolic landscapes surrounding the urban settlements, and the tipping points may not be far off. Most visually obvious is the urbanisation-by-stealth of the countryside: packing sheds are growing into distribution centres; shiny plastic is covering more and more agricultural land; wineries are becoming large bottling plants; and labourers' cottages are being transformed into holiday chalet compounds, in places even morphing into small, upmarket residential security estates.

Less obvious, unless experienced during rush hour, are growing traffic volumes. The landfill site outside Stellenbosch town – now legally full, yet still accepting waste while other affordable and acceptable alternatives are investigated – continues to tower upwards. The water quality in the main rivers is extremely poor – so much so that it might pose a threat to the export of produce to overseas markets that insist on acceptable standards of irrigation water. And, lurking in overcrowded rooms and informal shacks is the housing challenge. Stellenbosch Municipality's current waiting list is about the same as the current number of formal dwelling units in the municipality – around 20 000; in other words, about the same number of dwelling units as have been built in Stellenbosch Municipality during the entire past 300 years are needed. Where will they be built? And how did this untenable situation arrive?

A history of planning in Stellenbosch
It certainly didn't start like this ...

Planning in the first 250 years
Stellenbosch prides itself on being South Africa's first town (Cape Town is regarded as the first [mother] city.) It was established in December 1680 by the Governor of the Cape, 28 years after Cape Town was first established. While there was undoubtedly pre-colonial settlement of the land along the Eerste River before the arrival of the settlers, today's town has its roots in the original settlement pattern, whose main elements were in place by 1710 (see Figure 1.2).

Long streets were aligned at right angles to the contours and provided with furrows, so that water for irrigation and drinking could gravitate to each property. Buildings were built along the edges of the blocks so as to minimise their impact on land for cultivation.

This pattern reflected a very close relationship with the natural environment on which the town depended for its water, food and fibre and building material needs. Figure 1.3, WF Hertzog's famous map of Stellenbosch town in 1817, clearly illustrates the settlement's close relationships with the natural systems:

- the source of the water drawn from a tributary of the Eerste River at a suitably elevated point;
- Dorp Street laid out some 150 m to 200 m away from the river, so as to keep it out of the flood lines; and
- optimal and equitable use of good agricultural land on 'water erven' (long, thin plots leading down to the river banks) – a pattern also seen in Jamestown and Franschhoek.

Figure 1.2   The growth of Stellenbosch town: 1710, 1770, 1817, 1905 (all drawings at same scale) [Oude Meester Groep 1979]

Figure 1.3 Stellenbosch town and environs, 1817 [Oude Meester Groep 1979]

The energy requirements of such a settlement were very small. Water moved by gravity. Transport was supplied by animal traction, the animals deriving their energy from the same nearby food gardens that fed the residents. Water heating and cooking requirements were supplied by renewable energy, i.e. wood, which was chopped either in the natural forests in the *kloofs* of Jonkershoek and Blaauwklippen Valley or in the plantations and woodlots already in evidence on Herzog's map. Space heating relied on the passive insulation resulting from the prevailing approaches to building construction, material and orientation. All these features may still be seen and experienced at the Stellenbosch Town Museum.

There were no cars and trucks, pumps, generators, pipelines, cables or refrigerators. The main energy source was gravity, which meant that the need to travel and transport goods had to be minimised. Thus, the least fertile land was chosen for the location of buildings, while aligning the streets perpendicularly to the contours allowed water to be abstracted from the higher reaches of the main river via weirs above the town and distributed via kerb-side furrows to irrigate garden plots and provide drinking and washing water – a system that is still found in some rural towns and villages today (e.g. Prince Albert or Nieu Bethesda).

As people mostly moved on foot, the settlement remained compact even as retail and commerce developed; indeed, walking distance was a key constraint to the outward growth of towns and villages.

Important public activities dominated the main spaces of the town: a market place for periodic trading and a site of worship, with the church building of the most popular denomination by far the largest structure in the town.

In this way a large religious, administrative and agricultural service centre developed in the municipality of Stellenbosch town, while smaller villages emerged at Pniel, in the Dwarsriver Valley, and Franschhoek, in the Franschhoek Valley.

Then, in the 1870s, with the onset of a modern economy driven by the discovery of diamonds and gold, things began to change fundamentally in South Africa – changes that resulted in the emergence of new shape shifters acting on spatial development, such as:

- the arrival of an industrial economy;
- improvements in regional transport, starting with the railways, which facilitated urbanisation;
- the development of urban engineering to cope with health challenges, which required improved water supplies and sewage services and led to the development of reticulated service networks and bulk infrastructure plants; and
- the onset of modern methods of energy generation (thermal power stations, the internal combustion engine, and massive increases in the efficiency of steam traction).

These developments not only led to extended working hours and enabled greater specialisation; they also resulted in greater mobility of commuters and freight.

By the early 1900s Stellenbosch Municipality was still relatively unaffected by these shape shifters. Spatial constraints remained significant and the valleys of the municipality largely retained their closed economies. However, from the 1920s onwards these factors began to result in elemental changes, including:

- the arrival of affordable motor cars for the few who could afford them;
- the Post-Union transformation of local government and the emergence of scientific town planning;
- the impact of modernism on town planning (separation of functions as an antidote to unhealthy overcrowding and proximity to pollution; the concept of house and city as a machine, promoted by le Corbusier; and the garden city movement, with its promotion of the ideal urban lifestyle);
- improvements in urban engineering technology (including better pump performance; bigger and longer service networks; decreasing dependence on locally available energy; and new waste water treatment technologies); and
- improvements in hydraulic engineering, which enabled rivers to be canalised and piped and wetlands drained, thus increasing the amount of land that could be developed.

At this point, Stellenbosch town began to spread into the surrounding countryside as walking distance no longer imposed a spatial constraint (see Figure 1.4). Service industries and value-added manufacturing grew rapidly. Raw materials, water, building materials and electricity from coal were cheap and abundant. It appeared as though the urban economy could be more and more divorced from the local natural environment. The result of all this was an urban metabolism that had apparently thrown off environmental shackles and could begin to consume resources at an exponential rate.

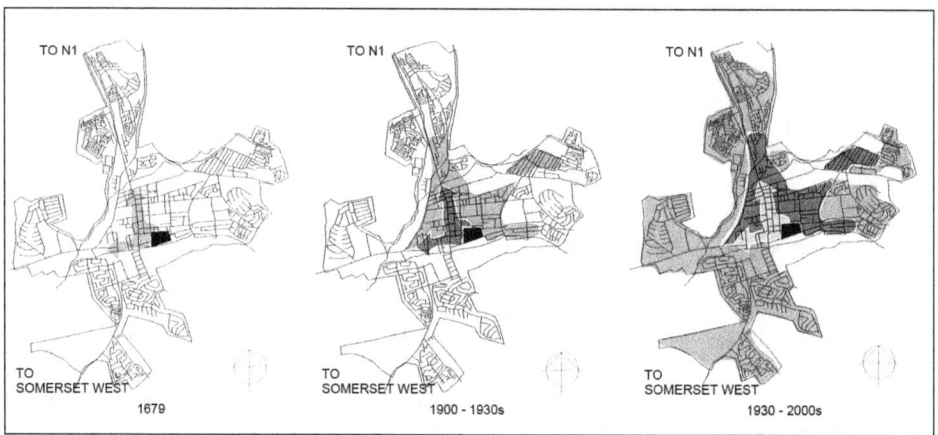

Figure 1.4   Urban growth of Stellenbosch [CNdV Africa, Urban Dynamics & GLS 2004]

Planning under Apartheid

The next major change in Stellenbosch Municipality's spatial planning history occurred during the apartheid era. The Group Areas Act was passed in 1950 and began to be implemented in the late 1950s and early 1960s. Stellenbosch was no exception to this process, with Franschhoek North (Groendal), Cloetesville, Idas Valley and Kayamandi emerging during this time.

▣ Stellenbosch Divisional Council Regional Development Scheme, 1967

During this period, the area demarcated today as Stellenbosch Municipality fell under three jurisdictions: Stellenbosch (town) Municipality, Franschhoek Municipality and the Stellenbosch Divisional Council (SDC).

The Stellenbosch Divisional Council covered a vast area – particularly to the west, where Kuilsriver formed its boundary – and included part of present-day Delft, Blue Downs, Eerste River, Gaylee, Kleinvlei, Scottsdene, and all of the Helderberg basin, including Gordon's Bay, Macassar and Sir Lowry's Pass Village – all settlements which it initially developed.

As the Divisional Council too became involved in implementing the Group Areas Act, people were moved from Stellenbosch town to these peripheral settlements, their numbers compounded by migration from the farms, which began in earnest in the 1970s.

Although the Divisional Council's Regional Development Scheme (SDC 1967) – a spatial plan similar to a present day spatial development framework (SDF) – explicitly earmarked land for the three main race groups, it was far ahead of its time with regard to environmental conservation and management. The scheme was underpinned by an analysis of heritage, geology and soils, topography, climate, vegetation (JP Acocks[3] was an important contributor here), demographics, transport, and the land use pattern. The conclusion of this study – informed partly by the fact that South Africa only had 14 million hectares of arable land, of which 3% is high value, and that the Stellenbosch division was relatively richly endowed in this regard (Adendorf 1984) – led to the policy

---

[3] JP Acocks was a scientist with the Department of Agriculture whose seminal *Veld Types of South Africa* (1988) was the first systems approach to understanding biodiversity as an ecology, and the impact of grazing in particular, and who also proposed new grazing methods that tried to emulate faunal grazing patterns' pre-partitioning of the range lands.

decision that there should be no urban development east of the proposed N7 route corridor, which ran through the middle of the division at the time. (In fact, this line still largely forms the City of Cape Town's eastern urban edge to this day.)

This policy was supported throughout this period by a united front constituted of officials led by Johan Adendorf,[4] chief town planner, and the council, particularly in the person of Stevie Smit, chairman of the Divisional Council for much of this period. It is likely that if the council had not enforced their policy, the western part of the municipality, between Kuilsiver and Stellenbosch town, would have been completely developed at present – at the time, the Cape Division took a very different position on land of a similar context between Bellville and Durbanville, and this area was completely built out from the 1960s to 1990s (see Figure 1.5).

Figure 1.5   40 year lateral growth comparison: Bellville to Durbanville (A and B), Kuils River to Stellenbosch (C and D) [Surveyor General 2000]

---

4   Johan Adendorf was the former chief town planner for the then Stellenbosch Divisional Council.

A comparison of these two policy trajectories – the one promoting large lot urban development; the other consolidating agriculture – over the past 40 years starkly raises key questions about sustainable development with respect to spatial planning:

- Which had the most benefit to society?
- Which had the most benefit to the environment?
- Are benefits to society and the environment mutually exclusive, or can they be mutually beneficial?
- How should these benefits and costs be measured?
    - In terms of rands? – Development and huge construction expenditure creates jobs, appreciates property, increases rates to the local authority, and mobilises against the loss of productive agricultural land and low-skilled jobs; and/or
    - In terms of environmental sustainability? – The conservation of nearby agricultural land (vegetable and animal protein close at hand) and the preservation and creation of jobs (low-skilled in agriculture and tourism).
- Is there a third way? Adendorf believed not. His experience suggested that the kind of finely detailed and sensitive response to upmarket rural development found in Europe and the UK – where development could be confined to small, carefully designed, isolated hamlets or the renewal of existing buildings – was not possible in the South African context, where the prevailing development mode was large lot subdivisions of entire farms, connected to bulk infrastructure services requiring large economies of scale to be viable. It could only be all or nothing. The debate continues to rage between developers, municipalities and the Provincial Government to this day.

▣ *Stellenbosch Guide Plan, 1988*

Guide plans – high level strategic planning documents – first began to be formulated in the Cape in 1980. They were approved in terms of the Physical Planning Act (Act 88 of 1967), and because they were under the jurisdiction of an Act, they could trump the provincial planning ordinances. They were also approved by government, unlike many of the other structural plans of the time. This gave them a legal standing that still endures today. The guide plans were intended to ensure compliance with the broad spatial planning policy of the National Government, which initially meant the spatial implementation of apartheid. They are currently administered by the Provincial Government.

Stellenbosch's 1988 Guide Plan covered the town and nearby agricultural areas to the north and south. It broadly delineated land uses and recommended preservation of agricultural and scenic land. A possible concern, although not explicitly stated, was to control the growth of Kayamandi, the only suburb of Stellenbosch town mentioned by name in the text. Although the planning policies from the late 1980s onwards did not feature any of the racist terminology that characterised plans from the 1960s and 1970s, their physical legacy had become solidly entrenched by this time.

Planning under the new democracy

Meanwhile, events on the political front were changing quickly. Influx control as a formal policy was abolished in 1986; Nelson Mandela was released from jail in 1991; and the first democratic elections were held in 1994. Clearly, there was a need – and an opportunity – for a more progressive approach to planning.

Stellenbosch Municipality at this time formed part of the Cape Metropolitan Area under the Western Cape Regional Services Council (WCRSC) and, at a conference in Caledon in 1991, it was agreed that an overall metropolitan development framework was needed. Somehow, instead of preparing a single framework, the WCRSC officials charged with this task partially disaggregated the metropolitan area into eight planning sub-regions, of which 'Stellenbosch and Environs' was one. A new planning lexicon arrived on the scene – 'activity corridors', 'spines and streets', 'nodes', 'open space systems' and 'densification' – to describe new approaches to restructuring the apartheid landscape.

▣ *Stellenbosch and Environs Sub-Regional Plan, 1995*

The Stellenbosch and Environs Sub-Regional Plan covered much of the area in the present-day municipality. Two main swings in planning policy were proposed in this document:

- The racist nomenclature for urban residential areas previously identified as 'white', 'coloured' and 'black' group areas was replaced with 'low-', 'medium-' and 'high-intensity' areas, respectively.
- The previously clear line between urban and rural areas, which had been maintained by the Divisional Council up to this time, was to be blurred with large 'rural development' areas around the existing rural nodes, where smallholdings and small farms were to be permitted (see Figure 1.6).

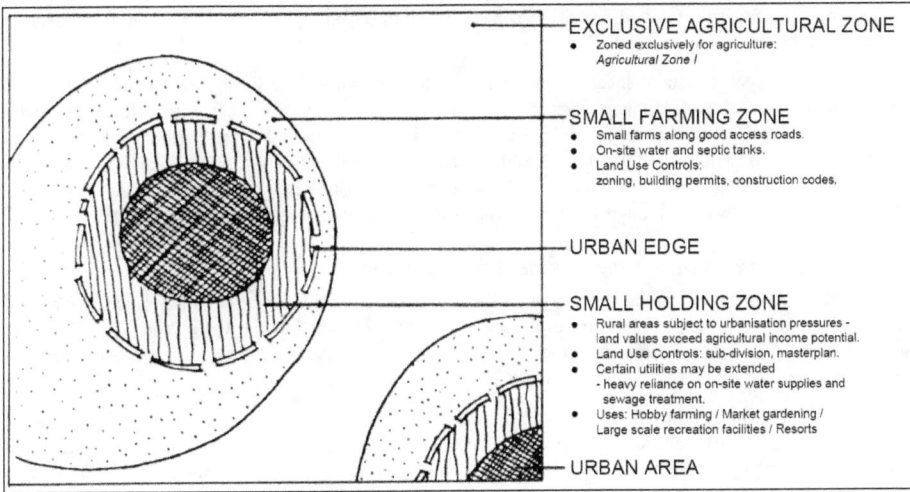

Figure 1.6 Grouping of small farms [WCRSC 1995]

The plan also contained an interesting definition of sustainability – "a characteristic of a process or state that can be maintained indefinitely" (WCRSC 1995:6) – and expressed concerns about the need to promote tourism, conserve the rural ambience and visual quality of the area, protect agriculture and promote densification, so as to minimise the need for lateral expansion of the towns and villages, and put forward proposals to address them.

On a philosophical level, the plan relied on the (chauvinistically-named) Man and the Biosphere (MAB) Programme of the United Nations Educational, Scientific and Cultural Organisation (UNESCO), which promoted a very clear approach to land use management, i.e. 'biosphere zoning' (see Figure 1.7).

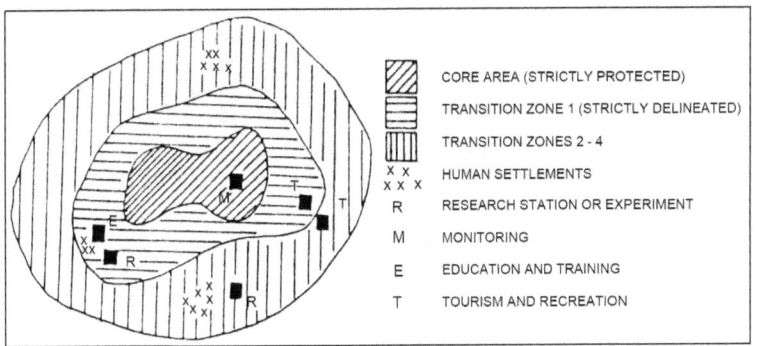

Figure 1.7 Schematic zoning of a biosphere reserve [WCRSC 1995]

This clear and simple approach to regional land use management – in terms of which protected wilderness areas and biodiversity hot spots are zoned as 'core areas' (where very little intervention should take place), through a series of more and more intense 'transition' zones culminating in urban settlements – has come to underpin broad land use management throughout the country, albeit with some important modifications, particularly as regards transition zones.

Early interpretations of transition zones gave rise to ambiguous situations, with little of the necessary clarity and consistency for effective policy planning and guidance. Thus, Transition Zone 4 could accommodate "small-scale market gardening, and space-extensive recreational activities such as golf courses", while urban development could be accommodated in Transition Zone 5, providing that areas of ecological, aesthetic and/or cultural value were preserved (WCRSC 1995:25). (At present, Transition Zone 1 is known as 'Buffer', Transition Zone 3 as 'Intensive Agriculture', and Transition Zone 4 as 'Urban Development', and so on; see the Winelands Integrated Development Framework.)

Meanwhile, the reorganisation of local government was in full swing. The former Divisional Council and Stellenbosch and Franschhoek Municipalities had become Transitional Local Councils in the urban areas, with the Winelands District Council (WDC) governing the rural ones. Towards the end of this period, the Demarcation Board drew its final local municipal boundaries and Stellenbosch Municipality (WC024), which has prevailed to this day with only minor changes, was born. However, in 2000, before these boundaries were finally proclaimed, the WDC produced a new plan, namely the Winelands Integrated Development Framework (WIDF).

▣ Winelands Integrated Development Framework (WIDF), 2000

Although the commissioning agent for this plan, the WDC, became defunct over this period, it approved the WIDF. Many of the principles and proposals of the plan are contained in the Cape Winelands Biosphere Reserve Framework Plan (Cape Winelands District 2009) currently in process with the Cape Winelands District Municipality.

The WIDF was a large and comprehensive plan, based on the philosophies of a number of theorists, including Norberg-Schulz, Heidegger, Trancik, Kelbaugh, Lynch, Alexander and Sharp. The Biosphere Reserve strategy was refined by replacing 'Transition Zone 1' (see Stellenbosch and Environs Sub-Regional Plan above) with the term 'Buffer', which continued to be a largely constrained area protecting the 'Core' areas. However, it contained a 'Rehabilitation' category to describe land that was severely degraded. The plan proposed that limited development rights could be considered within these previously disturbed areas in return for entrenching active conservation or rehabilitation of the buffer areas (WIDF 2000:202). The remaining 'Transition' areas were referred to as "flexible" areas of cooperation (WIDF 2000:55) and this sub-category was assigned intensive and extensive agriculture (*ibid.*:179); elsewhere in the text it is referred to as "Agriculture" (*ibid.*:178).

The WIDF also extended the concept of the 'Rural Development Areas or Zones', first raised in the Stellenbosch and Environs Sub-Regional Plan of 2000, to 'Special Management Zone' (SMZ). This was motivated by a plea to reconsider the prevailing strict approach to prevent the subdivision of agricultural land in order to "recognise the potential that exists in utilising the subdivision of agricultural land for non-agricultural purposes as a mechanism to support other important sectors of the district economy ..." (WIDF 2000:207). The SMZs were intended to comply with ISO 14001 certification and be controlled by a trust and trust fund seeded by developers, with an annuity income flowing from commissions on the sales of units in the SMZ scheme (WIDF 2000:232). In effect, it would comprise a kind of planning gain scheme to compensate for the development and create social or environmental benefits. This model has been proposed and implemented – although the extent of the implementation requires clarification – on large out-of-town projects such as De Zalze and Boschendal.

Proposals were also made for conservancies, the protection of scenic views, tourism promotion, waste water treatment management investigations, traffic calming measures, and a wide variety of other issues that required attention in the district.

Transformed local government – a new era?

In 2000 the transitional period came to an end and the new Stellenbosch Municipality (WC024) came into being.

New municipal boundaries were implemented, with Franschhoek, Pniel and Stellenbosch town (including Klapmuts) – that had previously operated as Transitional Local Authorities – forming the new Stellenbosch Municipality together with the southern third of the former WDC. This name was transferred to a new, much larger district council, including the valleys of Paarl and Wellington and extending across the mountains to the central Breede River Valley. After the November 1995 local government elections, the executive headquarters were moved to Worcester, while the administration sat in the former SDC offices.

However, despite the fact that the municipal boundaries were beginning to settle down, the Stellenbosch Municipal Council now entered an era of considerable volatility, with party political changes from the ANC to the DA and back again every 18 months to two years (see Table 1.1). Over time, this volatility extended to the executive director level of the administration, thus destroying much institutional memory on the one hand, and, on the other, not giving fresh incumbents with new ideas sufficient time to bed down.

Table 1.1 Changes in political power in Stellenbosch, 1996-2012

Date	Party in Power	Event
1996	ANC	Elections
2000	DA	Elections
2002	ANC / NNP	Floor crossing
2006	DA Coalition	Elections
2008	ANC Coalition	By-election
2009	DA Coalition	Motion of no confidence
2011	DA	Elections

Integrated Development Plans

The new Stellenbosch Municipality required a Spatial Plan (now called a 'Spatial Development Framework' (SDF) in terms of the new Municipal Systems Act) as an input into the Integrated Development Plan (IDP). IDPs represented a new initiative by central government to move local government onto a more regular and systematic footing. It was hoped that the process of producing annual budgets and reports would become a much more interdisciplinary and developmental activity to ensure integration and address the wide range of issues that municipalities face, but often don't address effectively if their line function departments operate in silos.

New overarching policy frameworks

The Provincial Administration's Department of Environmental Affairs and Development Planning (DEA&DP) completed a number of important planning policies during this period. In 2001 the department published a manual on biosphere planning as a guideline for municipal planning departments (Western Cape Provincial Government (WCPG) 2001), which – amongst other guidelines – set out similar spatial planning categories to those in the 1995 Stellenbosch and Environs Sub-Regional Plan.

In 2004, the DEA&DP completed the Provincial Spatial Development Framework (PSDF) (WCPG 2004) to support the Provincial Growth and Development Strategy and a number of developmental programmes the provincial government of the time was promoting under the banner 'Elihlumayo' and 'A Home for All'. The PSDF sought to coordinate provincial budget spending spatially, particularly in terms of the National Spatial Development Perspective; provide guidance to the preparation of district and local municipal SDFs; and set down a clear and consistent approach to land use management. It also sought to provide guidelines as to how the principles of the Development Facilitation Act and National Environmental Management Act (see Appendices 1 and 2) could be implemented. It revised some of the bioregional planning categories, in particular replacing the term 'Transition' with 'Agriculture' and referring to 'Intensive Agriculture', i.e. cultivation. Considering the potentially positive impact that proper veld management of grazing regimes could have on biodiversity, 'Extensive Agriculture' was moved to the 'Buffer' spatial planning category.

In an effort to develop a meaningful approach to sustainable spatial development, the PSDF also developed the Ecological Socio-Economic Relationship framework, which represented an attempt to understand, in an integrated and holistic way, a municipality or any other spatial or building entity as a metabolism with inputs and outputs with both spatial and aspatial constraints and opportunities.

The PSDF was adopted by the ANC provincial cabinet in 2005 and passed into law by the new DA provincial government in 2009.

Stellenbosch Municipality (WC024) Spatial Development Framework, 2005

Formulation of this plan began in 2002. Public participation was a much bigger feature of the post-2000 SDFs and this SDF went through three rounds over a period of four years. These were relatively well-attended compared to the average turnout for public participation in high-level policy plans, but the interest continued to emanate mainly from largely retired, white middle-class stakeholders, although there were efforts to access existing institutional structures, such as ministers' fraternals, in formerly black areas.

Project planners sought to break away from the previous spatial trajectory of limited internal change within settlements and *ad hoc* expansion of both high- and low-income housing on the urban periphery. Scenario planning was chosen as a vehicle to explore how the municipality might develop differently in order to accommodate its growing urban settlements.

Five scenarios were posited:

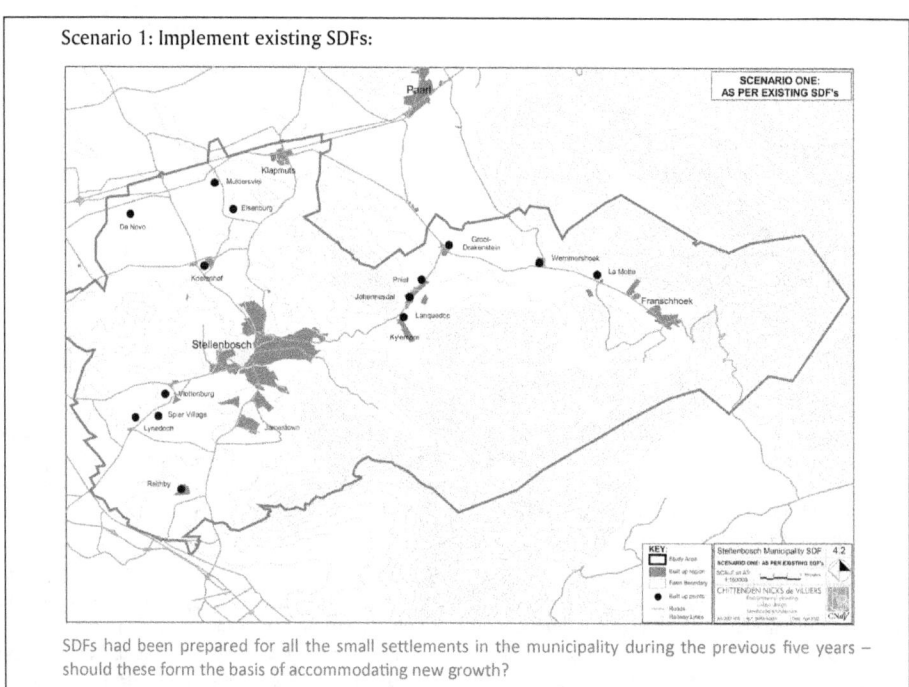

SDFs had been prepared for all the small settlements in the municipality during the previous five years – should these form the basis of accommodating new growth?

- Scenario 2: Locate development near employment growth nodes:

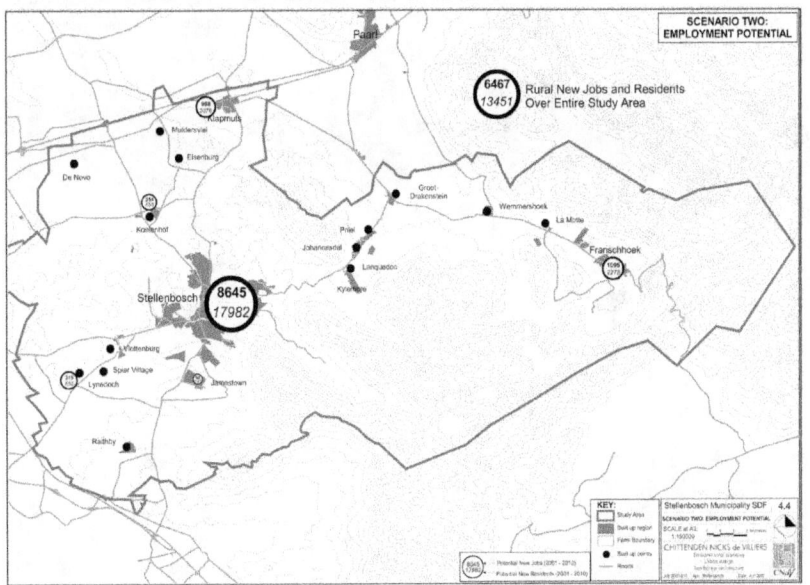

In terms of this scenario, an estimation was made about what employment generation was likely to be and where it would be located (Stellenbosch Municipality enjoyed stellar economic growth over this period); it then assumed that 50% of these workers should be able to walk to work, which has implications for finding land and housing very close to the employment centre.

- Scenario 3: Focus development at nodes on road transport corridors:

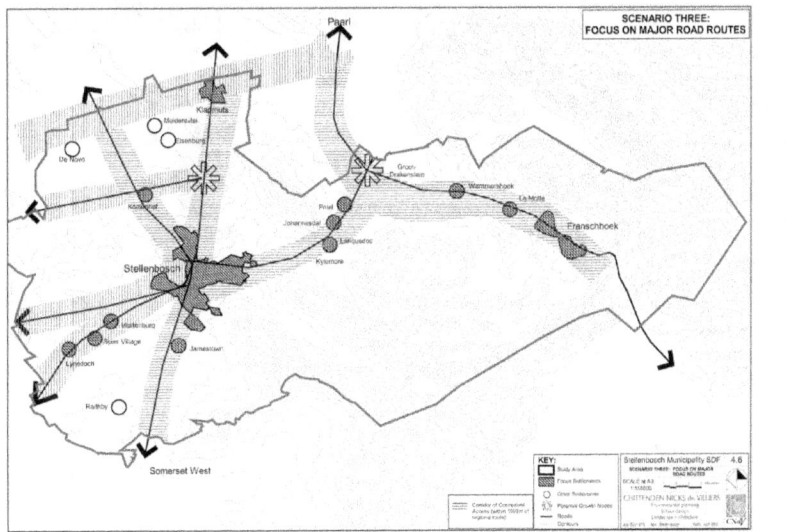

This scenario was based on the assumption that there would be a continuing need to commute and that this would be road-based public or private transport, therefore settlements on the road routes should enjoy development priority.

- Scenario 4: Focus development at nodes on rail corridors:

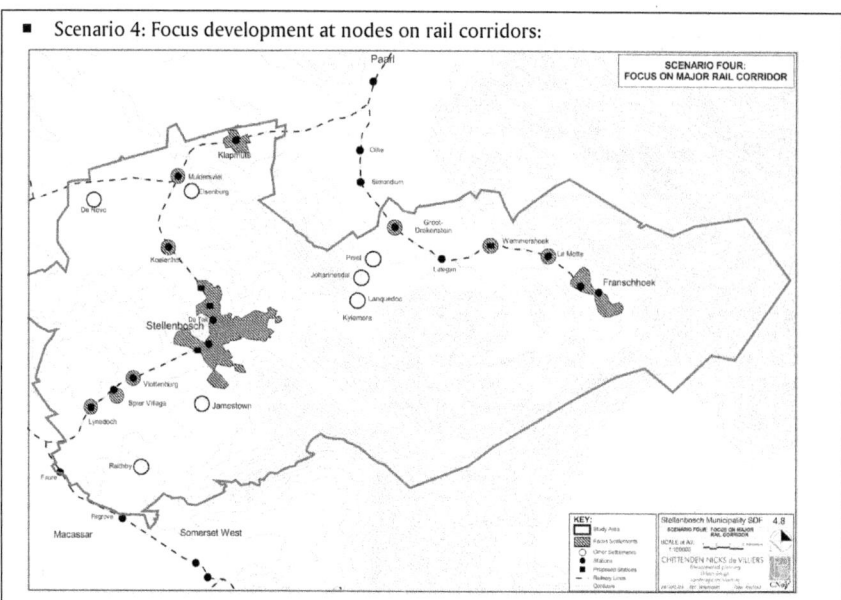

Scenario 4 was based on the assumption that commuting demand should be accommodated on the much higher-capacity, less environmentally-impactful rail system; settlements on rail routes should therefore enjoy priority (assuming also that a private or public operator runs a rail car service). The Franschhoek branch line could be brought back into service if correct operation, sufficient demand and safe protocol for level crossings were established.

- Scenario 5: Focus development at the three growth nodes:

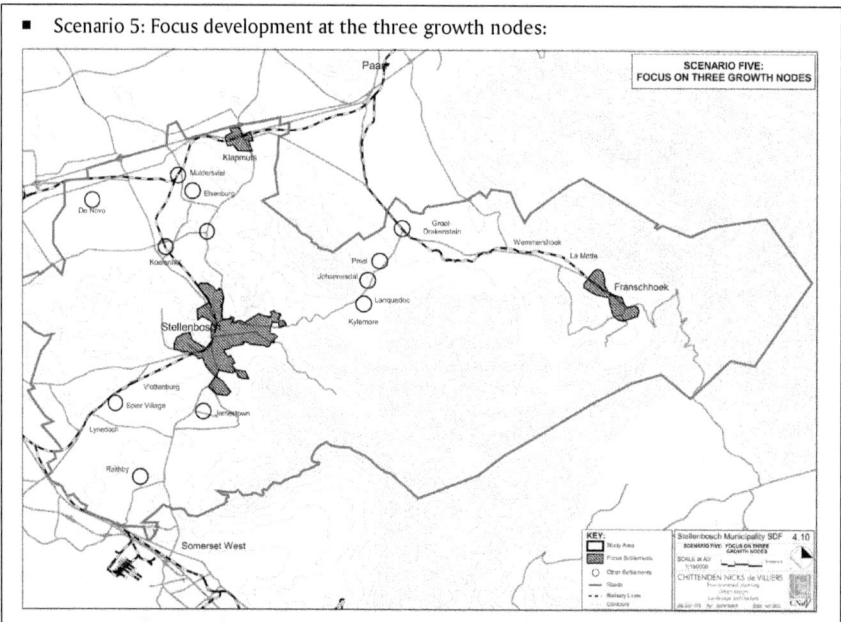

This scenario was based on the assumption that all new housing and jobs would be located in the nodes of Franschhoek, Klapmuts and Stellenbosch town, with no development anywhere else.

In addition to these scenarios, twenty implementation proposals and their required actions were identified, one of the most pressing being to assess the likely extent of the movement of farm labour to the nearest hamlet or village. This intra-migration was one of the most critical issues facing the municipality.

However, by 2006, when the political balance in council shifted and some of the senior line managers were replaced, few of these implementation proposals had been actioned.

Stellenbosch Town Spatial Development Framework, 2006

After an 18-month hiatus, another important planning initiative, which had started in 2004, was nearly brought to conclusion. This focused on Stellenbosch town and sought to prepare it for absorbing considerable urban growth with only limited lateral expansion, while at the same time ensuring that the essential character of the town be retained and strengthened. The initiative therefore emphasised urban restructuring and densification.

However, because of the natural conservatism of middle-class civic groupings – anxious about structural changes to the urban environment with which they are so familiar and whose positive elements they wish to retain – scenario planning was also used as a tool. Four scenarios were developed (see Figure 1.8).

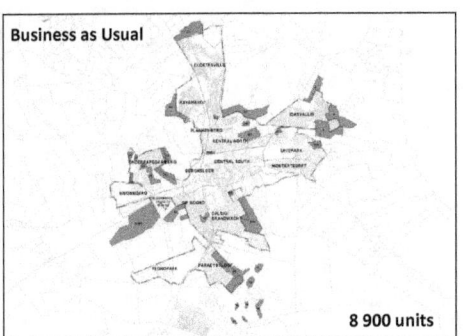
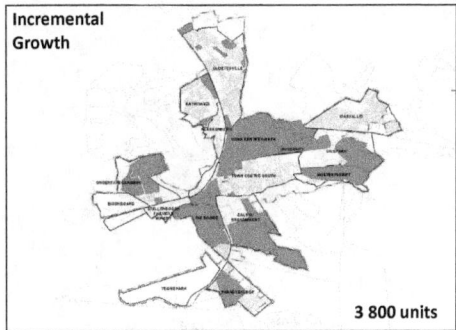
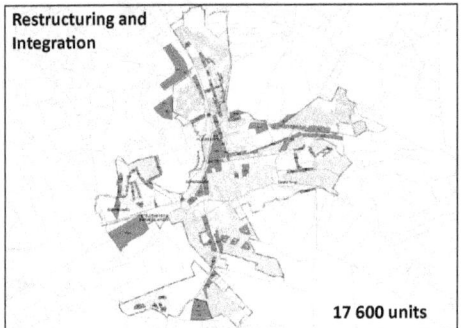
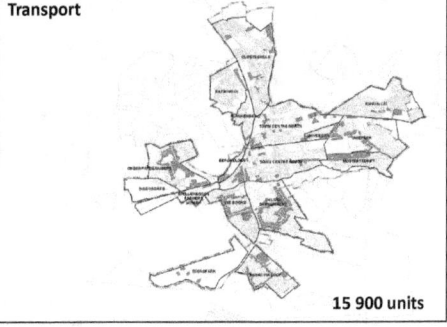

Figure 1.8 Target sites for urban restructuring and densification [CNdV Africa 2006]

An important aspect of these scenarios was that the interdisciplinary team of planners and civil and transport engineers who produced them also interrogated them with regard to their associated costs and benefits.

The scenarios were intensely workshopped and published in the media. In addition, a number of case studies of actual projects in Stellenbosch town demonstrating aspects of the 'Urban Restructuring and Integration' scenario were written up.

The outcome of the exercise was to be a large, integrated workshop in the town hall. One hundred key representatives from all communities in Stellenbosch were to be invited to debate the possibilities and choose a scenario, or to choose elements from each of the scenarios in order to make up a composite scenario. Clearly, for such a high-stake exercise to be successful, a stable, cohesive council and top management was essential. Unfortunately, this was not to be. Around the time of the critical workshop to bring closure to this stage of the project, there was another council change and the project was put on hold.

Finally, two years later, a technical in-house team put together what they considered to be the scenario most capable of addressing Stellenbosch town's future development pressures. Shortly after this, seven key sites were put out to tender for redevelopment (see Figure 1.9). This would have been a great opportunity to illustrate a new growth trajectory for the town using the projects as demonstrations. Although a few of the sites were controversial at the time, it was hoped that, if they were sensitively developed, many fears might be allayed. Unfortunately, there were a number of procedural issues with the tender process, including some grossly over-bulked proposals, and the programme was put on hold until the next council change, at which stage the whole process was shelved.

Figure 1.9   Stellenbosch Town: Tender sites in 2006 [CNdV Africa 2006]

Lessons for a future spatial plan for Stellenbosch

Urban reform in South Africa needs to achieve multiple objectives:
- move away from the apartheid spatial pattern;
- reduce carbon emissions;
- take climate change into account; and
- alleviate poverty.

Achieving these multiple objectives requires that a number of success factors be in place, including:

- courageous and strong political leadership that does not flinch from arguing key points against the support of its constituency – this requires a firmly entrenched and stable government;
- executive municipal management that can work in concert to develop a progressive vision from the departure point of each manager's key performance area;
- a supportive culture that welcomes carefully planned pilot policy projects as learning opportunities;
- continuous public participation and ventilating of new ideas through a range of platforms, from established IDP representative forums to the local mass media;
- a deepening of the systems approach to spatial planning, integrating it as fully as possible with the functions of different line departments and stakeholder groups; it cannot, and should not, be confined to the line activity of one particular department;
- mediate the necessary spatial compromises to facilitate short-term development needs with long term environmental, social and fiscal sustainability priorities; and,
- in particular, there must be a greater integration of the economic imperatives of growth and employment into the planning process, using techniques such as input/output models to achieve a deep, rather than shallow, understanding of the necessary preconditions and the way in which they may effectively be planned for.

However, due to the political and institutional flip-flopping that Stellenbosch Municipality has endured over the past decade, it has not been possible to embed these success factors.

To date, the two preconditions for successful urban and rural reform clearly have not been met: firstly, embracing a new urban and rural vision; and secondly, creating a sufficiently stable and courageous political platform to provide necessary, but sometimes non-populist, reforms. Only time will tell whether they will be satisfied following the 2011 local government elections.

REFERENCES

Acocks JP. 1988. *Veld types of South Africa*. 3rd Edition. South Africa: National Botanical Institute.

Adendorf J. 1984. Personal communication.

Calderwood DM & Connell PH. 1952. Minimum Standards of Accommodation for the Housing of Non-Europeans in South Africa. *Bulletin*, 8 (June 1952). Pretoria: National Building Research Institute.

CNdV Africa. 2004. *Stellenbosch Town Growth Management Strategy Report*. Stellenbosch: Stellenbosch Municipality: Department of Planning.

CNdV Africa. 2005. *Stellenbosch Municipal Spatial Development Framework*. Stellenbosch: Stellenbosch Municipality: Department of Planning.

CNdV Africa. 2009. *Stellenbosch Municipal Spatial Development Framework*. Stellenbosch: Stellenbosch Municipality: Department of Planning.

CNdV Africa, Urban Dynamics & GLS. 2004. *Towards a Growth Management Strategy for Stellenbosch*. Stellenbosch: Stellenbosch Municipality: Department of Planning.

Floyd TB. 1961. *Town Planning in South Africa*. Pietermaritzburg: Shuter & Shooter.

Miller M. 1989. *Letchworth: The First Garden City*. 2nd Edition. West Sussex: Phillimore & Co. Ltd.

NBRI (National Building Research Institute). 1951. *Minimum Standards of Housing Accommodation for Non-Europeans*. South Africa: National Housing Office.

Oude Meester Groep. 1979. *Stellenbosch 3 Eeue.* Stellenbosch: Stadsraad van Stellenbosch.

Republic of South Africa. Cape Winelands District. 2009. *Cape Winelands Biosphere Framework Plan.* Draft 2. Stellenbosch: Department of Planning.

SDC (Stellenbosch Divisional Council). 1967. *Regional Development Scheme.* Stellenbosch: Department of Planning.

Surveyor General. 2000. *Maps for Bellville in 1958, Kuils River in 1959, Durbanville in 2000, Kuils River in 2000.*

WCPG (Western Cape Provincial Government). 2001. *Manual for Bioregional Planning.* Cape Town: Department of Environmental Affairs and Development Planning.

WCRSC (Western Cape Regional Services Council). 1995. *Stellenbosch and Environs Sub-Regional Plan.* Draft 1. Cape Town: Department of Planning.

APPENDIX 1

Summary of the principles of the Development Facilitation Act (DFA) (Act 67 of 1995)

Although the DFA largely does not apply in the Western Cape, the principles espoused by the Act are applicable to all land development. The following principles are of particular relevance and require that land development should:

- provide for urban and rural land development and facilitate development of formal and informal, existing, and new settlements;
- discourage illegal occupation of land, with due recognition of informal land development processes;
- promote efficient and integrated land development through –
  - the integration of social, economic, institutional and physical aspects of land development;
  - integrated land development in rural and urban areas;
  - promoting availability of residential and employment opportunities in close proximity to each other;
  - optimising the use of existing resources;
  - promoting a diverse combination of land uses;
  - discouraging the phenomenon of urban sprawl and contributing to the development of more compact towns and cities;
  - contributing to the correction of historically distorted spatial patterns of settlement in the Republic; and
  - encouraging environmentally sustainable land development;
- allow active community participation in the land development process;
- develop the capacities of disadvantaged persons involved in land development;
- encourage and optimise the contributions of all sectors of the economy to land development;
- promote sustainable land development at the required scale through –
  - the promotion of land development within the means of the Republic;
  - the establishment of viable communities;
  - sustained protection of the environment;
  - meeting the basic needs of all citizens in an affordable way; and
  - the safe utilisation of land;
- promote speedy land development; and
- promote security of tenure and provide a possible range of tenure alternatives.

The following principles are of particular relevance to the Western Cape:

- The promotion of the availability of residential and employment opportunities in close proximity to or integrated with each other (Section 3(1)(c)(iii));
- The optimisation of the use of existing resources, including resources relating to agriculture, land, minerals, bulk infrastructure, roads, transportation and social facilities (Section 3(1)(c)(iv));
- The promotion of a diverse combination of land uses, also at the level of individual erven or subdivisions of land (Section 3(1)(c)(v)); and
- Discouraging the phenomenon of urban sprawl in urban areas and contributing to the development of more compact towns and cities (Section 3(1)(c)(vi)).

The above principles recognise that urban areas should utilise land more efficiently and effectively, particularly in locations close to employment opportunities, so as to discourage urban sprawl. The use of sites to accommodate multi-level apartments complements these principles, since such developments minimises pressure on the outward expansion of the city.

APPENDIX 2

Principles of the National Environmental Management Act, 1998 (Act 107 of 1998), Section 2

"2. (1) The principles set out in this section apply throughout the Republic to the actions of all organs of state that may significantly affect the environment and –

    (a) shall apply alongside all other appropriate and relevant considerations, including the State's responsibility to respect, protect, promote and fulfil the social and economic rights in Chapter 2 of the Constitution, and in particular the basic needs of categories of persons disadvantaged by unfair discrimination;

    (b) serve as the general framework within which environmental management and implementation plans must be formulated;

    (c) serve as guidelines by reference to which any organ of state must exercise any function when taking any decision in terms of this Act or any statutory provision concerning the protection of the environment;

    (d) serve as principles by reference to which a conciliator appointed under this Act must make recommendations; and

    (e) guide the interpretation, administration and implementation of this Act and any other law concerned with the protection or management of the environment.

(2) Environmental management must place people and their needs at the forefront of its concern, and serve their physical, psychological, developmental, cultural and social interests equitably.

(3) Development must be socially, environmentally and economically sustainable.

(4) (a) Sustainable development requires the consideration of all relevant factors, including the following:

        (i) that the disturbance of ecosystems and loss of biological diversity are avoided, or, where they cannot be altogether avoided, are minimised and remedied;

        (ii) that pollution and degradation of the environment are avoided, or, where they cannot be altogether avoided, are minimised and remedied;

        (iii) that the disturbance of landscapes and sites that constitute the nation's cultural heritage is avoided, or where it cannot be altogether avoided, is minimised and remedied;

        (iv) that waste is avoided, or where it cannot be altogether avoided, minimised and reused or recycled where possible and otherwise disposed of in a responsible manner;

        (v) that the use and exploitation of non-renewable natural resources is responsible and equitable, and takes into account the consequences of the depletion of the resource;

        (vi) that the development, use and exploitation of renewable resources and the ecosystems of which they are part do not exceed the level beyond which their integrity is jeopardised;

        (vii) that a risk-averse and cautious approach is applied, which takes into account the limits of current knowledge about the consequences of decisions and actions; and

        (viii) that negative impacts on the environment and on people's environmental rights be anticipated and prevented, and where they cannot be altogether prevented, are minimised and remedied.

    (b) Environmental management must be integrated, acknowledging that all elements of the environment are linked and interrelated, and it must take into account the effects of decisions on all aspects of the environment and all people in the environment by pursuing the selection of the best practicable environmental option.

(c) Environmental justice must be pursued, so that adverse environmental impacts shall not be distributed in such a manner as to unfairly discriminate against any person, particularly vulnerable and disadvantaged persons.

(d) Equitable access to environmental resources, benefits and services to meet basic human needs and ensure human well-being must be pursued, and special measures may be taken to ensure access thereto by categories of persons disadvantaged by unfair discrimination.

(e) Responsibility for the environmental health and safety consequences of a policy, programme, project, product, process, service or activity exists throughout its life cycle.

(f) The participation of all interested and affected parties in environmental governance must be promoted, and all people must have the opportunity to develop the understanding, skills and capacity necessary for achieving equitable and effective participation, and participation by vulnerable and disadvantaged persons must be ensured.

(g) Decisions must take into account the interests, needs and values of all interested and affected parties, and this includes recognising all forms of knowledge, including traditional and ordinary knowledge.

(h) Community well-being and empowerment must be promoted through environmental education, the raising of environmental awareness, the sharing of knowledge and experience, and other appropriate means.

(i) The social, economic and environmental impacts of activities, including disadvantages and benefits, must be considered, assessed and evaluated, and decisions must be appropriate in the light of such consideration and assessment.

(j) The right of workers to refuse work that is harmful to human health or the environment and to be informed of dangers must be respected and protected.

(k) Decisions must be taken in an open and transparent manner, and access to information must be provided in accordance with the law.

(l) There must be intergovernmental coordination and harmonisation of policies, legislation and actions relating to the environment.

(m) Actual or potential conflicts of interest between organs of state should be resolved through conflict resolution procedures.

(n) Global and international responsibilities relating to the environment must be discharged in the national interest.

(o) The environment is held in public trust for the people, the beneficial use of environmental resources must serve the public interest, and the environment must be protected as the people's common heritage.

(p) The costs of remedying pollution, environmental degradation and consequent adverse health effects, and of preventing, controlling or minimising further pollution, environmental damage or adverse health effects must be paid for by those responsible for harming the environment.

(q) The vital role of women and youth in environmental management and development must be recognised and their full participation therein must be promoted.

(r) Sensitive, vulnerable, highly dynamic or stressed ecosystems, such as coastal shores, estuaries, wetlands, and similar systems, require specific attention in management and planning procedures, especially where they are subject to significant human resource usage and development pressure."

Chapter 2

LAND

Farmland and tenure security

JUANITA PIENAAR

Introduction

To a large extent, the farmland surrounding Stellenbosch is its lifeblood: world renowned for its award-winning wines and a major attraction in the tourist industry. However, among other things, productivity and sustainability require sufficient fertile farmland of suitable quality held in secure tenure. On a national scale, in order to address the historical imbalances and the skewed land ownership and settlement patterns within a new constitutional dispensation, a three-pronged land reform programme aimed at (a) broadening access to land (redistribution), (b) upgrading of insecure rights (tenure reform programme), and (c) restoring land or land rights (restitution programme) (Pienaar & Brickhill 2007) was launched in 1994. All three of these sub-programmes resonated to some degree within the Stellenbosch jurisdictional area.

Seventeen years after the introduction of the new Constitutional dispensation, the effectiveness of the land reform programme is still under discussion (Pienaar 2011). The restitution programme can be quantified by, among other things, the number of claims that have been finalised and the amount of land restored, while calculating the hectares of land that have been redistributed since 1994 may give some indication of the success of the redistribution programme. However, the tenure reform programme differs from the other two sub-programmes in that it is extremely difficult, if not impossible, to quantify tenure security. What is clear, however, is that:

- more evictions have occurred post-1994 than before that date (Social Surveys and Nkuzi Development Association 2005);
- occupiers' and farm workers' rights are still tenuous (Pienaar & Geyser 2010);
- the housing crisis has been exacerbated by farm dwellers losing their homes on farmland (Liebenberg 2010); and
- an especially valuable tool in both accessing and securing land rights, namely the commonages, has not been utilised optimally (Anderson & Pienaar 2003; Legal Resources Centre 2010).

In light of the importance of farmland in the Stellenbosch area, this chapter will reflect on tenure security provisions within the constitutional and legislative context and the effectiveness of these measures. The discussion will also encompass the rights and responsibilities of the three main role players in this regard: the farmer (or landowner), the farm worker, and the local authority. Instead of extending security of tenure, as the tenure reform programme set out to do, other (perhaps unintended) consequences have come to the fore, and it is within this context that the key challenges to tenure security will be set out.

Tenure security within farming communities

Background

Historically, most farm work was performed by a permanent on-farm workforce, with 75% of jobs held by men (Shabodien 2007). Farm workers mostly resided on the farm, which led to a rather unique relationship between the landowner and the farm worker in that residential and labour issues were interwoven. Accordingly, the provision of housing was considered a production input cost and did not form part of social subsidies, since the building of houses was subsidised pre-1994 by the state as part of its broader system of agricultural subsidies. Apart from overarching, racially-based land measures relating to the occupation of land in general (Pienaar 1998), there was no legislation regulating the eviction of farm workers and the concomitant tenure and labour issues before the advent of the new constitutional dispensation and its approach to landholding and corresponding rights.

Constitutional and legislative context

Section 25 of the Constitution of the Republic of South Africa of 1996 stipulates that:

- "A person or community whose tenure of land is legally insecure as a result of past discriminatory laws or practices is entitled, to the extent provided for by an Act of Parliament, either to tenure which is legally secure or to comparable redress" (Section 25(6)); and
- "Parliament must enact the legislation referred to in subsection (6)" (Section 25(9)).

In accordance with the Constitution, various legislative measures relating to the upgrading of insecure rights in general, and the provision of more secure tenure in particular, were promulgated, including the Land Reform (Labour Tenants) Act 3 of 1996 and the Extension of Security of Tenure Act 62 of 1997 relating to agricultural land (Badenhorst, Pienaar & Mostert 2006). The Western Cape differs from the rest of the country in that it has neither traditional communal land nor labour tenancy areas, and has the largest number of farm dwellers (Republic of South Africa: Department of Land Affairs (RSA DLA) 2007). In light of these peculiarities, the discussion will focus mainly on the Extension of Security of Tenure Act (hereafter referred to as the ESTA).

The ESTA and tenure security

As is clear from the title, the main aim of the ESTA is to provide for more secure tenure for persons falling within the ambit of the Act. In essence, the Act applies to rural areas or land used for agricultural purposes and persons who occupy it with either tacit or explicit consent from the landowner or person in charge, and sets out the rights and responsibilities of all relevant parties.

Tenure security may be achieved in two ways: firstly, by strictly regulating eviction in accordance with constitutional values and principles; and secondly, by providing for long-term tenure security.

Effective regulation of eviction means that evictions may not be arbitrary and that strict requirements must be met. The ESTA provides for normal and urgent eviction proceedings and draws a distinction between persons who occupied the land before 4 February 1997 (when the Bill was first published for comment) and persons who became occupiers after that time. Strict procedural requirements relating to, among other things, the provision of notices, must be met. Section 9(3) provides for the submission of a probation report in order to bring the relevant circumstances of both the occupier and the landowner or person in charge before the court. Unfortunately, some judgements have been handed down without the submission of such a report, as waiting for the report to be submitted may lead to proceedings being prolonged beyond a reasonable period of time. Eviction may only occur on the legal ground stipulated in the Act and only after the court has considered all relevant circumstances. Once it is satisfied that all procedural and substantive requirements have been met and that the granting of an eviction order would be fair in the particular circumstances, the court may grant the order. Furthermore, all eviction orders granted by the lower courts (the magistrates' courts) are subject to an automatic review procedure in the Land Claims Court.

Chapter II of the Act provides for long-term security of tenure for occupiers by means of on- or off-site developments. The former entails a development on occupied land belonging to the landowner; the latter entails a development on land belonging to someone other than the owner of the occupied land. Although the Act formulates a clear preference for on-site/on-farm development, there has not yet been any large-scale development on privately owned farm land. Nor has off-site development been successful. In fact, hardly any developments – either on or off farms – have been implemented in the Stellenbosch jurisdictional area.

Role players in tenure security

Achieving tenure security requires a unique partnership between three main role players, namely landowners, persons residing and working on farms, and government, in particular local government.

Landowners

Section 25 of the Constitution stipulates that no one may be arbitrarily deprived of property (s 25(1)) and that expropriation of property may only occur "for public purpose or in the public interest" (s 25(2)(a) and only with the payment of "just and equitable compensation" (s 25(3)) as determined in accordance with the criteria set out in Section 25(3)(a-e). A landowner's ownership of his or her farm is therefore protected. However, the property clause also specifically provides for land reform (s 25(5), (6) and (7)) and the reform of all natural resources (s 25(8)). Accordingly, the exercise of ownership and the protection of property are governed by the broader transformational framework of the Constitution, which invariably calls for striking a balance between the rights of landowners and the rights of the landless (Van der Walt 2005; 2009).

The whole of Chapter III of the ESTA is devoted to the specific rights and obligations of landowners (or persons in charge) and occupiers (farm workers). Section 5 sets out a list of fundamental rights: the right to human dignity; freedom and security of person; privacy; freedom of religion, belief and opinion and of expression; freedom of association; and freedom of movement. All of these rights may, however, be limited, provided that such limitations are reasonable and justifiable in an open and democratic society based on human dignity, equality and freedom (s 36 of the Constitution). Apart from the list of fundamental rights, additional provision is made for landowners or persons in control of property with regard to the removal and impounding of trespassing animals in certain circumstances (s 7(1)). Furthermore, a landowner "may not prejudice an occupier if one of the reasons for the prejudice is the past, present or anticipated exercise of any right" (s 7(2),(3)). Provided that the substantive and procedural requirements are met, landowners are also entitled to terminate occupational rights and apply for an eviction order.

Occupiers and farm workers

An occupier is a person who resides on the land in question with the tacit or explicit consent of the landowner. In practice, the majority of occupiers are farm workers. The ESTA draws a distinction between a 'normal' occupier and a 'long-term' occupier. The difference is that a long-term occupier is a person who has been on the land for 10 years or longer and who has reached the age of 60 (s 8(4)). This category of occupiers has more rights and enjoys greater protection under the Act than persons falling within the category of 'normal' occupiers (Pienaar & Brickhill 2007).

Besides the fundamental rights listed in Section 5 of the ESTA, which apply to landowners, persons in charge, and occupiers alike, Section 6 also lists the specific rights and responsibilities of occupiers and/or farm workers, and stipulates that these rights are always "balanced with the rights of owners or the person in charge" (s 6(2)(a)). In addition to the rights emanating from the Act itself, parties are also free to include particular rights or any other privileges in the agreement that forms the basis of the occupier's status.

In view of the ESTA's underlying aim of promoting tenure security, the primary right of occupiers stipulated in Section 6(1) is the right to reside on and use the land, and to have access to such services as have been agreed upon with the landowner or person in charge. Depending on the circumstances, the latter may include a right to water, electricity, refuse removal, and educational or training facilities. Occupiers also have the right "to receive *bona fide* visitors at reasonable times and for reasonable periods"; the right "to receive postal and other communication"; a right "to family life" in accordance with their particular culture; the right to bury a deceased member of his or her family in certain circumstances; and the right "not to be denied or deprived of access to water … [or] to educational or health services" (s 6(2)(3)(4). In addition, the judgement handed down in the case of *Nkuzi Development Association vs Government of RSA* (2002 (2) SA 733 (LCC)) also established the right of occupiers and farm workers to legal representation and legal aid.

Long-term occupiers have the additional right not to have their residency terminated unless they have unlawfully or intentionally caused damage to property or threatened other persons residing on the land. In practice, this means that they may retire on the land they have been occupying without the obligation to work or render any other services. Provided that they do not commit any breach (as contemplated in section 10(1)), the long-term occupier's spouse and/or dependants may also remain on the land for up to 12 months after his/her death. Finally, a special provision allows for the burial of long-term occupiers on land in accordance with Section 6(5).

Local authorities

As representatives of the national government at a local level, local authorities or municipalities are automatically involved in land reform matters in general. Tenure security is usually associated with access to land for either residential or agricultural purposes, or both. As for the tenure security of farm workers, one may look at it from a short- and long-term perspective. In order to promote short-term tenure security, evictions must be strictly regulated. Long-term security, on the other hand, would entail farm workers acquiring strong tenure rights on the land they occupy (or in relation to other land if they are relocated in accordance with off-site developments). Tenure security is also necessary with regard to land being used for agricultural purposes. In this context, private or public land may be utilised in accordance with the Provision of Land and Assistance Act 123 of 1991. However, in this instance tenure security should not be limited to the acquisition of ownership alone.

In terms of short-term tenure security, the ESTA stipulates that all notices of eviction applications also be served on the municipality where the land is located, as well as the head of the provincial Department of Rural Development and Land Reform (s 9(2)(d)). By itself, this does not require any proactive conduct from local government. However, the joinder of local authorities in eviction applications under the ESTA recently came under the spotlight in an appeal from the magistrate's court to the Land Claims Court (*Pietersen vs Van Deventer* (LCC 158/2009); judgement delivered on 25 March 2010). As the local authority is the only body that is truly able to paint the full picture about the housing situation in a given area, it was argued that it has a direct and substantial interest in all eviction matters. Unfortunately, the necessity of joining the local authority in this particular instance was not investigated in detail. Since the legislature had already recognised the need to *notify* the role players of an eviction in the area "short of them being joined" (par 10 of the judgement), the court was satisfied that joinder in itself was unnecessary. Therefore, in terms of the ESTA, the precise role of local authorities in eviction applications is relatively passive at present and requires no specific activity apart from filing eviction notices.

The role of local authorities in promoting long-term security of tenure is likewise vague and falls short of proactive conduct (Chapter II of the ESTA); however, its role with regard to the use of municipal commonage, as a special category of state-owned land, may be explored in more detail.

Actors supporting tenure security in Stellenbosch area:

- The Centre for Rural Legal Studies
- Stellenbosch University Legal Aid Clinic
- Women on Farms
- Sikhule Sonke
- Lawyers for Human Rights
- The Legal Resources Centre

Municipal commonage: Land for tenure reform

It is clear that local authorities have a role to play in the identification and acquisition of land for redistribution and tenure reform purposes. Various mechanisms exist; for example, land may be either purchased or expropriated by the state, or donated by private owners to local authorities. It is also possible for existing categories of state-owned land to be designated for redistribution or tenure reform purposes, and it is in this context that commonage has an important role to play.

There are two kinds of municipal commonages, namely 'existing' municipal commonage and 'new' municipal commonage (Mostert, Pienaar & Van Wyk 2010). Existing commonage refers to land that was granted free of charge to (white) municipalities by the state for the use and benefit of (mostly white) town residents during the 1800s and later. In the case of Stellenbosch, about 2 000 hectares of Crown land was given free of charge to the local authority in 1883. New municipal commonage is former white-owned, commercial agricultural land that was purchased after 1994 as commonage by a municipality with state funds. The title deeds to these parcels of land are subject to special conditions to prevent municipalities from alienating commonage land by sale, donation, or land swaps. These conditions underpin local authorities' responsibility to make such land available to its residents – with an emphasis on the poor and less privileged – on a secure and equitable basis (Legal Resources Centre 2010).

In fact, commonage is not only intended to remain the property of the local authority in perpetuity, but also to be utilised in the best interest of the community at large. Historically, in the former Cape Province in particular, special sets of commonage regulations setting out the rights of inhabitants with regard to the use of commonages were developed (Mostert *et al.* 2010). By the mid-twentieth century, communities' wealth increased and technology changed to such an extent that commonage land was increasingly leased to (white) tenants for commercial farming and to generate income. Apart from the fact that the nature of commonage use changed from, for example, common pasture to commercial farming, this transition also excluded black people (Anderson & Pienaar 2003).

Under the new post-1994 political dispensation all residents became entitled to access to municipal commonages. However, in most instances, large portions of commonage land had already been leased to white commercial farmers – in some cases for up to 50 years – or had been sold in contravention of the title deed conditions. Most of the commonage land in South Africa set aside for land reform purposes is presently either still being used by commercial farmers in accordance with long-term lease agreements, or for grazing (Legal Resources Centre 2010). In fact, the use of commonage for small-scale agricultural purposes or subsistence farming, with the focus on the poor or vulnerable members of society, is rather rare, even though the land reform programme specifically provides for it to be used for exactly these purposes.

As in the rest of the country, Stellenbosch Municipality entered into various long-term lease agreements with large commercial farmers involving large portions of commonage land just before the start of the new constitutional dispensation. Given that most of these leases span 50 years and were entered into around 1991 (which means that they will be in effect until 2041), and that the local authority intends to honour these arrangements, the greater portion of municipal commonage will be tied up in existing lease agreements for decades to come, unless subletting is negotiated or agreements are terminated (Brummer 2010; Stellenbosch Small Farm Holdings Trust Project 2007).

In one instance only, a commercial farm, namely Spier, terminated an existing sublease contract with a neighbour and made this land, which comprises 65 hectares, available to upcoming farmers. Subsequent to this, the Stellenbosch Small Farm Holdings Trust (Farm 502 Land Reform Project) was established with the assistance of the Legal Resources Centre in Cape Town and the Trust entered into a lease agreement with the municipality at a nominal amount. Providing the necessary infrastructure and negotiating access to additional water involved long, time-consuming negotiations, while rendering the land viable and producing good crops on it proved challenging (Brummer 2010). Each upcoming farmer has five hectares on which strawberries and other fruit and vegetables are grown, and the individual farmers must each pay a subsidised rental fee, as well as for water. In the event of non-payment, a farmer has to vacate his portion of the land. Accountability and responsibility are therefore integral to the success of the venture (Stellenbosch Small Farm Holdings Trust 2007).

Two further elements are essential for the success of commonage ventures: tenure security and cooperation between all role players. As for the former, while ownership remains with the local authority, it is essential that participants' rights are set out clearly. Tenure security is thus not limited to ownership only, but can also include lease, leasehold, or even a permit-based system, provided that the scope of the tenure right and security is clearly defined and systematically specified. Effective cooperation between all stakeholders is also essential. For example, in order to secure more water for allotment, an effective working relationship between local government, farmers, the Department of Water Affairs, and the Department of Rural Development and Land Reform is required. If any one of these partners fails to contribute and cooperate as required, the venture is ultimately doomed.

Implications of existing legislation and policy

The importance of preserving valuable agricultural land within the Stellenbosch jurisdictional area is clear; however, it is equally important to ensure tenure security for those who own the land, use the land, and work the land. Attempts at striking a balance between the conflicting aspirations and rights of landowners and occupiers and applying the relevant legislative measures have yielded mixed (and perhaps unexpected) results.

Ten years after the commencement of ESTA, the agricultural work force had already shrunk by more than 22% nationally (Shabodien 2007) and over time landowners have reduced the number of persons residing on their farms and replaced them with temporary or contract workers. As the ESTA only applies to rural or peri-urban areas, it may be circumvented by employing farm workers residing in towns and transporting them to and from work. This has serious implications, in particular with regard to workers' housing needs and their safety while travelling. As much as 60% of agricultural work in the Western Cape is temporary, with two thirds of jobs performed by women (Shabodien 2007; UBISI 2011). A temporary female work force also has specific implications, for example, no provision for pension, sick or maternity leave, and generally lower wages compared to male employees (RSA DLA 2007; Sloan 2010; South African Human Rights Commission 2008). In response, interest groups such as the Women on Farms Project and Sikhule Sonke, an independent women-led trade union for farm women, were established to address the specific needs of women. Both organisations are active in the Stellenbosch area.

Despite the ESTA having been on the law books for more than a decade, evictions have increased since 1994 (Social Surveys & Nkuzi Development Association, 2005), which, in turn, has dramatically exacerbated the housing shortages in urban areas and towns and increased the urgency of identifying land for housing purposes. However, according to the Nkuzi Development Evictions Report, only 1% of all post-1994 was lawful (Social Surveys & Nkuzi Development Association, 2005) which means that legislative measures and the courts have played a role in only 1% of cases where people lost the roof over their heads. This raises serious questions about the enforcement of legislation in general, and the monitoring of evictions in particular. Clearly, it is nonsensical to have legislative measures in place if they are not enforced or respected. The situation is compounded by the beneficiaries' general ignorance of their rights and responsibilities: as is clear from case law, despite it being in force since 1997, occupiers and farm workers apparently still do not know and understand their rights under the Act (see eg *Herman Diedericks vs Univeg Operations South Africa (Pty) Ltd T/A Heldervue Estates* (LCC18/2011).

Neither have the existing legislative measures providing for housing developments – either on the farm or elsewhere – been implemented effectively. Some of the blame for this must be attributed to the legislature: Chapter II of the ESTA is vaguely formulated and fails to assign clear duties or responsibilities to the provincial Departments of Rural Development and Land Reform and local authorities. Moreover, successful developments require a unique partnership between the relevant stakeholders, including both the public and private sectors, to surmount the numerous problems related to them, such as insufficient funding and issues associated with transport, infrastructure and the administration of land, including the survey and subdivision of land.

Another impediment to effective tenure reform and land redistribution legislation is the fact that has been operating within an incomplete policy framework. The Rural Development Policy was only published in 2009 and clear, concise guidelines and frameworks remain lacking, while the continual amendments to grants, plans, and application procedures render some plans and strategies invalid or outdated, complicating matters for local authorities and beneficiaries who must perforce expend much time and effort just to stay abreast of developments.

Finally, commonage land – specifically earmarked for local use by indigent persons or beneficiaries who qualify under the overall land reform programme – has not been put to effective use for redistribution and tenure reform purposes.

Key challenges

Stellenbosch Municipality's Local Economic Development Strategy specifically highlights the imperative to reduce poverty and promote sustainable livelihoods (Stellenbosch Municipality 2008). The promotion of Broad-based Black Economic Empowerment and land reform has likewise been prioritised. In terms of total area and prime land, land ownership remains skewed in favour of residents formerly classified as white, thereby marginalising the majority of residents previously classified as coloured or black. Effective access to land (and water) – and the attendant access to markets – remains crucial, and associated to this is the need for capacity development and mentoring. Apart from restating the general need to secure land, the local authority's Land Reform Strategy also called for the provision of housing opportunities (close to rural settlements) for small-scale or subsistence farmers and viable commercial-scale land reform projects. The specifics regarding possible locations, support, and the roles of the various parties are outlined in the Cape Winelands Area-based Plan. Recommendations relating to the implementation of the Land Reform Strategy in Stellenbosch include the approval of the Area-based Plan prepared by the then Department of Land Affairs; increasing awareness of the need for land reform in the Stellenbosch area; and the institution of a local land reform process (*ibid.*).

Since the publication of the 2008 Economic Strategy, there have been a number of important developments with regard to land reform in general. Although the long-awaited Green Paper on Tenure Reform has not yet been published, a Draft Policy on Tenure Security and a Draft Tenure Security Bill were published for comment on 24 December 2010 (Government Notice 1118 in *Government Gazette*, 3894). The main thrust of these documents is to promote tenure security by repealing the two separate acts aimed at labour tenants and occupiers (farm workers) respectively, and combining the relevant measures into a single new Act, the Tenure Security Bill. Some of the proposals contained in the documents were informed by the Department of Rural Development and Land Reform's Strategic Plan for the period 2010-2013, as well as the National Summit on Vulnerable Workers in Agriculture, Forestry and Fisheries, which took place in July 2010 in Somerset West in the Western Cape.

All of these developments – most notably the restructuring of the Department of Land Affairs in May 2009 (resulting in the Department of Rural Development and Land Reform) – have impacted on the approval and implementation of the above plans and strategies (Du Plessis, Olivier & Pienaar 2009). However, most of them have been put on hold in expectation of the finalisation of the Recapitalisation and Development Programme (RADP) and, to a lesser extent, the Green Paper on Tenure Reform. These initiatives, in particular the RADP, also impacted on the Commonage Infrastructure Grant that formerly applied to commonages. Accordingly, most of the ideals and strategies envisaged by Stellenbosch Municipality have failed to come to fruition.

Various key challenges therefore remain. In a general sense, the prompt identification of suitable land for residential and agricultural purposes remains a major challenge. Improved synergy and cooperation between the relevant institutions and governmental departments is integral to resolving this issue. As far as long-term tenure security is concerned, more emphasis on the development of 'agri-villages' is required. The new Tenure Security Bill will impact directly on how these developments are to be approached (Pienaar & Kamkuemah 2011). Chapters 6 and 7 of the Bill are integral to the viability of Section 20(11), which stipulates that no eviction order may render a person homeless. Chapter 6 deals with agri-villages and land development measures, while Chapter 7 addresses the management of resettlement units and agri-villages. Essentially, Section 26 of the Tenure Security Bill allows the Minister to apply the provisions of the Provision of Land and Assistance Act 126 of 1993 to institute land development measures, including the establishment of agri-villages. Although the Bill is still rather vague, local authorities will be burdened with participating in the provision of settlement areas.

In terms of short-term security and evictions, particular challenges may be identified, including the necessity of promoting lawful evictions and strict compliance with eviction measures, which in turn requires the implementation of improved monitoring systems. Promotion of access to information is also crucial, as occupiers and landowners must know and understand their rights in order to exercise them – which brings one to the issue of effective enforcement and the need to further expand access to legal representation for all parties.

Conclusion

The actual use of both private and public land will, to a large extent, depend on legislative measures and policy developments still in the pipeline. As both the Tenure Policy and the Tenure Security Bill are still in draft form, their passage through parliament before promulgation and actual implementation are still a long way off. These measures will impact directly on the short- and long-term security of farm workers and occupiers and will have important implications for landowners and local authorities. Proactive conduct and greater participation are envisaged with regard to both landowners and local authorities, for example, by submitting specific plans to enhance security and identifying suitable land for relocation well in advance.

However, while tenure security developments remain in limbo, it is essential that municipalities do all and everything that they are able to at present. This means that local authorities must draft and adopt budget-linked plans for the ongoing, financially sustainable management of land as part of their Integrated Development Plans. Since lease agreements relating to commonage land are being upheld at present, it is important that new initiatives and their associated groundwork are in place at the time these leases lapse. Perhaps a committee, located within the local authority, can already be established at this stage to explore possible avenues for commonage usage. Such a committee would be well positioned to identify relevant stakeholders and initiate discussions that might outline options for when the leases lapse. As one small commonage farming venture has already proved successful, it may be worthwhile at this stage to already encourage subletting of existing leases on a greater scale, coupled with accountability and sufficient infrastructure and support.

Notwithstanding the uncertainties clouding the way forward with regard to land use in general, and land reform in particular, the need to preserve Stellenbosch farmland is undisputed; however, it is also essential that it be optimally utilised for the benefit of the Stellenbosch community as a whole.

REFERENCES

Anderson M & Pienaar K. 2003. *Evaluating land and agrarian reform in South Africa*. Occasional Paper No 5: Municipal Commonage (PLAAS).

Badenhorst PJ, Pienaar JM & Mostert H. 2006. *Silberberg & Schoeman's Law of Property*. Cape Town: Lexis Nexis.

Brummer W. 2010. Ons boer met geloof. *Die Burger*, 6 August. p. 2.

Du Plessis W, Olivier NJJ & Pienaar JM. 2009. Land Matters and Rural Development. *South African Public Law*, (2): 588-601.

Legal Resources Centre. 2010. *Municipal Commonage: How to access and use it*. Cape Town: June, Legal Resources Centre & Trust for Community Outreach and Education.

Liebenberg S. 2010. *Socio-Economic Rights – Adjudication under a Transformative Constitution*. Claremont: Juta.

Mostert H, Pienaar JM & Van Wyk J. 2010. *The Law of South Africa: Land*. Durban: Lexis Nexis.

Pienaar JM. 1998. Farm workers: extending security of tenure in terms of recent legislation. *South African Public Law*, (2):423-437.

Pienaar JM. 2011. Tenure reform in South Africa: Overview and challenges. *Speculum Juris*, (1):108-133.

Pienaar JM & Brickhill J. 2007. Land. In: S Woolman, T Roux & M Bishop (eds). *Constitutional Law of South Africa*. Cape Town: Juta.

Pienaar JM & Geyser K. 2010. 'Occupier' for purposes of the Extension of Security of Tenure Act: the plight of female spouses and widows. *Tydskrif vir die Hedendaagse Romeins-Hollandse Reg*, (2):248-265.

Pienaar JM & Kamkuemah A. 2011. Farmland and tenure security: new policy and legislative developments. *Stellenbosch Law Review*, (3):724-741.

RSA DLA (Republic of South Africa. Department of Land Affairs). 2007. *Farm worker eviction: Briefing by Women on Farms and Deputy Minister of Land Affairs*. [Accessed 3 February 2011] http://www.pmg.org.za.report/20070821-farm-worker-eviction-briefing-women-farms-deputy-minister-land-affairs

Shabodien F. 2007. Farm workers need dignified homes in South Africa. *Women on Farms Project*:1.

Sloan J. 2010. *Mayday! Mayday! Farm workers remain at the bottom of the pile*. News by Greater Good South Africa. [Accessed 20 February 2012] http://www.myggsa.co.za/news/2307

Social Surveys & Nkuzi Development Association. 2005. *Still searching for security: The reality of farm dweller evictions in South Africa*. Polokwane North: Nkuzi Development Association.

South African Human Rights Commission. 2008. *Progress made in terms of Land Tenure Security, Safety and Labour Relations in Farming Communities since 2003*.

Stellenbosch Municipality. 2008. *Local Economic Development Strategy*. Stellenbosch: Stellenbosch Municipality.

Stellenbosch Small Farm Holdings Trust Project. 2007. *Summary Report*. Stellenbosch: Stellenbosch Small Farm Holdings Trust Project.

UBISI. 2011. High HIV rate among female farm workers. *UBISI Mail*, March. p. 7.

Van der Walt AJ. 2005. *Constitutional Property Law*. Cape Town: Juta.

Van der Walt AJ. 2009. *Property in the Margins*. Oxford: Hart Publishing.

Legislation

Constitution of the Republic of South Africa No 108 of 1996.

Draft Tenure Security Bill (24 December 2010 in Government Notice 1118 in *Government Gazette*, 3894).

Extension of Security of Tenure Act (Act 62 of 1997).

Land Reform (Labour Tenants) Act (Act 3 of 1996).

Provision of Land and Assistance Act (Act 123 of 1991).

Policy Documents

Draft Policy on Tenure Security (24 December 2010 in Government Notice 1118 in *Government Gazette*, 3894).

Republic of South Africa. Department of Rural Development and Land Reform. 2010. *Strategic Plan of the Department of Rural Development and Land Reform for the period 2010-2013*. Pretoria: Government Publishers.

Case Law

Herman Diedericks vs Univeg Operations South Africa (Pty) Ltd T/A Heldervue Estates (LCC18/2011).

Nkuzi Development Association vs Government of RSA. 2002 (2) SA 733 (LCC).

Pietersen vs Van Deventer (LCC 158/2009).

Chapter 3

URBAN SPACES
Quartering Stellenbosch's urban space

RONNIE DONALDSON & JOLANDA MORKEL

Introduction

The human dimension makes a space a place. Matthews (2005:35) explains that 'place' is space that is informed by experience, because "ideas of place are informed both by the physical reality of the situation and by the idiosyncratic and culturally-biased perceptions of the individual experience of that space". A key characteristic of the apartheid city is the functional and racial separation of urban spaces. Drawing on Marcuse and Van Kempen's (2002) postmodern urbanism debates on the partitioned city, this chapter highlights some of the most significant dynamic 'quarters' produced in Stellenbosch as a result of spatial, socio-economic and counterproductive, politically-driven processes within the partitioning debate.

The discussion includes a debate on the significance of urban spaces, as well as the interrelationship between exclusion and race, class, and subcultures; the walling/hardening of boundaries between and among the quarters; and the central role of living spaces in these processes.

The chapter is divided into three sections. The first focuses on the significance of urban spaces within the context of sense of place, 'placelessness', and quality of urban space. In section two, we briefly discuss five salient processes actively shaping urban spaces, namely studentification, gating, neo-ruralism, gentrification and squatting. In the last section, some scenarios for dealing with the aforementioned processes are provided.

The Stellenbosch context

The university town has undoubtedly undergone a dramatic spatial and demographic transition since the 1970s. According to the 2001 census data, the town's population grew from 29 955 in 1970 to 96 250 in 2001, with the highest growth rate (595%) among blacks (although they constitute only the third largest group in terms of numbers).

Table 3.1   Population of Stellenbosch [Compiled from Statistics South Africa 1970; 1980; 1996; 2001]

Race	1970	1980	Population growth (%) 1970-1980	1996	Population growth (%) 1980-1996	2001	Population growth (%) 1996-2001	Population growth (%) 1970-2001
Black	2 848	2 846	-0.07	11 715	312	19 804	69	595
Coloured	13 269	18 182	37.03	26 224	44	51 662	97	289
White	13 778	19 510	41.60	22 063	13	24 553	11	78
Indian/Asian	60	71	18.33	226	218	231	2	285
TOTAL	29 955	40 609	35.57	60 228	48	96 250	60	221

The dramatic growth in population between 1980 and 2001 has expectedly impacted the town's urban space in various ways: pressure on accommodation for students, pressure on accommodation for migrants from the Eastern Cape, and pressure on accommodation for the affluent who desire secluded lifestyle environments. When the actual population growth per former group area is correlated with the actual urban spatial expansion, it is seen that there is a mismatch in terms of race-space relations (Figure 3.1). In 1938, the physically built footprint of Stellenbosch (excluding Kayamandi) was 263 415 ha; by 1983, it had increased to 1 083 355 ha; and by 2009, it had grown to 1 516 739 ha. At the two latter points in time, Kayamandi had a footprint of 3 132 ha and 99 984 ha respectively. This reflects a horizontal spatial expansion of 40% for Stellenbosch (excluding Kayamandi), compared to 3 092% for Kayamandi.

Figure 3.1   Spatial growth of the built-up area between 1938 and 2009
              [Stellenbosch University: Centre for Geographical Analysis 2010]

The evidence suggests that student recruitment exceeds strategic planning and provision of accommodation by higher education institutions in university towns worldwide (Smith, 2009). Stellenbosch is no exception. In 1990, Stellenbosch University registered 11 379 students; by 2010, this number had swelled by 137% to 26 964 (Figure 3.2). Of these, just over 14 000 lived in Stellenbosch (this excludes the students studying at the Bellville campuses), with the university housing approximately 7 580 of them – leaving over 6 000 students to find alternative accommodation in the private sector (Benn 2010).

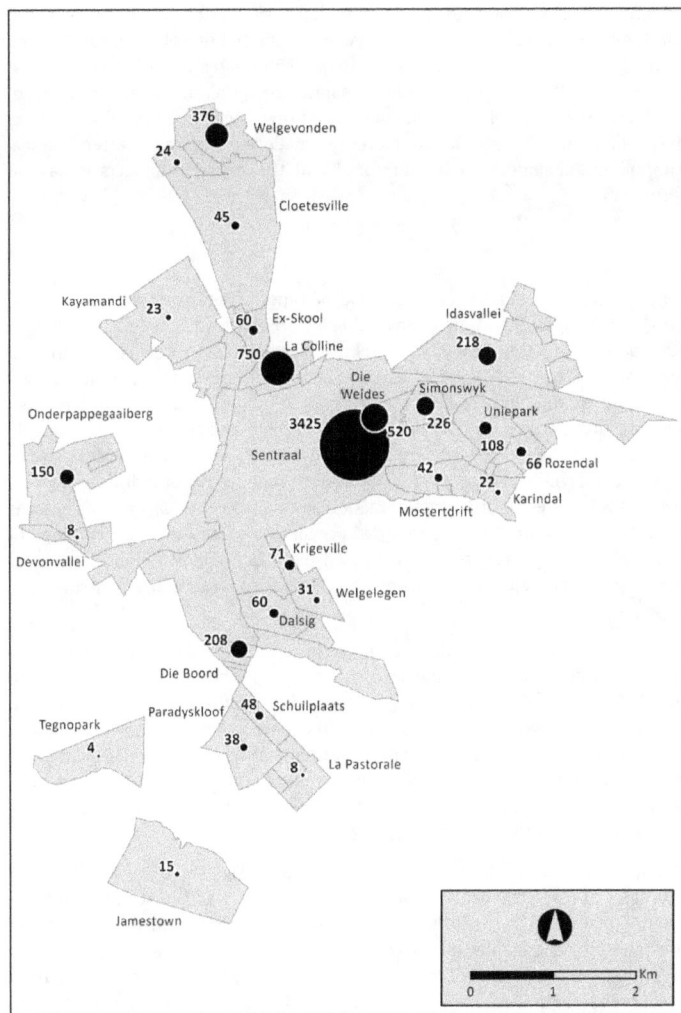

Figure 3.2　　Student population per suburb [Adapted from Benn 2010]

Stellenbosch is clearly a town bursting at its seams. Yet, the results of two consecutive studies conducted during the past decade (Van Der Merwe, Zietsman, Ferreira, Davids, Swart & Kruger 2005; Van Niekerk, Donaldson, Du Plessis & Spocter 2010) showed that it has the highest growth potential in the Western Cape, which will contribute to further pressure for its growth and more spending from the provincial government to sustain that growth. The town is at a crossroads and is in an institutional crisis for which urban politics is directly responsible. We call this an 'urban spatial identity crisis'. Marcuse and Van Kempen (2002:5) argue that "the state's [proactive rather than reactive] actions are crucial to implementing partitioning. By the same token, the state can be effective in countering the forces that lead to partitioning".

Seethal (2004) explored the politics of reaction, the politics of resistance (and reconstruction), and the cultural politics of difference and identity that characterised Stellenbosch Municipality from 2000 to 2002. Although Seethal's study is dated, the application of his analysis may have had similar results in 2012. He argued that political hegemony in Stellenbosch, as dominated by the Democratic Alliance regime, "advanced middle-class material successes, co-opted senior municipal officials and maintained cosy neo-traditional preserves into which they retreated (even though the majority poor residents were denied basic services and infrastructure)" (2004:n.p.). After 2002, the then dominating ANC-NNP Alliance "engaged in the politics of exclusion and the marginalisation of the opposition – pushing the 'others' to the margins of discourse and of physical space (for example in council debates, and in its efforts to prematurely terminate the appointments of senior white officials)". The net result of the seesawing political stage has been a political hegemony for urban space/place-making and the geography of difference in Stellenbosch; moreover, given this highly contested political environment, it is unlikely that political hegemony for urban space and the negative impacts it has for sustainability will dissipate before 2025.

Meaningful urban space

Complexity, surprise, diversity of activity and users, vitality, a sense of time, and historical continuity are some of the key requirements for a meaningful urban experience (Rowley 1998). Rowley suggests a number of considerations for meaningful place-making: functional and social use, sustainability, visual characteristics, and quality of the urban experience, while Pistorius (2002) identifies six qualities that contribute to the making of meaningful urban space: balance, freedom, equity, complexity, integration, and community. *Balance* refers to important relationships, including that between a city and its surroundings and nature – a sustainable one. *Freedom* is a necessary quality for a democratic society. *Equity* allows all people easy access to the opportunities the city provides. A certain level of *complexity* (as opposed to monotony), or richness, in terms of a diversity of activities and facilities is required to promote interaction among residents. *Integration* ensures accessibility to these activities and facilities, which allows the broad population to be exposed to a range of activities and events. Finally, *community* relates to a sense of identity (sense of place) and belonging. The issues of integrated urban space, sense of place, and authenticity as prerequisites for meaningful urban space are considered here.

Sense of place, unlike the historic or traditional sense of place, emerges from the everyday human experience of a place across time – the repetitious sense of time emphasising the daily reiterated stories of living in a given place. According to Liu (2009), the practice of everyday life today creates more significant cultural meanings and sense of place than does the history of the place itself. In conservation theory, sense of place is often understood as a set of subjective feelings for the overall traditional character, national identity, or existing sociocultural character of a place – that is, the sense of place informing urban conservation practice is mainly derived from the cultural traditions of a place (*ibid.*). Fortunately, there is a growing concern about the disappearance of sense of place as evidenced in the placeless approach of modernism.

The Stellenbosch Spatial Development Framework (SDF) formulates the goal to "conserve the architectural, historical, scenic and cultural character of the settlements, forms and areas in the Stellenbosch Municipality" (CNdV Africa 2010:7). It describes the main arterial roads linking the major settlements with the surrounding region as playing "a key role in establishing the municipality's sense of place because of the views they offer, the experience of travelling along them, and the nature of their detailed design" and warns that, "should their design speeds be increased, their curves straightened, their edges fringed with suburban-style precast concrete kerbing, urban development intensified and advertising boarding erected alongside, the sense of place of the municipality will be considerably diminished and one of its main tourist attractions destroyed". The spirit or sense of place has to be preserved to retain a particular local quality and protect the character of a place. In addition, the concept of sense of place has been much appreciated because of its respect for the conditions of the authenticity and integrity of heritage in contemporary conservation practice. In Stellenbosch, Die Vlakte was a district possessing these qualities for successful urban space, comparable in many ways to District Six in Cape Town. Unfortunately, this former vibrant and mixed-use townscape in the area east of Victoria Street, between Bird and Ryneveld Streets, was destroyed in the 1960s and early 1970s through a process of forced removals (Giliomee 2007). A good surviving example of such a meaningful open space (with the potential of being further enhanced) is Die Braak, where the landscaping reinforces the rural character and allows views to the mountains (balance); all people are free to act out daily life (freedom and equity); and quiet and busy, sunny and shaded spaces are provided, defined by harmonious, yet varying facades (necessary complexity). It is visible and accessible (integration) and provides a strong sense of belonging and identity (community).

Urban places without a sense of place are sometimes referred to as 'placeless' (or inauthentic) places. Placeless landscapes are those that have no special relationship to the places in which they are located – they could be anywhere, such as gated estates with pseudo-vernacular styles and monotonous, 'studentified' residential spaces. Recommodification of historical areas that have been purposefully 'touristified' (such as the inner historic core of the town) and standardised massification of housing developments (Academia and Concordia) result in such places losing their sense of place – or to quote Gloria Stein, "there is no there, there" (1937:289). In places with historical significance, the effect of modernisation, globalisation, modern consumer trends, the need for convenience, development pressure and property rights result in inappropriate generic building and space-making (Pistorius 2002).

Stellenbosch has not escaped this phenomenon. One of the many examples is the singularly out-of-character and out-of-place Star Trek (see Anonymous 1996) service station in Dorp Street, opposite the iconic Oom Samie se Winkel and De Akker. The call for adequate protection of the historic milieu – including not only conservation-worthy buildings, but also other tangible and intangible forms of heritage, such as oral history and cultural landscapes – by heritage watchdogs, is made for the sake of the touristification of urban space. Nonetheless, there is an appeal for an approach that allows for change to take place, instead of a doggedly conservative one – or, to quote Welch, "seeking a kind of fossilised past" (1993:27). Change should be appropriate; it should acknowledge the distinct identity and sense of place. In Stellenbosch, heritage has acquired an elitist reputation, based on the traditional association with the preservation of the white European heritage. The built heritage of the Cape Malay and slave traditions has unfortunately been swallowed by the former.

In 2011, the urban spaces of Stellenbosch as public life appearance, and urban space as an inner self, are opposites. The destruction of the 'true inner self' – in other words the built environment heritage of the town and 'ruralscapes' – is slowly, but surely, masking a persona of progression, class advancement, almost uncontrolled modernisation, and untouchability. Yet, for communities struggling to adequately feed, house and ensure the health of their citizens, the devotion of time and energy to heritage goals may appear to be a luxury which is not shared equitably. Why then search for authentic urban spaces in Stellenbosch? Architects and developers justify their applications for the demolition and unsightly restructuring of historic buildings by arguing that layering is just another mark left by a new generation. Conservation strategies for places mean preserving what is valuable from the past even while adapting to change (Ferris 2002).

Thinking through current urban sociospatial relations

In the context of Van Kempen's (2002:50) argument that "cities are not 'naturally' divided: they are actively partitioned", it is clear that introspection of urban spaces in Stellenbosch is necessary to provide a meaningful, imagined urban environment. Although there are multiple spatialities at play, we focus on five processes most prominently (re)shaping urban space.

Studentification

Given its university town character, the balancing of student life and urban space in Stellenbosch warrants attention. Studentification of suburbs causes spatial dysfunctionality, where, in most cases, only the needs of a student subculture are catered for. More often than not, planning policy is formulated to change the zoning from single residential to general residential, resulting in social protest by residents for fear that their suburb should be invaded by student accommodation. We argue that both dramatic percentage increases of student enrolment and resident-based totals directly contributed to a fundamental spatial transformation of the town in terms of sociospatial relations, urban spatial expansion, and the resultant spatial impacts as a consequence of these pressures. Neither the university nor the municipality has kept pace with this influx and alternative urban social-spaces have subsequently been created. One such spatial outcome is studentification – the process whereby students start inhabiting certain parts of a suburb or town, gradually displacing the original residents. The spatial consequences of studentification are immense and include an array of economic, social, physical, and cultural impacts (Smith, 2009; Fox 2008). As a consequence of the studentification of urban space, a particular spatial pattern has emerged over the past two decades, the result of which is seen in the current spatial distribution of students' places of residence. Figure 3.2 indicates a major cluster in the town centre and to its immediate north. Negative social, physical, and cultural impacts are experienced in the studentified neighbourhoods, resulting in problems such as noise, traffic congestion, high residential density, and the loss of neighbourhood character (Benn 2010). In Die Weides and Simonswyk, there is a lack of social cohesion between permanent residents and students, a situation which is exploited by property developers.

In the process, studentification has inflated the property market, thus denying low middle- and middle-income earners the opportunity of buying property in the town.

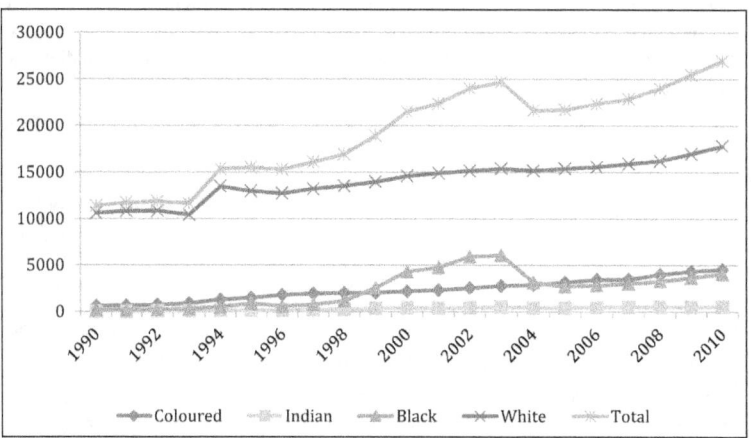

Figure 3.3   Student numbers according to population group: 1990-2010
[Compiled from Stellenbosch University Student Records 2010]

Gated developments

A second sociospatial impact derives from gated developments. In a country struggling to shed its apartheid legacy, it has been widely argued that gated developments inherently present undertones of re-segregation – a re-creation of urban spaces of separation. Critical viewpoints on gating revolve around three issues: lack of spatial integration, proposed policy on social mixing, and access to natural and urban public spaces (Republic of South Africa: Department of Housing 2004; Office of the Presidency 2004). According to the South African Human Rights Commission (2006:5), gated developments "cause social division, dysfunctional cities and lead to the further polarisation of our society" – a scenario very much evident in Stellenbosch. Whereas gated living has become a prominent feature in the urban landscape for a distinct class in society (mainly affluent and white), very little investigation has been done on the actual impact it has in non-metropolitan places. In the Western Cape, there has been a gradual increase in these developments since the late 1990s. Although Stellenbosch does not have many gated developments yet, the existing ones on the periphery of the town already dramatically scar the rural landscape and transform the rural spaces into urban quarters of homogeneity (Spocter 2010). It is a common trend achieved via the local authority's claim that "municipalities are in tough financial situations, and the privatisation of public property can help relieve budget pressure" (Nissen 2008:1141). Land for development is running out. It is our contention that the *de facto* local authority emerging from this is the property developer, who will insist on walling a new development that happens to be located in 'undesirable spaces' near a township or low-income area – Welgevonden, De Zalze, and La Clémence, as well as the newest addition, the 'Berlin-styled' Mount St Simon, wedged between Kayamandi and Cloetesville, are cases in point.

Post-productivist countryside

The tenets of post-productivism include a shift in focus to the emergence of non-food-producing farm jobs and activities for income; counter-urbanisation, leading to social and economic restructuring; the creation of a consumptionist countryside; the demand for amenity value from rural landscapes; and agriculture that has lost its central role in the countryside (Van Niekerk, Donaldson, Du Plessis & Spocter 2010). The inmigrants driving this demand for housing tend to be middle- and upper-class urbanites, with their key driving forces being the quest for improved lifestyle in exclusive housing units, the need to be close to nature and unspoilt areas, and for a higher degree of personal and property security compared to that in metropolitan areas. These spaces are slowly being transformed into a picture of urbanised rurality that may be embedded in the minds of the inmigrants and their vision of the services needed to cater for their lifestyle. It is significant that concomitant to the development of residential sites in non-metropolitan areas is the development of commercial and retail services, which further change the character of these tranquil spaces (Phillips 2005). In addition, non-metropolitan spaces are also targeted by hi-tech industries for their offices and operations.

Partitioning of space

This background context bears a striking similarity to the remaking of urban spaces in the Jamestown-Technopark node of Stellenbosch. As a fourth urban sociospatial process, i.e. the partitioning of space, the Jamestown node is a good illustration of the destruction of a once authentic urban space – an historical rural hamlet (for coloureds) – by urban spatial transformation. After 1910, individual families who leased property in Jamestown had an option to buy. Strawberry farming was the main rural activity and land was passed from father to son for the next 80 years. A strong sense of family and collective community evolved over these decades, but since the 1980s, pressure to subdivide increased as families grew in numbers (Perold 2010). The Jamestown interdependent node presents a strong juxtaposition of varying urban space best exemplifying the inherited spatial complexities of partitioned urban spaces. In a short period, the urban space of this hamlet has undergone a dramatic transformation from being a predominantly mono-functional, coloured, dormitory, rural agricultural village to an eclectic urban spatial mix of squatting, retirement, and lifestyle gated developments, a decentralised shopping centre, gentrification, and residential desegregation, all quartered. The question we pose is whether this intra-quartering of the town's nodes is desirable. Moreover, the danger of this model lies in its being replicated in other identified nodes (for example Klapmuts).

Abandoned spaces

No discussion about sustainability in South Africa can ignore the so-called 'abandoned spaces'. Bourdieu (2009:127) contends that "the ability to dominate space, notably by appropriating (materially or symbolically) the rare goods (public and private) distributed there, depends on the capital possessed. Capital makes it possible to keep undesirable persons and things at a distance at the same time that it brings closer desirable persons and things (made desirable, among other things, by their richness in capital), thereby minimising the necessary expense (notably in time) in appropriating them." Conversely, the "lack of capital intensifies the experience of finitude: it chains one to a place" (*ibid.*). Another characteristic of the quartering of the town is the 'abandoned city', that is, the spaces the indigent are 'chained to'. Here we have urban space for the "very poor, the excluded, the never-employed and the permanently unemployed, the homeless and the shelter residents" (Marcuse 1989:705). Table 3.2 shows that the greatest need for formal housing is concentrated in one of the town's former black townships, Kayamandi, with Jamestown and Cloetesville ranking lower in terms of the need for formalised housing.

Table 3.2   Housing needs analysis [Compiled from Stellenbosch Municipality* 2010]

Details	Total
Number of registered indigent households	7 535
Total number of municipally owned public rental units	607
Total number of staff rental units	161
Formal structures (overcrowded dwellings and backyard shacks) in WC024	±9 000
Estimated informal dwellings and shacks (total amount, by area) in WC024: Kayamandi Franschhoek (Langrug) Klapmuts Cloetesville (Slabtown) Pniel, Kylemore, Lanquedoc, Johannesdal Jamestown Devon Valley	**7 032** 5 531 1 234 149 28 19 37 34

*Note: WC024 represents the whole Stellenbosch municipal area (i.e. including Franschhoek)

Towards sustainable urban space

There is always tension between continuity and change in a city (Liu 2009). The dramatic population growth and pressure for urban expansion (especially for student accommodation, gated estates, and informal settlements) experienced over the past two decades call for an imaginative relook at how authentic spaces with a sense of place are to be conceived and produced over the next twenty years.

Creating more meaningful urban space

If left unchecked, the transformation of authentic urban spaces into new townscapes reflecting a current trend of a 'fantasising' lifestyle based on escapism – for example, converting a 1960s-style house into a Tuscan villa or a Byzantine-style home – could be complete within twenty years. It is ironic that the strong property market at the high end of the scale, driven by the exceptional scenic and small-town qualities to be found so close to the City of Cape Town (CNdV Africa 2005), has led to strong pressure to develop the exact scenic landscape that it is drawn towards, creating inauthentic spaces in the process.

The 2010 SDF emphasises the need to "guard against the erosion of the city's strong sense of place that has become an international brand" (quoted in CNdV Africa 2005:3). Over the past few years, there have been a number of urban conservation and design proposals formulated for Stellenbosch by consultants (for example Kruger Roos Architects, Urban Designers and Planners, and Chittenden Nicks de Villiers) appointed by the Stellenbosch Municipality, but implementation remains uncoordinated and ineffectual. Is there a plan for managing urban change that encompasses both the historical and everyday human experiences of the town? How may the *genius loci* be preserved under the pressure of new functional demands? What is to be conserved? What is to be changed? How much can be changed? In order to maintain a 'sense of place', while at the same time allowing for urban change, these questions should be incorporated into an urban policy-making vision. Changes to sensitive urban environments must respect the spirit, prevailing character, and atmosphere (*genius loci*) of a place; otherwise it will lose its identity (Norberg-Schultz 1984). However, respecting the *genius loci* does not equate to copying old models. The challenge is to determine the identity of the place and to interpret it in ever-fresh ways. Development is seen to be in direct opposition to conservation and vice versa. However, opportunities presented by appropriate development should not be overlooked in an attempt to resist development. Change is necessary, provided that it is appropriate and reinforces, rather than erodes, the local identity or distinctiveness.

Designing appropriate student living space

Clustered student accommodation results in dead urban space that lies uninhabited for over a third of the year. An emerging sterile and disjointed student urban landscape destroys the historical nature of urban space in the town. Replacing one monotonous residential landscape (for example Die Weides) with another monotonous landscape (studentification of Die Weides) with no regard for integrating the streetscape and mixing of land use, is foolish. The *ad hoc* densification of the emerging student housing node in Die Weides may only serve as a temporary reprieve for a university that is growing in student numbers without planning accordingly for better space utilisation. Likewise, the current student housing policy of the municipality is seriously flawed in its attempt to control studentification and is in desperate need of rethinking and reformulation. The university's greed in not capping student numbers, while at the same time failing to provide housing alternatives, is inexcusable. Student enrolment should be directed at innovative cyber learning opportunities for non-resident students. Another possibility is the creation of a dispersed university campus network, for example the relocation of the Faculty of Economic and Management Sciences to the Bellville Park campus (to build on existing synergy with the Graduate School of Business and the Medical Faculty at Tygerberg). Similarly, may the Faculty of Agriculture not be incorporated with Elsenburg? Further infill and intensification on the Stellenbosch campus is needed, whereby hostels can be remodelled without impacting the current meaningful open spaces. The partnership formed between the university and the municipality (astonishingly formalised only in 2009!) should provide a sustainable platform for addressing the crucial urban spatial problems created by increased student intake and the studentification of residential spaces, or creatively help shape the (re)structuring of knowledge-based economic urban spaces. Is it not time for the town to position itself rather as a knowledge-based town?

Protecting the ruralscape against rural gating
The spatial outcomes of gated developments directly impact the town and regional identity. Wilson (1995:n.p.) refers to Lewis Mumford's view of the relationship between the urban and the rural, arguing that Mumford's 1960s observation is applicable to Stellenbosch in 2011, in the sense that sprawling urban spaces will inexorably lead to necropolis, the death of the city. The original finite space (the containment of urban space) no longer exists. Wilson quotes Mumford's vivid, but depressing, description:

> As the eye stretches towards the hazy periphery, one can pick out no definite shapes ... one beholds rather a continuous shapeless mass ... the shapelessness of the whole is reflected in the individual part, and the nearer the centre, the less ... can the smaller parts be distinguished ... The city has absorbed villages and little towns, reducing them to place names ... [I]t has ... enveloped those urban areas in its physical organisation and built up the open land that once served to ensure their identity and integrity ... [Then] as one moves away from the centre, the urban growth becomes ever more aimless and discontinuous, more diffuse and unfocused ... Old neighbourhoods and precincts, the social cells of the city, still maintaining some measure of the village pattern, become vestigial. No human eye can take in this metropolitan mass at a glance.

One alternative to sprawl (such as in gated sprawl) into high-value agricultural areas in the Cape Winelands, is the reshaping of the urban settlement system for the development of agri-villages, ecovillages and green urbanism. These developments, if properly introduced, can counter the unsustainable and highly (segre)gated golf, polo, and other lifestyle estates and avoid a replication of the Jamestown node, where gentrification is displacing the poor. Such a system can be facilitated in nodes on land of varying quality, size and situation, and plan for sustainable urban agriculture and local food system movements, such as the Slow Food Movement, where the farmer's market becomes a basic town centre amenity. At the same time, new urbanism projects and smart growth initiatives have demonstrated the possibility of creating healthier, more liveable urban centres (Krause 2006). For the implementation of this paradigm to succeed, nodal developments should guard against the continued growth in mono-functional gated developments. The Lynedoch EcoVillage is an example of a socially mixed hamlet built around a school and crèche for poor people and non-partitioned ecologically designed housing. The Klapmuts node (sometimes referred to as the Midrand of the Boland) on the other hand, is in the process of being carved up into walled security estates with the institutional endorsement of Stellenbosch Municipality and the National Department of Housing – a 'new town', composed of, in the words of Swilling (2011:n.p.), "disconnected spatially coterminous gated communities ... an urban horror story". The Raithby case is a typical example of how developers exploit sustainable development ideals for their own personal gain. Planned and approved as an organic community with an ecologically designed, semi-gated community, this development was supposed to reflect an emergent middle-class green urbanism. However, since its approval, the focus has seemingly shifted from an ecologically sensitive development to just another gated estate.

Sustainable urban spaces or 'back to the future'?
The current 'back to the future' mentality of the municipality in 'dealing' with squatting is reminiscent of apartheid policies of segregation. Having a task force arresting all unemployed newcomers and repatriating them elsewhere is unconstitutional. Discarding innovative proposals for dealing with abandoned spaces is short-sighted, especially in the context of the National Spatial Development Perspective and the provincial policy on prioritising infrastructural funding in towns with a very high development potential, such as Stellenbosch (Van Niekerk *et al.* 2010). The inability to proactively and creatively think about undoing the abandoned spaces in the urban space and place-making process resonates with the notion of the geography of difference in Stellenbosch, where politicians and government officials are acting as disabling agents. Policy suggestions, such as that the "provision of low-income housing that is sustainable requires a new urban form, different to what has recently (the last 50 years) been accepted as the norm. This new urban form should reflect the uniqueness of the surrounding setting as well as being affordable and capable of rapid development" (CNdV Africa 2005:3) remain worthless unless they are implemented and urban spaces better utilised.

Towards 'technoburbia'

The quartering of the urban fabric has resulted in separated spaces. Modern telecommunications and information technologies provide us with a new social space, so-called 'technoburbs', which allows for interconnectedness between people and places. This is where new dimensions of social interaction, education, and even the facilitation of public participation can be achieved. A real challenge in future will be how to restructure public spaces to incorporate cyberspace. As Thompson (2002:68) observes:

> ... although the electronic revolution means we no longer need to go into the streets of our towns and cities to find out the news or to arrange to meet people and organise events, it does mean that we can now use those streets, squares and parks with much greater confidence that we will find what we want there, meet whom we want, be able to do what we choose. Given that we are social animals, and that we crave real contact with each other and with nature, perhaps public open space will be more, and not less, used in future than it has in recent decades.

Conclusion

In order to manage a sustainable sociospatial transformation in Stellenbosch, it must be acknowledged that some changes are inevitable. The process of sustainable transformation is ongoing, but needs to be skilfully managed in order to achieve appropriate urban sociospatial development. Spatial approaches for growth and development may include the following: avoiding the destruction of the traditional agricultural landscapes; guarding against transformation of the rural-urban linkages into a post-productivist countryside; preventing the erosion of the urban historical fabric; appropriately managing the process of studentification of urban space; guarding against the expansion of fortified spaces; and sensibly dealing with sociospatial urban processes such as gentrification and social integration. Restructuring partitions, or exclusions, either by design or decay, is a non-negotiable course.

REFERENCES

Anonymous. 1996. Die nare jaar van die middedorp. *Eikestadnuus*, 8 November:19.

Benn J. 2010. Studentifikasie in Stellenbosch. MA thesis. Stellenbosch: Stellenbosch University.

Bourdieu P. 2009. Site effects. In: P Bourdieu et al. *The Weight of the World. Social Suffering in Contemporary Society*. New York: Polity.

CGA (Centre for Geographical Analysis). 2010. Ortho-rectified aerial photograph of Stellenbosch: 1938 and 2009. Stellenbosch: Stellenbosch University.

CGA. 2010. Spatial growth of the built-up area between 1938 and 2009. Unpublished paper. Stellenbosch: Stellenbosch University.

CNdV Africa. 2005. *Integrated Development Plan: Preliminary spatial development framework*. Stellenbosch: Stellenbosch Municipality.

CNdV Africa. 2010. *Spatial Development Framework (09.1834)*. Stellenbosch: Stellenbosch Municipality.

Ferris J. 2002. *Conservation and Regeneration in the Nottingham Lace Market*. [Accessed 20 December 2010] http://www.arcchip.cz/w02/w02_ferris.pdf

Fox MJ. 2008. Near-Campus Student Housing and the Growth of the Town and Gown Movement in Canada. *TownGownWorldE-Journal:* January 2008. [Accessed 12 January 2011] http://www.towngownworld.com/images/town_and_gown_movement_in_canada.pdf

Giliomee H. 2007. *Nog altyd hier gewees – Die storie van 'n Stellenbosse Gemeenskap*. Kaapstad: Tafelberg Uitgewers.

Krause S. 2006. *A Call for New Ruralism*. University of California, Berkeley: Institute of Urban & Regional Development New Ruralism Initiative. Berkeley, CA: University of California.

Liu W-K. 2009. *Everyday Life, Morphology and Urban Conservation: An Interdisciplinary Approach to Urban Change*. [Accessed 20 December 2010] http://ace.caad.ed.ac.uk/ear2009/upload/pdfs/014-Liux.pdf

Marcuse P. 1989. 'Dual city': a muddy metaphor for a quartered city. *International Journal of Urban and Regional Research*, 13(4):697-708.

Marcuse P & Van Kempen R. 2002. States, cities and the production of urban space. In: P Marcuse & R van Kempen (eds). *Of States and Cities: The Partitioning of Urban Space*. Oxford: Oxford Geographical and Environmental Studies.

Matthews RS. 2005. *The Production of Sustainable Urban Space: A Comparative Analysis of Wallingford and the Carfree Reference District*. [Accessed 20 December 2010] http://www.carfree.com/papers/Matthews_CarfreeSpace.pdf

Nissen S. 2008. Urban Transformation from Public and Private Space to Spaces of Hybrid Character. *Czech Sociological Review*, 44(6):1129-1149.

Norberg-Schulz C. 1980. *Genius Loci: Towards a Phenomenology of Architecture*. London: Academy Editions.

Perold K. 2010. To space: exploring its representations, constructions and negotiations in Jamestown, Stellenbosch, South Africa. Unpublished paper. Stellenbosch University: Department of Fine Arts. Stellenbosch: Stellenbosch University.

Phillips M. 2005. Differential Productions of Rural Gentrification: Illustrations from North and South Norfolk. *Geoforum*, (36):477.

Pistorius P. 2002. *Texture and Memory: The Urbanism of District Six*. 2nd Edition. Cape Technikon: Sustainable Urban and Housing Development Research Unit. Cape Town: Cape Technikon.

Republic of South Africa. Department of Housing. 2004. *Breaking New Ground: A Comprehensive Plan for the Development of Sustainable Human Settlements*. Pretoria: Government Printers.

Republic of South Africa. Office of the Presidency. 2004. *Towards An Urban Policy / Strategy*. Pretoria: Government Printers.

Rowley A. 1998. Private-property decision makers and the quality of urban design. *Journal of Urban Design*, 3(2): 151-173.

Seethal C. 2004. Postmodern urban politics in South Africa: the case of Stellenbosch (2000-2004). Paper presented at the Centennial Convention of the Association of American Geographers. Philadelphia, USA: March 2004.

Smith D. 2009. Student geographies, urban restructuring, and the expansion of higher education. *Environment and Planning A*, 41(8):1795-1804.

South African Human Rights Commission. 2006. *Road Closures/Boom Gates Report*. Pretoria: South African Human Rights Commission.

Spocter M. 2010. The Theoretical Context on Non-Metropolitan Gated Developments in the Western Cape. Unpublished paper. Stellenbosch University: Department of Geography & Environmental Studies. Stellenbosch: Stellenbosch University.

Statistics South Africa. 1970; 1980; 1996; 2001. Census data. [Accessed 4 January 2011] http://www.statssa.gov.za

Stein G. 1937. *Everybody's Autobiography*. New York: Vintage Books.

Stellenbosch Municipality. 2010. *Annual Report*. Stellenbosch: Stellenbosch Municipality.

Stellenbosch University. 2010. Student Records. Stellenbosch: Stellenbosch University.

Swilling M. 2011. Personal communication.

Thompson CW. 2002. Urban open space in the 21st century. *Landscape and Urban Planning*, (60):59-72.

Van der Merwe IJ, Zietsman HL, Ferreira S, Davids AJ, Swart GP & Kruger D. 2005. *Growth Potential of Towns in the Western Cape*. Report commissioned by the Western Cape Provincial Government. Stellenbosch University: Centre for Geographical Analysis. Stellenbosch: Stellenbosch University.

Van Kempen R. 2002. The academic formalizations: Explanations for the partitioned city. In: P Marcuse & R van Kempen (eds). *Of States and Cities: The Partitioning of Urban Space*. Oxford: Oxford Geographical and Environmental Studies.

Van Niekerk A, Donaldson R, Du Plessis D & Spocter M. 2010. *A Revision of the 2004 Growth Potential of Towns in the Western Cape Study*. Research study for the Western Cape Provincial Government Department of Environmental Affairs and Development Planning. Stellenbosch University: Centre for Geographical Analysis. Stellenbosch: Stellenbosch University.

Welch CT. 1993. Historical Continuity in Town Planning. *Planning History*, 15(2):27-31.

Wilson E. 1995. The Rhetoric of Urban Space. *New Left Review (a)*:n.p.

Chapter 4

HOUSING
The challenge of informal settlements

LAUREN TAVENER-SMITH

Introduction

Housing in Stellenbosch is increasingly defined by informality. This trend is likely to continue as urbanisation drives growth in the population of urban poor, who will continue to house themselves in the most affordable way possible by building their own shack in an informal settlement, in the backyard of a low-income formal settlement, or on any other well-located, vacant urban space.

This chapter outlines, in a national context, the inadequacies of the conventional response to informal settlements (the provision of a house on a single stand) and the potential for and barriers to an alternative response (the upgrading of informal settlement infrastructure). Informal settlement upgrading has the potential to be a more realistic and humanistic response to the scale and dynamics of informal settlements. However, this will only happen if the state can be made to really 'see' and understand the longstanding informal practices of the poor in their attempts to improve their dwellings and surroundings, and to shape their official response around the everyday process of upgrading. This approach is often referred to as 'incrementalism'.

Following a contextualisation of the tensions between rhetoric and practice at a national level, the chapter turns its focus to the local situation and looks at a grassroots movement aimed at building support for incrementalism through social mobilisation. The emergence of this process, still in its early stages, is described and the tensions that have the potential to undermine the progress made are identified.

In conclusion, the chapter situates the challenge of informal settlement upgrading within the context of a future affected by climate change and resource depletion, and hypothesises hope for transforming these simultaneous challenges into mutual opportunities.

Informal settlements in Stellenbosch

The scale of informal settlements in Stellenbosch (as in many municipalities across South Africa) is difficult to measure due to the poor quality of available data. Demographic information used in official documentation and reporting varies widely and there is little agreement amongst officials (Stellenbosch Infrastructure Task Team (SITT) 2012) what the true size of the Stellenbosch population is. It is likely that both the latest Integrated Development Plan (IDP) (Stellenbosch Municipality (SM) 2011b) and Water Services Development Plan (WSDP) (SM 2011a) overestimate the number of individuals residing in Stellenbosch. Table 4.1 collates demographic and informal settlement statistics from Stellenbosch Municipality's annually updated IDPs and the 2011 WSDP and reveals the widely varying estimates across different years of the publication.

Table 4.1   Demographic and housing trends in Stellenbosch [SM 2007-2010; 2011a; 2011b]

Source	Total Population	Informal Settlement Residents		Proportion Informal	New Shacks
	Individuals	Individuals	Households	Individuals	Per Year
WSDP 2011/2012: 2010 estimates	222 575	41 488	10 352	19%	435
IDP 2011: 2010 estimates	222 575	91 413	16 928	41%	711
IDP 2010, IDP 2009: 2007 estimates	200 527	6 435	1 192	3%	50
IDP 2008, IDP 2007: 2006 estimates	135 874	23 830	4 413	18%	185

Based on the author's own estimates the size of the resident population in Stellenbosch as of mid-2012 is approximately 166 000. This figure is based on the number of billing units as per the records of Stellenbosch Municipality, as well as extrapolations (at an annual average growth rate of 2.89% pa) on the Census 2001 and Community Survey 2007 data. A study by Zietsman (2007) indicated that between 2001 and 2006 the population of Stellenbosch grew from 118 000 to 135 000 people. Growth rates were highest amongst the African population, with an estimated 43% of all population growth taking place in this demographic group during this period (Zietsman 2007).

Informal settlements have grown as the size of the African population has expanded. Low living conditions and poverty are concentrated amongst this group of people. The current size of the informal settlement population in Stellenbosch is estimated at approximately 27 000 people (16% of the total population) living in 9 000 households. These figures are based on informal settlement enumerations, aerial photograph and Google Earth shack counts, and interviews with key players in the informal settlement upgrading plans including David Carolissen (2011a), from Stellenbosch Municipality and Aditya Kumar (2011) from the Community Organisation Resource Centre (CORC).

It is likely that approximately 300 new shack-dwelling households form in Stellenbosch every year. Some of this new household formation is driven by emigration, predominantly from the Eastern Cape. Growth in informal settlements is also driven by those residing in backyard shacks in low income, formal neighbourhoods who choose to relocate to a shack in an informal settlement, often because the latter is rent free. Stellenbosch Municipality receives 300 housing capital subsidies annually (Division of Revenue Act (DORA) 2011) which, considering the estimated backlog in housing provision, is clearly inadequate. In the current economic climate it is unlikely that the municipality will be able to continue topping up the subsidy (between 2003 and 2008 an average of 140 additional units were constructed per year (Swilling & Hendler 2008, Davidson 2012)). Because the housing subsidy is structured to cover the cost of land purchases, the majority of subsidised houses are located in developments far from the nodes of economic opportunity (Carolissen 2011b).

Infrastructure challenges and opportunities in Enkanini, Stellenbosch

In January 2011, a group of Stellenbosch University post-graduate students based at the Sustainability Institute began a research project which aimed to co-produce knowledge with multiple groups in order to inform alternative implementation logics for infrastructure upgrading in informal settlements. Enkanini, a settlement 1.5 km northwest of the town centre of Stellenbosch, has been the core site for engagements between residents and researchers. The observations related to the challenges and opportunities in informal settlements are based on the time spent in and engaging with residents from Enkanini.

- *Infrastructure challenges in Enkanini*

Enkanini is home to an estimated 8 000 to 10 000 people, largely young job seekers who have urbanised and left their Eastern Cape homes to seek better opportunities in the city. Enkanini is a desirable place of residence due to its proximity to transport linkages and employment opportunities in central Stellenbosch, as well as the wine farms and tourism activities which converge around the town. Enkanini is a one of the few completely illegal settlements in Stellenbosch.[1]

---

1 A 2006 court order for eviction of Enkanini residents still stands, although it is due to be rescinded as the position of eviction is no longer viable given the current population.

In 2009 the municipality connected Enkanini to the water supply and sewerage networks, however, as the settlement has grown, the facilities are increasingly unable to adequately service the larger population. Free Basic Water and Sanitation is extended to residents through six centralised toilet block and tap facilities; in 2012 it is estimated that each of the 60 toilets is shared by 150 people and each of the 36 taps services 250 people. The newest areas of Enkanini are farthest from the public toilets and evidence of open defecation is most apparent in these parts of the settlement. Congestion at toilets and tap points manifests downstream, as shown by sewerage overflows reportedly due to piping diameters which are unable to accommodate the expanded user loads. During summer time and over weekends, the reservoir serving both Enkanini and Kayamandi is regularly drawn down, resulting in low pressure which compounds the congestion problems.

There is no grid electricity infrastructure in Enkanini, and the dominant position at the municipality is that the viability of installing the grid is compromised by the severe gradients and the low potential for cost recovery. The delivery of Free Basic Energy is contingent on a connection to the grid, thus residents of Enkanini forego their entitlement and expend significant amounts of gas, paraffin, wood, coal and candles for heating, cooking and lighting. Shacks fires related to open flames from these energy sources are common and there is high risk of loss of property and life. Risky energy access strategies also include the informal and technically illegal connections to the grid connected houses in adjacent Kayamandi. These lines can be easily exposed or torn down, for example by trucks navigating the settlement or during inclement weather, increasing exposure to live cables. Electricity from informal vendors is expensive, and residents in Enkanini are subject to their discretionary price setting. These transactions around informal electricity have also resulted in tensions between residents of Enkanini and Kayamandi.

Drainage is a severe problem in Enkanini. During the rainy season Enkanini is flushed out, but becomes dangerous to navigate as the top soil has been washed away, leaving only the bald and very slippery clay hillside. A perennial, but easily avoidable, drainage issue is caused by runoff from the taps at the toilet blocks, as the taps are used heavily, particularly for washing clothes. Although residents regularly walk up to 250 m to collect water, disposal of used domestic water is usually within a 15 m radius of the home onto the ground, into drainage channels, onto the road within 10 m to 15 m of the house. In the context of open defecation and solid waste accumulation, runoff in the settlement becomes contaminated, and is one of the main transmission mechanisms for disease, both chronic (such as the skin rashes present on many children) and acute (such as a high incidence of diarrhoea).

- *Resilience in Enkanini*

Common issues affecting neighbours, such as runoff channels passing through or in close proximity to houses (which usually lack proper foundations) or poor access due to gradients, are often points around which small group actions are undertaken. For example, there are a number of staircases carved into the hillside in order to improve mobility in the steep terrain. Neighbours have also dug intricate systems of drainage channels to divert water away from their houses and into the natural drainage system of the Plankenbrug River. In the densest part of Enkanini, residents sharing a common roadway collected money amongst themselves in order to purchase drainage piping to sink a drainage channel that had been blamed for the skin rashes and sores of their children who played near the drain.

In response to space constraints, some residents have added a second storey to their dwelling. Others are constructing brick walls around their shacks, the pace of their work determined by the rate at which their incomes permit them to purchase bricks. Some shacks are fitted with gutters to collect rainwater from the roof and a shebeen owner has constructed a rudimentary urinal to soak away the urine from his customers. Many residents have planted trees, shrubs, flowers and vegetables on the land directly outside their shacks, which in addition to creating a sense of homeliness and visual appeal, helps to bind the soil and assists in retaining steep slopes and to smooth food consumption during cash-constrained period (e.g. relishes made from spinach is commonly reported as an important household food-security strategy).

The importance of social networks as a coping strategy in cash-poor households is significant. There is a general knowledge among residents regarding the relative poverty levels of others in the settlement, and those who suffer the greatest hardship regularly rely on in-kind assistance from neighbours, particularly in order to obtain food and clothes for children. Social safety nets are particularly strong amongst extended family members, and individuals who belong to various social groups including churches, burial societies and savings clubs. Volunteer groups have self-organised in Enkanini to patrol the settlement on weekend nights, in particular to ensure that shebeens stop trading after a certain hour (Bronkowski 2011).

Figure 4.1 **A: *Grey water drainage.*** Even during the dry season a continual flow of grey water emanating from the communal taps means that drainage problems pose an all year round risk. Grey water is a carrier of pathogens, particularly in the context of open defecation. **B: *Informal connections in Enkanini.*** Informal connections are the means by which Enkanini residents access the national electricity grid. This mode of access imposes high costs, both financial (if payments to metered households are exchanged) and physical (in terms of the hazards from high-voltage live wires). **C: *Incremental building in Enkanini*.** An incremental approach to building a house in Enkanini: a shack is being converted to a brick house. The pace of construction is determined by the availability of surplus household income to purchase bricks. **D: *Urinal in Enkanini*.** A urinal constructed by the owner of a shebeen in Enkanini. The urinal allows his customers ease of access to ablution facilities and prevents a concentrated flow of urine from negatively impacting neighbours and contaminating the nearby wetland (the latter is an unintended impact). [Wessels 2011. Photographs reproduced with permission of Stellenbosch University: Sustainability Institute]

The conventional response to informal settlements in South Africa: Formalisation[2]

Post-transition housing policy has sought to remedy the fragmented social landscape engineered over the preceding decades by the architects of apartheid. A capital subsidy for housing was viewed as the best way to reduce the inherited housing backlog estimated at 1.4 million houses in 1993 (Gilbert 2004:20; Khan 2010:37,40; Wilkinson 1998:225). Under the Reconstruction and Development Programme (RDP), the subsidy aimed at providing eligible poor households[3] with a "permanent residential structure with secure tenure" and access to "potable water, adequate sanitary facilities, including waste disposal and domestic electricity supply ... situated in areas with convenient access to economic opportunities, health, educational and social amenities" (Department of Housing 1994:21).

Policy implementation (il)logic

Most RDP houses have been constructed by private developers who accessed the subsidy on a project-linked basis (Huchzermeyer 2001). This developer-led approach is as much due to local government capacity constraints as it is a reflection of the balance of power between constituents, structured as it is to favour the dominant interests of developers and political elites (Bolnick & Bradlow 2011; Khan 2010).

One million houses were pledged (Republic of South Africa: Department of Housing 1994) and "more or less" delivered (Gilbert 2004:28) during the first five years after 1994. However, the approach taken to achieve these roll out figures has been criticised as target-driven, with the housing problem reduced to a technical equation and private sector developers rolling out houses to minimum standards as fast as the subsidies were allocated (Khan 2010).

Given the scale of the challenge (the housing backlog is larger in 2012 than it was in 1996), this approach proved fundamentally inadequate to make a dent in the housing problem. Moreover, for those who have 'benefitted' from subsidised housing, the envisaged welfare gains have not always transpired. The dominant location of subsidised houses; the costs associated with occupying a house relative to a shack; and the technocratic implementation logic of developer approaches all negatively affect potential welfare outcomes.

▣ Location

Informal settlements usually arise as an expression of the urban poor's (very limited) choice in terms of residential location(s) allowing them access to urban centres and the associated jobs, transport, education, and social network opportunities at a very low cost (Misselhorn 2008). Project-linked subsidies include, where applicable, the cost of acquiring land – the price of which is a function of its proximity to town centres – but, as well-located land is scarce, a 'one house, one plot' development model is unachievable given affordability constraints. Thus, since land is cheaper on the urban periphery, RDP-style housing developments have usually located the poorest people the furthest away from employment centres and community facilities (Gilbert 2004).

The relocation of shack dwellers from often relatively well-located informal settlements subjects them to significant pecuniary (notably transport-related) and social costs. The latter relate to the disruption of social networks and safety support systems, as well as to the increased distance placed between the poor and the legal and political systems (Misselhorn 2008).

And so the legacy of apartheid planning lives on, with poor people of colour engineered to live on the edges of cities in settlements that burden residents with "appallingly inconvenient and expensive" lifestyles (Dewar in Khan 2010:37), so that there has "been no meaningful alteration of the inherited socio-spatial boundaries which include the privileged and exclude the poor" (*ibid.*:38).

---

2   It is not the intent of this essay to examine in depth the discourse on post-apartheid housing. For a more complete review please see: Khan 2010; Huchzermeyer 2001; and Wilkinson 1998.

3   Households with a monthly income of less than R3 500 may put their names on a list to become eligible to receive a minimum standard house when their area is targeted: "Register now and await your turn" (Cape Provincial Housing Chief 2000, in Gilbert 2004:27).

▫ *Affordability*

For the urban poor, the cost of living in a formal house as opposed to a shack is typically much higher. Not only do they incur additional costs associated with utilising the infrastructure services to which their new homes are connected, but they must also carry the cost of maintenance and repairs to the structure and fixtures.

Developer-built RDP houses are usually constructed to minimum specifications at the lowest possible cost. In a survey of Cape Town RDP housing developments, 48% of homeowners reported structural problems with their houses, while 58% reported a non-functioning toilet – and virtually all respondents declared that they were unable to afford to make necessary repairs (Govender, Barnes & Pieper 2011). As a result, the asset value of houses is eroded by inadequate maintenance and they come to represent a liability rather than an asset.

Formal housing offers improved access to infrastructure services, but it typically comes at a higher price. For example, research has shown that, subsequent to electrification, the electricity consumption of low-income settlements typically does not increase as much as anticipated, and connected households continue to use multiple fuel sources, mainly because of affordability constraints (Winkler, Simoes, La Rovere, Mozaharul, Rahman & Mwakasonda 2011). Most households do not have toilet paper, and toilets connected to waterborne sewerage cannot withstand commonly used items such as newspaper, packaging or rags (Govender *et al.* 2011).

A common response to liquidity needs is for RDP beneficiaries to generate income through backyard shack rentals. The resultant overcrowding puts pressure on infrastructure systems (especially water and sanitation) and significantly contributes to the spread of communicable diseases, including tuberculosis and diarrhoea (Govender *et al.* 2011).

▫ *Disempowerment*

The formalisation of informal settlements is a welfarist, as opposed to developmental, state approach (Freund 2007). According to Bolnick and Bradlow, this "paternalistic" approach transforms the urban poor "into entitlement-driven dependants, trapped in a dysfunctional system of bureaucratised delivery", as is evident from "the steady decline in community savings, collective action and self-reliance" (2010:38).

This statement is not equivalent to the oft-quoted aversion to grant dependency displayed by political elites and neoliberal apologists (Khan 2010) and should not be viewed as a general-purpose vilification of social grants, whose role in the context of endemic unemployment and poverty is not disputed here. However, using the promise of a house as political leverage and then failing to deliver keeps the poor waiting in anticipation, deterring them from opting to build their own houses. The problem is augmented by the bias of housing subsidies toward developer-driven projects, which effectively displaces the incremental efforts made by the poor on a daily basis to improve their living conditions and excludes them from making any meaningful contribution to their own development (Bolnick & Bradlow 2011:270-271).

A radical shift in development responses to informal settlements in South Africa: Upgrading

Changes in policy towards informal settlements

By 2004, the highest levels of government acknowledged that its conventional approach to dealing with informal settlements was not working (Republic of South Africa: Department of Human Settlements (RSA DHS) 2004:3-6). Faced with a larger housing backlog than that which was inherited 10 years earlier (despite large-scale delivery in the interim) and the slowdown in the pace of delivery due to private sector withdrawal from subsidy projects, government conceded that a change in approach was overdue and introduced the Breaking New Ground (BNG) plan for housing delivery.

The BNG plan included *in situ* informal settlement upgrading as an official response to informal settlements. The shift in policy from a conventional formalisation strategy (providing a house) to one focussed on upgrading water provision, sanitation, drainage, and storm water infrastructure, roads and pathways, refuse removal and street lighting, is detailed in Chapter 3 of the 2009 Housing Code (RSA DHS 2009:21,51,57). *In situ* upgrading is emphasised in order to avoid relocations, which disrupt fragile social and economic networks (Del Mistro & Hensher 2009).

The Upgrading of Informal Settlements Programme (UISP) was established to facilitate the translation of this policy into practice. In terms of the 2009 Housing Code, upgrading is seen as happening in four phases, the first three of which are eligible for support through the UISP subsidy. Phase 1 funding covers surveys, geotechnical investigations, land acquisition and interim services, while Phases 2 and 3 make provision for detailed town planning, land and contour surveying, and permanent engineering services (RSA DHS 2009).

Phase 4, the Housing Consolidation Phase, which includes actual construction and the proclamation of residential property ownership rights, is included in the UISP process, but excluded from its funding mechanism. The terms of the UISP subsidy clearly de-link services and land from housing, as seen in the following excerpt from the Housing Code (RSA DHS 2009:15):

> *Beneficiaries of the [informal settlement upgrading] programme will only receive access to land, basic municipal services and social amenities and services. To qualify for housing assistance benefits such as registered ownership and a consolidation subsidy, beneficiaries need to comply with the requirements of relevant programmes.*

In 2010 the Minister of Human Settlements committed to upgrading 400 000 shacks by 2014 (Tissington & Royston 2010). If this happens in practice – and in such a way that the urban poor are accommodated on well-located land – it will signify the first wave in undoing the decades of apartheid spatial engineering.

▣ The basis for change

The preamble to BNG includes notes from a self-reflective government, one that recognises that the subsidies of the day worked to reinforce the spatial inequalities of the apartheid era (RSA DHS 2004). The acknowledgement was coloured by undertones of a political reality of increasing service delivery protests and a growing discontent among the poor that government was failing to deliver on its promises. The "powder keg" (Misselhorn 2008:3) this created necessitated a radical response, and the logic behind upgrading from a political perspective is clear: by extending a lower level of service to a greater number of people, scarce resources may be spread more broadly (if thinly) across a greater number of beneficiaries in order to mollify the discontent of as many as possible. An econometric study confirmed this logic by showing a decreasing marginal utility of investment in upgrading and that "(w)ith limited budgets, the greatest total utility would be gained by implementing the lowest level (of service) for the most number of households" (Del Mistro & Hensher 2009:348).

▣ Upgrading in practice

The policy shift from formalisation to upgrading has not been significantly reflected in practice, with implementation still organised according to conventional delivery systems. As a consequence of the systems, mechanisms and regulations of the UISP subsidy, which have been criticised as mechanistic and inflexible (Misselhorn 2008), the upgrading process is still largely focussed on housing deliverables as the "*de facto* policy response" (*ibid.*:16) and *in situ* upgrading has been avoided in practice: "(a)fter the rhetoric, all that is left is a single solution: – a house on a fully serviced site in a greenfield development" (Del Mistro and Hensher 2009:350).

Misselhorn (2008) describes the chasm between the cost of this approach to upgrading and the available funding. If South Africa were to eliminate the entire housing backlog of 1.2 million by 2014, and assuming a 7.5% per annum growth in informal settlements and an R80 000 subsidy for infrastructure and a top structure, R27 billion would be required *per year*. This stands in sharp contrast to the amounts allocated in housing grant budgets, which ranged from R6.6 billion in 2006/7 to R8.3 billion in 2007/8. As a result, upgrading, under the implementation logic borrowed from the process of formalisation, seems likely to follow a similar path of "unfunded mandates" (Khan 2010:254).

Given the population densities of informal settlements, in most contexts a site and service model en route to a permanent structure on a single plot is not viable for *in situ* upgrading, because relocations would be necessary in order to accommodate everyone (Misselhorn 2008). Yet, housing typologies that are more suitable for *in situ* upgrading have been crowded out by a government that equates a fully serviced RDP-type house as the only legitimate form of housing (Bolnick & Bradlow 2011). In fact, the continued vilification of shacks and informality indicates that the government has neither recognised nor legitimised the strategies of the poor to house themselves.

Incrementalism: the logics of everyday actions underpinning development responses

As non-poor people try to understand poverty, different epistemologies according varying degrees of agency to the poor emerge (see Bayat 2000). If, instead of viewing the poor as marginalised to the degree that they are incapable of actively participating in their own development, one acknowledges their continual efforts to connect themselves "piece by scrappy piece" to the resources and services needed to maintain and improve their quality of life – to "translate toeholds into footholds into full scale (albeit often fragile) inclusions ..."(Swilling & Annecke 2010:126), then one may begin to recognise the effectiveness of informal responses to housing needs.

Incremental building of housing units is estimated to account for 50 to 90% of residential development in most developing countries. This strategy allows for heterogeneity in approach reflecting the needs of each household and permits households to climb the housing ladder at the pace of their own economic opportunities (Ferguson & Smets in Lizarralde 2011).

Incrementalism is the everyday process of slum upgrading undertaken by the poor. As an official response, incremental upgrading must first "seek to understand [the] individual and collective livelihood struggle" of the poor (Mitlin 2001:509) and then structure support in ways that replicate natural responses (Lizarralde 2011). In this manner, official development responses in support of everyday incrementalism can help to bridge the gap between the informal and the formal through extending the logics of longstanding informal practice (Bolnick & Bradlow 2010). This is in contrast to the conventional approach, which seeks to eliminate informality entirely.

Slum Dwellers International

Slum Dwellers International (SDI) is a network of federations of the urban poor and support NGOs active in 33 countries in the Global South. The SDI 'brand' represents a common identity for action and activism around informal settlement upgrading (Swilling 2005). This SDI brand is composed of three salient methods developed by and for low-income urban communities to self-mobilise; participatory enumeration and mapping constitute an important exercise in community organisation and is intended to produce in-depth information about residents' demographic, socio-economic and service-related situations. This information is intended to inform the engagements between organised communities and local authorities. The second methodology is the frequent exchange of informal settlement residents between different settlements. This is intended as a learning exercise, as well as a means by which to grow networking capacity of informal settlement residents. The SDI trademark of daily savings is the third ritual in their toolbox for mobilisation.

Local action for informal settlement upgrading is centred on mobilising the community (through rituals) and connecting mobilised communities to local officials (the responsibility of the support NGO). As a result of the mobilisation process, power is redistributed in favour of the poor in negotiations between them and planners, while the bridging of the formality/informality divide facilitates the "analysis of problems and emerging solutions reflect(ing) the needs and aspirations of the urban poor" (Patel, Burra & D'Cruz 2001:48).

SDI is renowned for its pragmatic and patient approach (Bolnick & Bradlow 2010), which includes collective risk taking in order to demonstrate alternative solutions to the state, and setting precedents through constructive action, which "forms a breach in the system and potentially offers opportunities for further policy reform" (Mitlin & Patel 2005:28).

Towards incremental upgrading in South Africa: Progress, pathways and partnerships

At the 2005 UN World Summit, the Millennium Declaration to improve the lives of 100 million slum dwellers by 2020 was formalised and institutionalised in the UN Habitat and Human Settlements Foundation Slum Upgrading Facility. In the same year in Africa, a Special Ministerial Conference on Housing and Urban Development (AMCHUD) put institutional mechanisms in place to translate the declaration into African governments' agendas.

In 2006, the then Housing Minister of South Africa, Lindiwe Sisulu, entered the state into a formal partnership with SDI and declared her ministry's commitment to promoting linkages between government and poor people's movements through the South African SDI Alliance, so that slum upgrading could become a community-led development process (Sisulu 2006).

The SA-SDI Alliance comprises a range of different organisations, including two parallel networks of urban poor, namely the Informal Settlement Network (ISN) and the Federation of the Urban Poor (FEDUP). Support NGOs include CORC, which plays a leading role in organising community mobilisation and connecting communities to government; the uTshani Fund, a finance and lending institution; and iKhayalami, a partner NGO that designs and manufactures improved, modular shacks (Bolnick & Bradlow 2010).

Cape Town

The SA-SDI Alliance has worked in partnership with the City of Cape Town Informal Settlement Unit since 2009 and is actively networking informal settlements in the Western Cape around an upgrading agenda (Ziervogel 2011). The salient objective guiding local action and activism is to include mobilised informal settlements in city-level planning processes that will determine how informal settlement upgrading will be put into practice in the future (Bolnick 2011).

The Cape Town ISN is gaining momentum, with thirty settlements and counting in the network at present (Kumar 2011: personal communication). Each of these settlements have completed enumerations, compiled reports, and developed geographic information system (GIS) maps to understand local needs and capabilities. These mobilisation actions have increased the bargaining power of the informal settlement leaders, who represent their community by conveying findings and recommendations to City of Cape Town line managers at monthly meetings.

Incremental upgrading in Joe Slovo (Nyanga) and Sheffield Road (Phillipi) uses household savings to improve sanitation and water services, and create public spaces and areas for emergency services by rearranging clusters of shacks in a process known locally as 'blocking out'. Households also pay 10% of the cost of improving their shack with more robust, fireproof wall materials.

The upgrading in Joe Slovo and Sheffield Road, although very much still in progress, is used as learning case studies for Cape Town ISN settlements in earlier stages of mobilisation. ISN networks in other provinces also take field trips to Joe Slovo and Sheffield Road, and international exchanges have afforded Cape Town ISN leaders the opportunity to travel to India, Uganda and Malawi to learn from the experiences of other mobilised communities.

The fact that very few settlements are implementing upgrading may, from the perspective of conventional development, be used to criticise it as having an ineffectual response rate to the urgency of the housing problem. However, such indicators of success do not reflect the strides being taken in the less-tangible social process. Upgrading will either happen as a people's process, with government participating, or as a government- and engineer-driven process in which the poor can (in theory) participate.

Stellenbosch

In 2010, a delegation from the Stellenbosch Municipality travelled to Uganda to witness SDI methods and outcomes in practice. Approximately a year later (December 2011), a Memorandum of Understanding (MOU) between the municipality and the alliance was signed and an Urban Poor Fund (which comprises contributions of R3.5 million from Stellenbosch Municipality and the SA-SDI Alliance and R12 000 of community savings) was established. The fund will be used not only for capital and to cover the operational costs of the physical upgrading of informal settlements, but is also intended to back investments in social processes such as enumeration and mapping, and to build the institutional capacity of communities to manage upgrading projects (Fieuw 2011).

The MOU and financial pledge in support of informal settlement upgrading in Stellenbosch follows a year of ground-level mobilisation of informal settlement communities in Stellenbosch. Langrug was the first informal settlement in Stellenbosch to elect a leadership and begin mobilising around an upgrading agenda using SDI methods. Since enumerations were conducted in early 2011, task team members and leaders have produced detailed maps to depict the demographic, socio-economic and infrastructure realities of life in the settlement. These maps, together with written reports, narratives and photographic accounts, have been used as a strategic communication and resource leveraging tool as community leaders engage with the municipality.

The importance of enumeration and mapping exercises is manifest in the increased bargaining power of residents in municipal engagements. Equally important, if not more significant, is the political, social and organisational capital developed in the process of undertaking enumerating and mapping. The SDI activities form a nexus around which communities may organise themselves, and this has been observed insofar as mobilisation in Langrug has triggered enumerations (as the entry point to mobilising) in other settlements in Stellenbosch. At present, all informal settlements (except Enkanini and the informal parts of Kayamandi) have been constituted as part of the Stellenbosch ISN and are in involved in various post-enumeration activities, including data analysis, report writing and mapping (both manual and digital).

The mobilising activities in Langrug have already borne fruit, where 16 households were relocated to turnkey, fire resistant shacks, so that space could be made for laying emergency vehicle servitudes and bulk reticulation systems. In addition, a drainage project was undertaken in partnership between the Langrug ISN leadership, the Stellenbosch Municipality and CORC, and visiting students and professors from the Worcester Polytechnic Institute in Massachusetts, USA. One of the key lessons learned from this drainage project was the need to ensure that consistent maintenance, in the form of removing build-up in the channels, is undertaken.

Challenges to informal settlement upgrading

The shift in national housing policy requires people to adjust their expectations away from a house delivered by the state to infrastructure delivered in partnership between themselves and government. However, this shift is hampered by a strategic inertia in the promises of politicians. In Stellenbosch, 'giveaway' houses are still used as a political leveraging tactic (Carolissen 2011a). For example, at an IDP meeting in Kayamandi during the pre-2011 local government election campaign, the previous mayor of Stellenbosch conflated upgrading and housing delivery in front of a room full of informal settlement residents. In doing so, he epitomised the use, whether intentional or through his own misunderstanding, of promising free housing as a tool for realising political agendas.[4]

The Director of Informal Settlements at Stellenbosch Municipality, David Carolissen, has first-hand experience of the mechanisms by which the expectation of a 'one house, one plot' model is entrenched. He relates his local, grassroots experiences to fundamental problems with the National Housing Code: Upgrading Informal Settlements policy. According to Carolissen (2011b:n.p.), "at the core, policy is still designed to deliver a house", with eligibility criteria for local governments to access provincial government subsidies requiring that upgrading plans include the housing and consolidation phase of the UISP. At this point, regulations regarding minimum standards and technical specifications related to the housing subsidy become applicable, which means that infrastructure upgrading is ultimately shaped according to the same implementation logic, albeit in a staggered manner, as formalisation. In this context, upgrading is still confined to be compatible with the 'one house, one plot' model and there is limited flexibility for alternative housing typologies that are congruent with local needs and resource constraints.

Stellenbosch Municipality faces multiple challenges in regards to infrastructure and there is concern that underspending in recent years has led to an infrastructure backlog that threatens the local economic base of the municipality. In response to these challenges, a group of municipal officials, politicians, academics and business leaders came together to establish the SITT at the end of 2010. The group meets twice a month. At one such meeting, on 2 May 2012, the larger SITT group and selected other experts met to define a shared understanding of the challenges and to brainstorm possible alternatives.

The information presented in this section is based on notes and records from regular meetings and the workshop held on 2 May. The quantitative information is based on a modelling exercise undertaken by the author and one of the developers of the Municipal Services and Finance Model (MSFM) who was contracted by the SITT. The results are based on the first run of the MSFM for Stellenbosch Municipality, and further refinement is required. However, the model estimates are in line with those obtained by the Director of Engineering Services, Mr Andre van Niekerk, in his preliminary quantification of the current backlog in Stellenbosch.

The MSFM calculates the capital expenditure required over 10 years to meet service delivery targets and assesses the capital finance sources available. It also calculates the operating expenditure required to operate and maintain infrastructure and determines whether expected operating revenues will be sufficient to cover this expenditure. The model is based on current demographic profiles and service levels, as well as anticipated

---

4   Author's notes taken at the meeting on 13 April 2011.

future demographic and economic growth trends, future servicing plans, and unit costs of service provision based on regional averages.[5]

Based on the initial results of the MSFM,[6] Stellenbosch Municipality needs to spend almost R3 billion between 2013 and 2022 in order to extend infrastructure to currently unserviced populations, to rehabilitate existing infrastructure, and to expand and upgrade infrastructure for a growing population and local economy. The average annual capital budget for Engineering Services for the 2012/3 to 2014/15-period is approximately R150 million. At this rate, only half of the required capital expenditure is budgeted for over ten years.

Some additional capacity to borrow is only marginally helpful – even if the municipality were to extend its loan book to maximum tolerable limits – after which no other borrowing is permitted – a funding gap would transpire in the third year (see Figure 4.3). Under the most optimistic borrowing scenarios and assumptions regarding developer contributions to bulk infrastructure and collection rates, the model predicts a total funding gap of approximately R1 billion by 2022. Furthermore, institutional capacity constraints, which have resulted in unspent capital budgets in recent years, impede the action required to overcome the infrastructure problems in Stellenbosch.

Of the R3 billion required capex, R1 billion is for social infrastructure, which translates into infrastructure spending with zero return for the municipality in the present paradigm. Social capex includes rehabilitating current infrastructure, extending infrastructure to those currently unserviced, and providing for the growing population into the future. The DORA allocations to Stellenbosch have been declining steadily over the past few years and the current capital grants are only sufficient to cover 28% of required social spending. The social funding gap of R800 million projected over the next ten years constitutes 76% of the total R1 billion funding gap.

Stellenbosch Municipality will have to make difficult decisions regarding how to allocate its expenditure infrastructure for users who either do not pay for services or who pay a price equal to that required to cover costs (social spending), and for high-income and non-residential users who pay cost plus a surcharge (economic spending). It is politically difficult to allocate to economic spending when there is great need in the social arena. Equally, it would be short-sighted for the municipality to allocate funds required for economic infrastructure if this will erode the revenue base of the municipality in the long term.

In addition to the capital expenditure requirements, there is a fear that by extending services to the poor, the municipality will lock itself into operating expenditure patterns that threaten the current favourable position of the municipality (currently ranked second in the country). This mindset is based on an immutable viewpoint that cost recovery in this segment of the population is not possible. This fear reinforces the popular conception that the poor are a drain on the reserves of the local municipality: they are not as yet viewed as an asset and there is no notion of their upward mobility or of their existing willingness and ability to contribute to service delivery costs.

---

5   Input data derived from Stellenbosch Municipality planning documents, including the Water Services Development Plan; the Comprehensive Integrated Transport Plan; Integrated Development Plans for the current and recent years; the asset registers and monthly reports of the municipal Engineering Services; NERSA D-forms; Stellenbosch Housing Strategy A-Tables; DORA gazette; Capital Budget 2012-2015; and Stellenbosch Municipality Annual Reports on interest repayments, provision for bad debt and depreciation.

6   The first iteration of the model used demographic information from the Integrated Development Plan (SM 2011b) and the Water Services Delivery Plan (SM 2011a). The model is in the process of being updated; lower population and household estimates will result in lower estimates for required capital expenditure.

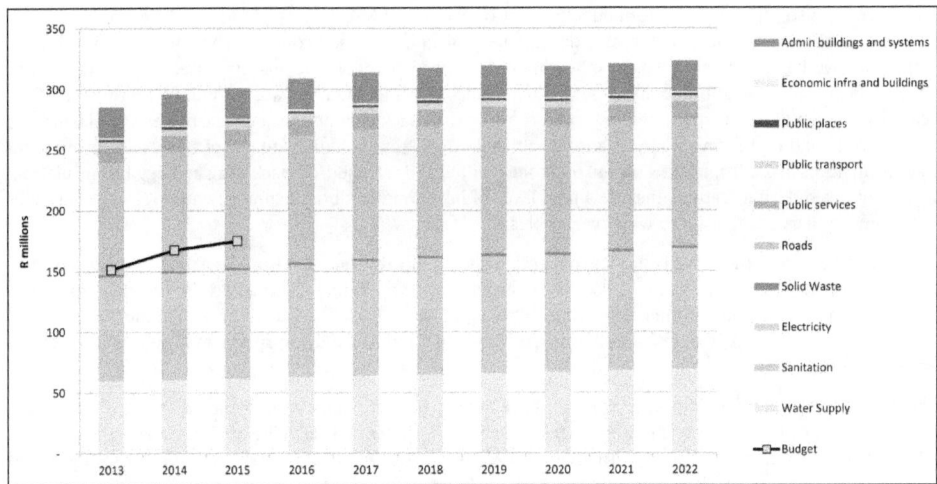

Figure 4.2   Projected capital expenditure requirements on infrastructure[7] for Stellenbosch, 2012-2022 [MSFM 2012]

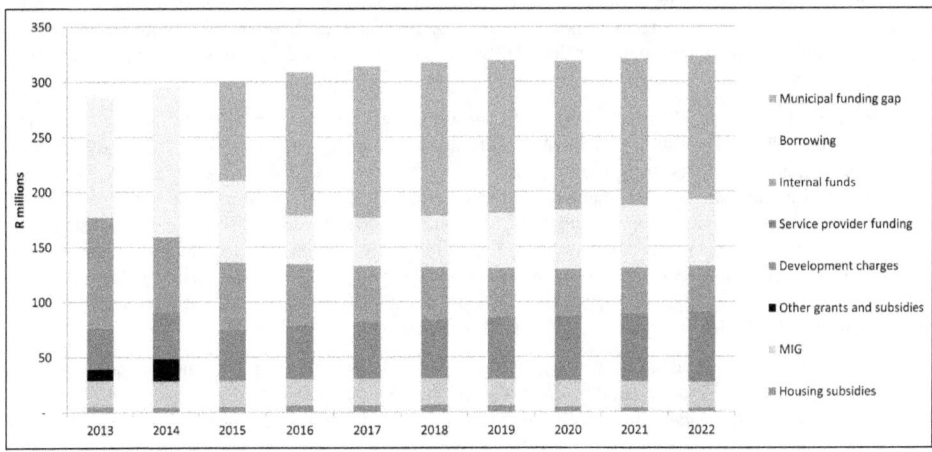

Figure 4.3   Sources of finance for infrastructure[8] in Stellenbosch, 2012-2022 [MSFM 2012]

Seeking workable infrastructure alternatives for informal settlements

The seemingly dire infrastructure situation projected for Stellenbosch is based on a model that assumes a continuation of business as usual, both in terms of technologies and delivery mechanisms. However, there are possible alternatives.

These alternatives form the focus of the research agenda of a group of post-graduate students based at Stellenbosch University's Sustainability Institute. Eighteen months of collaborative research between the students, Enkanini residents, officials from Stellenbosch Municipality, technology partners and local SDI affiliates, resulted in a tangible alternative to inadequate housing and energisation.

---

7   Excludes expenditure on land and top structures associated with housing delivery.

8   Excludes expenditure on land and top structures associated with housing delivery.

The iShack, constructed in 2011, formed a key part of one of the MA students' (Andreas Keller) research. The iShack is an improved shack, ecologically designed to maximise thermal comfort. The iShack is north-facing, with a roof overhang to maximise passive heating and cooling potential. The materials used to build the iShack were chosen based on affordability, accessibility, availability and their impact on indoor thermal comfort. Features that moderate temperature and comfort inside the shack include the strategic placement of windows on the north- and east-facing walls; an adobe half-wall on the south side, made out of straw and surrounding clay earth; floors made from bricks picked up at the landfill; and tetrahedron packaging and egg box insulation. Other improved design features include a final layer of insulating cardboard sprayed with fire retardant paint and rainwater harvesting used to water vegetables.

The performance of the iShack has been continually monitored and measured against a control shack. Preliminary results suggest that, at peak day-time outdoor temperatures, the iShack is 4 °C degrees cooler than the control shack; and at the coolest time of the night, the iShack is 1.5 °C warmer than the control shack (Keller 2012). The iShack is significant, because it shows that improved living spaces may be achieved at very little extra cost.

The iShack and three other shacks in Enkanini have been fitted with solar energy units developed specifically for the context by innovative technology partners, Specialised Solar Systems. The basic solar systems comprise a 20-watt photovoltaic cell, battery, distribution box, two indoor lights, an outside motion sensor light, and a cell phone charger. The key innovation is that the system functions on a direct current (DC) microgrid, thus avoiding alternating current (AC) inversion losses. This efficiency translates into capital savings, because the size of the required photovoltaic (PV) cells is minimised. The system is also designed to be affordable, robust and modular. It is thus well suited to the context and is amenable to incremental additions based on the income and expenditure patterns of poor households. With additional solar panels and greater battery capacity, the system can be upgraded to include DC appliances such as fridges, radios, hi-fis, televisions, DVD players and microwaves.

Without regular maintenance, it is likely that these solar systems will not function optimally into the future. The failure to account for the long-term functionality of technical systems is a major cause of the unsustainability of most development projects. In recognition of this fact, the technological change in Enkanini has been accompanied by a simultaneous investment in building the capacity of local operators to operate, maintain and repair the solar systems. Five Enkanini residents were trained in 2011 to undertake the physical installation, operation, maintenance and repair of the systems. In addition, their training included operating these systems as a small enterprise with revenue streams based on marginal user payments for the servicing. At present, the researchers, Enkanini residents, Specialised Solar Systems, and SDI affiliates are exploring alternative financing and asset ownership modalities as the next phase of the experiment.

Much energy has been expended and excitement generated around both the iShack and the solar systems. It is anticipated that this experimental research will set precedents with regard to what is possible, both technologically and institutionally, around alternative energisation and housing. Collaborative research around alternative sanitation options is also underway, and testing of new physical systems and cooperative business models is expected in late 2012.

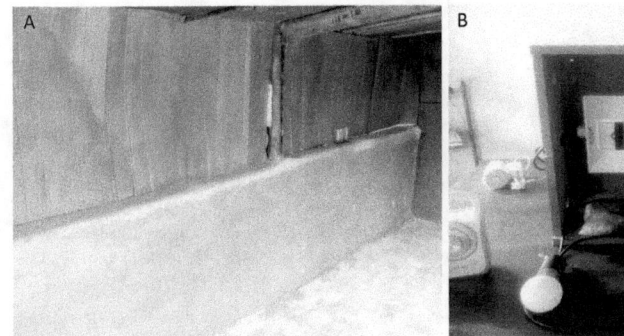

Figure 4.4 A: **Interior wall of the iShack, Enkanini.** *The interior adobe wall on the southern wall of the iShack. The reclaimed bricks are used a thermal heat stores. The cardboard, sprayed with green fire retardant paint, overlays the insulating layers made from egg cartons and tetrahedron packaging.* **B: Solar-powered light-emitting diode (LED) light in the iShack, Enkanini.** *Experiments around the solar-powered direct current microgrid system developed by Specialised Solar Systems are ongoing in Enkanini. The system includes a 12-watt photovoltaic panel, two internal LED lights, an outside motion sensor light and a cell phone charger.* [Keller 2011. Photographs reproduced with permission of Stellenbosch University: Sustainability Institute]

Conclusion

Action toward improving the lives of informal settlement residents in Stellenbosch is underway at a grassroots level, as is evident from the on-going enumerations and learning exchanges, the emergence of savings clubs, and the settlement-level mapping and planning undertaken by informal settlement community leaders. In addition, networks of informal settlement residents, decision makers in government, funders, and researchers are becoming stronger.

Radical shifts are necessary, considering that neither formalisation nor upgrading underpinned by conventional formalisation logic are viable responses to the challenges faced by residents of informal settlements. Moreover, given the scale of the backlog and the associated financial constraints, and the scarcity of resources (which will become even more pronounced in the future), conventional approaches to urban infrastructure development are not viable.

It is not constructive to devise a 'one-size-fits-all' plan for upgrading informal settlements that aspires to achieve a certain level of service delivery (measured in numbers) within a stipulated project period. Target-driven 'quick fixes' must be replaced with strategies founded on practical action, patience and perseverance. In this way, the solutions arrived at may reflect the needs *and* capabilities of each household. As they evolve locally, initiatives are likely to be heterogeneous across households and settlements.

The effort that is currently being invested in the social mobilisation process will serve to promote the everyday, incremental upgrading undertaken by informal settlement residents in the absence of official assistance. Mobilised informal settlement leadership and residents are critical in order to drive the upgrading approach in such a way that is supportive and consistent with incrementalism. If a movement for incremental upgrading can be consolidated across the whole Stellenbosch area, the momentum for upgrading may provide an opportunity to further empower the urban poor and create a more equitable urban social landscape.

REFERENCES

Bayat A. 2000. From Dangerous Classes to 'Quiet Rebels': Politics of the Urban Subaltern in the Global South. *International Sociology*, 15(3):533-557.

Bolnick J. 2011. Personal communication. Stellenbosch: 7 February.

Bolnick J & Bradlow B. 2010. Rather a better shack now than wait twenty years for a formal house. *Trialogue*, 104(1):35-41.

Bolnick J & Bradlow B. 2011. Housing, Institutions, Money: The Failures and Promise of Human Settlements Policy and Practice in South Africa. *Environment and Urbanisation*, 23(1):267-275.

Bronkowski J. 2011. Documenting experiences living in Enkanini. Unpublished narrative. Stellenbosch University: Sustainability Institute Informal Settlement Upgrading Research Group. Stellenbosch: Stellenbosch University.

Carolissen D. 2011a. Personal communication. Stellenbosch: 28 June.

Carolissen D. 2011b. Interview with Andreas Keller. Stellenbosch: 8 April.

Davidson B. 2012. Personal communication. Stellenbosch: 10 May.

Del Mistro R & Hensher DA. 2009. Upgrading Informal Settlements in South Africa: Policy, rhetoric and what residents really value. *Housing Studies*, 24(3):333-354.

Division of Revenue Act (DORA). 2011. Act No 6 of 2011: Transfers to Local Municipalities. [Accessed 12 June 2012] http://www.info.gov.za/view/DownloadFileAction?id=146048

Fieuw W. 2011. *Partnerships for Change: New Approaches to Financing Informal Settlement Upgrading*. [Accessed 10 April 2012] http://sasdialliance.org.za/partnerships-for-change-new-approaches-to-financing-informal-settlement-upgrading/

Fischer-Kowalski M, Swilling M, Von Weizsäcker EU, Ren Y, Moriguchi Y, Crane W, Krausmann F, Eisenmenger N, Giljum S, Hennicke P, Romero Lankao P & Siriban Manalang A. 2011. *Decoupling United Nations Environment Programme (UNEP). 2011 Natural Resource Use and Environmental Impacts from Economic Growth*. Report of the Working Group on Decoupling to the International Resource Panel. Nairobi: United Nations Environment Programme (UNEP).

Freund B. 2007. South Africa as a Developmental State? *Africanus*, 37(2):191-197.

Gilbert A. 2004. Helping the Poor through Housing Subsidies: Lessons from Chile, Colombia and South Africa. *Habitat International*, 28:13-40.

Govender T, Barnes J & Pieper CH. 2011. Housing Conditions, Sanitation Status and Associated Health Risks in Selected Subsidised Housing Settlements in Cape Town, South Africa. *Habitat International*, 35:335-342.

Hendler P & Swilling M. 2008. Stellenbosch 2017 Housing Strategy: A Final Proposal to Stellenbosch Municipality. Unpublished document. Prepared for Stellenbosch Municipality by the Sustainability Institute (Stellenbosch University) and Probitas Real Estate Finance Education CC. Stellenbosch.

Huchzermeyer M. 2001. Housing for the poor? Negotiated Housing Policy in South Africa. *Habitat International*, 25(3):303-331.

Informal Settlement Network (ISN) & Community Organisation Resource Unit (CORC). 2011. Langrug Settlement Enumeration Report. Presentation by Langrug community leaders to Stellenbosch Municipality. Franschhoek & Stellenbosch: June. [Available from CORC offices, Mowbray, Cape Town]

Keller A. 2011. Photographic narrative. Prepared for Stellenbosch University: Sustainability Institute Informal Settlement Upgrading Research Group. Stellenbosch: Stellenbosch University.

Keller A. 2012. iShack Preliminary Performance Results. Unpublished document. Prepared for Stellenbosch Municipality. Stellenbosch: Stellenbosch University.

Khan F. 2010. Critical Perspectives on Post-Apartheid Housing Praxis through the Developmental Statecraft Looking Glass. PhD thesis. Stellenbosch: Stellenbosch University.

Kumar A. 2011. Personal communication. Cape Town: 12 July.

Linde H. 2011. Personal communication. Stellenbosch: 16 February.

Lizarralde G. 2011. Stakeholder Participation and Incremental Upgrading in Subsidised Housing Projects in Columbia and South Africa. *Habitat International*, 35(2011):175-187.

Misselhorn M. 2008. *Position Paper on Informal Settlement Upgrading*. Part of the Strategy for the Second Economy for the Office of the South African Presidency. Compiled for Urban LandMark by the Project Preparation Trust. KwaZulu-Natal. pp. 1-54.

Mitlin D. 2001. Housing and Urban Poverty: A Consideration of the Criteria of Affordability, Diversity and Inclusion. *Housing Studies*, 16:509-522.

Mitlin D & Patel S. 2005. Working for Rights from the Grassroots. *Alliance*, 10(3):27-28.

Municipal Services and Finance Model (MSFM). 2012. MSFM: Run 0. Prepared for the Stellenbosch Infrastructure Task Team by the Palmer Development Group and the Sustainability Institute, Stellenbosch University. Stellenbosch.

National Upgrading Support Programme. 2010. Progress Report of the National Upgrading Support Programme (NUSP). Paper presented at the National Upgrading Forum. Pretoria: 22 September.

Patel S. 2008. Tools and Methods for Empowerment by Slum and Pavement Dwellers' Federations in India. In: N Kenton (ed). *Critical Reflections, Future Directions*. International Institute for Environment and Development. Participatory Learning and Action Series, 50:117-130.

Patel S, Burra S & D'Cruz C. 2001. Shack/Slum Dwellers International (SDI) – Foundations to Treetops. *Environment and Urbanisation*, 13(2):45-60.

Republic of South Africa. Department of Housing. 1994. White Paper on Housing. A New Housing Policy and Strategy for South Africa. *Government Gazette*, 345(16178). Notice 1376.

RSA DHS (Republic of South Africa. Department of Human Settlements). 2004. *Breaking New Ground: A Comprehensive Plan for the Development of Sustainable Human Settlements*. [Accessed 22 June 2011] http://abahlali.org/files/Breaking%20new%20ground%20New_Housing_Plan_Cabinet_approved_version.pdf

RSA DHS. 2009a. Incremental Interventions: Upgrading Informal Settlements. *National Housing Code*, Part 3, Vol. 4:1-72.

RSA DHS. 2009b. Incremental Interventions: Housing Subsidy Quantums. *National Housing Code*, Part 3, Vol. 4:1-15.

Sisulu L. 2006. Partnerships between Government and Slum/Shack Dwellers' Federations. *Environment and Urbanisation*, 18(2):401-405.

SITT (Stellenbosch Infrastructure Task Team). 2012a. Working Note on a Learning and Innovation Approach. Prepared for SITT public workshop by Mark Swilling on 2 May 2012. Stellenbosch.

SITT. 2012b. Infrastructure Backlogs in Stellenbosch. Presentation by André van Niekerk at SITT strategy meeting on 20 January 2012. Stellenbosch.

SITT. 2012c. Quantification of Backlog. Unpublished document accompanying presentation by André van Niekerk at SITT strategy meeting on 20 January 2012. Stellenbosch.

SM (Stellenbosch Municipality). 2007. *Integrated Development Plan: 2007-2008*. 2nd Generation, Original document. Stellenbosch: Stellenbosch Municipality.

SM. 2008. *Integrated Development Plan: 2008-2009*. 2nd Generation, Revision 1. Stellenbosch: Stellenbosch Municipality.

SM. 2009. *Integrated Development Plan: 2009-2010*. 2nd Generation, Revision 2. Stellenbosch: Stellenbosch Municipality.

SM. 2010. *Integrated Development Plan: 2010-2011*. 2nd Generation, Revision 3. Stellenbosch: Stellenbosch Municipality.

SM. 2011a. *Water Services Delivery Plan: 2011-2012*. Stellenbosch: Stellenbosch Municipality.

SM. 2011b. *Integrated Development Plan: 2011-2012*. 2nd Generation, Revision 4. Stellenbosch: Stellenbosch Municipality.

Swilling M. 2005. Hear the Forests Grow: SDI Evaluation Africa. Unpublished article. Stellenbosch: Stellenbosch University.

Swilling M & Annecke E. 2012. *Just Transitions: Explorations of Sustainability in an Unfair World*. Cape Town: UCT Press & Tokyo: United Nations University Press.

Tissington K & Royston L. 2010. Making up lost ground in SA's informal settlements. *Business Day*, 15 November.

Wessels B. 2011. Photographic narrative. Prepared for Stellenbosch University: Sustainability Institute Informal Settlement Upgrading Research Group. Stellenbosch: Stellenbosch University.

Wilkinson P. 1998. Housing Policy in South Africa. *Habitat International*, 22(3):215-229.

Winkler H, Simoes AF, La Rovere EL, Mozaharul A, Rahman A & Mwakasonda S. 2011. Access and Affordability of Electricity in Developing Countries. *World Development*, 39(6):1037-1050.

World Bank. 2011. *Accountability in Public Services in South Africa: Selected Issues*. [Accessed 12 June 2011] http://siteresources.worldbank.org/INTSOUTHAFRICA/Resources/Accountability in_Public_Services_in_Africa.pdf

Ziervogel C. 2011. *South Africa-SDI Alliance Cements Partnership with Mayor of Cape Town*. [Accessed 10 April 2012] http://www.sdinet.org/tags/town

Zietsman HL. 2007. *Recent Changes in the Population Structure of Stellenbosch Municipality*. Prepared for Stellenbosch Municipality. Stellenbosch. [Accessed 10 April 2012] http://www.stellenbosch.gov.za/jsp/util/document.jsp?id=778

Chapter 5

LOCAL ECONOMIC DEVELOPMENT
Promoting a productive Stellenbosch

WOLFGANG H. THOMAS

Introduction

At the time of writing this contribution on the local economic development of Stellenbosch, the 2011 local government elections are still fresh in people's minds, with the focus usually on the local economic development (LED) challenges of high unemployment, grinding poverty among far too many people, and extreme differences in income, wealth and quality of life in our towns and villages.

Stellenbosch can easily be included in this view of South Africa's LED dilemma or challenge. It is the purpose of this chapter to outline the process, success, and shortcomings of LED in the Stellenbosch area and to reflect on the likely (or desired) LED dynamics over the next two decades.

In focusing on LED in Stellenbosch, one must differentiate between the town itself – including all its suburbs, such as Kayamandi, Cloetesville, and Jamestown – and the municipal area of Stellenbosch, which also includes smaller towns like Klapmuts in the north (along the N1), Franschhoek in the east, and Raithby in the south. While our prime focus will fall on the town of Stellenbosch, we also have to reflect on the development momentum in these surrounding villages.

Before we look at the different dimensions of local economic activities and development processes in the area, we have to put Stellenbosch into the broader (comparative) perspective of more than 800 towns or places (and 278 municipalities) in South Africa. In order to obtain a development perspective of any place, its comparative advantages (or disadvantages) are extremely important, because they help us to better understand the development dynamics, opportunities, and challenges. Having reviewed these, as well as the different dimensions of LED in Stellenbosch, the main part of this chapter focuses on the unfolding process of LED looking ahead over the next decade or two.

Factors and forces shaping the economic development of Stellenbosch

When we compare different towns and their LED potential, a few factors can generally explain much about their respective development dynamics. These factors include the geographic location, available (natural) resources, growth sectors, infrastructure standards, socio-political stability, and global competitiveness. In this section, we briefly look at each of these in order to show that in many ways Stellenbosch is probably one of most privileged or advantaged towns in South Africa as far as its LED potential is concerned.

Location

Being located at the edge of a metropolis with a population of 3.6 million while retaining a distinct 'small-town character', gives Stellenbosch a strong competitive advantage – sharply contrasting with similarly-sized towns located 400 km or more from the nearest metropolis.

Aside from being a mere 50 km away from Cape Town's central business district (CBD), Stellenbosch is also just 30 km away from the sea (at Somerset West/Strand), and only a few kilometres away from one of the most attractive mountain ranges of the Boland. In addition, the town is a mere 28 km from Cape Town International Airport, one of South Africa's top (air) links to the global economy, and not much further away from Cape Town harbour, the shipping portal to both the Atlantic and the Indian Oceans. As far as inland transport is concerned, Stellenbosch is equally close (that is, less than 20 km) to the N1, N7 and N2, the three national road links between the Western Cape and the rest of the country (and the continent). Thus, for a medium-sized town in South Africa, Stellenbosch could hardly be better located geographically.

Resources

Historically, Stellenbosch is known for its agricultural production, with the emphasis mainly on vineyards, although vegetables and other specialised agri-products have also been significant in the past. Generally, agriculture has been shaped by the winter-rainfall pattern, the proximity of mountain ranges, and the relatively small size of farming units.

If we look further afield than agriculture (and the absence of any significant mining potential), other local resources include the mountains and unique (wine) estates, such as Spier, surrounding Stellenbosch (linked to the proximity of the sea), which makes the area ideal for tourism, especially if we link these physical factors to the rich history of the area and the proximity to Cape Town. It is therefore no wonder that Stellenbosch has become one of the iconic tourist destinations of South Africa.

Interpreting 'resources' in even broader sense, we realise that, due to its proximity to Cape Town and its history and attractive location, as well as the presence of the university, Stellenbosch is rather well-endowed with the 'human resource' factor of production, in particular higher-skilled labour (and retirees). Going one step further, we can say that Stellenbosch is also well-endowed with 'capital' as far as it relates to technological know-how and specialised skills.

Finally, if we also include world-class business leadership as a locally relevant (or exploitable) resource, we may mention well-known business leaders, such as the Ruperts, GT Ferreira, the Venfin people, Capitec's Michiel le Roux, and Christo Wiese as people who have strengthened the business image of Stellenbosch.

Taking all these resource factors together, it is clear that, in the sphere of resource endowment, Stellenbosch is much better off than most medium-sized towns in South Africa.

Infrastructure

In sharp contrast to the situation in many towns across South Africa, the different infrastructure dimensions of Stellenbosch are relatively well-developed. This applies to transport infrastructure (road, rail, and airline access) and electricity and water supply, as well as local services, such as refuse collection, street lighting, and sewage removal. While there are specific, localised gaps in the supply and distinct challenges with respect to future capacities, the overall level of current services is of a high (metropolitan) standard. However, this advantage may well be eroded if sufficient funds are not allocated to extend this infrastructure and properly maintain and upgrade it.

Viable economic sectors

Here again, closer scrutiny reveals an amazingly wide range of sectors in which the local economy is involved at a significant level. Without going into detail, we can list the following sectors and niches:

- agriculture (viticulture, horticulture, vegetables, poultry, cattle/dairy);
- forestry;

- fishing (including aquaculture on a limited scale);
- manufacturing (including agro-processing, wineries, timber processing, high-tech industries, clothing, construction material);
- construction, maintenance, and repairs focusing on local urban and nearby rural structures (with strong capacity in locally adapted architectural style and design);
- trade (including retailing, some wholesaling, import/export, and informal trade);
- financial institutions (including national headquarters of specialised firms);
- tourism-related establishments;
- education and training institutions (including Stellenbosch University);
- transport-related services;
- business and professional services (including corporate headquarters);
- social and health services; and
- public administration offices (including the Stellenbosch Municipality and Winelands District Municipality offices).

In the case of some of these sectors, such as the wine industry, the university, tourism facilities, forestry, and professional services, the local institutions, activities and/or exports are well-known countrywide.

Enterprise mix

The diverse economic sectors and niches of Stellenbosch reveal a wide range in terms of size and sophistication of enterprises, varying from local branches of large multinational firms, head offices of South African companies to medium-sized businesses and non-profit and social enterprises, as well as a large number of small, micro- and informal enterprises (operating from homes, along streets, and in lively markets). Thanks to the university and the town's rich cultural history, we also find a significant number of religious, cultural, sports, and other social bodies headquartered or active in Stellenbosch. In many cases, it is the pleasant business and living environment of Stellenbosch that encourages enterprises and non-business entities alike to set up shop in Stellenbosch rather than nearby Cape Town suburbs.

Socio-political stability

Compared to densely populated areas of larger towns and poverty-stricken urban areas, Stellenbosch has a relatively diverse and stable socio-political environment, with coloured, African, and white inhabitants each constituting a significant segment of local society. Similarly, there is strong support among local voters for each of the main political parties, resulting in some political instability as a result of the frequency of changes in leadership in the municipal government. A steadily increasing number of immigrants from the rest of Africa have raised the spectre of xenophobia in Stellenbosch as well, but its dimensions appear relatively limited at this stage.

As far as the global competitiveness of Stellenbosch is concerned, we have already referred to a few factors working in its favour: the university and sophisticated research and technology-linked local enterprises, its world-class wine industry, the tourism sector (which is related to the wine industry), and the growing financial services sector.

LED progress in Stellenbosch

Having outlined and stressed the comparative advantages (or growth opportunities) of Stellenbosch, we now have to look at the actual performance of the local economy over the past decade. If we interpret socioeconomic development in a broad sense, we would have to touch on many of the developments and issues covered in the other chapters of this book. More narrowly interpreted or demarcated, we focus on macro-growth indicators, general social development indicators, business and sector development, black economic empowerment (BEE) and inequality as particular challenges. We must also differentiate between trends and issues in the town of

Stellenbosch itself and the more complex development pattern of the whole municipal area, with the emphasis in this chapter falling on Stellenbosch town.

The main sources of information for this review section includes the available reports on the Stellenbosch Municipality's Integrated Development Plan (IDP), the Spatial Development Framework (SDF), and the local 2009 Economic Development Strategy (see references at the end of the chapter). Other important sources are the reports prepared by the Cape Winelands District Municipality, of which Stellenbosch is one of the two major towns (Paarl, which is slightly larger, being the other).

Before we look into any specific dimension of the LED process, we have to mention the data dilemma, which is not limited to Stellenbosch or the Western Cape; in fact, it may be less severe here than in other parts of the country. Starting with population estimates (which raise serious questions), we lack readily accessible, plausible data about important trend factors, such as annual population growth, unemployment levels, informal sector engagement, and more specific sector growth trends. The generation, dissemination, interpretation, and wider discussion of such data seem to present some of the greatest challenges for the LED process (and other aspects of socioeconomic development). This is not unique to Stellenbosch, but might be tackled here easier than elsewhere.

Population growth

If we take the annual population growth in a town as (*inter alia*) reflecting the overall growth of the local economy and its 'absorptive capacity', Stellenbosch appears to have done very well over the past decade. Based on the 2001 national census and the 2007 community survey, its population is said to have grown from 118 700 to 200 500, that is, at *an annual growth rate of 9.2%* (Stellenbosch Municipality IDP 2009)! Extrapolating this rate to 2012 would imply a total population of almost 246 000 (using a more modest annual growth rate of 4.2% for the last five years).

Annual growth rates of 4.2% to 9.2% (compared to South Africa's overall population growth rate of less than 2%) would not only suggest highly successful LED, but should also sound a strong warning about sustainability challenges.

- Aggregate growth is likely to be influenced significantly by the student population living in Stellenbosch. Their number (25 000+) has steadily increased over the past decade, but it is also partially misleading, since they do not necessarily reside in Stellenbosch full-time.
- During the past decade, the share of Africans in the total population of Stellenbosch has increased rapidly, from about 20% in 2001 to at least 28% in 2011. Some of this increase has been due to migration from the Eastern Cape, while the eviction of workers from Boland farms has also swelled their numbers.
- It seems likely that the 2007 community survey overstated the total population of the area, but that issue can only be solved on the basis of the 2011 national census. Once those results are available and have been verified, it seems most important for the municipality (together with other stakeholders) to start using these figures for a systematic and openly discussed demographic growth-monitoring process. This should reveal many of the particular characteristics and trends of this town, such as a relatively large older age group, a 'bulge' in the student age segment, relatively large white and coloured groups, and the rapid increase in the African segment.

Gross Domestic Product

Available data for Gross Value Added (GVA) in the Cape Winelands District suggest that, between 2002 and 2009, Stellenbosch Municipality's real GVA increased at an average annual rate of 4%, compared to 3.2% for the Cape Winelands District (Cape Winelands District Municipality 2010). This includes (in line with national trends) some relatively high growth rates of 5% in 2002 and 2008, but also a low 1% during the 2009 recession.

Once again these figures – when viewed within the broader context of national and provincial Gross Domestic Product (GDP) growth rates – would seem to confirm reasonably satisfactory growth. However, they also raise important questions.

- Compared to the alleged population growth rate of 4.2 to 9.2% per year suggested earlier, the real GVA growth of 4.0% per year would be disappointing, since it would indicate negative per capita growth.
- Considering the large student population, it may be difficult to actually measure annual GVA with respect to the town. The same applies to the role of Stellenbosch as a place of retirement, which means that it receives a lot of pension payments for its older residents from external sources.
- Being in many ways a suburb of greater Cape Town, Stellenbosch's 'own' GDP is rather difficult to quantify. For example, many of the town's commuting residents are contributing to the GDP of Cape Town.

Thus, here, as in the case of population growth, it will be necessary to develop more meaningful indicators of 'aggregate local economic growth'.

Employment, income and social development

In current discussions about the challenges faced by South African towns and municipalities, there is much emphasis on the reduction of unemployment and poverty levels and a narrowing of income and wealth gaps. From available statistics we note that, with respect to Stellenbosch, a rate of 17.1% unemployment is given for both 2001 and 2007, while the labour force is supposed to have grown at 9.5% per year over the 2001 to 2007 period. This would seem to be a remarkable performance, much in line with the relatively lower unemployment rates of the Western Cape (compared to other provinces). Yet, once again, these and related further data about the age distribution of the unemployed call for more in-depth analysis, taking into account the significant role of the university student body (*inter alia* in part-time employment) and the seasonality of some of the agricultural, agro-processing and tourism niches.

If we look at poverty levels, the absence of clearly defined local indicators becomes even more important, in particular if we take into account the rapid population growth of Africans, which would suggest a major influx of job and income-seeking migrants from the Eastern Cape and other parts of the country. The level of statistically unrecorded, informal earnings is likely to be substantial among this group. In a rather different context, this may also apply to parts of the student population, who rely on quite diverse sources of income and part-time earnings.

Media reports and the observation of living conditions in informal settlements (and backyard activities in formal settlements) make it clear that Stellenbosch faces serious poverty and deprivation challenges, which constitutes one of the reasons to create even faster LED. Yet, it seems to be more a case of 'a successfully expanding place in an ideal location attracts so many people that poverty remains an issue'.

Local business development

Disaggregated data on employment and value added per sector confirm that, during the past decade, all the sectors in the local economy expanded, including the largest, 'community services' (which includes educational institutions), manufacturing (with the emphasis on agro-processing), and financial services. During the early part of the decade, the growth in the construction sector was exceptionally high (11.2% per year), which reflects the expansion in residential, university, and infrastructure construction (Stellenbosch Municipality IDP 2009).

Not reflected in the available data, but visible to local observers, has been the increase in virtually all size categories of business and social enterprises, from large corporates and globally-linked firms to medium, small and micro enterprises, right through to the buoyant informal sector (in the townships, along some streets and markets, as well as in the homes of local residents, students, retirees, and migrants). Against this background, one wonders whether the municipality is fully aware of the significant size of the local informal sector, and whether municipal policies reflect an understanding of its significance in the fight against unemployment and poverty in the town (and sub-region).

Black Economic Empowerment and the reduction of inequalities

Available statistics are not very helpful for drawing clear conclusions about progress over the last decade with regard to these two strategic goals of LED in South Africa. The high population growth rate and the relative increase in the local number of Africans could be seen as a positive sign of black advancement. The changing student composition of Stellenbosch University might be seen as a further factor, to which the closer integration of the different suburbs of the town might be added. Yet, available data and trends are open to contradictory interpretations.

Sustainability of development

Local population growth rates of 4% to 9% per year, continued growth across a fairly wide spread of (sub-) sectors, and rising *per capita* income levels may be impressive signs of dynamic LED in Stellenbosch, but they raise the general concern about development sustainability along the present path to even higher population levels. What does the high growth and expansion imply for waste removal, electricity demand, land use and the availability of water? These topics are explored in several of the other chapters in this collection. Given the many factors stimulating growth here (highlighted previously in the '*Factors and forces shaping the economic development of Stellenbosch*' section), the bigger challenge for Stellenbosch may now be *how to reconcile its growth potential and growth momentum with sustainability concerns and long-term resource constraints.* Thus, the emphasis has to be on progress with *sustainable* LED, rather than merely the growth momentum. Here again, the availability of reliable, detailed data about the growth factors (population, local production, water usage, etc.) is so important that it warrants higher priority in LED planning.

Driving a holistic LED strategy for Stellenbosch

Against the background of its growth (as summarised in the previous section) and nationwide political and institutional developments, Stellenbosch (like all other towns in the country) faces the challenge of pursuing a systematic, holistic, and politically acceptable strategy of proactive local economic development for the town and the adjoining places within the municipal area. This strategy has to simultaneously focus on three distinct objectives, each calling for a multitude of policy steps:

- stimulating overall growth in line with population increases and expectations of rising living standards;
- reducing or alleviating poverty and other forms of social deprivation of segments within the local society; and
- tackling the sustainability challenges posed by the high population growth and environmental and other existing or emerging constraints, including infrastructure upgrading.

In the following section, we focus on the issues and strategy aspects that fall more directly into the sphere of LED, but without losing sight of the many broader sustainability issues. We first look briefly at the evolution of the current LED strategising process and then outline key challenges to an evolving LED process.

Historical perspectives

It would be wrong to assume that the municipal, business, and civil society leaders of Stellenbosch during the pre-1990 decades did not pursue an LED strategy aimed at stimulating local business and wider socioeconomic development with respect to Stellenbosch. However, these activities were limited by

- the deliberate exclusion of Africans in planning for the permanent settlement of people in Stellenbosch (as part of 'grand apartheid');
- the neglect of many of the social and economic interests and needs of coloured people in and around Stellenbosch (as part of local apartheid and separate development);
- the absence of a systematic, integrated LED strategy for the whole municipal area; and
- little ongoing support for LED efforts from the provincial authorities (the Western Cape provincial government) and national government (although *ad hoc* support was provided for specific projects or problem areas).

The creation of new coloured or African (Group Areas) suburbs did pave the way for new commercial and suburban development nodes (in Cloetesville, Idas Valley, and Kayamandi), but these were felt to be artificial creations and of little benefit in comparison to the damages caused by the implementation of Group Areas-linked removal policies and the exclusion of black entrepreneurs from businesses in the town centre.

Naturally, the effects of these discriminatory policies were even more damaging in the smaller towns and settlements within the municipal area (like Franschhoek, Jamestown, Kylemore, and Koelenhof) where 'separate' black business areas were just not viable.

The abolition of the Group Areas Act and other discriminatory policies after 1995 could not change urban business development patterns overnight. Besides, it took a few years to reform municipal legislation, rationalise municipalities (from more than 800 in the country to 278), and establish new frameworks for (*inter alia*) local economic development. Thus, it was only in 2004 to 2006 that a new approach to LED strategy was accepted nationally and put on the task list of municipalities. A first LED strategy document was accepted for the Stellenbosch municipal area in 2006, with a revised strategy prepared in 2009 and a further revision due in 2012.

These LED strategies were seen as a refinement of the much broader Integrated Development Plan (IDP), which municipalities are compelled to draw up on a regular (annual) basis, to show in some detail how public spending is to be effected in terms of the different activity areas and projects of the municipality. Thus, while the IDP is primarily spending and projects/programme orientated, the LED strategy is expected to be strategy and policy orientated.

During these years of municipal restructuring (1996 to 2006), most of the municipalities (including the Stellenbosch Municipality) also experienced drastic staff changes and internal restructuring processes. This implied, quite naturally, that relatively new sections such as 'LED strategising' suffered from a lack of experienced professionals and sufficient support staff. As a further consequence, most of the professional documents (like LED strategy papers) had to be prepared by outside consultants who did not always liaise closely with internal staff. Close interaction between LED staff, external consultants, municipal politicians, community, as well as business was expected, but much of the grassroots interaction was limited to a few public meetings with frank discussions, rather than extensive interaction around critical issues. During 2009, 2010 and 2011, there have been formal municipal workshops or conferences about Stellenbosch Municipality's LED strategy, with summaries of the discussion made available (see references). Nevertheless, the breadth and depth of interactions between municipal, business, and community stakeholders have so far been rather limited.

The remainder of this section puts forward a few critical steps to strengthen this interaction and the evolution of a viable LED strategy for Stellenbosch that addresses the three objectives mentioned earlier (that is, stimulating overall growth, reducing poverty, and overcoming sustainability challenges).

Critical steps in a future-orientated LED strategy

We can highlight five critical areas where systematic action appears to be required in order to achieve the overall goal of implementing a holistic LED strategy.

- *Strengthening partnerships in the LED process*

The LED literature is replete with references to the need, character, and dynamics of public-private partnerships (PPPs) for the successful unfolding of the LED process. Yet, in practice, these efforts are often limited or more tedious and formalistic, rather than really helpful to both sides. All too often, there is an underlying perception (on both sides, though usually for different reasons) that it is really the public sector, government, or, locally, the municipality who should tackle and solve LED issues. Yet, if we look at the LED process in its full breadth, the private sector is by far the greatest 'player'– through its dozens of large firms, hundreds of smaller enterprises, thousands of informal operators, dozens of NGOs, social enterprises, and community activities, as well as all the other stakeholders (such as retired people, tourists, and church leaders, to name but three).

South Africa has reached a point in its socio-political evolution since 1995 where we have to seriously change the perception that it is the sole responsibility of those in government (or in the local municipality) to deal with challenges. References to elected councillors cannot reduce the responsibility of the private sector to get more actively engaged in the day-to-day issues of LED.

Considering its position as a university town, as well as a place filled with retired people with lifetime experience and a satellite to greater Cape Town, Stellenbosch's private sector stakeholders and LED volunteers should be well qualified to participate in the process of forward planning, policy design, and leadership sharing for the LED challenges outlined earlier. However, to put it bluntly, attendance by a few private sector representatives of one or a few annual LED review workshops is just not good enough.

▣ Developing deeper insights into LED processes and challenges

We indicated earlier that the statistical basis from which we can judge overall progress with LED in and around Stellenbosch is quite shaky. This becomes even truer when we look at the full complexity of the three interacting and partially conflicting goals and objectives of LED (growth, poverty alleviation and sustainability). To overcome this dilemma, which impedes planning, as well as the implementation of strategies, we have to give more attention to:

- the effective utilisation and interpretation of national and provincial LED statistics and trend analyses;
- the production and revision of locally generated data and trend projections;
- the dissemination of such data in easily understandable and accessible ways; and
- the active discussion of this data and trend assessments among all relevant stakeholders (to help us reach consensus about the issues and challenges and help shape realistic policies).

This 'task' applies to all 278 municipalities in South Africa. However, with Stellenbosch University central to the Stellenbosch context, we seem to be in the advantageous position of having an institution right here that has (should have?) the capacity to help with this task. We shall return to this challenge later.

▣ Strengthening capacities for the implementation of LED policies

Once we realise the complexity of the LED goals (even for a well-endowed place like Stellenbosch) and the diversity of issues that have to be addressed through policies, negotiations between conflicting parties, and action by diverse institutions, it becomes clear that implementation capacities are crucial for the ultimate success of LED efforts. This relates not only to the LED core staff of the municipality, but also to:

- staff in other municipal departments, who should cooperate in policy implementation;
- municipal councillors, who have to represent the wider public and who should understand the diversity of dimensions of LED issues and the need to proactively search for compromise policies or action;
- leading staff of business associations (chambers of business, or *sakekamers*, etc.), who have to play a pivotal role in 'representing' the (often quite diverse) interests of the business community in tackling LED issues (for example, the 'control' of informal business activities); and
- other public leaders who should get involved in LED issues such as poverty alleviation, land reform implementation, youth development, and the marketing of Stellenbosch.

The simple answer to this capacity challenge would be the offering of LED-focused training programmes, as carried out by provincial or national LED offices or private consultants, or via in-house municipal initiatives. Yet, the challenge is much greater and calls for the appointment of (more) dedicated LED staff in the municipality, in business associations, and in other stakeholder bodies, the implementation of more extensive training programmes, the mobilisation of retired mentors as LED officials, and the widening and deepening of media coverage of LED processes and issues.

▣ Reconciling growth, equity, and sustainability goals

Efforts to pursue the three goals of LED intervention may easily lead to politicised, zero-sum confrontations. For example, more money is to be spent on low-income housing needs, which implies reduced funding for road maintenance or inner-city infrastructure; or health considerations and the protection of tax-paying small enterprises seem to call for drastic steps to keep out (illegal) street traders.

Closer focus on the full spectrum of LED issues and conflicts shows that win-win (rather than zero-sum) solutions are often possible – and are known to have succeeded in some places. However, this calls for thorough investigations, a search for compromises, and delicate negotiations. Case studies of other places must be studied and their lessons adapted to local conditions.

- **(More) Active engagement of Stellenbosch University in the LED process**

In each of the above four critical areas for strengthening the LED process we indicated that the resources of Stellenbosch University in terms of capacities, skills, and people might (if properly engaged) be of significant help. This applies as much to the direct training of public sector officials and the offering of part-time training or shorter courses as it might to the provision of consultancy services and the mobilisation of mentors. A more indirect (but no less effective) way in which Stellenbosch University could be of assistance would be the active involvement of US staff or researchers in the private sector or in civil society bodies engaged in LED planning or initiatives. On the research side, there is the opportunity to supervise dissertation research related to practical issues around LED in the Stellenbosch area or LED implementation in general. Case study research might be particularly fruitful for MA students who have to complete mini-dissertations (especially if the municipality cooperates on the data collection side). Another fruitful area could be the facilitation of local workshops between the PPP bodies, especially when they want to tackle difficult issues.

As indicated in some of the chapters in this publication, the role of Stellenbosch University could be particularly important around the critical issue of sustainability, where ignorance levels are still high and the absorption of worldwide experience is a major challenge for local authorities and other LED stakeholder groups. In this context, fairly simple processes (such as the production of 'fact sheets' on comparable experiences in other countries or towns) could already be of significance for the local municipality and could pave the way for the planning of broader tasks.

Looking ahead

Over the next two decades, many of the issues currently at the forefront of public attention may enjoy less priority, while others may be much more topical. In this final section, we look at two aspects in particular, namely:

- how the LED process is likely to unfold over the next twenty years, and
- what the more critical issues are likely to be in short to medium term (leaving aside totally new forces or factors).

How the LED process might unfold

From current critical discussions about LED in South Africa we know that, at this stage, issues around unemployment, basic service delivery to the poor, poverty relief, and the strengthening of South Africa's global competitiveness are the central issues in our LED processes and perceived LED progress. As discussed earlier, these issues are also relevant for Stellenbosch, even though the problems here seem less severe compared to other towns and other parts of the country. Reasons for this have also been suggested, namely the diversity of local growth forces and sectors and the proximity of Stellenbosch to the metropolis (where poor people are more likely to cluster in the larger townships than in the relatively small townships of Stellenbosch).

Looking into the future, it seems likely that dissatisfaction of the local poor is to continue, since 'solutions' are enjoying attention, but take time and seldom show uniform successes. Thus, it seems quite possible that Stellenbosch will also face grassroots protests. Similarly, unemployment is likely to remain significant, but at a rate much lower than the national (and provincial) average. Income and wealth inequalities are also significant, but the overall structure of income and wealth distribution remains quite complex in this university town, with its large number of retirees and its proximity to the metropolis.

If the LED-focused interaction between the municipality, the private sector, the university, and other stakeholder groups unfolds as suggested in the previous section, it should be possible for Stellenbosch to 'manage' LED conflicts and gradually reduce the extent and destabilising impact of the apartheid past and resulting conflicts.

As these social development issues are being addressed, we should, over the next two decades (probably more in the second than the first), expect sustainability issues to demand increasing attention, resources, and effective management. This may initially centre on the water storage and water supply position of Stellenbosch, but will probably also include greater focus on alternative electricity generation and other challenges linked to global warming (and its impact on local agriculture). At the same time, the growth of the metropolis is likely to further push up land prices on the metropolitan edge, which could lead to renewed frustration among low-income households intending to settle in the Stellenbosch area. Rising land prices combined with increased property rates (to fund the refurbishing of the ageing infrastructure) could also increase local frustration levels.

Future LED priorities

Speculation about the longer-run pace of changes taking place in the global world – rich and poor countries, technologically advanced and emerging countries in both the northern and southern hemispheres – is intense and includes the possibility of rather dramatic changes. Will global warming 'destroy' the Western Cape's viticulture and other profitable agricultural sub-sectors? Will a rising sea level have dramatic results along the Cape Town to Mossel Bay coastline? Will local political or social conflicts reduce South Africa's (and Stellenbosch's) appeal for retirees to settle here?

As far as Stellenbosch is concerned, it would seem that all these longer-term issues, challenges, or threats are relevant and should be looked into carefully. Yet, considering the particular location of the place, its diversity of growth sectors, the presence of the university (assuming its proactive engagement in the search for, and implementation of, solutions to these problems), and its proximity to the Cape Town metropolis seem likely to reduce the overall risk and magnitude of negative consequences.

This relatively reassuring conclusion, however, puts even greater emphasis on the need for the university to get involved in the deeper understanding of LED challenges, as well as the range and feasibilities of counteraction. This applies as much to the need for attention to local issues and challenges as it relates to the broader regional, national, and continental challenges that may unfold during these two decades.

REFERENCES

Cape Winelands District Municipality. 2010. *Regional Development Profile: Cape Winelands District 2010.* WC048. Stellenbosch: Cape Winelands District Municipality.

Phillips E. 2007. *Local Economic Profile of the Stellenbosch Municipal Area.* Cape Town: Western Cape Government Department of Economic Development and Tourism.

Stellenbosch Municipality. 2006. *Integrated Development Plan for the 2007-2011 Term.* Stellenbosch: Stellenbosch Municipality.

Stellenbosch Municipality. 2007. *Integrated Development Plan 2007-2012.* Stellenbosch: Stellenbosch Municipality.

Stellenbosch Municipality. 2008. *Social and Human Development Strategy.* Stellenbosch: Stellenbosch Municipality.

Stellenbosch Municipality. 2009. *Integrated Development Plan for the Municipal Area of Stellenbosch.* Stellenbosch: Stellenbosch Municipality.

Stellenbosch Municipality. 2010. *Local Economic Development Strategy – Version 2, 2009.* Stellenbosch: Stellenbosch Municipality.

Stellenbosch University: Sustainability Institute. 2009. *Stellenbosch Sustainable Human Settlement Strategy.* Stellenbosch: Stellenbosch University.

Thomas WH. 2011a. Challenges around Stellenbosch's LED process: Understanding LED to harness inclusive economic growth and competitiveness. Conference address at Stellenbosch Municipality, Stellenbosch: 15 April 2011.

Thomas WH. 2011b. Supporting Local Economic Development in the Stellenbosch Municipality Area. Feedback from a series of workshops held during March to April 2011. Unpublished report. Stellenbosch: Stellenbosch Municipality.

Thomas WH. 2010. Stellenbosch Municipality's LED Strategy: From drafting to implementation. Feedback from workshop held 6 May 2010. Unpublished report. Stellenbosch: Stellenbosch Municipality.

RESPONSE TO
LOCAL ECONOMIC DEVELOPMENT

Local economic development – myth or mission?

ANN HEYNS

Wolfgang Thomas's chapter provides an overview of local economic development (LED) in Stellenbosch and the measures required to achieve more sustainable LED. For the author, the bigger challenge for Stellenbosch may be how to reconcile its growth potential and growth momentum with sustainability concerns and long-term resource constraints.

Stellenbosch faces serious poverty and deprivation challenges. In order to remedy these problems, Thomas suggests pursuing a systematic, holistic and politically acceptable strategy of proactive and sustainable LED for all people living within the municipal area.

I would agree with the Thomas's argument that proactive and sustainable LED is only achievable through expanding the breadth and depth of interactions between the municipality, university, private sector and community stakseholders. However, what I find lacking in this chapter is the question of how these expanded interactions may be brought about.

The objectives pursued by LED are admirable. Yet, past experience has failed to articulate measures that might inspire every inhabitant of Stellenbosch to participate in the process. As a result, the majority of residents perceive LED as being the sole responsibility of the municipality. Herein lies the crux of the problem. Given the socio-economic disparities between the various communities in Stellenbosch, we have concentrated so hard on *community* (i.e. previously disadvantaged, as opposed to advantaged) that we have overlooked *commonality*. It is my belief that the key to unlocking the vast potential for economic development and transformation lies in the word 'local'.

All inhabitants of Stellenbosch have this in common: they are 'locals'. Should visitors to the town approach someone for directions, they have only one goal in mind: ask a 'local'. The Concise Oxford Dictionary (1990:695) defines 'local' as "belonging to or existing in a particular place, or neighbourhood, peculiar to or only encountered in a particular place."

It stands to reason that if all these 'locals' (whether by birth, vocation or address) would embrace, emphasise and promote their commonality, the way would be paved for a transformative mind shift that would see all stakeholders in town engaging in the LED process.

Stellenbosch is world-renowned for its spectacular natural beauty. Surrounded by magnificent mountain ranges and more than 150 award-winning wine estates, it is also famed for the preservation of its historical neo-Dutch and modern Victorian architecture. Further enhanced by (among other things) the vitality of a university campus, a vibrant café culture, culinary expertise, world-class sports facilities and the presence of corporate giants alongside a bustling informal economic sector, Stellenbosch bears witness to a proud and dignified heritage.

In this commentary, I will look at ways in which the development and promotion of Stellenbosch as a tourism destination may play a leading role in proactively engaging all stakeholders in the LED process. This process will, however, fail to be either meaningful or sustainable if the tourism industry in Stellenbosch remains predominantly in the hands of white owners in the town centre, and fails to include the marginalised communities on the outskirts of town. In Stellenbosch, the inequality of tourism-product infrastructure may be eradicated through the development of what communities on either side of the economic scale have in abundance: inherent talent and a rich cultural heritage. It is also necessary to clarify that the tourism industry is not merely defined by infrastructure created for leisure purposes. The presence of corporate businesses in town brings a flow of visitors who avail themselves of our highly esteemed academic, technological and business acumen.

Tourism as conduit for economic development

Tourism contributed 25% of the GDP of Stellenbosch in 2008/9, compared to 17% in Cape Town. It is estimated that 80% of new jobs globally are created in this (SMME) sector. Tourism is the only business sector that has the capacity to provide jobs and a platform for entrepreneurs to access the tourism industry through innate talent and natural resources.

In order to take full cognisance of the potential of tourism for reaching the objectives of LED, I am focusing on a niche market in tourism that will draw on the commonality of the Stellenbosch 'locals', without the barriers of socio-economic inequalities.

Cultural tourism

Cultural tourism refers to any travelling for the purpose of participating in or experiencing the culture or heritage of humanity in a destination of choice (Goeldner & Ritchie 2009). Ivanovic (2008) sustains that the greatest motivator for travel is to understand culture and heritage, both contributing to economic development.

People are hungry for unique and authentic experiences and there is an international demand for cultural tourism. Moreover, cultural tourism has a higher spend than any other sector in global tourism. 'Culture' and 'heritage' are singled out as the new buzz words in the tourism industry, with a higher expectation of revenue due to the international demand for responsible tourism. Statistics produced by Goldman Sachs predict that South Africa will be one of the markets dominating global tourism during the next 20 years, with the BRICS (Brazil, Russia, India, China and South Africa) countries it will accounting for 61% of the growing market. Part of the National Department of Tourism's development strategy for 2012 is the development of tourism opportunities in rural areas to ensure rural areas utilise their unique assets, basic resources and characteristics (Department of Tourism (RSA DoT) 2012).

Sustainable tourism

According to the World Tourism Organization's definition, sustainable tourism development:

> ... meets the needs of present tourists and host regions while protecting and enhancing opportunity for the future. It is envisaged as leading to management of all resources in such a way that economic, social and aesthetic needs can be fulfilled while maintaining cultural integrity, essential ecological processes, biological diversity and life support systems. [Earth Council 1995:30]

Expanding on this formulation, sustainable tourism should "operate in harmony with the local environment, communities and cultures, so that these become permanent beneficiaries and not victims of tourism development" (Van Egmond & Van Egmond 2007:150).

In terms of LED planning, the activity of route tourism is of special interest, for it often involves developing cooperative planning arrangements and relationships between different localities in order for them to collectively compete as tourism spaces. Spatial networks are constituted by 'packaging' rural tourism products into "inclusive and coherent routes through the use of themes and stories (such as folklore, working lives, food and drink routes, religious routes) which help to move the tourist around geographically dispersed attractions" (Clarke 2005:92).

Moreover, Briedenhann and Wickens point out that "the clustering of activities and attractions in less developed areas stimulates cooperation and partnerships between communities in local and neighbouring regions and serves as a vehicle for the stimulation of economic development through tourism" (2004:72).

Case studies

Route 360

Stellenbosch 360 is a non-profit organisation established in November 2011 with the vision to encourage collaboration and unity around the positioning of Stellenbosch as the most inspiring and innovative town in the world. In keeping with this vision, the organisation has launched an initiative to develop a cultural tourism route, Route 360, aimed at addressing the desperate need to grow the economy in the previously disadvantaged communities of Stellenbosch while simultaneously encouraging the sharing of knowledge and lifestyle experiences among locals. This initiative stands as a prime example of a unified, all-inclusive approach to job creation and poverty alleviation.

The Midlands Meander

The Midlands Meander is another good example of such an approach. The Meander focuses on a cluster of 120 small enterprises, including art studios, country hotels, flower farms, cheese makers, tea gardens, craft studios, pottery and weaving workshops, trout farms and golf courses. Collectively, this group of hospitality ventures and craft activities produce a network of quaint experiences that has put the Natal Midlands firmly on the national tourism map.

The Midlands Meander began as an unplanned, collaborative local-economic initiative among a group of local artists, potters and weavers in 1985. Later, a series of accommodation and other hospitality enterprises were added to the attractions in the area. An important step was the amalgamation of different routes in 1992 and the formation in 1993 of the Midlands Meander Association, a non-profit organisation that started as, and has remained, a voluntary association dedicated to marketing the Midlands Meander. Each year, the organisation produces a Midlands Meander map and brochure, with production costs shared among its membership, who in turn receive coverage in the brochure. In addition, the Association also supports the Meander by purchasing exhibition space at travel and trade shows.

The Ghoema Route

The Ghoema Route, founded in 2007 by the Ghoema Route Association, provides a third case study on route development or collective marketing. It meanders along the R44, from Paternoster on the West Coast, through the Cape Winelands and Cape Metropole, and ends at L'Agulhas in the Cape Overberg region, with the aim of promoting wine and cultural tourism. The Association is especially interested in a specific sector of cultural tourism among farm workers and believes that resources such as innate talent and a rich heritage may provide both the farm owner and the farm worker the opportunity to grow their economy to their mutual benefit. There are approximately 12 000 farm workers in the Stellenbosch region, for whom Afrikaans is their mother tongue – this provides endless opportunities for the domestic tourism market, as storytellers, performing artists, crafters and skilled labour provide the visitor with unique and authentic experiences. By developing cultural tourism, the farm owner adds value to an existing offering of wine tasting, cellar visits, restaurants and accommodation. Solms Delta, near Franschhoek, is a prime example of this winning recipe (Von Meijenfeldt 2012).

Conclusion

In conclusion, given the complex challenges that LED faces, Stellenbosch Municipality will have to look at fresh and innovative ways of capturing the attention of all major stakeholders in Stellenbosch and implement solutions that are both proactive and sustainable. The golden thread running through each community, corporate or leisure establishment is our rich cultural heritage. By investigating and investing in our natural and innate resources, we stand a chance to bridge the divide between our fragmented societies. The pursuit of cultural tourism as a socio-economic activity aimed at job creation and general well-being is an attainable, measurable, creative – and pleasurable – way to achieve LED objectives.

REFERENCES

Briedenhann J & Wickens E. 2004. Tourism routes as a tool for the economic development of rural areas – vibrant hope or impossible dream? *Tourism Management*, 25:71-79.

Clarke J. 2005. *Effective Marketing for Rural Tourism: Rural Tourism and Sustainable Business*. Clevedon, UK: Channel View.

Earth Council, World Travel and Tourism Environment Research Centre, World Tourism Organisation, World Trade Organisation. 1995. *Agenda 21 for the travel and tourism industry: towards environmentally sustainable development*. Madrid: World Tourism Organisation.

Goeldner CR & Ritchie JRB. 2009. *Tourism: Principles, practices, philosophies*. New Jersey: John Wiley & Sons.

Ivanovic M. 2008. *Cultural Tourism*. Cape Town: Juta & Company Ltd.

RSA National Department of Tourism. *Rural Tourism Strategy*. April 2012:9. [Accessed 14 August 2012] http://www.tourism.gov.za/AboutNDT/Branches1/domestic/Documents/National%20Rural%20Tourism%20Strategy.pdf

The Concise Oxford Dictionary. 1990. New York: Oxford University Press.

The Ghoema Roete. 2012. [Accessed 26 May 2012] http://www.ghoemaroete.co.za

The Midlands Meander. 2012. [Accessed 26 May 2012] http://www.midlandsmeander.co.za

World Tourism Organization. [Accessed 26 May 2012] http://www.sdt.unwto.org/en/content/about-us-5

Van Egmond T & Van Egmond AN. 2007. *Understanding Western Tourists in Developing Countries*. Wallingford, UK: CABI International.

Von Meijenfeldt L. 2012. Cultural Tourism to the Wine Farms in Stellenbosch. Bachelor thesis. Diemen, the Netherlands: InHolland University.

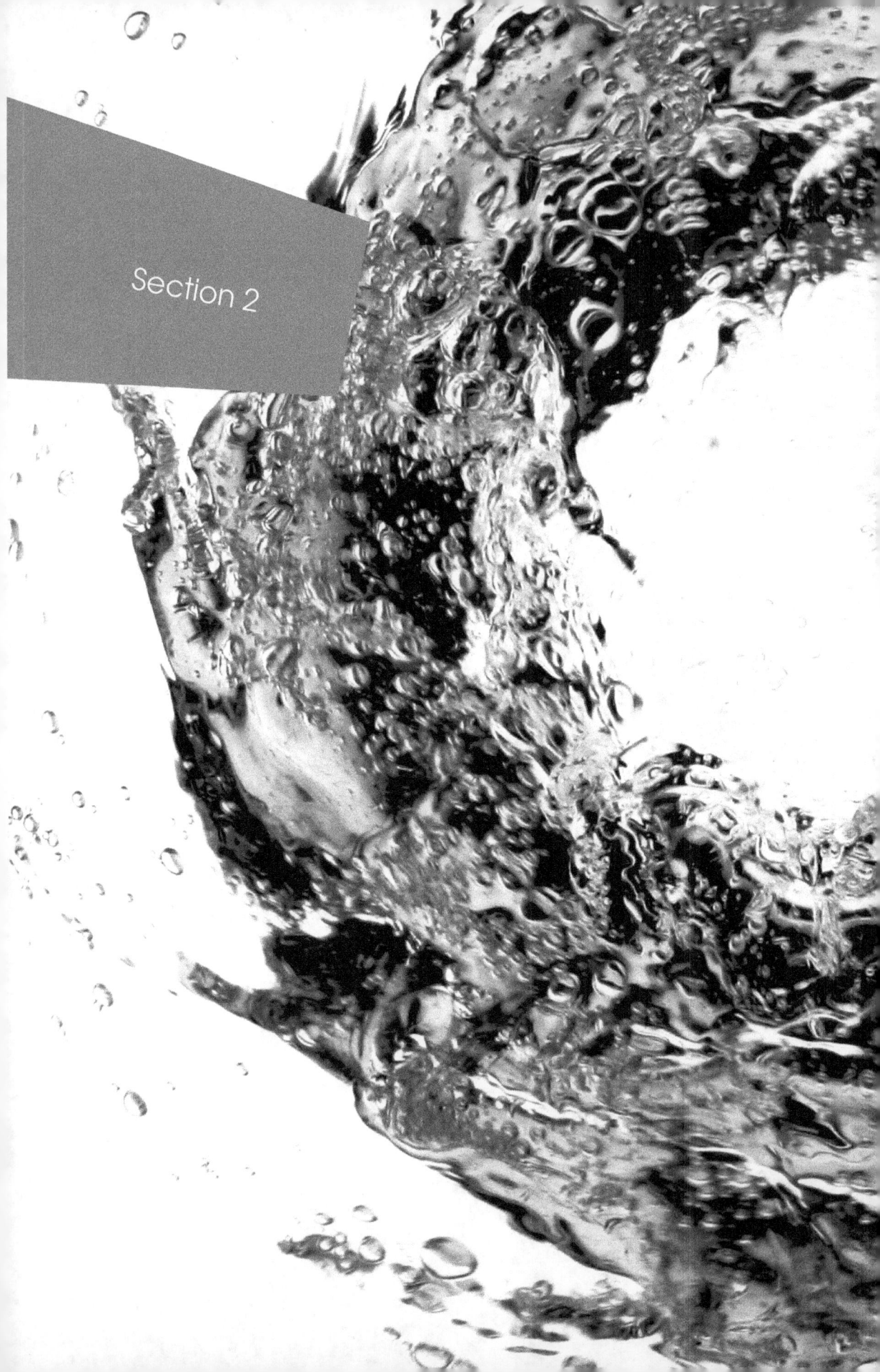
Section 2

Flows

Flows

Section Introduction

The fundamental lifeline of Stellenbosch is its network of flows that facilitates the conveyance of food, water, waste, energy, people and goods. However, this network is critically challenged at present.

Food production is focused on export markets, dominated by large-scale commercial farms and conventional farming methods that are damaging to the environment. Distribution involves long distances and is dominated by national retailers. Linear thinking of 'extract, use once, and dump' dominates, especially where conventional waste and sanitation systems are concerned. The landfill site is at capacity and very little recycling takes place. For example, nutrients locked up in dumped waste flows are not returned to local soils; on the other hand, farmers purchase increasingly costly nutrients for distribution on their lands.

Water scarcity is already a limiting factor in socio-economic development. Water quality is negatively impacted by over-extraction, pollution from poorly functioning sewerage management systems, winery and food processing effluents, urban run-off, and storm water. Local waste water treatment plants currently receive loads that are close to, or in excess of, their rated capacities. Both water supply and sanitation services are challenged by poorly maintained and aging infrastructure.

Also heavily reliant on an aging and capacity-constrained infrastructure, the supply of **electricity** continues to shrink against a backdrop of escalating demand. This electricity is almost exclusively generated from coal, and thus carries a heavy carbon footprint. In addition, outdated fiscal models make it impossible for Stellenbosch Municipality to support well-tested strategies and practices for the rational use of energy.

Congestion and poor road safety are due to a lack of public transport and provision for non-motorised transport. At an annual growth rate of 3.7%, traffic volumes are expected to double between 2009 and 2030. Existing infrastructure will not cope.

The municipality lacks sufficient qualified and experienced staff, as well as sufficient funding for extending service delivery and upgrading and maintaining ageing infrastructure. However, solutions can be found if the community, NGOs and the private sector collaborate in decision making. The **decentralisation of services is crucial** to the sustainable use and management of resources. Localisation allows consumers to participate in the system and also reduces the high capital costs of distribution infrastructure. Pricing and tariff structures directly impact human behaviour, and should be redesigned to reflect resource scarcity and the true cost of supply, while making provision for the poor. However, what is most needed, is a change in infrastructure design to allow both human and natural systems to flourish and to close the loop of urban resource flows.

Some of the solutions discussed in the chapters that follow include the following:

- Food production could be localised via land reform and urban agriculture, and distribution through farmers' markets or consumer cooperatives. Food and human waste could be composted.
- Recycling rates need to increase, with separation at source of all solid waste. Non-recyclable materials must be phased out.
- Water services in areas without Blue Drop Status need rigorous risk assessment, based on proper monitoring. Aging infrastructure must be repaired or replaced, water pollution addressed, and alternative water resources and storage systems investigated.
- Decentralised, alternative waste water systems, such as biolytic or composting systems, must be considered.
- Converting electricity consumers to electricity producers could allow Stellenbosch to become technically energy neutral and resilient to external shocks. The largest consumers have a responsibility to make the largest investments in both energy efficiency and renewable energy. However, replacing electric geysers with solar hot water heaters could release for new developments all the electricity that Stellenbosch would need for many years into the future.
- Public and non-motorised transport systems, offering mobility, access and safety, must be established.

Chapter 6

FOOD

A sustainable system for Stellenbosch[1]

CANDICE KELLY, JESS SCHULSCHENK, ANRI LANDMAN & GARETH HAYSOM

Introduction

The main aim of any food system[2] is to ensure food security[3] for the groups it serves (Ericksen 2007). Indeed, as Landman explains:

> ... food systems function to feed people. Effective food systems would produce and equally distribute sufficient quantities of nutritious food among the global population. If the continued operation of these systems relies on basic ecosystem services such as biological diversity, freshwater, climate regulation, pollination and pest control, we can expect them to operate within nature's capacity to provide these services.
>
> [Landman 2011:35]

It is clear that the global food system is not achieving these aims and is, therefore, operating neither effectively nor sustainably. Nine hundred and twenty-five (925) million people were reported to be suffering from hunger in 2010, "even after the recent food and financial crises have largely passed, [which] indicates a deeper structural problem that gravely threatens the ability to achieve internationally agreed goals on hunger reduction" (United Nations Food and Agriculture Organisation (FAO) 2010:4). Moreover, the global food system's impact on ecosystems has also not been sustainable: the food system uses 70% of all fresh water and 25% of the earth's land surface for food cultivation (Millennium Ecosystem Assesment (MEA) 2005). It also has significant consequences for climate change: about 70% of anthropogenic nitrous oxide emissions are attributable to agriculture, mostly from land conversion and the use of nitrogen fertilisers (*ibid.*). In other words, the food system is both causing, and being increasingly constrained by, environmental destruction.

Given the many different measures used to assess food security, it is problematic to describe the status of food security in Stellenbosch. According to Statistics South Africa, 14.5% of Western Cape residents have inadequate or severely inadequate access to food (2010), while the Urban Food Security Baseline Survey conducted in late 2008 found that food insecurity among the urban poor in the cities of Johannesburg, Cape Town and Msunduzi may be as high as 70% (in Frayne, Battersby-Lennard, Fincham & Haysom 2009). Gericke and Labadarios (2007) point out that, according to the 2005 National Food Consumption Survey-Fortification Baseline, as many as 51% of households experienced hunger, while those at risk of hunger was at 28%. Put differently, only 1 in 5

---

1   This chapter is based largely on Jess Schulschenk's 2010 MPhil thesis (Stellenbosch University), which has also been adapted as an article for *Development South Africa*, 28(4).

2   A food system is "the assembly of all components and activities that produce, distribute, consume and waste food. The boundaries of a food system can move from including only the Stellenbosch food system, to the South African food system and eventually to the global food system" (Landman 2011:18).

3   Food security exists when all people, at all times, have physical and economic access to sufficient, safe and nutritious food to meet their dietary needs and food preferences for an active and healthy life (United Nations Food and Agriculture Organisation (FAO) 1996).

households were food secure. Closer to home, a recent survey of food security conducted by the Department of Human Nutrition at Stellenbosch University identified pockets of food insecurity in Stellenbosch (Van Niekerk 2009). Despite these differing figures and measures, the reality is that a large portion of the poor experience hunger, which is also reflected in Stellenbosch.

Stellenbosch faces further resource challenges, including intensifying water shortages due to climate change and population growth (Hewitson 2006); limits on growth as a result of electricity restrictions (Stellenbosch Municipality 2008); and an overloaded landfill and sewerage treatment plants. Furthermore, Stellenbosch lies within the Cape Floral Kingdom, a biodiversity hotspot. Yet, less than 10% of the endemic and critically endangered Renosterveld remains (Von Hase, Ragout, Maze & Helme 2003).

The degraded status of Stellenbosch's supporting ecosystems will be further discussed in other chapters in this book, and we will focus more narrowly on food security in this chapter. Whatever the actual figures for hunger in Stellenbosch, it is clear that they are too high and that the food system in Stellenbosch is not functioning as effectively as it should. Yet, access to food is a fundamental human right and we should never allow anyone in our society to go hungry.

This chapter discusses the macro trends influencing food systems and food security, before moving on to analyse the Stellenbosch food system by looking at the flows of food (and energy, nutrients and other inputs) in food production, distribution and consumption, in order to determine where there may be room for making Stellenbosch's food system more effective and sustainable. The main problems with the Stellenbosch food system are discussed and recommendations given for how the food system may be improved.

Macro trends influencing food systems

It should become clear from a complete reading of this book that human society has radically altered the conditions on the planet and set off a 'polycrisis' – "a multiple set of nested crises that tend to reinforce one another" (Swilling 2009). Globally, there are a number of key issues that will affect, or are already affecting, ecosystems and the people who are supported by them. The major issues we face are highlighted by a set of key global documents, summarised in Table 6.1.

Table 6.1   Key components of the polycrisis

Issue	Reference	Key message
Climate change	Intergovernmental Panel on Climate Change (IPCC) 2007	Climate change is now unavoidable; it is driven by anthropocentric actions and mitigation is essential
Inequality	United Nations Development Programme (UNDP) 1998	Policy focus must move beyond poverty to include inequality; consumption must be decoupled from development
Urbanisation	United Nations Department of Economic and Social Affairs (UN DESA) 2008	By the end of 2008, over half the world's population lived in cities concentrated predominantly in developing countries
Slums	United Nations Human Settlements Programme (UN-HABITAT) 2003	One-third of the world's population lives in slums and the 'locus of poverty' is shifting to cities
Peak oil	Association for the Study of Peak Oil and Gas (ASPO) 2008 & International Energy Agency (IEA) 2008	The era of cheap oil is officially over (IEA) and oil production will soon peak (ASPO)
Ecosystem degradation	MEA 2005	Human activity has damaged ecosystems to the extent that their ability to support human life is threatened
Agriculture	International Assessment of Agricultural Science and Technology for Development (IAASTD) 2008	Food production must find ways to improve the livelihoods of small-scale farmers and increase productivity without damaging the environment

It is clear that many of these crises will impact agriculture, food systems and food security in Stellenbosch. One of the biggest impacts on food security, given the current structure of the food system, will be peak oil. As we move towards peak oil, oil will become more and more expensive. Given the heavy dependence of almost every stage of our food system on oil – from the inputs of conventional agriculture, such as synthetic fertilisers and pesticides, to the transport of food in increasingly globalised food chains – we may expect food prices to rise dramatically. This means that the poor will become even more food insecure. It also means that more sustainable farming practices will become more economically competitive, in addition to their ecological advantages (Worldwatch Institute (WI) 2011).

There is a worldwide movement calling for increased localisation of food systems (Norberg-Hodge, Merrifield & Gorelick 2002; Winter 2003; Hopkins 2008), which means more local production, processing and distribution, rather than relying on corporate-controlled national and global food networks: "Locally and regionally produced food offers greater security, as well as synergistic linkages to promote local economic development" (Rosset 2002:xix). The proponents of such an approach see it as not only imperative in order to mitigate vulnerability to present and future impacts of the global polycrisis, but also as a way to make the current food system more equitable. This is due to the multiplier effect of localising the food system, which both stimulates the local economy and builds resilience. A recent study by Sonntag (2008) found that "locally directed spending by consumers more than doubles the number of dollars circulating among businesses in the community" (Sonntag 2008:v). In this way,[4] money spent on locally produced food generates twice as much income for the local economy (New Economics Foundation 2001) and promotes stronger social cohesion for the entire local community (Taylor, Madrick & Collin 2005).

Local economies tend to address inequalities in wealth by spreading their gains evenly through the entire community; in this way, "improving the economic welfare of farmers, farm workers, small producers and shopkeepers benefits entire local economies, providing in turn deep social benefits to communities as a whole" (Norberg-Hodge *et al.* 2002:31).

Keeping food flows local also allows for increased money circulating locally, thereby increasing communities' resilience to shocks, especially among the poorer members of society. In other words, it is important to understand not only the flows of food in Stellenbosch, but also who owns those flows and who captures the benefits from them. We now move on to an analysis of the existing Stellenbosch food system in order to get a sense of how the current flows of food and money operate.

Status of food security in Stellenbosch

The images in Figure 6.1 show local Stellenbosch families with the food they consume in one week and the amount of money they spend to acquire it. The high level of inequality between diets is clear, as well as the importance of increasing the availability of affordable fresh produce for optimal nutrition.

Figure 6.1 indicates that Stellenbosch reflects the national patterns: high levels of inequality[5] in both income and food consumption. While the exact figures for food insecurity in Stellenbosch are contested, the scale of faith-based and NGO feeding schemes and food support programmes are indicative of a society experiencing food insecurity. Research by Van den Berg (2010) revealed that, on weekdays (largely due to school feeding schemes), as many as 25 000 meals were served daily in Stellenbosch; one organisation served as many as 4 000 meals per day, with half the beneficiaries under the age of 18. Similar to the rest of South Africa, Stellenbosch has a high incidence of dietary-related disease (Mash 2010), and it must be borne in mind that the impacts of poor nutrition manifest in a range of health challenges, which, coupled with the wider impacts of food and livelihood insecurity, will lead to increased vulnerability by entrenching poverty and inequality if not addressed through targeted programmes (McCullum, Desjardins, Kraak, Ladipo & Castello 2005).

---

4   A further study by Ward and Lewis (2002) in the United Kingdom found that for every £10 spent with a local producer through an organic box scheme, the money was circulated within the local economy to make a total contribution of £25 to it (in comparison to £14 if spent at a local supermarket).

5   Stellenbosch has massive discrepancies between poverty and wealth (Stellenbosch Municipality 2009), while South Africa is in the top ten most unequal countries globally (UNDP 2008) with a Gini co-efficient of 0.72 (Statistics South Africa 2008:03). (The Gini coefficient is a measure of inequality in a population, where 0.0 expresses perfect equality and 1.0 expresses maximal inequality.)

Figure 6.1   Comparative weekly food consumption in Stellenbosch and associated cost [Photographs by Luke Metelerkamp 2010. Concept used with permission of Menzel & D'Aluisio 2007]

Worryingly, despite Stellenbosch Municipality being a key stakeholder in regional food security, it has not shown any intention of devising a comprehensive food security strategy. The 2010 Integrated Development Plan (IDP) for Stellenbosch recognises food insecurity as a challenge, but fails to include the improvement of food security among its strategic objectives. The plan cites food security as a key element in reducing poverty with municipal resources, but stipulates no clear performance indicators for food security projects (Stellenbosch Municipality 2009). The 2010 IDP's only other references to food security relate to its being judged a plausible outcome of the municipality's land reform programme and the creation of a single community garden, which can support three families, in November 2008. The local economic development (LED) strategy also recognises food insecurity as a challenge, but assumes that successful land reform will begin to address it (Stellenbosch Municipality 2008). This approach shows a lack of understanding on the part of the municipality about the complex nature of the food security problem and the fact that food insecurity is not just a result of low agricultural production.

This is clearly not a desirable situation – nor a sustainable one – for a region looking to improve the lives of its citizens and grow its economy. It falls outside the scope of this chapter to consider the wider economic, social or political factors that contribute to food insecurity; besides, these factors cannot necessarily be changed by the residents of Stellenbosch. However, it is possible to influence the existing local food system, and so we move now to an overview of the present food system in Stellenbosch in order to assess which elements may be unsustainable and might be changed by local residents.

Analysis of the Stellenbosch food system

The information in this chapter was compiled from various sources and agencies. The aim is to provide a general overview, rather than a detailed account.

Production

Stellenbosch has a strong agricultural sector, dominated by grape and deciduous fruit production, which account for 87.5% of total production by volume for Stellenbosch (Statistics South Africa 2006). With an estimated 60% of produce from the region currently being exported (*ibid.*; Louw 2009), the bulk of food consumed in Stellenbosch is imported from other areas[6] and most of the cultivable land is dedicated to crops that do not directly support local food security.

Farms are predominantly of commercial size (Statistics South Africa 2006) and privately-owned by wealthy individuals or businesses.[7] Despite the national government land reform programmes and the municipality's own stalled land reform[8] plans (Stellenbosch Municipality 2010), most land remains in the hands of those previously classified as white and land reform has not taken place at the scale required to redress the imbalances of the past. Farmers use mostly conventional agricultural methods that rely heavily on purchased inputs such as fertilisers and pesticides. At the same time, a handful of farmers in the region are engaging in sustainable agricultural practices that preserve and nurture soil integrity and promote optimal nutrition in the produce grown (Landman 2010).

The manufacturing sector is the largest employer in Stellenbosch and is closely linked to the agricultural sector, which itself is not a large employer, but indirectly supports a number of other sectors. The proportion of workers in the agricultural sector has declined dramatically from 24% of total employment in 2001 to 6.9% in 2009 (Stellenbosch Municipality 2009). (While this may be due to shifts to other sectors, no further research could be found to justify this decrease.)

Distribution

Tracking the flows of food in and out of Stellenbosch proves to be challenging, as much of the information required is either not recorded or not readily available. What is clear, however, is that most residents rely on supermarkets for food (nationally, the four largest retailers, Pick n Pay, Woolworths, Shoprite Checkers and Spar, account for over 60% of sales (Van Rooyen 2009)).

Selected interviews suggest that the flow of food from local food producers is directed, almost entirely, away from the Stellenbosch region (Schulschenk 2010). Produce is often transported to central distribution points before being sent back to large retail outlets or informal traders, with the main motivation being to achieve the economies of scale and diversity required to meet consumers' demand for a wide variety of affordable food (Becker 2009; Coetzee 2009; Lubbe 2009).

---

6 Cereals and sugars are entirely sourced from other regions, as Stellenbosch does not produce any of its own. Vegetables, fruit and livestock products may originate from Stellenbosch or elsewhere, but as Stellenbosch consumes more than it currently produces, it may be inferred that most of Stellenbosch's food is brought here from elsewhere.

7 Nationally, fewer than 40 000 commercial farmers occupy 85% of agricultural land, with six hundred and seventy-three (673) farmers producing 33.5% of gross farm income in 2002, while fewer than 2 500 farmers produced more than half the gross income (Vink & Van Rooyen, 2009). It is argued that this same typology is reflected in Stellenbosch.

8 Detailed within the 2010 Stellenbosch Municipality Integrated Development Plan under Key Performance Area 5.

Most major retailers in Stellenbosch source their produce from central distribution warehouses nationwide, which in turn receive their produce either directly from farmers or from local packing sheds that pack produce acquired nationally. Some Spar and Pick n Pay Family Store outlets allow store managers the option to source their produce directly from local farmers (Engel 2009; Espos 2009). An interview with a local Spar owner and case studies of other Spar initiatives reveal that this support for local producers stems from a commitment to supporting local communities, rather than from considerations such as food miles or carbon foot prints (Louw, Vermeulen & Madevu 2006; Espos 2009).

Although it currently contributes to only a small portion of food distribution, the network used by local emerging farmers was found to be more diverse and localised. From the interviews conducted with emerging farmers, it was found that they sell their produce through a variety of markets, including directly to local vendors (known locally as *smouse*), pack sheds, retailers, restaurants, community initiatives, or central distribution markets such as the Cape Town Market.

Community food gardens are also considered as forming part of the distribution network. A survey by the Western Cape Department of Agriculture identified eleven community food gardens supported by government initiatives in the region (Murdoch 2009), while the authors' investigations revealed a further ten community-based initiatives. The volume of food generated through these initiatives is marginal in comparison to the food bought from major retailers or informal vendors, yet plays a critical role in supporting food security for the families involved (Mbambalale 2009; Menze 2009).

Consumption

As no primary data was available for the Stellenbosch or Western Cape regions, total consumption was inferred by weighting the findings of the National Food Consumption Report (Nel & Steyn 2002) for the size of the relative age groups in the Stellenbosch population, based on the most recent available Community Survey (Statistics South Africa 2007). The consumption thus inferred indicates that the broad Stellenbosch population do not enjoy an optimally nutritious diet (substantially insufficient volumes of fresh fruit, vegetables and dairy, coupled with marginally excessive levels of meat and grains; refer to Figure 6.2), which is supported by the high incidence of dietary-related disease (Mash 2010).

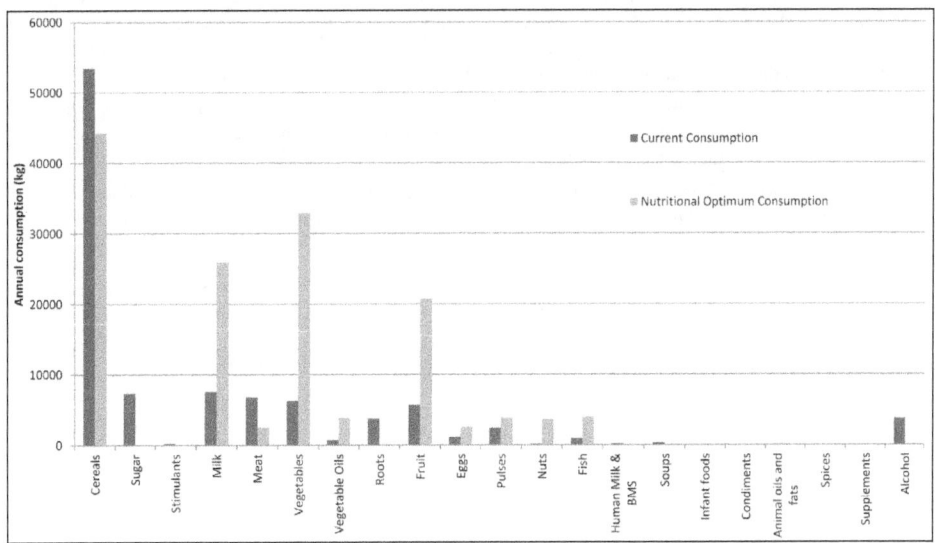

Figure 6.2 Actual current consumption vs. nutritionally optimal consumption by food group for Stellenbosch [Schulschenk 2010]

It is estimated that poor people in South Africa may spend up to 60 to 80% of their income on staple foods; hence, constantly rising food prices force them into ever-deepening poor nutrition (Naylor 2008). It is also important to bear in mind that households have a number of different objectives and may choose to forgo food in order to preserve other assets or choices; moreover, people's food choices are not always rational: "... prices are influential, as are income levels, cultural traditions or preferences, social values, education and health status" (Ericksen 2007:5).

Conclusions about the current food system in Stellenbosch

This section will address some of the problems inherent in the present structure of Stellenbosch's food system. The problems that have been identified may be attributed to one or more of the following distinct factors:

- they contribute to the current inequality in the food system (who controls and benefits from food flows?);
- they contribute to undermining the ecosystems on which the food system depends; and/or
- they reduce food security or increase vulnerability to future shocks such as peak oil.

Production

- *The focus on export markets:* While many are quick to point out the benefits of earning foreign exchange, one must interrogate how fairly these are shared (as pointed out earlier, Stellenbosch is one of the most unequal places on earth). Many will also highlight the job creation benefits of the export market, but again, one must question the security and quality of those jobs. An Oxfam study of the employment conditions of women in the South African fruit and wine export industry found that their jobs were insecure and underpaid, often due to the tight margins received by producers themselves (Oxfam 2004). And the authors' personal experience of living and working in Stellenbosch has shown us that farm workers' children who attend a local primary school often go hungry.

  The other problem with this focus on export markets is that these markets will become increasingly less attractive as oil prices continue to rise. This leaves those who depend on earnings from this sector extremely vulnerable, as they need money to purchase food.

- *Dominated by large-scale commercial farmers:* The ownership of land in Stellenbosch is inequitable and has not shifted much since 1994. This means that the poor do not have access to land from which to earn a living, thus further perpetuating inequality. These large-scale farms are also usually heavily mechanised, which reduces the need for labour and translates into fewer jobs in the agricultural sector.[9]

- *Conventional farming methods predominate:* The conventional approach to farming relies heavily on artificial fertilisers and pesticides derived from oil. As peak oil approaches, the cost of these inputs will rise, rendering farms that rely on them less competitive.[10]

  These inputs have also been shown to have adverse impacts on the environment, causing pollution in rivers and groundwater, degrading soils and damaging biodiversity. Moreover, they may have negative health impacts on the workers who handle them.

Finally, many of these inputs are produced outside Stellenbosch, which means that money spent on acquiring them is largely lost to the local economy.

---

9   Employment in the agricultural sector declined by 15% during the 1990s (Vink & Van Rooyen 2009).

10 "Given the degree to which the modern food system has become dependent on fossil fuels, many proposals for de-linking food and fossil fuels are likely to appear radical. However, efforts towards this end must be judged not by the degree to which they preserve the status quo, but by their likely ability to solve the fundamental challenge that will face us: the need to feed a global population of seven billion with a diminishing supply of fuels available to fertilise, plough, and irrigate fields and to harvest and transport crops" (Heinberg & Bomford 2009:10-11).

Distribution
- *Food travels long distances:* As was discussed earlier, most of the food produced in the Stellenbosch area flows away from it, although some is transported back by various distributors. At present, this model may make the most sense in terms of efficiency for the individual distributors; however, it will become increasingly less competitive as peak oil approaches.
- *Dominated by national retailers:* Most national retailers are owned by shareholders and based outside the Stellenbosch area. This means that the money paid by Stellenbosch residents for food mostly leaves the local economy. Moreover, this distribution network, requiring consistent bulk supply of high quality fresh produce, often excludes small-scale farmers who cannot meet these requirements.

Consumption
- *Poor nutrition:* Due to poverty and poor food choices, the average Stellenbosch resident is not consuming an optimally nutritious diet. This results in health problems that negatively impact not only the individuals concerned, but society as a whole.
- *Waste not recycled:* In a truly sustainable system, food and human waste would be composted and put back into the soils from which it originated, thereby returning nutrients to these soils. At present, most Stellenbosch waste ends up either in the landfill or the sewage works and is not fed back into the soils of the region.

The above points demonstrate that the flow of food in the Stellenbosch community is neither sustainable nor equitable. In addition, it is also clear that the current structure of the food system sees the flows of money associated with the flows of food benefiting a small minority only, or leaving the Stellenbosch region entirely, to the benefit of external agents.

Moving forward

This section focuses on possible interventions that may increase food security and keep more of the flow of money associated with the food system within the region, while reducing the vulnerability of the system to existing or impending shocks such as climate change and peak oil. The concept of localisation is introduced and explained, followed by suggestions for moving Stellenbosch towards a more sustainable and equitable food system.

Localisation may be defined as the prioritisation of the flow of resources within a network of community-based enterprises involved in the production and distribution of food at the local scale for the purpose of local consumption, including, but not limited to, financial, human, social and environmental capital (Schulschenk 2010). Local food economies thus refer to local food initiatives at a community level, within the context of a predominantly modern global food system.

The movement towards localisation is driven by a growing concern for the food security of local communities. It is informed, one the one hand, by an anti-globalisation sentiment (Feenstra 1997); and on the other, by the contribution of the long-distance transport of food (and the modern food system in general) to greenhouse gas emissions and climate change[11] and a view to the future, with peak oil and the increasing costs of food transport in mind (Peters, Bills, Wilkins & Fick 2008).

A focus on geography alone, however, will not address all the challenges currently presented by the modern food system. Oversimplified localisation strategies, where a local – instead of sustainable – food system is promoted as the end goal, must be avoided. The fact that a food system is local, does not it render it automatically sustainable or equitable. Sustainable food systems must be continuously reflexive, inclusive, and strategically positioned in order to successfully address key contextual challenges (Landman 2010).

---

11 In theory, advocating a reduction in food miles by localising agricultural production has its merits. However, the transport of food is often less significant than other contributors to climate change in the modern food system when considered in the context of the entire carbon footprint or life cycle assessment of most food products.

Critically considered localisation not only improves communities' resilience to external shocks (food crises, peak oil), but also builds local food economies that create opportunities for local small-scale producers (increasing their share of the food 'dollar') and distributors, while creating opportunities to add value – essentially, it has the potential to stimulate entire economies around local food.

On the other hand, the limitations of localisation include the inadequate capacity of a given region to meet all the food requirements of the local community and the inability of local food initiatives to compete with the largely subsidised and cost-externalised modern food system.

The potential of a local or community-based approach to food systems to promote greater sustainability lies in that it is in the interest of local communities to promote their own food, livelihood and environmental security and that it connects local communities more closely to the impacts of their decisions (Pretty 2002). Local food systems encourage relationships with the food system that promote more responsible decision making (Feenstra 1997; Hinrichs 2003).

Stellenbosch has the potential to create a sustainable food system, but this will require engaging all stakeholders through community participation. Resilient social capital and a community knowledgeable about sustainable food systems will contribute to informed strategies for restructuring production, distribution and consumption systems within the larger food system. This could be achieved, in part, through the set of recommendations for the Stellenbosch food system presented below.

Production

Stellenbosch has the potential to grow all the food required to meet local demand, in terms of both current and nutritionally optimal consumption, but is limited in its capacity to grow certain crops currently consumed (roots, sugar cane, maize, rice or tropical fruit). In addition, water scarcity, poor soil quality and degraded soils (in some areas), and the domination of wine production also impose severe constraints to increasing local food production (Schulschenk 2010).

Despite these limitations, increasing local food production is critical. Increasing the production output of key food produce – specifically vegetables, fruit and nuts, as well as chicken, fish, milk and eggs – through targeted programmes to assist local communities and emerging farmers with access to land, resources and support is critical. This includes the prioritisation of food production in spatial planning (taking into account urban growth), including:

- *Land reform to increase local production*: This would serve to secure food for local consumption, while also creating employment and stimulating the local economy. It will require extensive support to be successful and investment should be considered in the context of the 'hidden' subsidies that other forms of agriculture currently receive. The cooperative model, where small-scale and emerging farmers pool resources and share support, has proved successful in many countries, but requires support in the form of building management capacity. There should be investment in centralised packaging and transport nodes as public infrastructures to assist emerging farmers in getting their produce to market. These could be coupled with learning centres to build new systems of agricultural skills and knowledge. Targeted skills development programmes for emerging farmers, with an emphasis on agro-ecological methods (to reduce dependency on fossil fuels and build resilience to climate change), marketing skills (in terms of local markets), and business management, are critical (Pretty 2001).
- *Productive use of urban space for urban agriculture*: This includes raised beds, rooftop gardens, common greens, sections of parks, schools, communal centres, and backyards or allotment gardens, as well as self-provisioning at factories, offices and businesses, and suburban farms around urban perimeters. If Stellenbosch were to allocate only 10% of urban space towards urban agriculture, a minimum of 4 620 tonnes of vegetables per year[12] could be produced within the urban boundary. Urban agriculture activities may range from community gardens to productive enterprises and market gardens that contribute to both local production and the local economy.

---

12 The calculation is based on a conservative estimate of the available urban land with potential for food production and the most conservative productivity rates (Schulschenk 2010).

Distribution

Local economic development can be achieved by increasing local food production, but also through the localisation of food processing, distribution and retail, which forms part of a wider local food movement for Stellenbosch. In addition to being a system component, distribution acts as an interactive link between the various other system components, which includes production, processing, consumption and waste (Landman 2011). Even if a production or consumption component becomes completely sustainable, it cannot ensure a sustainable food system without a sustainable distribution network already in place.

As argued by Marsden and Smith, "... sustainable wealth creation and local economic development require new entrepreneurial initiatives that focus on investing in the local environment, creating/strengthening local institutions and employing people and their resources" (2005:441).

Various different local food economy initiatives such as community-supported agriculture (CSA) initiatives,[13] local markets and consumer cooperatives are critical in ensuring that locally grown produce reaches consumers directly and effectively. Farmers will also receive the best price for their produce through this marketing system, but it requires community support and awareness. More local restaurants could also buy high quality, sustainable produce from emerging farmers.

A local distribution hub for locally produced food should be established to assist farmers and consumers to connect more directly and maximise the benefits of central coordination. By connecting local consumers more directly to the food economy that supports them and the individuals behind that food economy, closer bonds may be forged within a community around local food.

Another, often overlooked, distribution network with the potential to act as an effective entry point to make the flow of food in Stellenbosch sustainable, is that of the informal traders known locally as *smouse* (hawkers). They sell loose fresh produce and small amounts of basic food stuffs sourced mainly from central distribution markets such as Epping or Kraaifontein, where the prices are cheapest (Lebo 2009). Other than price, the freshness of produce is an important procurement consideration for *smouse*, because it gives them a competitive advantage over supermarkets. To ensure the freshness of their produce, they procure as much produce as possible from an intricate network of regional small-scale farmers (Linders 2010). The *smouse* could be encouraged to buy locally and be supported by providing them with dedicated stall space in the town centre.

There are also opportunities for local value-adding enterprises (such as canning or frozen foods), which will not only stimulate the local economy further, but also assist in maintaining a supply of local produce throughout the year.

Consumption

While Stellenbosch residents are not expected to suddenly shift towards a nutritionally optimal diet (and bearing in mind barriers such as cultural preferences, accessibility and affordability), there is potential for targeted programmes to increase the relative volume of nutritious, fresh produce consumed in Stellenbosch, including (but not limited to) school feeding schemes and community food gardens.

There is also a dire need to educate the poorer members of society about food choices and nutrition.

Community participation

A key recommendation in preparing the broad Stellenbosch community for a more sustainable, low-carbon future – and making the transition as gently as possible – is to build knowledge systems that promote learning for change (Pretty 2002). Sustainable food systems are deeply rooted within the community and cannot function without community principles, values, participation and partnerships (Feenstra 2002). As Garrett and Feenstra (n.d.:6) state:

---

13 CSA initiatives aim to ensure a better and fair price for the farmer, and fresh, affordably-priced, sustainable produce for the consumer. Members pay a fee upfront or buy a share in a harvest at the beginning of the season, which provides the farmers with the necessary input capital to grow their food. In return, members receive a weekly delivery of produce (Henderson & Van En 2007).

> *Perhaps one of the most important elements of designing community food systems projects is that it is a collaborative process. This means that it includes the participation of multiple formal and informal organizations, associations and individuals with a variety of backgrounds and expertise. The participation of a broad cross-section of the community is essential for the project to be representative and contribute to the growth of a community food system.*

They also explain the benefits of collaborations: It allows the group to tackle more complex issues; improves the coordination of services; supports policy development through support of a variety of constituencies; allows for more effective leveraging of resources; and better outreach in the community (Garrett and Feenstra, n.d.). Feenstra (1997) identifies leadership, collaboration and civic renewal as crucial in building stronger local food economies linked to equitable and sustainable communities. Learning through experience has been highlighted as one of the most meaningful methods of shifting behaviour, which points again to the importance of connections with local food systems that allow opportunities for such engagement. In this sense, local food economies are as much about the flow of knowledge and social capital as resources, as about the flow of food itself.

In order to facilitate the building of a stronger local food economy, a wider local food movement should be promoted in Stellenbosch. A campaign premised on 'local is *lekker*'[14] is being proposed by shop owners in Stellenbosch to support local produce from a community development, cultural heritage and local tourism perspective. This campaign could be extended further to target households, businesses and institutions in the Stellenbosch region. Procurement policies for locally produced food would be strategic in establishing a growing demand for local produce and is a core recommendation for consideration by the local tourism sector and major institutions that dominate the Stellenbosch economy. Initiatives commonly associated with vibrant local food economies that could support a campaign of this nature include guides to local and seasonal produce, school education programmes, chef and farm partnerships, harvest festivals and local food events (Pinkerton & Hopkins 2009).

There is also an opportunity to strengthen the local food economy through the promotion of agri-tourism, to provide locals and visitors alike with an experience of Stellenbosch's unique 'flavour' that goes beyond the famous wines of the region. Agri-tourism has the potential to both stabilise the incomes and increase the revenue streams of farming communities (Nowers 2007). Providing communities with the opportunity to engage with farmers and the production system by means of farm visits increases awareness about critical food system (and broader sustainability) issues.

Conclusion

It is clear that Stellenbosch's current food system is both inequitable and unsustainable, as well as highly vulnerable to the coming polycrisis. In the context of peak oil, climate change and growing demand resulting from population growth, Stellenbosch must prioritise local food production in order to build resilience against these shocks. The transition from fossil fuel dependency to post-oil communities will carry significant social and environmental costs if preparations are not made now to promote strong and resilient local communities. Such preparations will require small but strategic investments in the short term to avoid devastating human and financial cost in the long term. Considering the impacts of the modern food system, on the one hand, and the benefits of local food economies on the other, Stellenbosch must become more sustainable, equitable and resilient, and could become so by building a stronger local food economy.

---

14 Or: 'local is cool'.

REFERENCES

ASPO (Association for the Study of Peak Oil and Gas). 2008. [Accessed 26 September 2008] http://www.peakoil.net

Becker E. 2009. Personal interview. Cape Town: 31 August.

Coetzee W. 2009. Personal interview. Stellenbosch: 2 September.

Engel W. 2009. Personal communication. Stellenbosch: 5 June.

Ericksen P. 2007. Conceptualizing food systems for global environmental change research. *Global Environmental Change*, 18(1):234-245. Oxford: Elsevier.

Espos T. 2009. Personal interview. Stellenbosch: 24 August.

FAO (Food and Agriculture Organisation). 1996. *World Food Summit*. [Accessed 22 April 2010] http://www.fao.org/wfs/index_en.htm

FAO. 2010. *The State of Food Insecurity in the World 2010*. Rome: Food and Agriculture Organisation of the United Nations.

Feenstra G. 1997. Local food systems and sustainable communities. *American Journal of Alternative Agriculture*, 12(1):28-36.

Feenstra G. 2002. Creating space for sustainable food systems: Lessons from the field. *Agriculture and Human Values*, 19:99-106.

Frayne B, Battersby-Lennard J, Fincham R, Haysom G. 2009. *Urban Food Security in South Africa: Case study of Cape Town, Msunduzi and Johannesburg*. Development Planning Division Working Paper Series No 15. Midrand: Development Bank of Southern Africa.

Garrett S & Feenstra G. n.d. *Growing a community food system*. Community Ventures: Partnerships in Education and Research Circular Series Topics. A Western Regional Extension Publication WREP0135. [Accessed 16 August 2012] http://smallfarms.wsu.edu/wsu-pdfs/WREP0135.pdf

Gericke G & Labadarios D. 2007. A measure of hunger. In: D Labadarios. *National food consumption survey: fortification baseline*. Stellenbosch: Stellenbosch Municipality.

Heinberg R & Bomford M. 2009. *Report: The Food and Farming Transition*. [Accessed 13 October 2009] http://www.postcarbon.org/food

Henderson E & Van En R. 2007. *Sharing the harvest: A citizen's guide to community supported agriculture*. Vermont: Chelsea Green.

Hewitson B. 2006. Climate Change in South Africa. PowerPoint presentation. Cape Town: 24 October.

Hinrichs C. 2003. The practice and politics of food system localisation. *Journal of Rural Studies*, 19:33-45.

Hopkins R. 2008. *The Transition Handbook: From oil dependence to local resilience*. Devon: Greenbooks.

IAASTD (International Assessment of Agricultural Knowledge, Science and Technology). 2008. *Executive Summary of the Synthesis Report of the International Assessment of Agricultural Knowledge, Science and Technology for Development Programme*. [Accessed 3 June 2008] http://www.agassessment.org

IEA (International Energy Agency). 2008. *World Energy Outlook 2008: Executive Summary*. [Accessed 27 September 2009] http://www.iea.org/textbase/nppdf/free/2008/wea2008.pdf

IPCC (Intergovernmental Panel on Climate Change). 2007. *Synthesis Report – Summary for Policymakers*. [Accessed 12 June 2008] http://www.ipcc.ch/pdf/assessment-report/ar4/syr/ar4_syr_spm.pdf

Landman A. 2010. Agro-ecological farms in the Stellenbosch region: an overview report. Unpublished report prepared for the Sustainability Institute. Stellenbosch: Stellenbosch University.

Landman A. 2011. Growing sustainable food systems: a study of local food distribution initiatives in Stellenbosch. MPhil thesis. Stellenbosch: Stellenbosch University.

Lebo N. 2009. Personal interview. Stellenbosch: 27 August.

Linders G. 2010. Personal interview. Stellenbosch: 23 June.

Louw A, Vermeulen H & Madevu H. 2006. Integrating Small-Scale Fresh Produce Producers into the Mainstream Agri-Food Systems in South Africa: The Case of a Retailer in Venda and Local Farmers. Paper presented at the Regional Consultation on Linking Farmers to Markets: Lessons Learned and Successful Practices. Cairo, Egypt: 28 January-2 February.

Louw D. 2009. *Stellenbosch Agricultural Sector Overview*. Draft report. Cape Town: CNdV Africa.

Lubbe H. 2009. Personal interview. Stellenbosch: 25 August.

Marsden T & Smith E. 2005. Ecological entrepreneurship: sustainable development in local communities through quality food production and local branding. *Geoforum*, 36:440-451.

Mash R. 2010. Chronic diseases, climate change and complexity: the hidden connections. *South African Family Practice*, 52(5):438-445.

Mbambalale E. 2009. Personal interview. Stellenbosch: 27 August.

McCullum C, Desjardins E, Kraak VI, Ladipo P & Castello H. 2005. Evidence-based strategies to build community food security. *Journal of the American Dietetic Association*, 105(2):278-283.

Menze C. 2009. Personal interview. Stellenbosch: 27 August.

Menzel P & D'Aluisio F. 2007. *Hungry Planet: what the world eats*. Berkeley: Ten Speed Press.

Millennium Ecosystem Assessment (MEA). 2005. *Ecosystems and Human Well-being: Synthesis*. Washington DC: Island Press.

Murdoch J. 2009. Personal communication. Stellenbosch: 27 May.

Naylor R. 2008. The global food crisis exposes the fragility of sub-Saharan economic progress. *Science*, 323:239-240.

Nel J & Steyn N. 2002. *Report on South African food consumption studies undertaken amongst different population groups (1983-2000): Average intakes of foods most commonly consumed*. Pretoria: Department of Health.

New Economics Foundation. 2001. *Local Food Better for Rural Economy than Supermarket Shopping*. Press release. London: 7 August.

Norberg-Hodge H, Merrifield T & Gorelick S. 2002. *Bringing the Food Economy Home: the social, ecological and economic benefits of local food*. Devon: International Society for Ecology and Culture.

Nowers R. 2007. The socio-economic dimension of agri-tourism. *Elsenburg Journal*, 4:04-07.

Oxfam. 2004. *Trading away our rights*. Oxford: Oxfam International.

Peters C, Bills N, Wilkins J & Fick G. 2008. Foodshed analysis and its relevance to sustainability. *Renewable Agriculture and Food Systems*, 24(1):1-7.

Pinkerton T & Hopkins R. 2009. *Local Food: How to make it happen in your community*. Devon: Greenbooks.

Pretty J. 1995. Regeneration Agriculture: The Agro-Ecology of Low-External Input. In: J Kirkby *et al*. (eds). *Sustainable Development*. The Earthscan Reader. London: Earthscan.

Pretty J. 2001. Some benefits and drawbacks of local food systems. Briefing note for TUV/Sustain AgriFood Network, 2 November. London: City University

Pretty J. 2002. *Agri-culture: Reconnecting People, Land and Nature*. London: Earthscan.

Rosset PM. 2002. Lessons of Cuban Resistance. In: F Funes, L Garcia, M Bourgue, N Perez & P Rosset (eds). *Sustainable Agriculture and Resistance: Transforming Food Production in Cuba*. Havana: ACTAF (Asociacion Cubana de Tecnicos Agricolas y Forestales).

Schulschenk J. 2010. Benefits and Limitations of Local Food Economies in Promoting Sustainability: A Stellenbosch Case Study. MPhil thesis. Stellenbosch: Stellenbosch University.

Sonntag V. 2008. *Why Local Linkages Matter: Findings from the Local Food Economy Study*. Seattle: Sustainable Seattle.

Statistics South Africa. 2006. *Report on the survey of large- and small-scale agriculture*. Pretoria: Statistics South Africa.

Statistics South Africa. 2007. *Community Survey 2007 findings*. Pretoria: Statistics South Africa.

Statistics South Africa. 2008. *Income & Expenditure of Households: 2005/2006*. Pretoria: Statistics South Africa.

Statistics South Africa. 2010. *General household survey 2009*. Pretoria: Statistics South Africa.

Stellenbosch Municipality. 2008. *Stellenbosch 2017 Housing Strategy*. Stellenbosch: Stellenbosch Municipality.

Stellenbosch Municipality. 2009. *Integrated Development Plan for the municipal area of Stellenbosch*. Stellenbosch: Stellenbosch Municipality.

Stellenbosch Municipality. 2010. *Stellenbosch Municipality IDP 2nd Generation – Revision 3, April 2010*. [Accessed 12 October 2011] http://www.stellenbosch.gov.za/jsp/e-lib/list.jsp?catid=219

Swilling M. 2009. *What Africa's leaders are ignoring*. [Accessed 19 July 2009] http://www.sustainabilityinstitute.net/newsdocs/footprints/item/what-africas-leaders-are-ignoring

Taylor J, Madrick M & Collin S. 2005. *Trading places: the local economic impact of street produce and farmers' markets*. London: New Economics Foundation.

UN DESA (United Nations Department of Economic and Social Affairs). 2008. *World Urbanisation Prospects: The 2007 Revision. Executive Summary*. [Accessed 28 September 2009] http://www.un.org/esa/population/publications/wup2007/2007WUP_ExecSum_web.pdf

UNDP (United Nations Development Programme). 1998. *Human Development Report 1998: Overview*. [Accessed 27 September 2009] http://www.hdr.undp.org/en/media/hdr_1998_en_overview.pdf

UNDP. 2008. *The Human Development Report 2008*. [Accessed 27 September 2009] http://www.hdr.undp.org/en/reports/global/hdr2007-2008/

UN-HABITAT (United Nations Human Settlements Programme). 2003. *The Challenge of Slums: Global Report on Human Settlements 2003*. [Accessed 28 September 2009] http://www.unhabitat.org/pmss/getPage.asp?page=bookView&book=1156

Van den Berg W. 2010. Food Aid Intervention Study Research Recommendations. Research component of the Stellenbosch Food Security Initiative. Stellenbosch: Sustainability Institute.

Van Niekerk A. 2009. Personal communication. Stellenbosch: 30 October.

Van Rooyen K. 2009. The Great Food Price Rip-off. *The Sunday Times*, July 12:1-2.

Vink N & Van Rooyen J. 2009. *The economic performance of agriculture in South Africa since 1994: implications for food security*. Development Planning Division Working Paper Series 17. Midrand: Development Bank of Southern Africa.

Von Hase A, Rouget M, Maze K & Helme N. 2003. *A Fine-Scale Conservation Plan for the Cape Lowlands Renosterveld: Technical Report*. Claremont: Cape Conservation Unit.

Ward B & Lewis J. 2002. *Plugging the Leaks: Making the most out of every pound that enters your local economy*. London: New Economics Foundation.

Winter M. 2003. Embeddedness, the new food economy and defensive localism. *Journal of Rural Studies*, 19:23-32.

Chapter 7

WASTE

The long walk to sustainable waste management

THYS SERFONTEIN & COBUS KOTZÉ

Introduction

It is often said that there is no waste in nature, because everything is interconnected. Within this delicate and complex web, the outputs from one system become the inputs for another.

Though this may be the case in nature, humanity's waste systems are predominantly linear. Instead of processing waste into other usable system inputs, municipal waste services are characterised by an overemphasis on waste removal and disposal by municipalities and a general 'throw and forget' attitude on the part of ratepayers, who demand a cost-effective service with little regard for the environmental cost. The impact of this approach to waste management may be seen in many South African municipalities – as is the case with Stellenbosch – in the form of large landfill sites that bring about a complex set of environmental liabilities and often negatively impact on the health and safety of surrounding communities.

Stellenbosch Municipality is currently facing some serious waste disposal issues. It has a landfill site that does not meet all the regulatory requirements and is fast reaching capacity. Fortunately, these facts have resulted in an increased drive to find better alternatives to dealing with its waste.

The municipality has been actively working towards a more integrated approach to waste management during the past few years. This approach includes involving the general public, promoting waste minimisation, and diverting certain waste streams, such as garden chippings and building rubble, to be composted or utilised elsewhere. The municipality is also planning an integrated recycling facility – to be established at the old landfill site within the next five years (Stellenbosch Municipality 2010a) – in an attempt to minimise the cost of waste management and the quantity of material that needs to be landfilled. As part of its Integrated Development Plan, the municipality has also committed itself to the implementation of a Recycling at Source project (Stellenbosch Municipality 2010b).

Limited records were kept of the amount of solid waste disposed of at the landfill site between 2002 and 2009. As there is no weighbridge on site, the quantity of waste received is estimated in cubic metres (m^3).

Records of disposal (m^3) have been kept since April 2009. The average monthly disposal rates are as follows:

- General waste: 8 900 m^3
- Garden waste: 3 917 m^3
- Covering material: 4 162 m^3
- Number of vehicles visiting the landfill site: 3 200 vehicles

Based on the municipality's monthly statistics and information received from recycling companies, about 11.4% of the total waste received has been recycled and 88.5% landfilled since February 2010.

This chapter aims to blend the conceptual with the practical in order to find alternative pathways to the business-as-usual approach to waste management. In the first part of the chapter, we will look at the current situation at the Stellenbosch landfill, as well as future developments at the site, based on the five-year Integrated Waste Management Plan (IWMP). Then we will present the Recycling at Source project as a case study, in order to draw out critical clues as to how to go about changing a waste culture and driving these changes forward. Finally, we will close with how to direct urban flows to create closed-loop metabolisms and more healthy environments in Stellenbosch in the future.

Non-compliance of the Stellenbosch landfill

During April 2009, the Department of Environmental Affairs and Tourism (DEAT) submitted a formal letter to Stellenbosch Municipality regarding the landfill site's non-compliance with permit conditions. This situation was the result of a period of poor management practices, insufficient budget planning, and political and administrative instability in the municipality.

According to letters received from DEAT regarding the landfill site, which had been issued with a permit on 29 January 1999, it did not comply with several permit conditions, including:

- suitable notice boards;
- fencing to prevent unauthorised entry;
- a movable fence;
- access control;
- litter and dust control measures;
- covering of daily disposed waste;
- Monitoring Committee;
- evidence and results of internal and external auditing;
- site management plan; and
- recording minutes of Monitoring Committee meetings.

In order to prevent the withdrawal of its operating permit, the municipality had to motivate why DEAT should not issue such a directive and had to implement, as a matter of urgency, complying measures at the landfill site within a short period of time. If DEAT had withdrawn the operating permit, the Stellenbosch landfill site would have been closed, leaving the Vissershok landfill site, some 92 km away from Stellenbosch, as the only option for disposal. The direct impact to Stellenbosch Municipality would have been an increase of R1 million per month for disposal only (i.e. excluding transport) (Jan Palm Consulting Engineers 2008). It would also have meant that the waste service for residents of Greater Stellenbosch would have been the most expensive one in South Africa.

Council therefore resolved to immediately implement compliance measures at the site and to proceed with investigations towards sustainable future waste management for Greater Stellenbosch.

Implementing the Integrated Waste Management Plan

Complying with permit conditions

Stellenbosch Municipality acknowledged the shortcomings at the landfill site and immediately addressed the highlighted issues after the official letter from DEAT was received. During June 2009, municipal officials visited DEAT in Pretoria to discuss the shortcomings. A public meeting was also held during the same time to obtain people's views on future plans for the landfill site.

The public requested officials to:

- close down the landfill site;
- implement access control (fencing all round and gates);
- manage scavengers on site;

- implement waste minimisation projects;
- improve the visual impact of site; and
- investigate future solutions.

Though the landfill site was not closed down, several of the other public requests mirrored the permit conditions and were subsequently implemented. Funding to the value of R6.5 million was applied for and approved from the MIG (Municipal Infrastructure Grants) for storm water channels and a concrete palisade fence around the landfill site (MIG 2009). Access control was improved and information boards, dust, litter and odour were successfully addressed. Figure 7.1 shows the site prior to and after these improvements. Disposed waste is now also covered daily to prevent any further nuisance.

Figure 7.1   Stellenbosch landfill. **A:** Before compliance – 2009. **B:** After compliance – 2011 [Photographs by Thys Serfontein]

Managing the landfill site

In order to comply with the environmental legislation and permit conditions required of a landfill, a dedicated person must be present on site during operating hours. It is clear from the history of the Stellenbosch landfill site that it has not been managed properly in the past. The latest solution seems to be outsourcing the management of the new extended landfill to the private sector. This is being done elsewhere with success, for example the Karwyderskraal landfill site in the Overstrand Municipality (DEADP 2009). However, to date Stellenbosch landfill is still managed by Stellenbosch Municipality.

Banning green waste and builders' rubble

Green waste and builders' rubble may account for up to 60% of the total waste stream in a community. During the 2009/10 financial year, a ban on dumping green waste and builders' rubble was introduced at the Stellenbosch landfill site (Stellenbosch Municipality 2010b). Funding was made available by Council to appoint a contractor for the chipping of bush waste at the site and a contractor was also appointed for the crushing of builders' rubble. These waste minimisation measures have saved the landfill approximately 6 months' capacity (Jan Palm Consulting Engineers 2008).

Instituting double street refuse bins

Funding was made available by Council to install double street refuse bins for wet and dry waste in all public parks, the CBD and public areas.

Implementing a 'Pay as you throw' system

During November 2009, communication via municipal accounts, local newspapers and advertisements on the municipality's website informed residents that a 'Pay as you throw' system for waste management would be implemented (Stellenbosch Municipality 2010b). Users are only allowed to use the standard WCO24 240-litre wheelie bin issued by the municipality. Residents are allowed one wheelie bin per household and only waste that is inside these containers is collected. Businesses and residents have to pay an additional cost if they have more than one bin for collection.

After all these compliance measures were implemented, the biggest challenge – minimising the waste disposed of on the nearly full landfill site – still remained. The need for waste minimisation led to the implementation of the Recycling at Source project, which was launched during the 2009/2010 financial year.

Minimising waste: Implementing the Recycling at Source project

The Recycling at Source project was initiated in response to the non-compliance of the landfill site, the implications of the new Waste Act (Act 59 of 2008), and the realisation that the landfill site had less than 18 months' capacity left. The project encourages the separation of recyclables and non-recyclables at source in certain neighbourhoods in Stellenbosch and is enhanced by the presence of double street refuse bins for wet and dry waste separation in the CBD. The project is ongoing at present.

Planning the project through workshops

During the planning phase, there seemed to be numerous obstacles to the collection of recyclables and the practical implementation of such a project in the different areas. However, waste management workshops arranged by the Sustainability Institute and Stellenbosch Municipality during 2009 helped to identify the shortcomings and (main) stakeholders in waste management in Stellenbosch, as well as the potential roles and responsibilities of stakeholders. These workshops helped to renew the public's trust in the service delivery of Stellenbosch Municipality. They also resulted in new commitments by politicians and newly appointed officials to create and maintain positive relationships with communities and stakeholders.

From the outset of these workshops, a number of potential problems and concerns were identified, which had to be carefully considered before finalising the new waste management processes and waste minimisation projects. Where possible, these problems were addressed and contingency plans formulated. Some of the major roles and responsibilities identified are described in Table 7.1, while Table 7.2 summarises more specific challenges within the municipality itself.

Table 7.1 Major roles and responsibilities of stakeholders in waste management [Sustainability Institute 2009]

Role Players	Responsibilities
Stellenbosch Municipality	<ul><li>Provide political, financial and technical support</li><li>Provide capacity, distribute information and provide staff and public support</li></ul>
Sustainability Institute	Arrange waste workshops for all stakeholders and communicate best practices
Recycling organisations:<ul><li>CL Waste</li><li>Star Metals</li><li>Huis Horison</li></ul>	<ul><li>Manage activities and pay informal recyclers</li><li>Separate, measure, store, transport and sell recyclables to markets</li><li>Provide feedback on demand and viability of recyclable material</li></ul>
Participating residents:<ul><li>Ratepayers Association</li><li>Chamber of Commerce</li><li>Community and business leaders</li><li>Schools</li></ul>	<ul><li>Support the project through promotion, participation and commitment</li><li>Provide feedback and ideas for improvement</li><li>Promote environmental responsibility and waste administration</li></ul>
Project Coordinator	Coordinate all activities, including project design, education, documentation, layout of artwork and publicity

Table 7.2 Waste management issues identified by the Engineering Directorate and submitted to Council [Stellenbosch Municipality 2011]

Issues	Detail
Political and administrative instability	Regular change in Council with new priorities; 6 Waste Managers in six years
Financial	Budget constraints
Capacity	Capacity limitation of municipal staff and collection services
Student population	Creates large volumes of waste; challenge to get co-operation for recycling
Affluent vs. poor areas	Different collection services are offered in different areas
Social issues	The placement of recyclables on kerb sides may attract undesirable elements
Market prices for recyclables	Market prices and transport to Cape Town affect the viability of the projects

Rolling out the project

Seven Stellenbosch suburbs, representing a cross section of the suburbs of Greater Stellenbosch, were identified for Phase 1 of the project, as summarised in Table 7.3. The remaining suburbs and other smaller hamlets in the area were supposed to be included in the project during the 2011/2012 financial year, but to date this has not happened.

Table 7.3 Suburbs identified to participate in Phase 1 of the Recycling at Source project [Stellenbosch Municipality 2011]

Suburb	Number of houses
Raithby	±100
Jamestown	±550
Brandwacht, Dalsig, Die Boord	±900
Onderpapegaaiberg	±600
Paradyskloof	±450
Total	±2 600

All households identified for Phase 1 were provided with a 240-litre wheelie bin and a clear recycling bag, with a once-a-week collection service. Bags full of recyclable material are collected on the same day as wet waste. This means that the same service is initially provided at the same collection costs, but residents are provided with an opportunity to reduce their waste by separating at source.

Encouraging consumer participation

Large clear bags, made from recycled plastic and bearing the printed 'Stellenbosch Recycles' logo, are delivered, free of charge, to all participating households in the suburbs identified. On recycling collection days, households are rewarded with a free bag if they actively participate by putting out a bag full of recyclables. Free replacement bags are either placed in letter boxes or fixed to the handle bar of the wheelie bin. The project will eventually be extended to all users, but this has not yet been implemented due to a shortage of suitable recycling centres in Stellenbosch.

Evaluating the project

For recycling and waste minimisation initiatives to be truly successful, they must become part of the general culture. Similar initiatives in other parts of the world have shown great success, with many developed cities like Seoul and London recycling a large percentage of their annual waste. The culture, history and populations of these places differ significantly from Stellenbosch, but the same goals are achievable.

The Recycling at Source project has already taught us the following lessons:

- Informing the public about how recycling at source works, together with information about what can or cannot be recycled, is vital to the success of such a project. Ongoing feedback and encouragement on project successes are also essential.
- Special attention must be given to the language medium and the quality of the information brochures and their visual impact. The strategy guiding the project implementation was to make recycling at source fun, inclusive and as simple as possible. Pamphlets, notice boards, posters on lamp poles, and the campus radio station, MFM 92.6, are only a few examples of high-impact media used to inform the Stellenbosch public about the project.
- The direct saving to Stellenbosch Municipality of removing one tonne of waste from the waste stream is ±R95. However, the true value of this figure is much greater if one factors in the value of future airspace at the landfill site, the generation of employment, and the saving on overall environmental costs (e.g. the impact of traffic). The extension of the Recycling at Source project to all other areas in the municipality, and the refinement of its operation, will further contribute significantly to these savings.
- The return on investment or the direct financial benefits of initiating such a project will always be difficult to justify in the short term. However, there are no alternatives. A recycling-at-source and waste-minimisation ethos has to be fostered from the grassroots level, with dedicated support and guidance, so that it may eventually become the accepted way of life over the long term.

Recognising the project

For the past ten years, DEAT has held the National and Provincial Cleanest Town Competition for municipalities in order to create awareness about how waste and waste material is managed by South African municipalities. This competition has become a very important benchmark for municipalities in terms of their compliance with international waste management and waste minimisation standards. Stellenbosch Municipality has been participating in the competition for over a decade, but only achieved a position in 2010, when it was named second runner-up in the Western Cape competition – for which it received approximately R60 000 in prize money. It maintained this position in 2011. The prize effectively recognised Stellenbosch as the third-best municipality in terms of waste management practice in the Western Cape, and the town's residents should be proud of this achievement. It is clear that forming partnerships between municipalities, relevant organisations and the public is a winning recipe for change.

Planning for the future

Due to the non-compliance of the landfill site, which required urgent action, the Recycling at Source waste minimisation project was implemented in a very short period of time; yet, it has already resulted in saving valuable landfill capacity. The implementation cost was about R2.5 million (Stellenbosch Municipality 2009).

By implementing such drastic compliance measures successfully, Stellenbosch Municipality earned a reprieve from DEAT. In 2010, DEAT issued an official notice indicating their satisfaction with the outcome of the measures implemented and their intention to withdraw the landfill operating permit was set aside. This notice, together with the support and participation of all stakeholders, paved the way for Stellenbosch Municipality to proceed with future planning. Crisis management has now been replaced with normal day-to-day operations and future planning for waste management.

The Stellenbosch landfill site officially reached capacity at the end of February 2012. A new cell is currently under construction; once it becomes operational (in September 2012), the rest of the original landfill site will be closed.

Redeveloping the landfill site

In 2008 already, Stellenbosch Municipality appointed waste specialists, Jan Palm Consulting Engineers, who have been involved at Stellenbosch landfill site since 1990, to assist the municipality with future planning. They conducted a study to identify a new landfill site in the Greater Stellenbosch area and investigate the option of transporting waste to the Vissershok landfill, as well as explore the possibility of establishing a waste-to-energy plant on site.

Their report concluded that it will benefit Stellenbosch ratepayers if:

- *a transfer station is built next to the existing landfill site;*
- *a Material Recovery Facility (MRF) is built next to the abovementioned transfer station. This facility would be used to accommodate some of the informal traders on site after closure of the existing landfill site;*
- *the existing site is extended for a five-year period;*
- *a waste-to-energy plant is not built – the quantity of waste disposed of per day (±180 to 200 tonnes) at the Stellenbosch landfill is not sufficient for the establishment of a waste-to-energy plant. A sustainable waste-to-energy plant project will only be potentially viable if the Drakenstein and Stellenbosch municipalities combined their waste streams (±450 t/day);*
- *the possibility of a combined regional landfill site for the Drakenstein and Stellenbosch municipalities is explored. The Cape Winelands District Council should investigate the establishment of such a site near Wellington for future use;*
- *a transfer station is built near Wemmershoek for Franschhoek residents.*

[Jan Palm Consulting Engineers 2008]

The cost of designing, launching and managing the initial projects and facilities, as recommended by the firm, amounted to approximately R40 million (MRF, transfer stations, extending the landfill site and expanding the Recycling at Source project). The recommendations were accepted and submitted in a 15-year capital budget to Council, and will be implemented in phases. Phase 1 was the construction of a new cell for disposal purposes, while Phase 2 will be the construction of the transfer station and MRF.

Figure 7.2 illustrates the future development plans for the Stellenbosch landfill site. The existing landfill site can be seen in the middle of the photo. A new cell, which will be operational for four to six years only, will be constructed north of the existing landfill site. A new material recovery facility and transfer station will be constructed south of the existing landfill site. This will be done as soon as the regional landfill site near Wellington becomes operational. Space for the chipping of green waste and crushing of builders' rubble will also be provided to the south of the existing landfill site.

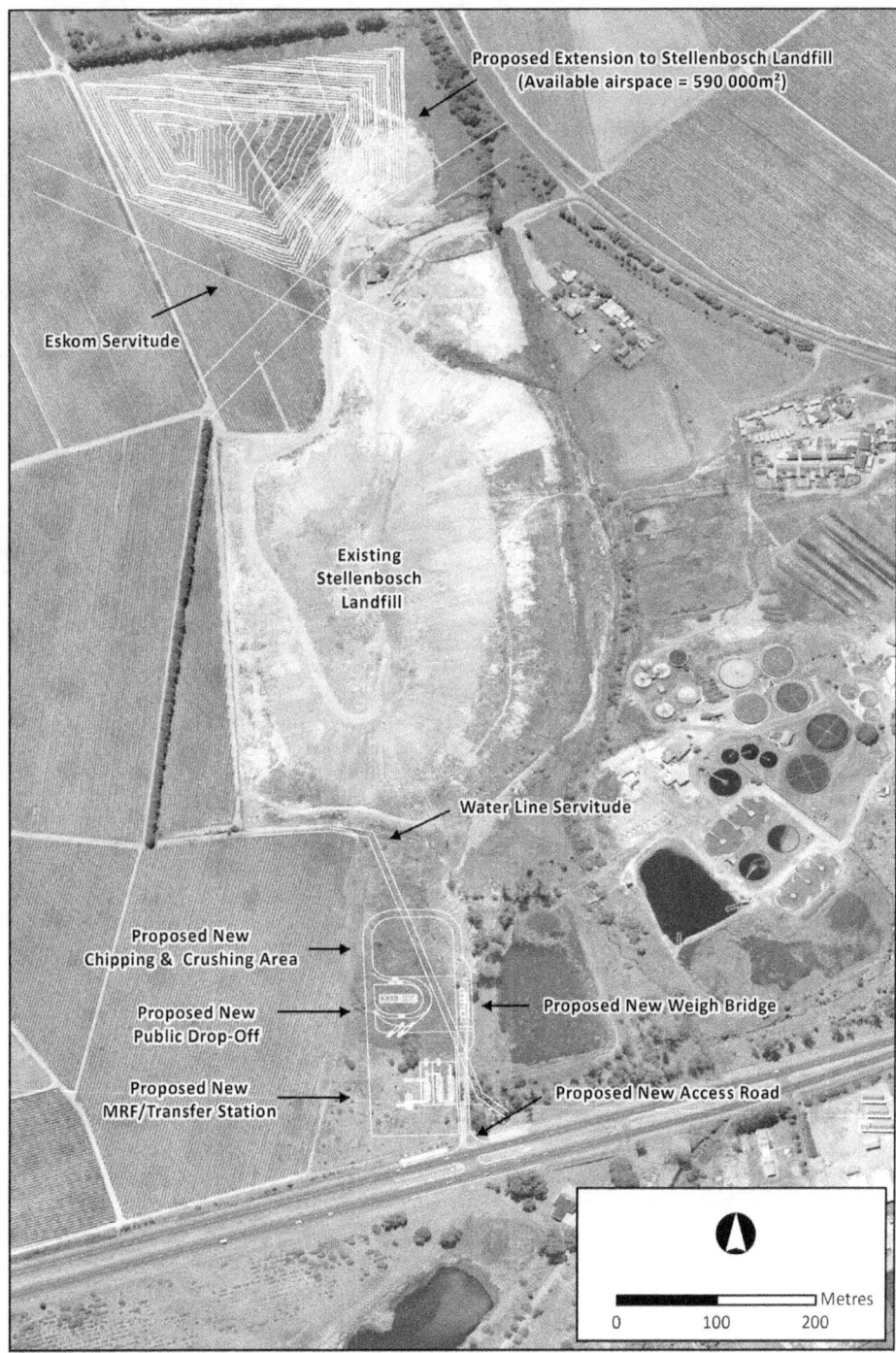

Figure 7.2   Future development plans for the Stellenbosch landfill site [Jan Palm Consulting Engineers 2008]

Closing the loop on waste

The slow adoption of the Recycling at Source project indicates how tough it is to affect behavioural change in South Africa, and that it will not be easy to create a sustainable Stellenbosch without considerable changes to mindsets and infrastructure. The conventional way of looking at resource flows in urban environments is to separate the technical systems that facilitate the flow of resources and the social systems of consumption. Instead, Guy and Marvin (2001) suggest a new paradigm, which views these systems as interdependent socio-technical networks with various technical, social, economic, environmental and political linkages.

Since the old paradigm is clearly not working, it is time for utilities to reinvent themselves and focus on network integration management, instead of the usual network expansion within the 'predict and provide' paradigm. By creating participating consumers – as opposed to resource users, utilities may respond more effectively to localised demand pressures within a broader regional context of resource scarcity and environmental innovation. In addition, consumers will be compelled to take more responsibility for their choices and become more active consumers. Evidence of this type of thinking may be seen in electricity demand-side management initiatives aimed at redirecting resources to flow in a more sustainable manner (Guy & Marvin 2001).

The reality is that people's habits will not change overnight and that resource flows in urban environments are a function of the socio-technical systems that direct these flows. The first step towards sustainable waste management is to design infrastructure that reflects closed-loop thinking. Birkeland (2009) suggests an alternative design paradigm that allows both human and natural systems to flourish by designing infrastructure that mimics natural eco-services, instead of trying to replace these services with mechanical imitations. An example of this is the construction of on-site waste water treatment systems, such as anaerobic digesters and artificial wetlands, at all new developments, instead of just building larger sewage infrastructure with large central waste water treatment plants.

The second step is to inform consumers. Recycling is normally the first aspect municipalities focus on as part of 'reduce, re-use and recycle' campaigns, mainly because there is a clear financial incentive to do so. However, if one scrutinises this waste warriors' slogan more closely, the overemphasis on recycling is a symptom of the old mindset, which essentially replaces "you may waste as much as you like – as long as you pay your rates" with "you may waste as much as you want – as long as you recycle". If one really wants to create societies without landfills, the slogan should be: "**reduce** your waste as much as possible by **re-using** what you have; and if you really do not have a choice, you must **recycle**." Recycling should be seen as a last resort, instead of the ultimate goal. This is the mind shift required of the consumer if we want to build a sustainable Stellenbosch.

Conclusion

Dealing with South Africa's municipal waste problem is not a simple issue. In the current political and organisational environment, there is often a large discrepancy between the theoretically possible and the practically feasible. However, the 'impracticality' defence may lead to stagnation if it is used to justify resistance to change. As always, the answer lies in a combination of efforts aimed at addressing both the social and technical aspects of the urban metabolism in a coordinated way.

Stellenbosch Municipality has not only assumed responsibility for complying with permit conditions at the landfill site and the new Waste Act (Act 59 of 2008); it has also formed a partnership with the public and committed itself to a more sustainable Stellenbosch.

Due to the low or uncertain return on investment, initiating projects for waste minimisation and compliance with legislation will always be difficult to justify in the short term. However, there are no real alternatives when the local landfill is at capacity and using the nearest other landfill is too costly. A new attitude towards waste management, based on closed-loop principles, must be fostered from the grassroots level with the dedicated support and guidance of the municipality and other role players, so that it will eventually become the cultural norm for Stellenbosch.

ACKNOWLEDGEMENTS

The authors wish to extend special thanks to the Stellenbosch Municipal Council for its vision – which strives to protect the interests of both the community and the environment – and for allowing us to contribute to this book.

We would also like to thank Eve Annecke of the Sustainability Institute for initiating waste management workshops for stakeholders in 2009 and, in doing so, helping to pave the way for sustainable waste management in Greater Stellenbosch.

REFERENCES

Birkeland J. 2009. *Positive Development: From Vicious Circles to Virtuous Cycles through Built Environment Design*. London: Earthscan.

DEADP (Department of Environmental Affairs and Development Planning). 2009. *Cleanest Town Competition Report 2009*. [Accessed 23 August 2012] http://www.westerncape.gov.za/Text/2010/6/cleanest_town_report_2009_opt.pdf

Guy S & Marvin S. 2001. Urban Environmental Flows: Towards a New Way of Seeing. In: S Guy, S Marvin & T Moss (eds). *Urban Infrastructure in Transition: Networks, Buildings, Plans*. London: Earthscan.

Jan Palm Consulting Engineers. 2008. *Investigation into future waste management scenarios*. Report No A114/1, March. Stellenbosch: Stellenbosch Municipality. [Accessed 21 August 2012] http://www.stellenbosch.gov.za/jsp/util/document.jsp?id=2297

MIG (Municipal Infrastructure Grants). 2009. Non-compliance with conditions of waste permit: Stellenbosch Municipal Disposal Site. Official letter from MIG to Stellenbosch Municipality: 20 April. Stellenbosch: Stellenbosch Municipality.

Stellenbosch Municipality. 2009. *Stellenbosch Landfill site non-compliance to permit conditions: possible closedown of site by DEAT*. Report to Council for approval of projects: 17 June. Stellenbosch: Stellenbosch Municipality.

Stellenbosch Municipality. 2010a. *Engineering 15-year Capital Budget 2010/2011*. Stellenbosch: Stellenbosch Municipality.

Stellenbosch Municipality. 2010b. *Integrated Waste Management Plan: Review and Update*. Stellenbosch: Stellenbosch Municipality.

Stellenbosch Municipality. 2011. *Recycling at Source Project in Stellenbosch*. Report to Council: 7 December. Stellenbosch: Stellenbosch Municipality.

Sustainability Institute. 2009. *The Stellenbosch Landfill is Full. Stakeholder Workshop, 9 July*. Lynedoch: Sustainability Institute. [Accessed 21 August 2012] http://www.sustainabilityinstitute.net/newsdocs/si-news/item/stellenbosch-landfill-is-full

Chapter 8

WATER SERVICES
Stellenbosch Municipality water services

EUGENE CLOETE, MARELIZE BOTES & MICHÉLE DE KWAADSTENIET

Introduction

Water is a renewable resource, but there are limits to its quality and quantity. Land development, polluted run-off from agricultural, commercial, and industrial sites, and aging waste water infrastructure threaten the quality of drinking water sources. In many areas of the country, ground water is being pumped faster than aquifers are being recharged and depleted aquifers are causing reduced ground water contributions to surface water flow (Robbins, Chilton & Cobbing 2007). Surface water withdrawals are diminishing in-stream flows to the point that habitat, as well as water supply uses, are threatened. Planning and taking action to protect sources of drinking water may also protect them for a multitude of other uses.

Drinking water systems have an enormous impact on public health, and the public health benefits of a well-run system cannot be overstated (Craun, Hubbs, Frost, Calderon & Via 1998). High-quality drinking water is a major contributor to a high standard of living and health. Customers rely on their water systems to provide safe water for drinking, bathing, cleaning, and cooking. Drinking water, which may be from ground water, surface water, or both, is vulnerable to being contaminated. If the source is not protected, contamination may cause a community significant expense and put people's health at risk. Cleaning up contamination or finding a new source of drinking water is complicated, costly, and sometimes impossible. Preventing the contamination of source water is often cheaper than remedying its effects, and is the first barrier to the outbreak of waterborne illnesses. Anyone engaged in producing safe drinking water should know that microbial pathogens represent the most pervasive and consistent proven risk to drinking water safety (Cloete, Rose, Nel & Ford 2004).

Water treatment provides a key barrier in minimising the risks from hazards that arise in raw water, whether from ground or surface sources or from private or municipal supplies. Identification of the hazards and risks while developing a Source Water Protection Plan is an essential step in determining the requirements for treatment at a plant. For municipal supplies, there may be additional hazards introduced during treatment and distribution that must be taken into account, such as disinfection by-products. Water treatment includes raw water or municipal water storage where such disinfectants are used. Storage may also act as a treatment step, depending on the nature of the storage, and may also be a potential point at which hazards are introduced; for example, pathogens from bird droppings if the floating cover of the storage tank is torn or ripped, allowing contamination of the finished water (Kirmeyer, Friedman, Martel, Howie, LeChevallier, Abbaszadegan, Karim, Funk & Harbour 2001), or a closed reservoir that leaks. Treatment will clearly vary according to the source and nature of the water; for example, natural mineral waters should only receive treatment approved by the local regulatory authority. The water treatment steps at municipalities include coagulation, flocculation, sedimentation, stabilisation, filtration, disinfection and chlorination (Rand Water 2012).

Water services supplied by Stellenbosch Municipality

The Stellenbosch Municipality provides piped water to the majority of inhabitants to their residences, either on site or at public taps. In 2006, only 58% of all people received water in their homes, with 17% relying on public taps as a main water supply (Stellenbosch Municipality 2008). This situation had drastically improved by 2007, with 87% of the population receiving water in their homes (Stellenbosch Municipality 2009; 2010). During the 2009/2010 term, a further 1 732 households were provided with potable water within at least 200 m of the dwelling for the first time (Stellenbosch Municipality 2010). During the 2011 term, 17 721 households had access to piped water inside a dwelling; 380 households to piped water inside a yard; and approximately 4 500 households to piped water on a community stand less than 200 m from the dwelling (Stellenbosch Municipality 2012a).

Table 8.1   Main water supply in Stellenbosch [Adapted from Stellenbosch Municipality 2010]

Population group	In dwelling	On site	Borehole	Other
African	87.3	9.9	0.0	2.8
Coloured	98.7	0.0	0.2	1.1
Asian	68.4	31.6	0.0	0.0
White	98.6	0.0	1.4	0.0
Total	95.3	3.0	0.3	1.4

The Civil Engineering Department of Stellenbosch Municipality provided the following water services during the 2008/2009 term: 11 850 074 kilolitres (kl) of water were abstracted and/or purchased by the municipality and 9 269 269 kl of water were supplied and metered. During the 2009/2010 term, 9 032 397 kl of water were supplied and metered (Stellenbosch Municipality 2010). Although the municipality supplied water to a greater number of households during the 2009/2010 term, there was a reduction in the volume of water metered when compared to the 2008/2009 term; hence it may be concluded that less water was used by end users during the 2009/2010 term than during the previous term.

An integral part of providing safe water is to identify and manage the risks. The following section will cover the process of risk management that is necessary for a safe municipal water supply.

Monitoring and managing water services

Risks to water quality

Risk is the probability that a hazard will give rise to unacceptable consequences, such as a breach of local water standards, waterborne diseases or chemical contamination.

A hazard may be considered to be:

- a microorganism or chemical with the potential to cause ill health or render water unacceptable for products delivered to consumers;
- the potential to cause a failure of local standards, even if there is no significantly increased risk to health or acceptability;
- circumstances such as susceptibility to heavy rainfall, leading to an increased microbiological challenge or disruption of coagulation in the water treatment process;
- other factors that may impact water quantity in the short or long term (for example, a burst water main, extended drought and significant pollution of a source).

Risk assessment

A comprehensive risk assessment for water should include a consideration of likely events in the production of potable water (Gale 2001; Haas 1983; Haas, Rose, Gerba & Regli 1993). Risk assessment considers both the inherent severity of the hazard and the probability of exposure to the hazard in a given situation.

The components of risk assessment are therefore:

- hazard identification (identification of the biological, chemical or physical agent, or disease or adverse health outcome);
- exposure assessment (determination of the frequency of disease and the number of people exposed to contaminated water, and the prevalence, growth, contamination, survival or destruction of pathogens in water);
- hazard characterisation (identification of the adverse health effects associated with the hazard); and
- risk characterisation (estimation of the risk and the number of cases, as well as the severity of the outcome).

Significant hazards must then be controlled within acceptable limits. Although it is possible to set quantitative microbiological limits for contamination, survival and proliferation, microbiological monitoring is expensive and retrospective. It is therefore preferable to decide beforehand the means by which the hazard is to be prevented or minimised, and to set limits or standards at these control points (see Figure 8.1), which may be more readily monitored (Gale 1998).

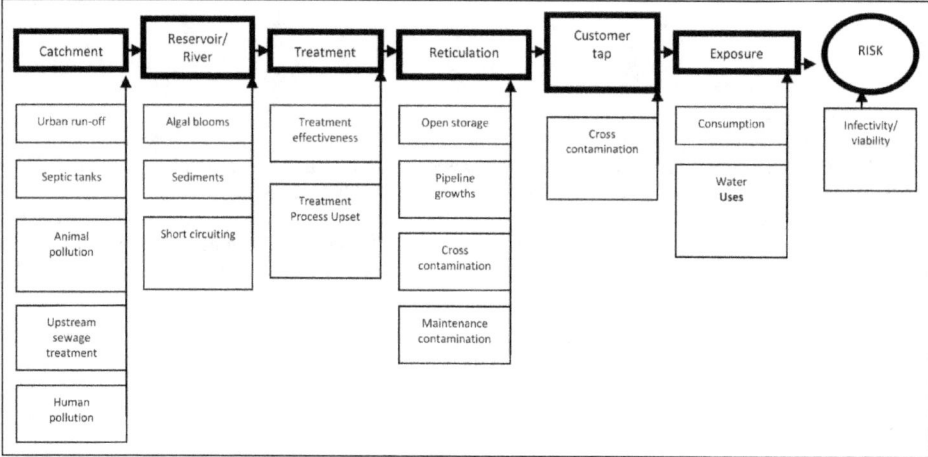

Figure 8.1   Sources of microbial risk in a municipal water supply system [Cloete, unpublished]

It may not be cost-effective to exercise tight control over all the identified hazards. Careful consideration of each hazard, with special regard to its position in the sequence of operations, will allow the identification of those that are truly critical, i.e. where loss of control would result in an unacceptable water safety risk.

In some cases, assessments are needed when there is significant public perception that the risk associated with a hazard is very high or unacceptable, even when the data does not support this and no standard exists. The assessment of risk helps to prioritise the actions that might be required and the investments or costs associated with particular interventions.

Approaches to risk assessment include:

- an analysis of the historic water quality data for a source or supply, which measures the likely exceedence of the standard with actual data;
- the evaluation of trends in the concentration of a contaminant, which allows extrapolation to determine whether a breach in standards is likely in the future;
- a simple evaluation of the likely presence of pathogens or a chemical; for example, a surface water catchment with livestock farming is most likely to have pathogens present in significant numbers in the raw water, and coagulation using aluminium will lead to aluminium being present in the treated water, and so both would require to be considered in more detail;

- the evaluation of extraneous risks that may impact water quality, such as infrastructure failure, sourcing raw water from new resources, floods or spills of chemicals, or human or animal wastes, etc.

The questions to be asked are:

- What are the contaminants and what are the concentrations or numbers?
- Do these represent unacceptable concentrations if they come into contact with the product, i.e. will they result in a breach of any of the standards directly or by adding to the concentration in potable water?
- What are the flows and what are the pathways to the final product?
- Can the contaminated water reach the product or come into contact with surfaces with which the product will also come into contact?
- If so, is it possible that a standard may be breached?

For a chemical contaminant, one issue is whether it is permanently or intermittently present. However, it is vital that the barriers are put in place to mitigate the risk at any time, because it could result in a breach of the standard. Therefore, for chemicals, the question is whether or not it is present and what the concentration is. For municipal drinking water supplies, it is often the practice to establish a health-based trigger value for chemicals, which is above the standard or guideline, but represents the maximum acceptable erosion of the margin of safety.

Managing risk

Assuring drinking water safety is an exercise in risk management, which inevitably demands making important decisions without complete or compelling evidence. Some key elements of a risk management approach to drinking water safety were summarised by the Walkerton Inquiry:

> ... being preventive rather than reactive; distinguishing greater risks from lesser ones and dealing first with the former; taking time to learn from experience; and investing resources in risk management that are proportional to the danger posed ...
>
> [O'Connor 2002:504]

The multiple barrier concept also helps to manage risk through several robust and effective measures:

> • "source protection and selection to keep the raw water as clean as possible, to reduce the risk of contamination breaching the drinking water system;
> • treatment, often involving more than one process, to remove or inactivate contaminants, to be effectively designed, operated and maintained;
> • distribution system security to protect against intrusion of contaminants and disinfectant residual use to assure delivery of safe water to consumers;
> • monitoring to control treatment processes and detect contamination in a timely manner to inform risk management responses;
> • response capabilities to adverse conditions that are well-conceived, thorough and effective."
>
> [Hrudey 2004:486]

Treatment should be based on the multiple-barrier approach, as in the case for Stellenbosch Municipality (see Figure 8.1). Stellenbosch Municipality was awarded with Blue Drop Status, which is based on a multiple-barrier approach, in 2009, 2010 and 2011.

Monitoring water services

It is important to establish properly documented procedures and appropriate operational monitoring procedures. There should preferably be continuous monitoring to provide real-time information of process stability and operation. For example, there should be continuous monitoring of turbidity (or other parameters such as particle numbers or aluminium) before and/or after coagulation, sedimentation and filtration. If specific chemical removal processes are in place, monitoring would need to identify any deterioration in the efficiency of the process. It is therefore essential that suitable operational limits are established to ensure that the processes are operating optimally. There should also be clearly documented procedures for correcting the system when monitoring indicates a deviation from the established limits.

- *Types of monitoring*

Operational monitoring is the process of gathering information to make decisions regarding the different component activities of producing municipal water, for example raw water monitoring or monitoring of drinking water treatment. This may take the form of continuous monitoring or off-line monitoring by sampling at a predetermined frequency or times. Some of the monitoring will be for surrogate parameters, for example turbidity, chlorine residual, or conductivity, as appropriate; while some will be the same as or similar to monitoring for parameters for which there are standards, for example *E. coli*. Frequencies of monitoring and the choice of parameters to be monitored must be carefully chosen to reflect the hazards of concern and the potential risks, but also need to be properly targeted to optimise the costs.

In the case of **microbial monitoring**, this is frequently achieved by examining indicator organisms. *E. coli* (or faecal coliforms if necessary) show that there is a possibility of the presence of faecal contamination. Any detection therefore requires a rapid response, initially to investigate and, if necessary, take appropriate remedial action to correct the process. This could be achieved by stopping the source of the contamination, or correcting treatment and checking or withdrawing any product that has already been prepared since the last clear data. The investigation should also be informed by other operational monitoring data, such as turbidity or chlorine residual. However, it should be remembered that some organisms, such as *Cryptosporidium*, are resistant to chlorine; therefore, operational monitoring data showing the integrity of particle removal barriers is also essential.

A clear distinction must be made between operational monitoring, investigation, and verification and audit monitoring. Operational monitoring is for informing process control in real time, or at least within a timeframe that allows the data to inform immediate decisions. Investigation provides initial data for assessing hazards and risks or for determining the source or cause of a problem. **Verification and audit monitoring** provide the data for management or an external authority to show that the process is working properly and is delivering safe and acceptable water, and is usually retrospective in nature. It would not be appropriate to rely entirely on verification monitoring of the final product, because if problems are only detected at this point, the cost of lost production may be substantial. In addition, the risk of a defective product being released would be much higher and this could be very damaging, if not catastrophic.

It is vital that all monitoring or analytical determinations are carried out to a high degree of accuracy and with appropriate quality control.

- *Blue Drop Certification Programme*

The Department of Water Affairs (RSA DWA) initiated the Blue Drop Certification Programme on 11 September 2008, mainly to regulate and ensure efficient management of drinking water quality (DWQ). The public is also informed of the DWQ performance of all the water supply systems, and closer working relationships are facilitated between Water Services Authorities and Water Service Providers. The Blue Drop Status award, indicated by a flag, may also be used by authorities for marketing purposes, so that citizens and tourists may use tap water with confidence. Towns without Blue Drop status should be perceived as being in the process of improving drinking water treatment. In the case of health failures, the department will communicate warnings to the public (eWISA, 2012).

Blue Drop status is only awarded if a Water Services Authority complies with 95% of the weighted criteria in the biannual assessment. Every drinking water system, from catchment to consumer, is evaluated and aggregated for each Water Services Authority in South Africa. In order to ensure continuous improvement in drinking water management practices, comprehensive and stringent criteria (from year 1 to year 3 and ongoing) are applied (eWISA, 2012).

The Stellenbosch Municipality serves the towns of Stellenbosch and Franschhoek, as well as the following settlements: Jamestown, Klapmuts, Koelenhof, Kayamandi, Lynedoch, Kylemore, Pniel, Simondium, Vlottenburg, Johannesdal, Jamestown and Raithby. The water supply systems of Paradyskloof, Idas Valley and Franschhoek are evaluated every year according to the Blue Drop Certification Programme. The 2012 report is summarised in Table 8.2.

Table 8.2 Blue Report Card of Stellenbosch Municipality [Adapted from RSA DWA 2012]

	Water treatment system				
	Blackheath	Faure	Franschhoek	Stellenbosch	Wemmershoek
Water safety planning (35%)	97	97	100	100	97
Treatment process management (10%)	50	50	75	88	50
DWQ compliance (30%)	100	100	45	100	100
Management, accountability (10%)	100	100	92	92	100
Asset management	100	100	97	97	100
Blue Drop score 2012	95.28	95.28	84.21	98.25	95.28
Blue Drop score 2011	96.34	96.79	75.02	97.11	93.40
Blue Drop score 2010	Not assessed	Not assessed	94.11	95.02	Not assessed

In 2009, Stellenbosch Municipality was awarded Blue Drop status and the individual supply systems, Paradyskloof and Idas Valley Treatment were certified as Excellent Drinking Water Quality Management Performers by the Department of Water Affairs (Stellenbosch Municipality 2010). Stellenbosch Municipality was awarded for the second year in row with Blue Drop status in 2010 by achieving an average score of 94.9%, which is an improvement of 0.9% from 2009. However, improvement in microbiological and chemical compliance is still needed. The status for Franschhoek could not be maintained due to compliance records that did not meet the standard. Stellenbosch Municipality once again received Blue Drop status in 2011, with an average score of 95.74%, which shows a gradual increase from 2009 and 2010. Three of the five water supply systems (Stellenbosch, Blackheath and Faure) were awarded Blue Drop status, with scores of 97.11%, 96.34% and 96.79%, respectively. The water supply system in Wemmershoek also achieved an increased score of 93.40%. However, with a score of 75.02%, the Franschhoek water supply system requires urgent improvement, as the water is of substandard quality, both microbiologically and chemically. In 2012, Stellenbosch achieved a Blue Drop score of 95.56% and was once again rewarded with Blue Drop status (RSA DWA 2012).

Challenges to water services

Limited water resources

Water resources in Stellenbosch Municipality consist of two major river systems, namely the Eerste River and the Great Berg River, and the newly built Berg Water Dam (see Figure 8.2a). Due to the impact of pollution, the status of the rivers in the municipality is of great concern. Rivers were evaluated in terms of the South African National Biodiversity Initiative (SANBI) classification and the majority was classified as critically endangered (Figure 8.2b).

The only river streams that fared better, i.e. that could be classified as vulnerable, rather than critically endangered, were situated in the Jonkershoek Valley (Stellenbosch Municipality 2010). Water resources in the Stellenbosch area are therefore under threat and steps should be taken to conserve them in order to ensure the sustainable supply of water to consumers. If the water quality in the rivers does not comply with certain standards, the amount of water available will be irrelevant. For example, water used for irrigation purposes must comply with the Department of Water Affairs and Forestry's[1] (DWAF) South African Water Quality Guidelines for Agricultural Use (RSA DWAF 1996b). Therefore, the state of the rivers is not only important for protecting the biodiversity of the river systems, but may also have an impact on the agricultural activities around Stellenbosch.

---

1 In 2009, the forestry responsibility was transferred to the newly formed Department of Agriculture, Forestry and Fisheries, and the former Department of Water Affairs and Forestry was renamed the Department of Water Affairs.

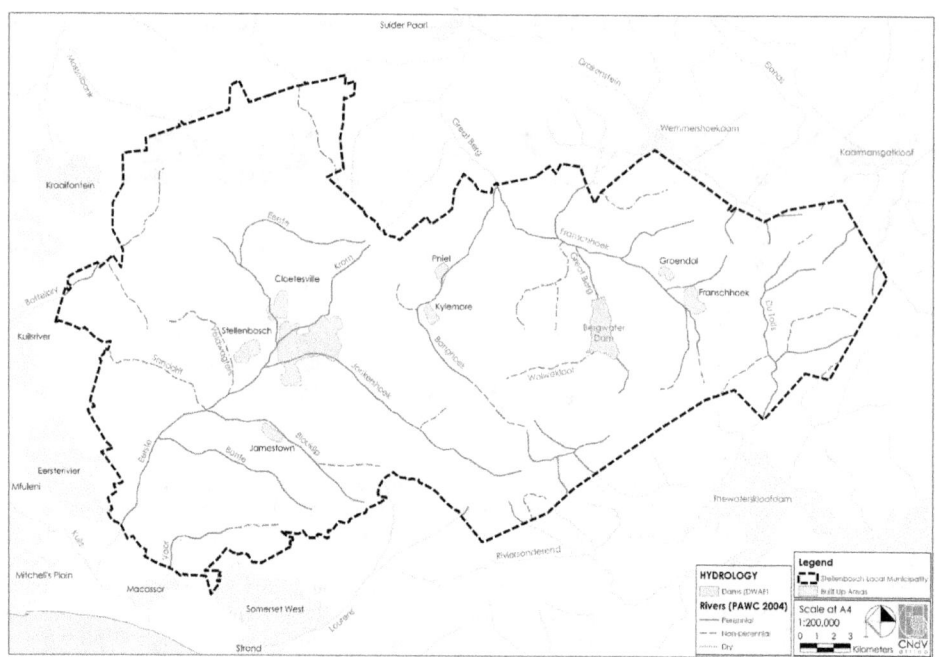

Figure 8.2a   The river systems and major dams in the Stellenbosch municipal area [CNdV Africa 2010]

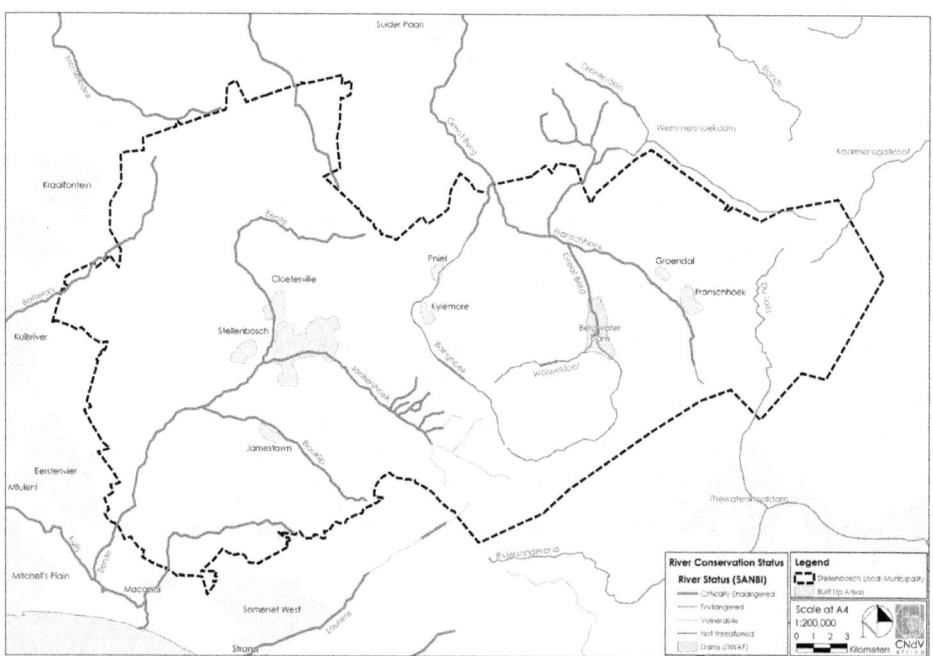

Figure 8.2b   Critically endangered rivers in the Stellenbosch municipal area [CNdV Africa 2010]

Stellenbosch Municipality has revised its Spatial Development Framework (SDF) to include support for sustainable development in the area, in accordance with the Draft Spatial Development Frameworks Manual prepared by the Western Cape Department of Environmental Affairs and Development. It is encouraging to note that the conservation of water is one of the key strategic issues in this long-term spatial framework.

The following issues have been outlined in the water section:

- *All rivers above a minimum size shall be protected by a 10 m to 30 m river conservation zone.*
- *No ploughing, urban development or intensive agriculture shall be permitted in these zones; instead, they should be restored to their natural riparian status.*
- *Other measures to save water, such as green roofs and water-wise gardening, should be encouraged.*
- *Waste Water Treatment Works should achieve minimum DWA water quality standards and should therefore be upgraded.*
- *Conservation areas should be protected to ensure the quality and quantity of the water in the upper reaches of the river system.*
- *Support should be continued for the eradication of alien vegetation in non-agricultural areas.*

[Stellenbosch Municipality 2010]

With regard to alien vegetation, the Stellenbosch Municipality should also determine the impact of invasive alien plants to catchment areas. The Working for Water (WfW) programme, funded by the South African government, was established together with the Payments for Ecosystem Services (PES). Through this programme, mountains and catchments are cleared of invasive alien plants to improve hydrological functioning and therefore water delivery. The PES awards contracts to unemployed individuals to restore the public or private lands that form part of the WfW poverty relief initiative. The clearing of invasive plants is not only government funded, but is also covered by fees paid by water users. The aim is to levy all 19 water management areas, based on the cost estimates of WfW divided by the total volume of registered agricultural, domestic and industrial sectors' water use, and weighted according to affordability, assurance of supply and equity.

Self-governing companies, such as the Trans Caledon Tunnel Authority (TCTA) in the Western Cape, fund water utilities that supply water to urban users in South Africa. The TCTA provided R1.6 billion for the construction of the Berg River Dam to improve water supply to the local farms and the City of Cape Town. Funded by water sales, the company paid WfW R8 million over three years to improve water supply to the Berg River Dam by clearing mountain catchment areas. Thus far, it has been estimated that if the invasions had not been cleared, an increase of between 4.3 and 6.2 million m^3 of water per year would have been lost (Turpie, Marais & Blignaut 2008). The WfW programme is regarded as highly successful in improving water supply by clearing catchment areas of invasive plants in a holistic manner, while having a positive impact on biodiversity and water (Hobbs 2004; Woodworth 2006).

In terms of saving water, Stellenbosch Municipality uses a block rate tariff structure to curb wasteful water consumption. The first block rate of 6 kl per month represents the lifeline volume and is supplied at a rate well below cost. Contributions from the higher tariff categories cover losses that occur in the first block. The draft sliding scale for households and a flat tariff for other consumers are presented in Table 8.3. These prices exclude VAT and a basic service charge of R39.55 for every domestic user and R42.54 for every other user (Stellenbosch Municipality 2012b).

Table 8.3 The blocks for water tariff charges in Stellenbosch municipal area [Adapted from Stellenbosch Municipality 2012b]

Amount of water (kilolitres)	Tariff: Normal consumption periods (Rand per kilolitre)	Tariff: 20% Water restriction periods (Rand per kilolitre)
Domestic: 0 – 6	3.45	3.45
7 – 20	4.90	5.38
21 – 40	8.19	9.83
41 – 60	12.19	16.73
>60	16.42	23.39
Commercial and industrial	10.27	-
Municipal and domestic leakages	5.43	-
Miscellaneous and all other users	10.27	-
Bulk users*: Up to 2 000	5.87	7.27
Over 2000	6.83	9.88

* Water consumption for irrigation of sports grounds (schools), sport clubs (Council property) and parks (Council Department of Parks and Recreation)

Other countries have similar mechanisms in place to encourage water saving. Australia, for example, has implemented various regulatory incentives for sustainable water management due to increasing water demand from a growing population and economy, as well as severe droughts. This includes the installation of rainwater harvesting tanks for domestic purposes (Coombes 2006).

Ageing and inadequate infrastructure

One of the major concerns in the water services network is the ageing infrastructure and a backlog of maintenance and replacement programmes. The increase in population growth increased the demand for bulk infrastructure to a point that exceeds the provision for capital funding. During 2010, 222 network bursts were repaired in the water network; R1.325 million was spent on improvements in bulk water supply; and R1 million on water pipe replacement (Stellenbosch Municipality 2010). The Paradyskloof water treatment works is currently being upgraded. The Water Master Plan (Stellenbosch Municipality 2008) is also being updated and guides all future infrastructure planning, including upgrading of bulk water supply systems, water treatment works infrastructure, water pump stations and reservoir infrastructure (Stellenbosch Municipality 2011).

The value of the water supply infrastructure has fallen by 52.3%. This figure was calculated by Stellenbosch Municipality from the depreciated replacement costs of the entire infrastructure (see Table 8.4).

Table 8.4 The replacement cost of the water infrastructure of Stellenbosch Municipality [Adapted from Stellenbosch Municipality 2011]

Asset type	Current replacement cost (CRC) (R)	Depreciated replacement cost (DRC) (R)	% DRC/CRC
Bulk water pipelines	99 352 866	51 050 866	51.38
Dams	54 963 625	26 389 276	48.01
Water pump stations	45 956 442	23 249 723	50.59
Reservoirs	132 970 330	66 924 861	50.33
Water consumer connections	139 755 000	52 151 236	37.32
Water reticulation pipelines	312 084 416	152 287 646	48.80
Idas Valley	29 485 288	17 620 798	59.76
Paradyskloof	26 178 467	11 147 691	42.58
Totals	840 746 434	400 822 097	47.67

The progressive deterioration of the water supply infrastructure has led to maintenance activities being focused on reactive measures, with less attention paid to preventative measures. However, the Stellenbosch Municipality Water Services plan for 2011/20112 contains a suggestion for the development and implementation of an Infrastructure Asset Management Plan, which makes provision for the protection of assets through preventative maintenance and their rehabilitation and/or replacement before the end of their economic life (Stellenbosch Municipality 2011).

Water and Sanitation Services has a capital budget of R70.907 million for 2011/2012; R96.73 million for 2012/2013; and R51.926 million for 2013/2014. However, the conditional backlog is in the order of R324.8 million, because 38.6% of the water supply infrastructure is in a poor or very poor condition. Reticulation pipeline assets make up the bulk of the backlog (Stellenbosch Municipality 2011). The Stellenbosch Infrastructure Task Team (SITT) calculated that around R55 million in capital expenditure is required annually to support water supply services in Stellenbosch (SITT 2012).

The challenges faced by Stellenbosch Municipality are the provision of potable water and waterborne sewerage for all; the employment of qualified and experienced staff; sufficient funding for service delivery, upgrading, refurbishment, expansion and maintenance of the network; and the upgrading of sewage treatment plants in Stellenbosch and Franschhoek. These challenges may be met by updating existing master plans and compiling new master plans that reflect current and future service delivery requirements and are aligned with the service master plans; planning for the upgrading and maintenance of all services; and drafting business plans to form the basis of the tariff structures for all services (Stellenbosch Municipality 2010). Targets for the next five years include the provision of additional raw water supply for the Jonkershoek and Idas Valley storage dams; a plan for a strategic single storage facility for Stellenbosch and complete planning for storage capacity for the Dwars River Valley; the refurbishment of the five most critical pump stations in the Stellenbosch municipal area; and the construction of regional treatment works to meet the requirements of the Franschhoek Valley (Stellenbosch Municipality 2010).

Pollution from human activity

The Eerste and Plankenbrug rivers flows through Stellenbosch and the surrounding region and its water is used for agricultural purposes. The origin of the Eerste River is in the mountains near the Jonkershoek Nature Reserve. The river is unpolluted upstream of the nature reserve, but human activity on the river bank increases between the reserve and central Stellenbosch due to agricultural irrigation and aquaculture. The highly polluted Plankenbrug River runs into the Eerste River downstream of central Stellenbosch. It is polluted by waste water and storm water from Kayamandi, an informal settlement, as well as industrial effluent from Stellenbosch's industrial areas and agricultural effluent from farms upstream and downstream of Kayamandi (Barnes 2003). Researchers have investigated the impact of human activity on the Eerste and Plankenbrug rivers by monitoring for metals (Jackson, Paulse, Odendaal, & Khan 2009), estrogenic endocrine disrupting chemicals (EDCs) (Swart, Pool & Van Wyk 2011), and microbial contaminants (Barnes 2003; Paulse, Jackson & Khan 2009). Water pollution in the Plankenbrug River may negatively impact agricultural activities in the region, leading to a loss in productivity and even contaminated food products.

Although trace amounts of metals are present in the environment, it is important that they do not exceed the permissible concentrations, as stipulated by DWAF (RSA DWAF 1996a; 1996b). Exposure to high metal concentrations may have detrimental short- and long-term effects on human health (Florea & Busselberg 2006). Jackson and his co-researchers investigated metal contamination in the Plankenbrug River over a 12-month period (May 2004 to May 2005) at four different sampling sites in and around Stellenbosch (Jackson et al 2009). Aluminium, iron, copper and zinc concentrations in the water samples exceeded the recommended concentrations as stipulated by DWAF for domestic use (RSA DWAF 1996a), but the concentrations of metal levels recorded were within the guidelines for agricultural water use, except for iron (Fe), with levels detected at 48 mg/l, which is well in excess of the recommended guideline of 20 mg/l (RSA DWAF 1996b).

EDCs present in water may have adverse human health effects, although the exact risks are still being determined by researchers (Snyder & Benotti 2010). The presence of EDCs in the Eerste River were investigated on samples collected at Jonkershoek Nature Reserve, the Stellenbosch sewage works, and Spier wine estate (downstream from central Stellenbosch). The water collected at Jonkershoek Nature Reserve served as a control, with low estrone concentrations (between 1.4 and 2.2 ng/l) detected and no estrogenicity observed in the bioassays. Water analysed from the sewage works effluent that enters the Eerste River and from the Spier collection site was shown to be estrogenic, with concentrations of estrone detected ranging between

14.7 and 19.4 ng/l.[2] These concentrations were similar to estrone concentrations detected in sewage effluent from Britain (Desbrow, Routledge, Brighty, Sumpter & Waldock 1998); Italy (Baronti, Curini, D'Ascenzo, Corcia, Gentilli & Samperi 2000); Germany (Ternes, Stumpf, Mueller, Haberer, Wilken & Servos 1999); and Canada and the Netherlands (Belfroid, Van Der Horst, Vethaak, Schafer, Ris & Wegener 1999). However, concentrations of ethynylestradiol, an EDC, were shown in laboratory conditions to have estrogenic effects on caged fish at concentrations as low as 0.1 to 0.5 ng/l (Purdom, Hardiman, Bye, Eno, Tyler & Sumpter 1994). This raises the potential risks of the presence of EDCs in the Eerste River.

Microbial pathogens present in water are usually responsible for low-level incidences of waterborne infections that are rarely debilitating to humans; however, severe illness and even death may occur among young children, the elderly and those with immune-compromised systems, such as HIV/AIDS patients (Abong'o & Momba 2008). Microbial agents responsible for waterborne infections include bacteria (*Escherichia* spp., *Salmonella* spp., *Campylobacter* spp.); viruses (hepatitis A and E); and protozoa (*Giardia* and *Cryptosporidium*) (Leclerc, Schwartzbrod & Dei-Cas 2002). Microbial contamination of the Plankenbrug River has been investigated by two research groups. Paulse and co-workers (2009) observed that, between June 2004 and June 2005, *E. coli* levels exceeded the maximum allowable level as stipulated by DWAF (RSA DWAF 1996a; 1996b). Barnes and Taylor (2004) also observed *E. coli* levels exceeding allowable levels over a four year period (1998 to 2002). A range of other potential pathogens were also isolated from the river, including the bacteria ß-haemolytic *Streptococcus* Group A, *P. aeruginosa, K. pneumoniae, Proteus mirabilis,* and *Enterobacter* spp., and viruses including enteroviruses, adenoviruses, human rotaviruses and human astroviruses (Barnes & Taylor 2004).

The findings above demonstrate the importance of monitoring potential hazards, and warrants further discussion. Stellenbosch Municipality is in partnership with the Adopt-a-River organisation. The goal of this partnership is, through contributions from local organisations and companies, to clean and maintain the Eerste River. Future planning of the municipality includes the roll-out of the Adopt-a-River programme to all rivers within the Stellenbosch municipal area; the development of an Adopt-a-River website link on the municipal and Department of Water Affairs websites; the implementation of seasonal alien clearing projects for identified rivers; and the implementation of seasonal river maintenance and clean-up (Stellenbosch Municipality 2012).

Conclusion

Stellenbosch Municipality has made considerable progress during the past few years in terms of the provision of safe water to both formal and informal communities. Nevertheless, not all the areas serviced by the municipality have Blue Drop status (for example, Kayamandi, Simondium and Raithby), which indicates the need for rigorous risk assessment based on proper monitoring in these areas. One objective of Stellenbosch Municipality is to implement projects for retaining and improving Blue Drop status (Stellenbosch Municipality 2012).

The risk of pollution of rivers (for example, the Plankenbrug River) and water sources that serve Stellenbosch communities and local agriculture must be addressed. If pollution is not controlled, the hazards and risks will increase in the future and may even make it impossible to provide a sustainable and safe water supply.

Although climate change is predicted to have little impact on rainfall in the region, it is expected that climate variability and the frequency of extreme events will increase with increased temperature and wind, which, in turn, could have an impact on water sources, especially surface waters. Future water systems therefore need to be more robust, and the use of alternative water resources and storage systems may need to be investigated. Stellenbosch Municipality acknowledges the need for the implementation of comprehensive Water Demand Management interventions to keep future water demand as low as possible. These interventions include frequent monitoring of the water supply system; regular and adequate system maintenance and repairs; and increased water efficiency. However, to date such interventions have been limited due to budget constraints and a lack of human resources (Stellenbosch Municipality 2012).

Lastly, the backlog in the repair and maintenance of infrastructure in the existing water network could have a severe impact on water supply and quality in the future. The objectives and milestones of Stellenbosch Municipality to ensure a reliable water supply include the completion of Phase 1 of the water infrastructure

---

2   The authors suggest that the *in vivo* and *in vitro* tests used in this investigation should be considered as a first tier of estrogenicity screening, and not applied for monitoring river samples (Swart *et al.* 2011).

replacement programme; the completion of the Paradyskloof Water Purification Works upgrade; the completion of Franschhoek Reservoir planning; the planning and design of various reservoirs; the establishment of an annual pipe replacement programme; and the implementation of medium- to long-term projects as part of the Water Services Master Plan and Water Sources Study (Stellenbosch Municipality 2012).

REFERENCES

Abong'o BO & Momba MN. 2008. Prevalence and potential link between *E. coli* O157:H7 isolated from drinking water, meat and vegetables and stools of diarrhoeic confirmed and non-confirmed HIV/AIDS patients in the Amathole District – South Africa. *Journal of Applied Microbiology*, 105(2):424-431.

Barnes JM. 2003. The impact of water pollution from formal and informal urban developments along the Plankenbrug River on water quality and health risk. PhD thesis. Stellenbosch: Stellenbosch University.

Barnes JM & Taylor MB. 2004. *Health risk assessment in connection with the use of microbiological contaminated source waters for irrigation*. Report prepared for the South African Water Research Commission. Report No 1226/1/04. [Accessed 4 June 2012] http://www.wrc.org.za/Knowledge%20Hub%20Documents/Research%20Reports/1226-1-04.pdf

Baronti C, Curini R, D'Ascenzo G, Corcia A, Gentilli A & Samperi R. 2000. Monitoring natural and synthetic estrogens at activated sludge sewage treatment plants and in receiving river water. *Environmental Science and Technology*, 34(24):5059-5066.

Belfroid AC, Van Der Horst A, Vethaak AD, Schafer AJ, Ris GBJ & Wegener J. 1999. Analysis and occurrence of estrogenic hormones and their glucuronides in surface water and waste water in the Netherlands. *Science Total Environment*, 225(1):101-108.

Cloete TE, Rose J, Nel LH & Ford T. 2004. *Microbial Waterborne Pathogens*. London: IWA Publishing.

CNdV Africa. 2010. Stellenbosch Municipal Spatial Development Plan – Strategies Report. Stellenbosch: Stellenbosch Municipality.

Coombes P. 2006. *Guidance on the use of Rainwater Harvesting Systems*. Callaghan, Australia: University of Australia. [Accessed 24 October 2012] http://waterplex.com.au/resources/

Craun GF, Hubbs SA, Frost F, Calderon RL & Via SH. 1998. Waterborne outbreaks of Cryptosporidiosis. *Journal of American Water Works Association*, 90(9):81-91.

Desbrow C, Routledge EJ, Brighty GC, Sumpter JP & Waldock M. 1998. Identification of estrogenic chemicals in STW effluent: 1. Chemical fractionation and in vitro biological screening. *Environmental Science and Technology*, 32(11):1549-1558.

eWISA. 2012. *Blue Drop Certification Programme Introduction*. [Accessed 4 June 2012] http://www.ewisa.co.za/misc/BLUE_GREENDROPREPORT/blue_drop_certification.htm

Florea AM & Busselberg D. 2006. Occurrence, use and potential toxic effects of metals and metal compounds. *Biometals*, 19(4):419-427.

Gale P. 2001. Microbiological Risk Assessment. In: S Pollard & J Guy (eds). *Risk Assessment for Environmental Professionals*. Lavenham: Lavenham Press.

Haas CN. 1983. Estimation of risk due to low doses of microorganisms: A comparison of alternative methodologies. *American Journal of Epidemiology*, 118(4):573-582.

Haas CN, Rose JB, Gerba C & Regli S. 1993. Risk assessment of virus in drinking water. *Risk Analysis*, 13(5):545-552.

Hobbs RJ. 2004. The Working for Water Programme in South Africa: The science behind the success. *Diversity and Distributions*, 10(5):501-503.

Hrudey SE & Hrudey EJ. 2004. *Safe Drinking Water – Lessons from Recent Outbreaks in Affluent Nations*. London: IWA Publishing.

Jackson VA, Paulse AN, Odendaal JP & Khan W. 2009. Investigation into metal contamination of the Plankenbrug and Diep Rivers, Western Cape, South Africa. *Water SA*, 35(5):289-299.

Kirmeyer GJ, Friedman M, Martel K, Howie D, LeChevallier M, Abbaszadegan M, Karim M, Funk J & Harbour J. 2001. *Pathogen intrusion into the distribution system*. Report No 90835. Denver, CO: AwwaRF (American Water Works Research Foundation) & AWWA (American Water Works Association).

Leclerc H, Schwartzbrod L & Dei-Cas E. 2002. Microbial agents associated with waterborne diseases. *Critical Reviews in Microbiology*, 28(4):371-409.

O'Connor DR. 2002. The Events of May 2000 and Related Issues. In: The Walkerton Commission of Inquiry. *Report of the Walkerton Commission of Inquiry, Part 1*. Toronto: The Walkerton Commission of Inquiry. pp. 502-504.

Paulse AN, Jackson VA & Khan W. 2009. Comparison of microbial contamination at various sites along the Plankenbrug- and Diep Rivers, Western Cape, South Africa. *Water SA*, 35(4):469-478.

Purdom CE, Hardiman PA, Bye VVJ, Eno NC, Tyler CR & Sumpter JP. 1994. Estrogenic effects of effluents from sewage treatment works. *Chemistry and Ecology*, 8:275-285.

Rand Water. 2012. *Water purification*. [Accessed 4 June 2012] http://www.randwater.co.za/WaterAndInfastructureManagement/Pages/WaterPurification.aspx

Robbins NS, Chilton PJ & Cobbing JE. 2007. Adapting existing experience with aquifer vulnerability and groundwater protection for Africa. *Journal of African Earth Science*, 47:30-38.

RSA DWA (Republic of South Africa. Department of Water Affairs). 2012. *Blue Drop Status 2012: Chapter 11 – Western Cape*. [Accessed 4 June 2012] http://www.dwaf.gov.za/Documents/RSP.aspx

RSA DWAF (Republic of South Africa. Department of Water Affairs and Forestry). 1996a. *South African Water Quality Guidelines. Volume 1, Domestic use*. 2nd Edition. Pretoria: Government Publishers.

RSA DWAF. 1996b. *South African Water Quality Guidelines. Volume 4, Agricultural Water Use*. 2nd Edition. Pretoria: Government Publishers.

SITT (Stellenbosch Infrastructure Task Team). 2012. Stellenbosch Municipality Services and Finance: Palmer Development Group Model. PowerPoint presentation. Stellenbosch: 2 May.

Snyder SA & Benotti MJ. 2010. Endocrine disruptors and pharmaceuticals: implications for water sustainability. *Water Science and Technology*, 61(1):145-154.

Stellenbosch Municipality. 2008. *Integrated Development Plan for the 2007-2011 term*. 2nd Generation, Revision 1: July. Stellenbosch: Stellenbosch Municipality.

Stellenbosch Municipality. 2009. *Integrated Development Plan for the 2007-2011 term*. 2nd Generation, Revision 2: May. Stellenbosch: Stellenbosch Municipality.

Stellenbosch Municipality. 2010. *Integrated Development Plan for the 2007-2011 term*. 2nd Generation, Revision 3: April. Stellenbosch: Stellenbosch Municipality.

Stellenbosch Municipality. 2011. *Stellenbosch Municipality Water Services Development Plan for 2011/2012 – Executive Summary*. Stellenbosch: Stellenbosch Municipality.

Stellenbosch Municipality. 2012a. *Integrated Development Plan 2-12-2016*. 3rd Generation: April. Stellenbosch: Stellenbosch Municipality.

Stellenbosch Municipality. 2012b. *Draft Tariff Policy – Appendix 3 2012-2013*. Stellenbosch: Stellenbosch Municipality.

Swart JC, Pool EJ & Van Wyk JH. 2011. The implementation of a battery of in vivo and in vitro bioassays to assess river water for estrogenic endocrine disrupting chemicals. *Ecotoxicology and Environmental Safety*, 74(1):138-143.

Ternes TA, Stumpf M, Mueller J, Haberer K, Wilken RD & Servos M. 1999. Behaviour and occurrence of estrogens in municipal sewage treatment plants – Investigations in Germany, Canada and Brazil. *The Science of the Total Environment*, 225:81-90.

Turpie JK, Marias C & Blignaut JN. 2008. The working for water programme: Evolution of a payments for ecosystem services mechanism that addresses both poverty and ecosystem service delivery in South Africa. *Ecological Economics*, 65:788-798.

Woodworth P. 2006. Working for water in South Africa – Saving the world on a single budget? *World Policy Journal*, 23:31-43.

Chapter 9

SANITATION

Alternative solutions for sustainable waste water management

BEN SEBITOSI

Introduction

This chapter looks at the sanitation systems in Stellenbosch and is developed on the premise that the provision of water and sanitation is a basic and fundamental human right. In this context, sanitation will refer to the removal of human excreta and dirty water.

A brief background of the traditional central sewage extraction and treatment system is given and its advantages, as well as its shortcomings, are explained before moving on to a description of the current performance of Stellenbosch's sanitation system. A number of suggestions are made for a way forward, including the concept of cluster-distributed systems, which seem to be gaining currency around the world, with a case study of an alternative cluster system that has been implemented successfully at the Lynedoch EcoVillage outside Stellenbosch. Lessons are also extracted from the energy sector, as well as from organisational strategies for sanitation in informal settlements in other developing country contexts. Finally, it is suggested that a sustainable future solution for Stellenbosch will not be provided by a single approach. Rather, multidisciplinary teams working collaboratively across institutional boundaries must explore a combination of approaches.

The central waste water management system: A background

The mainstream model for waste water management in use in South Africa is the centrally operated extraction, treatment and discharge system. This design emerged at the dawn of the industrial revolution in the nineteenth century to address a surge in urban populations. The model has served the world well for the past 150-odd years. Back then the utility had a single mission, namely to provide a service to the consumer. Hardly any ecological or economic constraints were considered. Modern imperatives such as environmental sustainability, economic efficiency and social responsiveness mean that the model has reached its limits, and alternatives must now be adopted for a sustainable future (Perkins 1989; Jenkins 1999; Del Porto & Steinfeld 2000; Van der Ryn & Berry 1978; Van der Ryn 2008).

From the outset, waste water treatment plants were designed to discharge to surface waters. The implicit assumption was that the environment had unlimited capacity to absorb waste. The purpose of treatment was to ensure that the body of water into which the sewage would be discharged could cope with it. Any treatment beyond that, it was argued, would be a big waste of money. It has in fact been suggested that this approach has over time promoted the adoption of mediocre waste water treatment technologies in preference to more effective ones (Perkins 1989).

In reality, receiving water bodies cannot sustain the effluent loads indefinitely. For example, the United States Environmental Protection Agency (EPA) has raised concerns about the amount of nitrates, which may enter groundwater and are implicated in causing certain birth defects and other disorders (Republic of South Africa Department of Water Affairs and Forestry (RSA DWAF) 2001). Certain nutrients may also promote the growth of algae, which reduce oxygen levels in water and may result in fish deaths. Consequently, communities with centralised sewers discharging to surface waters have been identified as sources of numerous pollution problems.

From a physical infrastructure perspective, the removal process was aimed at extracting waste as fast as possible. As in the case of any other utility traffic, the discharge volumes varied during the course of the day, with high and low consumer activity periods. Therefore, in order to ensure that system capacities were not exceeded during peak periods, sewage conveyance systems were traditionally designed on the basis of their peak flow capacity. This meant constructing bigger and consequently more costly tunnels than the required average. This cost factor continues to escalate with the progress of system upgrades and bigger pipes occasioned by rising demand. Moreover, as urban areas expanded, the distance between use and treatment increased exponentially. Consequently, the cost of an actual treatment plant is now much less than the cost of the piping required to move the material – in fact, Newman estimated that up to 85% of system costs are incurred from piping (Newman & Mouritz 1996). Thus, rather than being a facilitator, network piping has become the biggest obstacle to the provision of sewage services. This, more than any other reason, underpins the argument in favour of decentralisation.

The South African Department of Water Affairs (RSA DWA) has adopted the Green Drop Certification Programme as a quality standard for waste water management across the country. The initiative, which was launched in 2008, is aimed at ensuring the sanitation compliance of waste water treatment works and, in particular, at minimising their impact on the water bodies into which they discharge. The first Green Drop report (*Green Drop Report 2009: South African Waste Water Quality Management Performance*) was published in 2010 (RSA DWA 2010). Among other findings, it reported that "… [a mere] 7.4% of all waste water systems [in South Africa] can be classified as excellently managed, but the reality remains that various levels of improvement are required in about 55% of the systems assessed". The other prevalent waste water management problem in South Africa is the undercapacity of treatment plants, with DWA reporting a large number of waste water treatment works (WWTWs) receiving volumes of up to double their installed capacities.

The Minister of Human Settlements, Tokyo Sexwale, has said that a lack of sewage systems is hampering the progress of national government housing projects for the poor (South African Press Association 2010). Overseas researchers have reported project delays of up to five years while developers wait for municipalities to complete sewer connections before constructing houses. In addition, developers are often required to pay substantial sewer tap connection fees and fees to cover the cost of the extension of sewer lines, which they subsequently pass along to the new residents. In fact, the Spier Hotel (on the Spier wine farm outside Stellenbosch) opted to construct a private waste water treatment system after the bulk connection fee required by Stellenbosch Municipality was considered to be too high.

Issues of backlogs in the provision of sanitation are intricately linked to housing backlogs. Wallace and Hallahan (2005) cite the City of Cape Town as an example of this situation. According to an interview with a City official (Armitage, Winter, Spiegel & Kruger 2008), there was a housing backlog in the order of 300 000 units at that time, including some 150 000 in an estimated 220 informal settlements. In other words, nearly half a million people – out of an estimated total population of 3.27 million in 2007 – lived in shacks. It has been observed that while local authorities are generally able to provide potable water from the municipal network to communal taps scattered around settlements, there is usually no corresponding sanitation service, including the removal of grey and storm water. Communal sanitation may be provided (generally on the periphery of the site), but is frequently dysfunctional for a variety of reasons.

South African housing development regulations clearly specify water and sanitation as mandatory requirements. Lack of sanitation could, in that context, be seen as a symptom and its causes enumerated in terms of what ails the delivery of housing. However, informal settlements are a reality; and given the absolute necessity of sanitary services in the present rather than the future, the consideration of formal housing alone would clearly misrepresent the gravity of the situation. Moreover, problems with sanitation are not confined to poor areas alone. The only difference is that the problems related to middle-class settlements are externalised to the environment and remain hidden away as overloaded treatment plants are forced to discharge raw sewage. Undoubtedly, protracted underinvestment in the sector has played a major part in arriving at the present state as described in the *Green Drop Report*. Yet, even then, one might still legitimately ask whether underinvestment itself is a cause, or just another symptom of a larger cause.

Problems ailing the provision of sanitary services seem to be a complex web of socio-economic and political factors. According to Wallace and Hallahan (2005), one of the issues frequently raised by municipal officials during interviews was the lack of adequate funding. Municipalities seem to perennially struggle to collect enough revenue to provide the services they are mandated to deliver. Matters are made worse by the fact that certain income groups in informal settlements are exempt from any levies and that the national and provincial government subsidies intended to compensate for this loss of income appear to fall short of the required amounts – perhaps partly because a large number of people not exempt from levies seem to find loopholes to default on payment. Consequently, municipalities may feel more inclined (or obliged) to service the middle class, which appears to generate virtually all municipal revenues. However, given a past history of state-sanctioned service denial, such action has predictable consequences.

In the context of Stellenbosch, the generally sad state of sanitation described above, characterised by poor communication between communities and authorities and dilapidated infrastructure, was dealt a further blow by the recent history of poor cooperation in the municipal council as a result of a hung legislature shared by bitterly opposing political parties.

The current performance of Stellenbosch's sanitation system

Geographically, the district of Stellenbosch comprises numerous rivers, low-permeability soils, high water tables and high gradients. Under these circumstances water-borne sewage is traditionally the preferred technology. The shallow water table is particularly susceptible to contamination from underground disposal options such as pit latrines. Sanitation in Stellenbosch is provided through a network of underground piping that collects waste water from individual user premises and conveys it to a central treatment plant. The high gradients exploit gravity as a renewable energy resource to operate the sewer lines. Treated waste water is subsequently discharged into a convenient water body.

Table 9.1, adapted from the 2010 *Green Drop Report* (RSA DWA 2010), shows the respective design and actual flow capacities at waste water treatment stations in Stellenbosch. It is reasonable to assume that actual plant capacity is always less than what was originally installed due to ageing components and other factors. Therefore, one may infer that the daily capacity of 17 million litres that flows into the main Stellenbosch treatment plant (most likely) exceeds the existing capacity and hence the verdict of the 2010 *Green Drop Report*.

Table 9.1   The state of WWTWs in Stellenbosch [Adapted from RSA DWA 2010]

Waste water treatment plant	Design plant capacity in megalitres per day (Ml/d)	Average inflow in megalitres per day (Ml/d)	Surplus (+) or deficit (-) capacity (Ml/d)	Average flow as percentage of design capacity
Raithby	0.66	0.66	0.00	100
Stellenbosch	20.40	17.50	+2.90	86
Franschhoek	2.52	1.38	+1.14	55
Pniel	2.50	1.00	+1.50	40
La Motte	0.40	0.11	+0.29	28
Klapmuts	2.50	0.60	+1.90	24
Wemmershoek	0.63	0.11	+0.52	17

The *Green Drop Report* awarded Stellenbosch Municipality a poor average Green Drop score of 53%:

> *In terms of the overall Green Drop Assessment, the municipality is performing less than satisfactory. Generic improvement areas for all the works are compliance in terms of:*
> - *Registration and classification of the works, as well as the operating staff;*
> - *Waste Water Quality (WWQ);*
> - *Monitoring Programme Efficiency;*
> - *Regular submission of WWQ information to DWA;*
> - *Management planning relating to the WWTW capacity.*
>
> [RSA DWA 2010]

Many of these issues were echoed for most municipalities across the country, with expert witnesses citing a number of generic reasons for this state of affairs, including skills shortages, ageing and overloaded infrastructure, and budget constraints.

The above analysis is, however, limited in scope, as it focuses largely on traditional mainstream society, which constitutes only part of the population. In fact, over 60% of the residents of Greater Stellenbosch live in townships, informal settlements and low-income housing developments. Due to South Africa's unique history, past national policies and regulations discouraged or prohibited local authorities from providing adequate services to settlements that were considered temporary. This situation formally no longer applies, but in practice there are well-documented, massive backlogs (Goodlad 1996; Bond & Tait 1997; Pillay, Tomlinson & Du Toit 2007; Roelf 2010). In addition, there is the spread of informal settlements like Enkanini (near the Stellenbosch industrial area), which are deemed 'illegal', which means that the municipality is prohibited from providing services in that area.

Kayamandi is a township in Greater Stellenbosch with an estimated population of 33 000 (The Legacy Community Development Corporation 2010). One-fifth (20%) of its residents live in brick houses with in-house water connections and flush toilets (RSA DWAF (Republic of South Africa. Department of Water Affairs and Forestry)[1] 2001; Cousins 2004; Stellenbosch Municipality 2007; Cisneros & Blodig 2009); the remainder live in informal dwellings (shacks), with the population density often rising to as high as sixty people per hectare. These shack dwellers are serviced by sets of so-called 'bus toilets'. Each set comprises of a row of toilets, a water tap and a shower. Up to ten families share a toilet and quite often the ablutions are far from individual households. As a result, people resort to more convenient, if inappropriate, alternatives to dispose of faecal matter, such as surrounding fields, the storm water system and miscellaneous containers. Consequently, raw pollutants flow into the storm water system and thence into the rivers that flow through the town and surrounding area. The neighbourhood river (the Plankenbrug River) is severely contaminated with *Escherichia coli (E. coli)*,[2] with counts of up to 13 million per 100 ml recorded – in fact, Cousins (2004) mentions that the municipality has previously had to buy out a downstream farmer who was no longer able to irrigate his land with water from this river due to the high level of contamination. But it is not only downstream users that are affected. Though most middle-class households are generally oblivious to these waste water management issues (as waste extraction from their premises is almost invariably faultless), they have significant impacts on the communities living in the informal settlements.

From the above, it is apparent that urgent sanitation solutions are required for Stellenbosch residents from all strata of society. Traditionally, a good utility was one that expanded infrastructure capacity to meet projected demand. The Stellenbosch Infrastructure Task Team (SITT) has calculated that capital expenditure of R100-110 million per year will be required to upgrade Stellenbosch's sanitation system (SITT 2012). In recent years, the most notable capacity-constrained utility in South Africa has been the electricity provider, Eskom. Expert diagnosis of the widespread blackout problems that broke out in 2008 was unanimous: if Eskom had built enough plants there would have been no power cuts. Implicit in this verdict is the suggestion that the consumer carries no responsibility whatsoever. On the contrary, it is a well-established fact that wasteful consumer behaviour results in unnecessary and avoidable capacity demand. Consumer behaviour change is also the most cost-effective measure to implement.

Solutions for infrastructure development in the developing world need not follow the same path as that traced by the developed world since the advent of the Industrial Revolution. Instead, relevant technical solutions for advanced applications in the developed world may be used to leapfrog intermediate technologies and applied directly to the benefit of developing countries. The cell phone, for example, which bypassed the need for cumbersome, time-consuming and very expensive landline infrastructure, has over the past decade made impressive inroads and transformed communication in sub-Saharan Africa.

---

1   In 2009, the forestry responsibility was transferred to the newly formed Department of Agriculture, Forestry and Fisheries and the name of the department changed to the Department of Water Affairs (DWA).

2   *E. coli* is a type of bacteria used as an indicator to determine the extent of contamination of a water body from human waste. It does not, however, mean that this is the only pollutant present.

Alternative waste water management systems[3]

The challenges faced by traditional waste water utilities is neatly summarised by Niemczynowicz:

> The traditional approach to water-related problems must change drastically; waste water treatment technologies applied at present need to be complemented, and eventually replaced, by novel, economically efficient and environmentally sound technologies.
>
> [Niemczynowicz 1992:134]

Decentralised systems offer an alternative to the conventional centralised one. They are easy to install, maintain and upgrade, and typically comprise modular, interconnected and easily replaceable parts. In addition, they are easily scalable, take up considerably less space than centralised treatment options, and require less operation, which means that they are also less costly to maintain. Another added benefit is that the treated waste water may be reused for certain purposes, which represents further cost savings. In places such as Stellenbosch, where the centralised system is already stressed, partial decentralisation may alleviate the pressure on overloaded treatment plants, provide communities with water for reuse, and, depending on the technology used, generate methane gas as an alternative energy source. Current decentralised systems include septic tanks, aerated waste water treatment systems, composting systems and biolytic filtration systems.

Septic tanks

Septic tanks receive waste water directly from user premises. They are often designed with several chambers, from where there they may discharge to a secondary treatment stage or final disposal system. The primary function of the receiving chamber(s) is to prevent suspended solids from entering the disposal system, as the uncontrolled discharge of solids may clog the disposal system and lead to failure.

Basically, heavier solid particles settle to the bottom of the receiving tank, where anaerobic bacteria (the kind that live without oxygen) begin to act on them, thereby reducing the build-up of sludge in the tank, while the lighter materials float to the surface and proceed to the outlet stages. The process also results in the formation of biogas, a fuel gas which comprises mainly methane (approximately 60%) and carbon dioxide (approximately 40%) that may be used for cooking and other heating and lighting applications. However, this is often ignored and the gas is allowed to escape into the atmosphere.[4]

A major advantage of septic tanks is that they require little or no regular maintenance. No external power is required either, and they mix black and grey water. The main disadvantage is that they become dysfunctional as densities rise. This fact may, however, be exploited positively as a means of devolving some responsibility to communities and mitigating the capacity shortfalls at the central treatment plant: septic tanks not only force the costs of misuse onto users; they also enable the use of small-bore pipes to treatment works, as in the case of the Lynedoch EcoVillage.

Final disposal systems come in a variety of designs. The simplest is a soakage trench or bed system, where waste water is fed through perforated pipes laid in gravel-filled trenches; the waste water leaks through the pipe holes and seeps into the surrounding soil.

The above design may be augmented with an additional evapotranspiration seepage (ETS) system. Here, the disposal area is planted with selected plants that thrive in soggy environments. Apart from absorbing the liquid, the plants also extract the mineral nutrients from the waste water, thus rendering the remaining water reusable for certain applications. The removal of nutrients may, however, be a disadvantage for crop irrigation.

---

3   It is not possible to provide an exhaustive discussion of all the possible alternatives within the scope of this chapter. Other options that are gaining ground but are not discussed here include integrated, zero discharge and waste water reuse strategies, as well as the development and dissemination of viable alternatives for urban waste water reuse.

4   One tonne of methane gas has the equivalent greenhouse impact of twenty-one tonnes of carbon dioxide. This underscores the multi-pronged benefit of using the gas for cooking.

Aerated waste water treatment systems

The discharge from the septic tank may also be sent to an Aerated Waste Water Treatment System (AWTS) for further treatment. A number of small on-site treatment systems may be categorised as an AWTS and based on the following principles: Waste water from the outlet chamber of a septic tank (containing the small solids) is pumped into a new chamber, where aerobic bacteria (bacteria that require oxygen to live) break the solids down further (in some systems, a mechanical pump supplies the oxygen required by the bacteria). Then the waste water is allowed to settle again prior to being pumped to a final disposal system. One shortcoming with this design is that the pumps in an AWTS require a continuous power supply. Regular maintenance is also required to keep the system running properly.

Composting toilets

Composting toilet systems work by composting the solid waste on site and do not use conventional flush toilets. The resulting composted product may be used as a fertiliser, but should not be used on vegetable gardens. Two notable shortcomings of composting toilets are that grey water has to be treated separately and that solids are continually added, which makes it difficult to separate and remove the old, composted solids from the new, uncomposted material.

Biolytic systems

Biolytic systems are another technology that is becoming increasingly popular. Biolytic filtration treatment systems receive waste water directly from user toilets. They comprise a chamber where waste water is passed over a bed of continually accumulating organic matter. The waste water filters down through the bed, leaving the bulk of the solid material behind. Bacteria and earthworms break down this organic material. The activity of the earthworms also assists in aerating the organic waste, thus helping to suppress offensive odours. The residual waste water is discharged into a disposal system. Two major advantages are that these systems require no power and what little regular maintenance is required may be easily performed by the homeowner. The output water is also of relatively good quality, although pathogen removal will not have taken place.

Biolytic filters are specifically designed to receive both black and grey water. Alternatively, the filter size may be reduced to treat grey water separately. In many instances, biolytic filters are used exclusively to treat grey water in order to reduce the total quantum of waste water deposited into the public sewer. One shortcoming is that the system does not handle fatty waste water properly. Consumers must also be careful not to kill the worms with high concentrations of toxic chemicals contained in common cleansers and bleach. However, the biggest problem is flooding due to outlet blockages, which may kill the worms.

Ecosan systems

Another evolutionary concept in sanitation has been promoted by the Ecosan (short for 'ecologically-sound sanitation') movement (Eco-Solutions 2012). Basically, the toilet pan is designed to handle solids separately from urine: the solids fall straight into the composting container, while the urine is drained off to a urine tank or into the grey water system. The organic solids are decomposed with the help of fresh air, natural heat, micro-organisms and earthworms.

It is claimed that separating the urine from the solids increases the efficiency of the decomposition process. Thus, even in the case of centrally treated waste water the process would be far easier and consequently less energy-intensive in the absence of the urine component. Urine could be collected from households and delivered directly to fertiliser manufacturing plants, which would in turn eliminate substantial sections of the traditional waste water treatment plant (urine accounts for around 70% of the phosphorus and around 90% of the nitrogen in standard black waste water from household flush toilets).

Sweden is one country that has a long history of urine separation toilets dating back to the 1970s (EcoEthic 2012). The initial focus was on holiday houses, presumably because they were located far from the standard urban sewer lines and municipal water mains. However, serious production of these toilets only picked up in the 1990s.

Other options

There is also the possibility of encouraging private sector participation. For example, a local company called Water Rhapsody has installed water-restricting systems in all the toilet cisterns at the University of Cape Town. This not only translates into savings of fresh water, but also reduces the volume of waste water destined for treatment plants.

Another area worth examining is waste management in seagoing vessels. The strategy used in this area is to target specific elements in the waste. For example, marine regulations emphasise minimising the entry of carbon into the seas in order to support the visible components and eliminate pathogens. This is usually achieved using an aerobic biological procedure. Fat and flotation separators are used to separate oils, fats and food particles from the rest of the waste water. The biologically cleaned waste water is separated from the activated sludge either in a settling chamber or by membrane filtration. The combination of aerobic biological waste water treatment and membrane filtration is called membrane bioreactor (MBR).

Other options gaining ground include integrated, zero discharge, and the development and dissemination of viable alternatives for urban waste water reuse.

A case study of a cluster waste water management system – Lynedoch EcoVillage, Stellenbosch

The use of clustered decentralised waste water management is an emerging trend, as mentioned earlier. This model is bigger than an individual household septic system and smaller than the centralised system. The main attraction of cluster systems is that they combine some of the advantages of centralised systems and individual household ones. For example, with all the aforementioned advantages, the cost of individual household systems still remains high, making them a problematic proposition for low-income groups in South Africa. However, a study has shown that combining a certain number of households on a common community cluster system spreads these costs and minimises an individual household's capital cost (Armitage *et al.* 2008). This, combined with the prohibitive expenditure required for the traditional large piping networks, make clustering the model of choice for a sustainable future.

The waste water management system at the Lynedoch EcoVillage outside Stellenbosch is an example of a cluster. The ecovillage was founded in 1999 to promote (among others) the principles of sustainability and self-containment. Two waste water management technologies are used there, namely a biolytic system (marketed under BIOLYTIX) and a Vertically Integrated Wetland system (VIW).[5]

The main administration building and a guesthouse that can accommodate twenty-five guests are connected to the biolytic system (the operating principles of this technology are described earlier.) Also included on the system are the ablutions for a primary day school with five hundred children. The installed system capacity is 10 000 litres per day, or the equivalent of twenty households with an average of four people each. The effluent from the biolytic filter is used for irrigation purposes, because it has the advantage of nutrient retention. Biolytic technology is a standard technology, but the Lynedoch EcoVillage installation is based on a proprietary license purchased by the Spier Group from the Australian entrepreneur who developed the technology and holds the patents.

The VIW is currently servicing the rest of the complex, which comprises seventeen households, with another twenty-three to be built. The installed capacity of the system is 20 000 litres per day, which can service forty households.

Two to three households are connected to a dedicated septic tank, which retains the heavier solids for subsequent digestion by anaerobic bacteria, as described earlier. These tanks feed to a common disposal system through outlet piping. One notable aspect is the drastic reduction in pipe diameters compared to what they would have been in the absence of these tanks, which results in capital cost savings. The tanks also buffer the outlet discharge pipes from user load fluctuations and thus keep the flow rates at a constant average.

---

5   For more details, see http://www.sustainabilityinstitute.net/about/overview

Another very important feature is the fact that individual households are responsible for keeping their dedicated septic tanks in proper running order. While irresponsible dumping may occur on occasion (for example, children throwing balls into it), the resultant clogging only affects that particular tank and not the whole system. The affected households are also tasked with covering the cost of unclogging their system, so that a culture of responsible consumer awareness becomes the norm. The partially clarified waste water from the septic tanks is fed into the VIW, and the residual water from the VIW, which has a very low nutrient load, is then fed back to the individual houses, where it is reused for flushing toilets (which in itself reduces the normal household water requirements by 40%).

One pilot trial has five households clustered together on a biogas digester instead of a septic tank. The effluent from the biogas digester goes into the wider Lynedoch sanitation system. Although hardly mentioned in the literature, this arrangement has resulted in the successful harvesting of significant quantities of biogas. Generic literature suggests that the volume of biogas that may be generated by a household of four people would only be adequate for the cooking needs of one person (at most). In contrast, the five households currently connected to the biogas digester have been able to generate sufficient gas for nearly all their cooking needs. (This is, however, only possible if grass cuttings and organic kitchen waste are added to the biogas digester.)

Lessons from the energy sector

The field of electricity power infrastructure design is currently undergoing major rethinking driven by the 'smart grid' concept. The vision of the smart grid is underpinned by three key elements. The first is enabling active consumer participation in the management of their power requirements. In essence, it recognises and exploits the consumer as an energy resource. Consumers are empowered to understand their capacity to responsibly engage in the system. In the process, they enable the utility to recover supply capacity through conservation, while they enjoy improved service and savings. The second element of the smart grid is the adoption of alternative power generation technologies, including energy efficiency and demand-side management (EEDSM) and renewable and distributed power generation. This principle is based on the recognition that infrastructure capital and maintenance costs are major constraints to sustainable operations and thus present opportunities for reducing costs and improving performance. Eskom perennially reports power losses of some 8 to 10% (or the equivalent of a 3000 MW plant) during transmission. Therefore, cutting down on long and costly networks could lead to significant savings of this scarce resource. The third smart grid component is the implementation of a fully automated, remotely configurable, and self-correcting power distribution network allowing grid-wide demand or load management, such as rerouting demand from congested network branches to lightly loaded ones in real time. It also entails real-time utility-to-consumer feedback to inform consumers about the state of supply, which enables them to make informed decisions about their power usage. In this way, successful exploitation of consumer capacity is not limited to a passive reduction in consumption, but allows and enables consumers to actively feed power back into the grid and even obtain credit for it.

There are striking similarities between the challenges faced by administrators of power and sanitation utilities. For example, as in the case of power transmission, increased demand for sanitation services requires increased network capacity, with its attendant capital and maintenance costs. Moreover, the waste water from both these utilities constitutes a major environmental hazard. And, just as with electricity networks, sewage networks not only incur energy losses, but studies have shown that 10% of sewage leaks from piping *en route* to treatment plants, thus constituting a major source of pollution that has hitherto gone unnoticed (Stellenbosch Municipality 2007). At the household level, researchers also found that flush toilets may account for 20 to 40% of domestic water usage (Rose 1999), which is analogous to inefficient electrical appliances and provides yet another opportunity for conservation.

From the above, it would appear that the experiences of these seemingly disparate systems may be consolidated to their mutual benefit. It would be particularly prudent to recognise the fact that sanitation and power utilities serve a common consumer. Smart grid concepts could be easily adapted and applied to the benefit of water and waste water management. In fact, there appears to be a growing tendency towards more localised, community-scale systems for water and waste water management around the world (Rajbhandari n.d.; Parten & D'Amato 2009; Lim 2010; Slaughter 2010).

What would perhaps be even more novel for sanitation systems is the introduction of information technology concepts, similar to the smart grid concept of electricity provision, to the way sewage flows are conducted through a reconfigured sanitation network. This would be akin to 'smart metering', where consumers are informed about traffic congestion issues in real time and respond accordingly. Essentially, this represents a shift from the traditional top-down utility command-and-control paradigm to one where there is a gradual participation by consumers in the production and delivery of the service.

Fortunately, the South African local government policy advocates community participation through its Integrated Development Planning (IDP) policy; hence, it would appear that only implementation modalities remain to be determined and refined. However, from a South African social perspective, this will take no less than a radical cultural change from a society of dominant utility service providers and a consumer base traditionally expecting service as an entitlement, to one where the utility is willing to cede some responsibility to consumers, and consumers, in turn, are willing to accept it.

Lessons from projects in other developing countries

As a result of rapid migration to urban areas and the consequent proliferation of peri-urban informal settlements, many African and developing countries are plagued by water and sanitation challenges. In response, a number of international aid agencies have initiated projects aimed at devising solutions to this problem. One such project was jointly funded by the Swedish International Development Agency (SIDA), the Danish International Development Agency (DANIDA), and the United Kingdom Department for International Development (DFID), and carried out by the International Institute for Environment and Development (IIED) in Angola, Argentina, Ghana, India and Pakistan in liaison with the local agencies in the respective countries (Osumanu, Abdul-Rahim, Songsore, Braimah & Mulenga 2010).

Low cost was obviously a key consideration in the selection of toilet technologies for the project, so in this regard there are no fundamental lessons to be learned from it. However, what undoubtedly anchored the success of the project were their administrative, socio-political and business models, which offer invaluable lessons for Stellenbosch.

The first point emphasised by the project researchers was the importance of understanding the dynamics of the different country contexts in terms of actors, issues and approaches. They argue that traditional systems of service delivery (akin to the South African model), would seem to treat the poor as objects, rather than drivers, of their own development. In contrast, the proposed project strategy revolved around a self-help approach, aided (but not superseded) by outside technical assistance. This stands in sharp contrast to traditional, overly technical paradigms that rely on prescriptive approaches.

In terms of the new proposition, governments and/or municipal authorities are encouraged to align their investment with community effort, rather than ignore it. In other words, authorities (and donors) should find ways to support and link up with community initiatives. Communities should also be supported by means of (among others) public education about efficient water use, hygiene and HIV/AIDS, and the identification of economic opportunities.

Evidently, this proposition assumes the existence or creation of community associations and hence the subsequent establishment of partnerships between municipal authorities (or governments) and these associations. Fortunately the prevailing environment in Stellenbosch would appear to support such partnerships. Firstly, a high level of community cooperation seems to prevail in informal settlements. For example, it is a well-known fact that in support of a new community entrant, a whole neighbourhood undertakes the construction of his or her shack dwelling, which almost invariably ensures that the task is completed within a day. Secondly, as mentioned earlier, the new Integrated Development Plan policy advocates precisely the type of community-municipal authority liaison that is advocated here.

What might prove somewhat challenging is the practical implementation of such an arrangement. Osumanu and his co-researchers reported community resistance to making any financial contribution, however modest, towards any of the community projects. One of the problems identified as a disincentive for such investment was the lack of any form of land tenure or deed – an all too familiar issue in Stellenbosch and elsewhere in South Africa. For its part, South African consumer culture must strive to evolve from a service-demand mode to a more participatory one. If each individual of the millions of immigrants were looked at as a resource rather than a liability, the speed of service delivery would improve with every new entrant.

Conclusion

This chapter has attempted to shine some light on the state of sanitation in Stellenbosch. The evidence presented, including the DWA's *Green Drop Report*, points to a multitude of problems. A background literature study of the conventional central waste water management system, similar to the one currently deployed in Stellenbosch, shows that this model initially designed in the nineteenth century can no longer meet twenty-first century needs and challenges, including cost-effectiveness, environmental sensitivity and social responsiveness. And in the case of Stellenbosch, these general problems have been exacerbated by years of underinvestment and political instability.

A number of alternative solutions that are currently in use overseas have been described. A brief case study of the Lynedoch EcoVillage outside Stellenbosch points to the potential positive attributes of a community cluster system. In particular, it demonstrates the value of the consumer as a resource that is grossly underutilised in South Africa. Moreover, one household with non-technical occupants discovered that they could exploit their own ingenuity to obtain adequate supplies of biogas by introducing grass and organic kitchen waste as additional feedstock.

There are also innovative concepts from other fields that could be adopted to address the structural modelling of sanitation. In this regard, the adaptation of the smart grid model, where consumer responsibility provides the first line of capacity expansion, was suggested. It should, however, be borne in mind that the subject of human waste management may still be a taboo in much of our society, so that it would require a skilful approach to get consumer buy-in.

It would appear that the fundamental problem lies not so much in the choice of an individual technology, but in the administrative system. Appropriate system approaches must include both structural and non-structural elements, as opposed to a narrow-minded technological focus. This will require multidisciplinary cooperation. The most effective way to deal with waste water might very well be to manage the whole freshwater-to-waste water-and-back cycle as a single process.

The final solution lies in setting up a portfolio of performance indices that must not be compromised, including security of service delivery, economic efficiency, human well-being and environmental sustainability. To begin with, despite the challenges bedevilling the traditional central system, one should recognise the need to retain the economically significant – even if functionally deficient – present infrastructure and view the suggested alternative solutions as a complement to, rather than a replacement for, traditional systems. The question that must be explored soon is how much of which system should be adopted in which context.

The crucial first steps must be the recognition by society that a problem does indeed exist and that it needs to be addressed collectively and differently from a business-as-usual paradigm.

REFERENCES

Armitage NP, Winter K, Spiegel A & Kruger E. 2008. Community-focused greywater management in some selected informal settlements in South Africa. 11th International Conference on Urban Drainage. Edinburgh, Scotland, UK.

Bond P & Tait A. 1997. The failure of housing policy in post-apartheid South Africa. *Urban Forum*, 8(1):19-41.

Cisneros J & Blodig A. 2009. Sustainable Alternative to the Centralized Paradigm. *Bio-Microbics, Inc.* [Accessed 20 May 2012] http://www.biomicrobics.com/images/media/Bio_SLOTonline_Sustainable%20Alternative.pdf

Cousins D. 2004. Community involvement in the provision of basic sanitation services to informal settlements. MA thesis. Belville, Cape Town: Cape Peninsula University of Technology.

Del Porto D & Steinfeld C. 2000. *The Composting Toilet System Book: A Practical Guide to Choosing, Planning and Maintaining Composting Toilet Systems.* Burlington, VT: Chelsea Green Publishing.

EcoEthic. 2012. [Accessed 20 May 2012] http://www.ecoethic.ca/products_wl.html

Eco-Solutions. 2012. [Accessed 20 May 2012] http://www.eco-solutions.org

Goodlad R. 1996. The Housing Challenge in South Africa. *Urban Studies*, 33(9):1629-1646.

Jenkins J. 1999. *The Humanure Handbook – a Guide to Composting Human Manure*. 2nd Edition. Grove City, PA: Jenkins Publishing.

Lim KY. 2010. Evaluation of a community-scale drinking water treatment system. *International Journal for Service Learning in Engineering*, 5(2):17-31.

Newman P & Mouritz M. 1996. Principles and planning opportunities for community-scale systems of water and waste management. *Desalination*, 106(1-3):339-354.

Niemczynowicz J. 1992. Water management and urban development: a call for realistic alternatives for the future. *Impact of Science on Society*, 42(2):133-147.

Osumanu KI, Abdul-Rahim L, Songsore J, Braimah FR & Mulenga M. 2010. *Urban water and sanitation in Ghana: How local action is making a difference*. Human Settlements Working Paper Series: Water and Sanitation. London: International Institute for Environment and Development.

Parten SM & D'Amato V. 2009. *Analysis of Existing Community-Sized Decentralized Wastewater Treatment Systems*. [Accessed 20 May 2012] http://www.ndwrcdp.org/research_project_04-DEC-9.asp

Perkins RJ. 1989. *Onsite Wastewater Disposal: Designing, Constructing and Maintaining Septic Systems*. 1st Edition. Boca Raton, FL: CRC Press.

Pillay U, Tomlinson R & Du Toit J. 2007. *Democracy and Delivery: Urban Policy in South Africa*. Cape Town: HRSC (Human Sciences Research Council) Press.

Rajbhandari K. n.d. *Sunga constructed wetland for wastewater management: A case study in community based water resource management*. Shanta Bhawan: WaterAid Nepal. (See also: http://www.wateraid.org/documents/plugin_documents/sunga_constructed_wetland_for_waste_water_management.pdf)

Roelf W. 2010. South Africa battles huge housing backlog. *Reuters*, 21 April. [Accessed 21 April 2010] http://www.reuters.com/article/2010/04/21/safrica-housing-idAFLDE63K1YA20100421

Rose GD. 1999. *Community-based Technologies for Domestic Wastewater Treatment and Reuse: Options for urban agriculture*. IRDC (International Development Research Centre) Research Program: Cities Feeding People, Report Series 27. [Accessed 16 May 2012] http://www.washdoc.info/docsearch/title/125784

RSA DWA (Republic of South Africa. Department of Water Affairs). 2010. *Green Drop Report 2009: South African Waste Water Quality Management Performance*. [Accessed 20 May 2012] http://www.dwaf.gov.za/Documents/GreenDropReport2009_ver1_web.pdf

RSA DWAF (Republic of South Africa. Department of Water Affairs and Forestry). 2001. *Managing the water quality effects of settlements: economic impacts of pollution in two towns*. Pretoria: Department of Water Affairs and Forestry.

South African Press Association. 2010. Infrastructure problems slowing housing provision – Sexwale. *Engineering News*, 24 November. [Accessed 24 November 2010] http://www.engineeringnews.co.za/article/infrastructure-problems-slowing-housing-provision-sexwale-2010-11-24-1

SITT (Stellenbosch Infrastructure Task Team). 2012. Stellenbosch Municipality Services and Finance: Palmer Development Group Model. PowerPoint presentation, Stellenbosch: 2 May.

Slaughter S. 2010. Improving the Sustainability of Water Treatment Systems: Opportunities for Innovation. *Solutions*, 1(3):42-49.

Stellenbosch Municipality. 2007. *Integrated Development Plan for the municipal area of Stellenbosch*. 2nd Generation, original document. Stellenbosch: Stellenbosch Municipality.

The Legacy Community Development Corporation. 2010. *Kayamandi, Stellenbosch*. [Accessed 20 May 2012] http://www.legacykayamandi.com

Van der Ryn S. 2008. *The Toilet Papers: Recycling Waste and Conserving Water*. White River Junction, VT: Chelsea Green Publishing.

Van der Ryn S & Berry W. 1978. The Toilet Papers: Designs to Recycle Human Waste and Water: Dry Toilets, Greywater Systems and Urban Sewage. 1st Edition. Santa Barbara: Capra Press.

Wallace SD & Hallahan DF. 2005. Onsite Water Treatment: Cost-effectiveness of cluster systems in use today. *The Journal for Decentralised Wastewater Treatment Solutions*. [Accessed 20 May 2012] http://www.foresterpress.com/ow_0509_cost.html

FURTHER READING

Ayaz SÇ. 2008. Post-treatment and reuse of tertiary treated waste water by constructed wetlands. *Desalination*, 226: 249-255.

Beder S. 1989. From Pipe Dreams to Tunnel Vision: Engineering Decision-Making and Sydney's Sewerage System. PhD thesis. Sydney, Australia: University of New South Wales.

Crites R, Brown D & Caldwell. 2006. *Integrated planning for wastewater treatment and recycling for a small community development*. Davis, California: Water Environment Research Foundation.

Dowling TJ. 2007. Sustainable Development in Water and Sanitation – A Case Study of the Water and Sanitation System at the Lynedoch EcoVillage Development. MA thesis. Stellenbosch: Stellenbosch University.

Kennedy Smith TN. 2010. The rise of the phoenix or an Achilles heel: Breaking New Ground's impact on urban sustainability and integration. MA thesis. Stellenbosch: Stellenbosch University.

Morari F & Giardini L. 2009. Municipal wastewater treatment with vertical flow constructed wetlands for irrigation reuse. *Ecological Engineering*, 35:643-653.

Stefanakis AI & Tsihrintzis VA. 2009. Performance of pilot-scale vertical flow constructed wetlands treating simulated municipal wastewater: effect of various design parameters. *Desalination*, 248:753-770.

Xiao W, Bao-ping H, Ying-zheng S & Zong-qiang P. 2009. Advanced wastewater treatment by integrated vertical flow constructed wetland with vetiveria zizanioides in North China. *Procedia Earth and Planetary Science*, 1:1258-1262.

ENERGY

Towards sustainable energy flows for Stellenbosch

Chapter 10

ALAN BRENT, RIAAN MEYER & WIKUS VAN NIEKERK

Introduction

South Africa has the major challenge of closing the gap between its 'first' (developed) and 'second' (developing) economies, while decoupling the growth of the economy as a whole. The concept of decoupling refers to maintaining the growth with declining material throughput, with associated benefits such as improving the carbon emissions balance of the economy. To this end, the transformation of the energy system is a key intervention to facilitate the country's move towards a decarbonised economy (Brent, Hietkamp, Wise & O'Kennedy 2009).

At a national level, the industrial sector presents enormous opportunities for this transformation process, given that this sector accounts for nearly a third of the energy and over fifty per cent of the total electricity consumption of the country (Republic of South Africa. Department of Minerals and Energy (RSA DME)[1] 2009). However, for the local economy of Stellenbosch, which is less industry-intensive, the demand profile is more equally spread across the industrial, residential, transport, and the combined commercial, public services and agriculture sectors. Thus all these sectors will have to be targeted in parallel if the transformation of the energy system is to be realised.

This chapter provides an overview of the energy system of Stellenbosch Municipality and discusses possible technological strategies for transforming it. In closing, it suggests a number of interventions to make Stellenbosch energy neutral.

The current energy system in Stellenbosch

Electricity is the main energy carrier that is controlled internally by the Stellenbosch Municipality and the majority of the population is said to have access to it. However, paraffin use is still prevalent in the residential sector, especially in poor households in informal housing structures. In 2009, it was estimated that the backlog for basic electrical services stood at 400 households. At that time, approximately 25 000 households had access to electricity and 2 000 households had none, while 7 555 received a free basic service provision. The total operating cost was R136 187 694 (Stellenbosch Municipality 2010a).

The industrial, commercial and public services sectors utilise mainly liquid fuels as a backup source of energy, such as diesel generators. The agricultural sector uses a mix of electricity and liquid fuels with seasonal variations. The transportation sector uses primarily liquid fuels, with limited electricity use for public transport (rail), but this flow is not controlled internally by the municipality.

---

1 In 2009, the Department was split into the Department of Mineral Resources (DMR) and the Department of Energy.

Stellenbosch Municipality buys electricity in bulk from Eskom and is responsible for its distribution within the licensed area of supply of the municipality, as determined by the National Energy Regulator of South Africa (see Table 10.1). The municipality receives its electricity via a 66-kiloVolts (kV) transmission line for a maximum demand of 68 megawatt (MW), which is stepped down to a 11-kV distribution network and then a 400/230-V reticulation network. The network is perceived to be relatively stable at municipal level. Stellenbosch Municipality provides electricity services in the towns of Stellenbosch and Franschhoek, while Eskom supplies electricity (directly) to Klapmuts, Vlottenburg, Jamestown, Raithby, Kylemore, Lynedoch, and many of the economic activities within the Stellenbosch municipal area, particularly agriculture-oriented activities.

Table 10.1   Bulk electricity purchases, sales and losses for the 2008/2009 financial year [Stellenbosch Municipality 2010a]

	Quantity (kWh)	Cost (R)
Purchases	372 791 265	96 266 560
Sales	343 710 757	183 215 488
Losses	29 080 508	14 563 995

* kWh = kilowatt hour

Based on the limited available data, the following is deduced in terms of the indicative, current status of the energy system of Stellenbosch Municipality: it is almost entirely dependent on energy carriers, or flows, from outside the municipality. To begin with, the 66-kV capacity limit of the transmission line into the municipality is a constraint. In addition, the stability of the local electricity grid relies on the stability of the national grid, and all the other carriers, such as liquid fuels, are produced outside the municipality. This means that, given the characteristics of the national energy system, both Stellenbosch Municipality and the economic sectors within it are heavily reliant on fossil fuels and are, overall, major contributors to carbon emissions.

The figures in Table 10.1 highlight the fact that electricity sales generate significant income for the municipality. Thus, the municipality is not likely to incentivise the reduction of electricity consumption by its (electricity) customers, unless driven by other issues, such as the stability of the local electricity network or operating close to its power allocation from Eskom, which may constitute an incentive for the municipality to diversify the supply of electricity to its local network.

Energy security and escalating prices of fossil fuels will be the likely drivers of change in the various sectors that comprise the Stellenbosch economy. This means that alternative energy sources will become more attractive in the near future, or already are in some cases (Cowan & Daim 2009).

Strategies and technologies to transform the energy system

The Integrated Development Plan of Stellenbosch Municipality (2010b) stipulates two main strategies to transform the energy system in the residential sector:

- all new housing, whether low- and high-income housing, should have solar hot water heating installed; and
- all non-subsidised housing should install photovoltaic panels, wind generators, or other off-grid generators to meet demands above 300 kWh per month.

Apart from this residential sector focus, there are a number of opportunities for transforming the energy system across all sectors, of which only a few are highlighted here.

Energy efficiency

By 2008, the International Energy Agency (IEA) reported that globally, energy efficiency measures had already saved up to 58% of what would have been consumed if the measures were not put in place (IEA 2008). In terms of the South African context, Winkler and colleagues describe the potential benefits of industrial energy efficiency as follows:

> [T]he benefits of industrial energy efficiency in South Africa include significant reductions in local air pollutants; improved environmental health; creation of additional jobs; reduced electricity demand; and delays in new investments in electricity generation. The co-benefit of reducing carbon emissions could result in a reduction of as much as 5% of South Africa's total projected energy CO_2 emissions by 2020.
>
> [Winkler, Howells & Baumert 2005:20]

Table 10.2 lists a number of energy-saving measures, as well as a breakdown of the potential energy savings represented by each.

Table 10.2   Energy-saving measures and the potential electricity savings they represent [Adapted from Winkler et al. 2005]

Measure	Percentage of total electricity savings
Compressed air management	26%
Variable speed drives	18%
Efficient motors	17%
Efficient lighting	12%
Load shifting*	9%
Heating, ventilation and cooling	6%
Other thermal measures	6%
Refrigeration	4%
Steam systems	1%

* The practice of altering the pattern of energy use so that on-peak energy use is shifted to off-peak periods. Load shifting is a fundamental demand-side management objective.

Solar water heaters

The first five years of the millennium saw a number of investigations into the potential market for solar water heating in South Africa (Holm 2005). These studies all highlighted the significant potential of solar water heating for saving energy, and provided several scenarios for its implementation.

Solar water heaters (SWHs) could effectively provide 80% of the electricity consumed by conventional electrical geysers, electrical kettles, and other water heating methods used for household cleaning and personal hygiene in both high- and low-income homes. By replacing these devices, SWHs would reduce the demand for electricity generated in coal-fired power stations, which in turn would reduce emissions of harmful pollutants into the environment and would mitigate the critical electricity shortage in South Africa by reducing peak demand. On the demand side, SWHs would not only significantly reduce electricity bills, but would also supply readily available hot water in low-income homes where it was previously only available after time-consuming heating processes.[2]

Various types and models of solar water heaters are available on the market, with prices ranging from approximately R12 000[3] for a 150-litre system to around R30 000 for a 300-litre system.[4] Apart from flat plate collectors (see Figure 10.1), evacuated tube collectors are also a popular choice, as they are better suited for locations that have a higher number of overcast days and low ambient temperatures (Meyer & Gariseb 2009). At present, flat plate collectors dominate the South African SWH manufacturing sector, while virtually all evacuated tube collectors are imported (mostly from China).

SWHs are not only practical for the high- and middle-income sectors, but also offer a viable solution for low-income households (Meyer & Gariseb 2009). In April 2008, 184 SWHs were installed in Kwanokuthula,

---

2   As an added bonus, large-scale implementation of SWHs will also create more employment than conventional coal-fired power stations provide at present.

3   Prices are for commercially available units at the beginning of 2012.

4   These prices include the rebate of the Eskom programme. A complete list of approved suppliers can be found on the Eskom website: http://www.eskomidm.co.za/form/find/index.php

a low-income settlement on the outskirts of Riversdale in the Western Cape. The installed units comprised an 80-litre tank and a 1.5 m² flat plate collector panel, at a cost of R7 000 per unit. The systems were installed without any electrical backup, but owners were given the option of adding this element at an additional cost of around R200.

Figure 10.1　　**A:** Flat plate solar water heater at Lynedoch EcoVillage outside Stellenbosch. **B:** Flat plat solar water heaters in Kwanokuthula [Meyer & Gariseb 2009]

Alternative solar thermal applications

Experience of solar thermal solutions in Europe has shown that such applications may supply 3% to 4% of the European industrial sector heat demand (see Table 10.3). Given that South Africa has a better solar resource than Europe, and provided that there are suitable support initiatives in place, the penetration of solar process heat technologies into the industrial sector in Stellenbosch may be expected to yield comparable, if not better, results.

Table 10.3　　Industrial sectors and processes with the greatest potential for solar thermal uses [Adapted from European Solar Thermal Industry Federation (ESTIF) 2006]

Industrial sector	Process	Temperature level (°C)
Food and beverages	Drying	30-90
	Washing	40-80
	Pasteurising	80-110
	Boiling	95-105
	Sterilising	140-150
	Heat treatment	40-60
Textile industry	Washing	40-80
	Bleaching	60-100
	Dyeing	100-160
Chemical industry	Boiling	95-105
	Distilling	110-300
	Various chemical processes	120-180
All sectors	Pre-heating of boiler feed water	30-100
	Heating of production halls	30-80

▣　*Parabolic troughs and linear Fresnel systems*

Parabolic troughs probably represent the greatest possible contribution that could be made by solar industrial process heat systems; firstly by driving double-effect absorption chillers (see Table 10.4) for cellar operations, for example; and secondly, by providing process steam. (Air conditioning systems in commercial buildings may require single-effect absorption chillers with stationary collectors, as the roofs of these buildings may not be ideal for parabolic troughs. However, a small linear Fresnel system, driving a double-effect absorption chiller (see next section), would be ideal.)

Table 10.4 Chilling processes using thermal energy [Adapted from Renewables Academy (RENAC) 2010]

Process	Driving Temperature (°C)	COP	Size range	Sorbent	Comments
Absorption (1 stage)	75-115	0.6-0.75	50 kW-5 MW	Liquid (e.g. Li-Br)	Technically mature
Absorption (2 stage)	130-180	1-1.3	50 kW-5 MW		
Absorption	60-90	0.3-0.7	70 kW-1 MW	Solid (e.g. silica gel)	Limited number of suppliers

* kW = kilowatt

The use of solar collectors to drive large-scale thermal desalination plants (such as multi-effect desalination or multi-stage flash) also offers a solution to the challenge of potable water supply in the region.

Smaller parabolic troughs, with aperture widths ranging from 0.5 m to 2.3 m, can operate at temperatures between 100°C and 250°C. The advantage of these small troughs is that they are relatively lightweight and easy to handle. Some of them may even be installed on roofs.

Due to the concentrating nature of parabolic trough collectors, they are best used in climates with a high share of direct solar radiation. However, in moderate climates, such as in Central Europe, parabolic trough collectors have the same advantage over flat plate or evacuated tube collectors as in sunnier climates, provided that operating temperatures are generally above 130°C. At lower temperatures, and because they do not use diffuse solar radiation, the overall solar energy yield is still smaller than that of an evacuated tube collector. As always, careful system design, which takes into account the correct operating temperatures and load characteristics, and reliable climate data are paramount.

Parabolic troughs can be operated in a pressurised circuit, where the heat transfer medium does not evaporate in the collector field (indirect mode), or in a direct steam generating mode. In the indirect mode, thermal oil or water is the typical heat transfer media. In the direct steam generation mode, water is the best solution in regions where there is no danger of freezing, such as Stellenbosch.

It is also important to note that the manufacturing and installing of collectors for solar process heat for the industrial, commercial, and agricultural sectors is an industry in its own right, and that this industry satisfies the imperatives of labour-intensive employment, climate change mitigation, and energy security.

- *Concentrating solar power (CSP)*

A power cycle may be added to the abovementioned concentrating systems to produce electricity. Other configurations include central receivers or 'power towers', and dish Stirling systems. Economies of scale usually dictate that these installations be large – in excess of 20 MW. However, smaller sizes (in the 1 to 5 MW range) have been demonstrated in the United States, Europe and elsewhere, and may be considered for generating electricity at the municipal level. To this end, combining CSP with another source of energy, such as electricity generation from biogas produced in the treatment of sewage, has been proposed; this is referred to as hybrid systems. For off-grid, distributed power, a system configuration in the kW range is an option that is being demonstrated in Lesotho (Orosz, Mueller, Quolin & Hemond 2009). However, these CSP solutions are still in the development phase and their cost effectiveness remains to be demonstrated in the local context.

- *Photovoltaic panels*

There are several important reasons why photovoltaic (PV) systems are expected to play a highly significant role in South Africa (and Stellenbosch).

- **Ease of use:** PV modules represent one of the easiest ways to implement renewable energy. The power source itself is extremely robust, requires no maintenance (apart from occasional optional cleaning), and is one of the few products in the world that regularly carry a 20-year warranty. It is relatively simple to install many grid-integrated systems on the rooftops of domestic and commercial buildings. PV technology is also extremely well-suited to off-grid applications. For example, Figure 10.2A shows a solar PV panel installation on a farm near Ceres. The installation consists of four PV panels of 180 watt (W) each, with the ability to produce a peak electric power output of 720 W. On average, the installation can generate 3.6 kWh per day.

- **Electricity is generated at point of use:** PV systems can be installed almost anywhere, and can be readily integrated into the grid distribution system. They represent an opportunity for individuals, commercial enterprises, and other parties to take steps at their own premises to contribute directly to a more sustainable energy future.
- **Technologies are rapidly being developed** that allow the integration of PV cells or modules into dual-purpose coverings and fabrics, such as roof tiles and roofing membranes for flat roofs. Figure 10.2B shows a solar roof tile installation at the Drie Geuwels Guest House, located at the Sustainability Institute outside of Stellenbosch (Meyer & Gariseb 2009). The PV roof tiles are of a similar shape and size as normal tiles and cover only part of the total roof area, as may be seen in the picture. The peak output from this installation is 1,7 kW.[5] The electricity generated is fed directly into the guest house through a grid-tied inverter and is used to supplement the electricity currently provided by Eskom. No batteries are installed.
- The solar resource is well distributed and it is relatively easy to estimate how much energy will be produced at a particular location.
- The long-term marginal cost of electricity generated by fossil fuel resources is expected to increase, while PV systems are expected to become more affordable. Solar PV panels currently retail at around R20/W (hence the 180-W panels pictured in Figure 10.2B are priced at around R3 600 each).

Small wind turbines

Wind turbines offer a potential alternative energy solution for all sectors. Small horizontal axis wind turbines (HAWTs) are available off-the-shelf in various sizes, ranging from 500 W to 5 kW and with rotor diameters ranging between 1.5 m and 3 m (Meyer & Gariseb 2009).

Figure 10.2C shows a 1-kW horizontal axis wind turbine. It has a rotor diameter of 1.5 m and is used to power a stand-alone water filter at the Sustainability Institute in the Lynedoch EcoVillage outside Stellenbosch. The typical cost of a 1-kW turbine is around R30 000, including the turbine and controller box. A complete off-grid residential installation, including the turbine and pole for the turbine, batteries, cables, 4-kW inverter, and installation costs around R100 000.

Biogas digesters

A biogas digester manufactures biogas from organic waste materials, such as sewage and cow dung (Meyer & Gariseb 2009). Biogas is generated when bacteria degrades biological material in the absence of oxygen. The process is known as anaerobic digestion. Biogas consists mainly of methane (CH_4) and carbon dioxide (CO_2), which respectively make up two-thirds and one-third of the gas produced. In small-scale biogas digesters, the gas can be used directly for (mainly) cooking. Typically, 1 m³ of biogas will provide two hours of cooking. Two and a half cubic metres of biogas is equivalent to 1 kg of liquefied petroleum (LP) gas.

The Stanford Valley Farm and Conference Centre is located 10 km outside Stanford in the Western Cape (Meyer & Gariseb 2009). A 13-m³ biogas digester was installed there in September 2006. The biogas digester is fed by thirty toilet connections, restaurant food waste, and animal manure. During the process, liquid effluent overflows into the horizontal planted gravel filter, which overflows into an ultraviolet (UV) polishing pound, where it is aerobically treated before being recycled for irrigation. The energy generated by the biogas digester is used for cooking purposes in the restaurant.

At the Lynedoch EcoVillage, such a digester is being used to produce gas for cooking in two houses, each inhabited by three adults. The experience thus far is that, provided that organic kitchen waste and grass cuttings are added to the biogas digester on a weekly basis, the biogas digester meets 80% of the cooking requirements of both houses. Similar but larger systems may be integrated into sewage treatment works, thus offering a combined energy/waste solution at the municipal level.

---

5   The energy generated is monitored daily and may be viewed at http://www.solarworld.co.za

Figure 10.2 **A:** 720 W stand-alone PV installation at Nollie se Kloof in the Ceres district. **B:** Solar roof tile installation at the Sustainability Institute in the Lynedoch EcoVillage outside Stellenbosch. **C:** Small 1-kW wind turbine and water filter at the Sustainability Institute [Meyer & Gariseb 2009]

An energy neutral Stellenbosch?

Economic growth is usually associated with increased energy consumption. However, global trends and projections indicate that energy-efficiency measures can offset the increase in required energy. In other words, it is possible that the Stellenbosch economy can grow without a concomitant increase in energy consumption. The challenge is to decrease the demand for energy, something that cannot be achieved by simply implementing energy-efficiency measures.

In fact, in terms of electricity, Stellenbosch's dependence on the national grid, and thus Eskom, could conceivably be removed altogether by the implementing just two measures:

- Converting from conventional electric equipment: An obvious route is to replace all conventional electric geysers with solar water heaters or more effective heat pumps. Studies conducted at the Sustainability Institute at the Lynedoch EcoVillage outside Stellenbosch have estimated that at least one-third of electricity consumption in the residential and education sectors can be removed in this manner. Based on the global benchmark, this is a conservative estimate for those commercial and industrial sectors where hot water is required. With respect to chilling and cooling, similar savings can be made – especially in the agricultural, commercial and retail sectors – by utilising solar thermal systems directly, without electricity generation. At the household level, non-electric stoves should be considered.

- Converting electricity consumers to electricity producers: A number of studies in the Stellenbosch area have highlighted the feasibility of producing electricity on different scales; when done collectively, this could offset the remaining two-thirds of the electricity demand.

For example, in the agricultural sector, at Spier, a 2-MW concentrated solar power system, covering about 5 ha, could offset approximately half of the entire estate's yearly electricity consumption. Alternatively, about 200 ha could be planted with fast-growing biomass, which, through a gasification system, could offset all of the electricity consumption. And at another local winery, the minimum 300 kW of electricity that is utilised during any given operational day could be produced by wind or solar power, or a combination of both, with a payback period of around 10 years.

In the commercial sector, at InnovUS, Stellenbosch University's wholly-owned technology transfer company, solar panels are utilised to meet the peak electricity demand at midday, which amounts to substantial savings. Similar 2-kW to 3-kW PV systems could be installed on all residential rooftops and combined with small-scale wind turbines in the 1-kW range. All flat roofs in the public, commercial, retail and education sectors could host a variety of solar systems. Provided that astute demand-side management systems are deployed, these systems could, collectively, become net producers of electricity at certain times of the day.

The municipality could increase the energy savings by making effective use of the sewage treatment works and waste handling facilities to produce power. For example, the 5.5 billion litres of sewerage water that is treated annually, could, by utilising biogas digesters, generate in the order of the total annual electricity consumption of the municipality – around 400 gigawatt hour (GWh).

In terms of liquid fuels, our greatest dependence is associated with transportation. If public transport in Stellenbosch could be improved and expanded, especially by using electric vehicles that may be charged with the technologies discussed in this chapter, this would go some way towards addressing this challenge.

In summary, the ideal of energy neutrality is, technically, entirely possible to achieve; in fact, it has already been achieved elsewhere by means of innovative financing schemes and business models.

Conclusion

In order to achieve the goal of transforming the energy system of Stellenbosch across all sectors of the local economy, a number of municipal (policy) interventions are required. These include:

- enabling private and public entities to generate their own electricity and sell excess supply to the grid;
- enabling the creation and implementation of a framework for Power Purchase Agreements (agreements that allow these entities to sell electricity to a buyer such as the municipality or a private entity);
- combining heat and power, particularly in the industrial and agricultural sectors through solar thermal systems;
- removing [electricity] subsidies, penalising inefficiency, and incentivising low-carbon initiatives;
- upgrading/enforcing energy efficiency standards across all sectors, for example, building codes;
- binding renewable energy targets with strong regulatory certainty;
- providing public education programmes about energy efficiency and renewable energy; and
- implementing demand-side management.

As a strategic first step, the municipality could focus on the top ten energy consumers in the area and find ways to render them both energy and carbon neutral through investment in local energy alternatives, including the development of such alternatives, through the local system of innovation.

REFERENCES

Brent AC, Hietkamp S, Wise RM & O'Kennedy K. 2009. Estimating the carbon emissions balance for South Africa. *South African Journal of Economic and Management Sciences*, 12(3):263-279.

Cowan KR & Daim T. 2009. Comparative technological road-mapping for renewable energy. *Technology in Society*, 31(4):333-341.

ESTIF (European Solar Thermal Industry Federation). 2006. *Solar industrial process heat – State of the art*. [Accessed 12 April 2012] http://www.estif.org/fileadmin/estif/content/policies/downloads/D23-solar-industrial-process-heat.pdf

Holm D. 2005. *Market survey of solar water heating in South Africa*. Report prepared on behalf of Solasure for the EDC (Energy Development Corporation) of the CEF (Central Energy Fund), Johannesburg. [Accessed 12 April 2012] http://www.cef.org.za/solar_market_survey.pdf

IEA (International Energy Agency). 2008. *Worldwide trends in energy use and efficiency. Key insights from IEA indicator analysis*. Paris: OECD/IEA.

Meyer AJ & Gariseb G. 2009. Renewable energy options for domestic use. In: L Thompson-Smeddle (ed). *Sustainable Neighbourhood Design Manual*. Stellenbosch: Stellenbosch University.

Orosz MW, Mueller A, Quolin S & Hemond H. 2009. *Small-scale solar ORC system for distributed power*. [Accessed 12 April 2012] http://orbi.ulg.ac.be/bitstream/2268/24847/1/12156-Orosz.pdf

RENAC (Renewables Academy). 2010. *Solar thermal – Heating and cooling by the power of the sun*. [Accessed 12 April 2012] http://www.solarthermalworld.org/node/1155

RSA DME (Republic of South Africa. Department of Minerals and Energy). 2009. *Digest of South African energy statistics*. [Accessed 12 April 2012] http://www.energy.gov.za/files/media/explained/2009%20Digest%20PDF%20version.pdf

Stellenbosch Municipality. 2010a. *Annual report: 2008/2009 financial year*. [Accessed 12 April 2012] http://www.stellenbosch.gov.za/jsp/e-lib/list.jsp?catid=128

Stellenbosch Municipality. 2010b. *Integrated Development Plan*. 2nd Generation, Revision 3. [Accessed 12 April 2012] http://www.stellenbosch.gov.za/jsp/e-lib/list.jsp?catid=52

Winkler H, Howells M & Baumert K. 2005. *Sustainable development policies and measures: Institution issues and electrical efficiency in South Africa*. Cape Town: University of Cape Town.

Chapter 11

TRANSPORT

Improving traffic flows in Stellenbosch

MARION SINCLAIR, CHRISTO BESTER & ESBETH VAN DYK

Introduction

No community can reach its full potential without an effective transport system. Be it the daily commute to or from work, school or university, or other activities such as business, shopping and recreation, transport has a direct bearing on most human activities in most communities. The economic viability of a town is highly dependent on the efficient movement of goods and the delivery of essential services by means of a transport network. Moreover, the safety of all users of transport has become a critical issue in many developing countries, and especially so in South Africa.

In the global village of the twenty-first century, the sustainability of communities and even nations is dependent on the efficient use of energy in the transport sector. Not only do we see the end of the availability of fossil fuels in the not too distant future, but we have also become painfully aware of the consequences of the use of fossil fuels, particularly the pollution of the atmosphere and resultant global warming.

For many years, the focus of 'sustainable transport' internationally has been on developing new forms or systems of transport with low environmental impacts, and these priorities will remain important in future. Transport is the largest end-user of energy in developed countries and the fastest growing one in most developing countries. In terms of developing sustainable communities, efficient and appropriate transport systems are important to ensure access to markets, employment, education and basic services. As such, efforts to improve traffic flows through urban areas in order to reduce emissions; develop appropriate and workable public transport systems and non-motorised transport to reduce dependence on private car ownership; and provide a road network that maximises efficiency while optimising environmental protection, will continue to require coordinated effort.

In the local context, issues around sustainability are framed largely in terms of what the current transport conditions are, and how they can be redesigned locally to achieve the desired long-term improvements. In this chapter, we will be discussing the *status quo* of transport in Stellenbosch; the specific problems and challenges that transport issues present for the town; and the way forward towards the implementation of a sustainable, integrated and intermodal transport system.

The current transport situation in Stellenbosch

In his book, *Transportation for liveable cities*, Vukan Vuchic (1999) noted that the era of projects aimed at maximising vehicular traffic was coming to an end, with the focus shifting towards the broader goal of achieving liveable cities, i.e. economically efficient, socially sound and environmentally sustainable. Considering that our goal here is also the development of a sustainable and liveable Stellenbosch, we need, as a first step, to understand the characteristics of the current traffic situation in the town and the problems facing its residents and visitors.

The two primary problems are those of congestion and road safety, both of which result in significant economic and social costs; and both of which must be addressed if our goal is to be realised. While congestion and road safety have specific dimensions, both are compounded by three common factors: lack of public transport, poor provision for non-motorised transport, and patterns of transport operation and enforcement that often undermine efficient transport flows.

The problem of congestion

General issues

Considering the size of its population and the extent of its road network, Stellenbosch carries a disproportionately high number of vehicles on its roads and parking areas (legal and otherwise). There has been considerable growth in traffic volumes over the past ten years, to the extent that the majority of intersections on the main roads operate at Level of Service F.[1] For example, the annual average daily traffic on the R44 at the Blaauwklippen Road (just outside of town, towards Somerset West) increased from 20 500 in 2000 to 35 400 in 2009 – an average growth rate of 6.2% per year (see Table 11.1), compared to the average Western Cape growth rate of 3.7%. If this 6.2% growth rate is maintained, traffic volumes will triple by 2030; even at 3.7%, they will still double by 2030.

Apart from the simple fact of a steep increase in the number of vehicles on the roads, there are a number of reasons for the high rate of vehicle use in Stellenbosch, the primary one being the lack of public transport within the town. In addition, many students from surrounding towns make use of private transport to commute to campus, mainly because of the unreliability of the rail system, but also other problems related to safety and security on trains. There are also a large number of people who travel to, from and through Stellenbosch on their way to or from other towns to the south and north. Finally, there are the heavy vehicles using the R44 through Stellenbosch to avoid the weighbridges on the N1 and N2 freeways.

Freight and congestion

Freight transport in the Stellenbosch Municipality consists primarily of local traffic and through traffic to neighbouring municipalities, since none of the major corridors runs through this municipality. Local freight relates mainly to the agricultural industry, construction in the area and the needs of the inhabitants and business sector, for example Fast Moving Consumer Goods (FMCGs), fuel and other consumables.

Table 11.1 shows the 2009 and projected 2030 Average Annual Daily Traffic (AADT) volumes[2] on the respective roads around Stellenbosch. The projected volumes are approximately double the current volumes.

Table 11.1   Daily traffic flows around Stellenbosch

Counting station	Total vehicles		Heavy vehicles	Total vehicles	Heavy vehicles
	Observed 2000	Observed 2009	Observed 2009	Projected 2030	Projected 2030
R44 Blaauwklippen	20 510	35 406	1 266	73 224	3 213
R44 Cloetesville	12 928	19 339	1 106	39 995	2 824
R304 Kayamandi	14 151	18 247	867	37 737	2 203
Polkadraai	13 641	19 207	1 216	39 722	4 360
Helshoogte	5 358	6 893	373	14 256	1 131

Given the current level of congestion in Stellenbosch, it is clear that the estimated growth in freight volumes cannot be accommodated by the existing infrastructure and traffic management systems.

---

1   Level of Service is defined by the Highway Capacity Manual (Transportation Research Board, USA, 2000) in terms of the average delay per vehicle with Level A being the best (less than five seconds) and Level F the worst (greater than 80 seconds).

2   The average annual Western Cape traffic growth rate of 3.7% was used to calculate the volumes for 2030.

◘ *The consequences of congestion*

Traffic congestion, especially during the morning peak, has a number of negative effects. Saturated traffic flows are financially costly, because they contribute significantly to loss of productivity and economic inefficiency of business and pose huge problems for service delivery in the local area. In urban areas, a reduction in the average speed to below 30 km/h has a marked effect on time spent on the road. It has also been shown (Pienaar 1981) that, in terms of fuel and tyres, the cost of vehicle operation during congestion is much higher than during free-flow situations. This is exacerbated by the accelerations and decelerations necessitated by stop-start driving. Table 11.2 shows the daily fuel consumption of vehicles on the major roads in and through Stellenbosch.[3] These are the only roads for which recent traffic volume are available, and on them alone the total daily fuel use is in excess of 40 000 litres.

Table 11.2 Calculation of fuel consumption on major Stellenbosch roads[3]

From	To	Distance (km)	AM volume	PM volume	Daily volume	Peak speed	Off-peak speed	Peak fuel	Off-peak fuel	Total fuel (l)
R44 Strand-Adam Tas										
100 km/h	Webersvallei	0.350	3 392	3 263	33 275	60	80	201.2	844.8	1 046.1
Webersvallei	Technopark	0.682	3 360	3 400	33 800	45	80	447.0	1 672.2	2 119.3
Technopark	Blaauwklippen	0.915	3 524	3 594	35 590	30	80	814.4	2 362.3	3 176.7
Blaauwklippen	Van Reede	1.800	3 889	4 038	39 635	20	70	2 333.9	4 918.5	7 252.4
Van Reede	Safraan	0.439	2 703	2 814	27 585	20	50	396.2	890.3	1 286.5
Safraan	Dorp	0.487	3 304	3 209	32 565	20	40	518.8	1 317.9	1 836.7
Dorp	Adam Tas	0.310	2 095	2 094	20 945	20	40	212.4	539.6	752.0
Adam Tas	Merriman	0.556	3 258	2 869	30 635	20	40	557.2	1 415.4	1 972.7
Merriman	Bird	0.931	1 221	1 484	13 525	20	50	411.9	925.7	1 337.7
Bird	Cloetesville	2.530	1 968	1 900	19 340	50	70	899.3	3 373.3	4 272.7
Bird Street										
100 km/h	2nd Avenue	0.450	1 447	1 869	16 580	20	35	244.1	674.6	918.7
2nd Avenue	Adam Tas	0.758	1 112	1 765	14 385	20	35	356.7	985.9	1 342.6
Adam Tas	Molteno	0.494	1 120	1 170	11 450	20	35	185.0	511.4	696.4
Molteno	Merriman	0.390	1 099	1 194	11 465	20	35	146.3	404.3	550.5
Merriman	Dorp	0.832	1 214	1 356	12 850	20	35	349.8	966.6	1 316.4
Merriman										
Adam Tas	Bird	0.550	1 130	1 167	11 485	25	40	178.3	524.9	703.2
Bird	Ryneveld	0.312	1 026	1 201	11 135	25	40	98.0	288.7	386.7
Ryneveld	Marais	1.080	988	1 105	10 465	25	40	319.0	939.2	1 258.2
R310 Adam Tas										
Strand	80 km/h	3.17	1 978	1 887	19 325	30	50	1 532.0	4 503.9	6 035.8
Helshoogte Road										
R44	80 km/h	3.02	701	678	6 895	45	55	403.8	1 473.1	1 876.9
TOTAL (Litres/day)										40 138.1

The environmental consequences of traffic congestion are also a major concern. While South Africa has not yet implemented any robust monitoring of traffic-related pollution, international research (for example, Carslaw, Ropkins & Bell 2006) has shown that gaseous and particulate traffic-related pollutants increase in direct relation to traffic congestion. Increased air pollution, in turn, has been causally linked to respiratory illness, particularly among children, and so poses a direct health risk to local populations.

Congestion has also been shown to be a factor in the growing incidence of transport-related stress, potentially resulting in depression, aggression and disaffection, which have the potential to impact on driving behaviour and hence to undermine road safety. The psychological consequences of repeated exposure to traffic congestion

---

3 The fuel consumption was calculated from the distances travelled and the average speeds of the vehicles (Pienaar 1981).

are only now beginning to be understood. What is significant is that traffic congestion may have a direct impact on the psychosocial health of a population.

While congestion is most commonly conceived of as a problem specific to the through movement of vehicles at specific times of the day, it is also related to the general density of vehicles within a town and the facilities provided for these vehicles. One particular problem in this regard in Stellenbosch is the lack of sufficient provision for parking, which is seriously exacerbated by the high level of illegal parking that occurs in the town. Apart from adding to the challenges for through traffic, as well as to environmental and social problems, the parking problems create major obstacles for pedestrians and often have a marked impact on sight distance at intersections. It has been estimated (Vela VKE 2010) that the current shortage of parking spaces is between 4 000 and 5 000 on the university campus alone. To provide these spaces on open land (if available) would cost about R80 million. In parking garages and basements, the cost would be R500 million.

The problem of road safety

▣ International features of the problem

Past debates on sustainability have often overlooked the impacts of transport on the actual road users and, in particular, have ignored the problem of poor levels of road safety. Indeed, at the World Summit on Sustainable Development held in Johannesburg, South Africa, in 2002, there was not a single agenda item that addressed road safety, in spite of the fact that this is one of Africa's biggest and fastest-growing public health concerns.

In 2002, road traffic injuries ranked as the tenth leading cause of death in the world (United Nations World Health Organisation (WHO) 2002). By 2004, that ranking had risen to seventh, and it is expected that road injuries will rank as the fifth highest leading cause of death by 2030 (WHO 2005). While other causes of premature mortality worldwide are showing signs of improvement, road injuries are instead increasing in importance.

A lack of reliable data across the continent makes it difficult to pin down an accurate estimate of lives lost, but the number of fatal traffic injuries in Africa was estimated to be around 200 000 in 2002 (WHO 2009). As a continental average, the WHO has estimated that the aggregate rates of road traffic fatality per 100 000 members of the population in Africa are 28.0, compared to 11.0 for Europe (Peden, Scurfield, Sleet, Mohan, Hyder, Jarawan *et al*. 2004). At 32.0 per 100 000 people, South Africa's aggregate rate exceeds that of the continental average. Between 13 500 and 15 000 people are reported killed in traffic collisions in South Africa annually. Around 100 000 people are seriously injured, often resulting in significant long-term financial and social challenges and the perpetuation of poverty cycles.

There can be no sustainable development when the prevailing transportation systems allow for loss of life and lack of human safety at such a scale. According to a 2007 national Medical Research Council of South Africa (MRC)/University of South Africa (UNISA) study, which collected mortuary and police data related to all types of non-natural deaths across South Africa, 48% of all non-natural deaths occurred on roads, surpassing all other potential locations by a factor of at least 2.4 (Donson 2008). But if roads are the locus of many fatal and serious injuries, one should also point to the broader existing transport systems – of which roads are just one element – as being largely responsible for creating circumstances and road-user behaviour that precipitate collisions and hence injuries.

▣ Road safety trends

South Africa shares a number of road safety trends with other developing countries. These include:
- Road collision deaths affect the young and economically active disproportionately.
- Significantly more males than females are killed annually: the percentages in most developing countries equate to approximately 25% female: 75% male, and this is true for South Africa as well (see Figure 11.1).
- Pedestrian deaths account for a significant proportion of traffic fatalities. In South Africa pedestrians accounted for 35% of traffic fatalities in South Africa between 2007 and 2009 (Road Traffic Management Corporation (RTMC) 2008), and this figure was higher in the Western Cape, at 44% (see Figure 11.2).
- Buses, minibuses and trucks are over represented in accidents.
- Single-vehicle collisions or loss-of-control collisions are common.

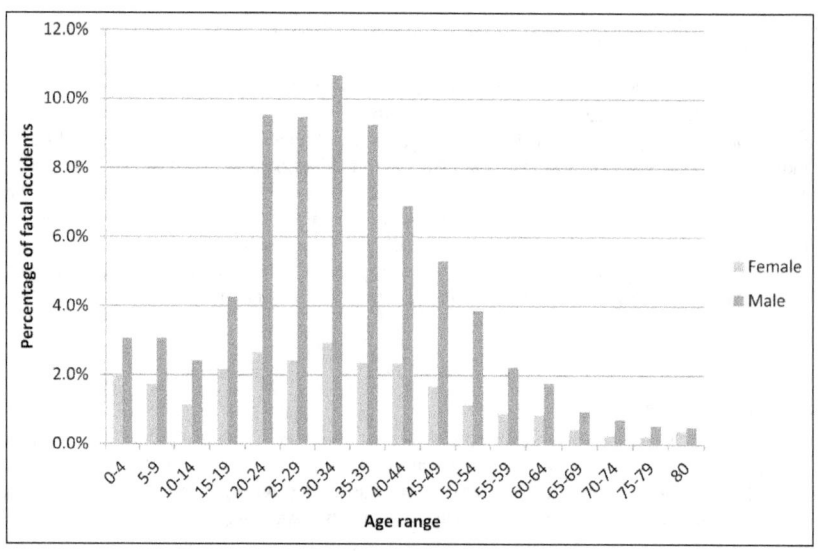

Figure 11.1   Age/gender comparison of fatal road accidents in the Western Cape, 2007-2009[4]

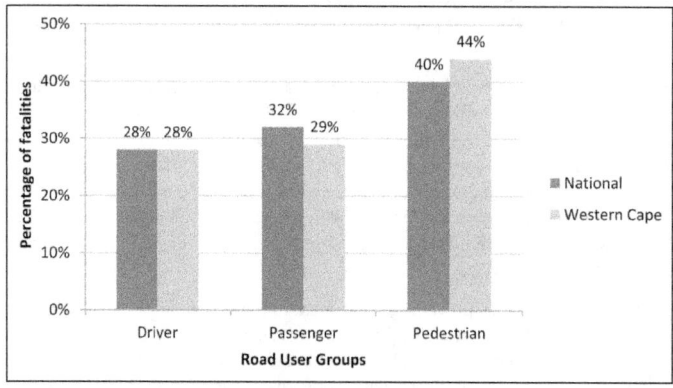

Figure 11.2   Comparison of fatalities by road user type, 2007-2009[4]

- *Problems specific to South Africa and the Stellenbosch area*

South African road safety problems are a result of a combination of factors that are common to Africa, including poor road infrastructure, poor road maintenance, lack of facilities for non-motorised transport, behavioural issues related to road users (including aggression, lawlessness, etc.), alcohol abuse and drug use, and the fact that police enforcement of traffic laws is limited and sporadic. However, two further features have shaped current road safety problems in South Africa.

Firstly, apartheid planning created spatial inequalities of access for different race groups, leaving poor communities on the outskirts of urban areas and forcing them into long commutes to work and school, while at the same time providing no safe pedestrian facilities or effective public transport. This is arguably the most crucial safety issue facing commuters in Stellenbosch today. Stellenbosch is a town where just less than 25%

---

4   Data for Figures 11.1 and 11.2 sourced from the RTMC National Collision Database, a copy of which was provided to Stellenbosch University by the RTMC in April 2010.

of its population is 15 years old or younger, indicating a high dependency ratio and a high percentage of the population relying on non-motorised or public transport. School buses, serving predominantly rural areas around Stellenbosch, are the primary means of school transport for approximately 13 000 children in the area. Recent efforts by the traffic police to regulate these buses have seen a crackdown on large numbers of unroadworthy buses and unlicensed drivers; yet such problems persist. The safety of school children is also compromised on a daily basis by the location of pickup and drop-off sites along main arterials and the fact that children as young as six years of age regularly cross fast-flowing arterials at informal crossing points.

Secondly, over the past decade South Africa has experienced an unprecedented increase in the registration of new vehicles, with the number of vehicles registered increasing by 38.8% between 2000 and 2009 (from 670 000 to 930 000 vehicles, respectively) (RTMC 2008). Increased traffic flows are undoubtedly a factor in the levels of traffic collisions, yet more work is needed to determine the specific nature of this relationship.

Stellenbosch is unusual in the sense that it is a university town, and its traffic flows (and collision levels) are significantly affected by the presence or absence of students. This is most noticeable in a monthly breakdown of collisions, which shows a marked decline in collisions during the month of December, when students are away on summer vacation.

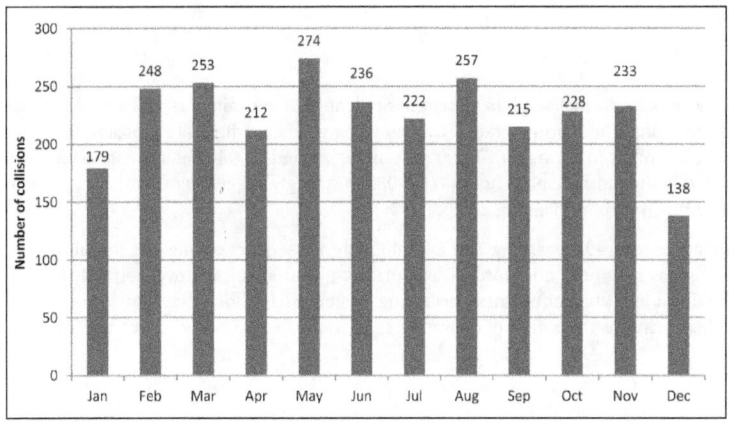

Figure 11.3 Stellenbosch collisions reported by month, 2009 [Data provided by Stellenbosch Municipality Traffic Department]

For the remaining eleven months of the year, the traffic police recorded an average of 232 collisions per month, 20% of which result in injuries. Over the three-year period from 2007 to 2009, 38 collisions generated 45 fatal injuries, while a further 1 534 people were injured, 224 of whom sustained serious injuries. During this time, 72 collisions with pedestrians were recorded, 22% of which involved hit-and-run drivers.

Issues contributing to congestion and poor road safety

Non-motorised transport (NMT)

Many of the trips made in Stellenbosch are still made on foot. Here one thinks of students, tourists, people from the poorer communities who do not have access to motorised transport, and all the motorised transport users who have to go to and from their parking areas and public transport interchanges. Very few trips are made by bicycle, which is largely ascribable to the narrow streets in Stellenbosch and the lack of a network of safe cycle routes.

The largest flow of pedestrians outside the university campus is from Kayamandi, along Bird Street to the central business district (CBD) – a route with a high number of pedestrian accidents.

In June 2008, Stellenbosch Municipality appointed consultants to draw up the Development of a Non-Motorised Transport (NMT) Network Plan for Stellenbosch, which included the identification of the various NMT generators and trip purposes, such as commuting, shopping, school and university, and recreational and competitive sport. Priorities for implementation were defined in terms of the needs for the 2010 World Cup (SSI Engineers & Environmental Consultants 2009).

Prior to this, the Stellenbosch Town Transport Master Plan (STTMP) identified the following routes and areas for priority attention (Arup Transport Planning 2007):

- Kayamandi – CBD;
- Pedestrianisation of Church and Andringa Streets;
- Town-wide NMT network upgrade;
- CBD street network;
- Stellenbosch Station – CBD;
- University – CBD;
- Kayamandi – Plankenbrug; and
- Kayamandi – Cloetesville.

Public transport

Apart from school buses, there is no formalised bus system in Stellenbosch and minibus taxi services cover only the low-income residential areas and farm workers. On any given weekday, there are typically 100 taxis operating on 27 routes from five ranks in town, with 7 500 passengers carried on 940 vehicle trips. One-third of daily passengers are carried in the morning peak hour (07:00-08:00) strongly suggesting that the majority of passengers are commuters (Arup Transport Planning 2007).

The Stellenbosch railway station sees 492 boarding and 843 alighting passengers during the morning peak hour, with 8% fewer during the evening peak hour (Arup Transport Planning 2007). Rail travel certainly has the potential to be a more significant and popular means of accessing Stellenbosch, though investment is needed to improve the amenities and increase the safety of passengers, as well as to diversify routes and increase train numbers.

Transport operations and law enforcement

Traffic and transport regulations and control are important for regulating the traffic flow in an orderly and sustainable way, thereby reducing congestion. It has been shown that the delay to vehicles may be reduced by optimal phasing and coordination of traffic signals. In Stellenbosch, traffic signals on the main congested routes are not part of a single system – some operate on an isolated basis, while others are controlled from Cape Town. To complicate matters even further, some intersections that are not signalised are controlled by traffic officers. This makes coordination particularly challenging.

When traffic regulations are not enforced, the safe and orderly flow of traffic is not possible. The enforcement effort in Stellenbosch is concentrated around the use of speed and red light cameras. Recent research (Roux, GB 2010) has shown that speed cameras in Stellenbosch reduce the speed of vehicles only in the immediate vicinity of the cameras. Another study (Verlinde 2010), which looked at the behaviour of drivers at stop-controlled intersections in Stellenbosch and Somerset West, indicated that fewer than 15% of drivers come to a complete stop under free-flow conditions. The 85th percentile of minimum speeds through these intersections was 19 km/h. In a survey of students, only 7.9% indicated that they always stop at these intersections; and when questioned about the reasons why they do not stop, 11.6% cited the lack of law enforcement as one reason. A study on pedestrian behaviour (Roux, AW 2010) also indicated a large disregard for traffic regulations by both drivers and pedestrians.

Rethinking transport with a view to the future

Principles

The guiding principles of the STTMP are an excellent starting point for the development of an altered transport system for the town. These principles are as follows:

- *Promote choice for mobility:*
  - *support those who cannot afford private transport;*
  - *provide public transport as a viable alternative for car users; and*
  - *manage congestion.*
- *Ensure access to:*
  - *opportunities for all;*
  - *land and developments; and*
  - *transport for the disadvantaged.*
- *Develop great streets (liveable areas):*
  - *provide a safe environment;*
  - *provide spaces and facilities of dignity; and*
  - *consider aesthetics principles and the conservation of historical and cultural assets.*
- *Achieve sustainability through:*
  - *securing funding for capital investment and the maintenance of infrastructure;*
  - *financial viability of public transport operations;*
  - *preservation of the natural environment; and*
  - *support of social and economic objectives.*

[Arup Transport Planning 2007]

These principles highlight the rights of Stellenbosch residents to safe and effective transport opportunities, while at the same time acknowledging the need to protect the natural and cultural environment of the town. The achievement of these principles must occur through the development of a vision of what the town can and should become, as well as addressing existing problems and challenges that are already pressing. The paragraphs that follow contain some suggestions for achieving this end.

Addressing congestion and freight issues

- *Changing infrastructure*

It has been shown worldwide that the problem of congestion cannot be solved by the provision of more road infrastructure (Vuchic 1999). However, this does not mean that bottlenecks in the infrastructure should not be improved. The intersection between the Strand Road and Adam Tas Road and the northern section of Bird Street are two examples where changes in the physical layout of roads could bring about an immediate improvement in congestion levels. The possibility of different bypass routes may also be considered.

- *Changing employment practices*

Crucially, the data shows that the peak 15-minute period around 08:00 in the morning carries 40% more traffic than any other 15 minutes during the day (Arup Transport Planning 2007). This is a clear indication that there is considerable scope for transport demand management actions, such as variable working hours at different employers, or even flexible working hours at the same employer. Small changes in the employment practices of local businesses and even educational institutions could result in greater efficiency of traffic flows.

- *Managing freight*

In terms of managing the needs of freight traffic, new ways will have to be found to organise freight flows in an efficient and sustainable way. Freight flows in the town centre include deliveries to supermarkets, shops, restaurants and offices. Therefore, the efficiency of freight logistics has a direct impact on product prices and on-shelf availability.

According to the *BESTUFS*[5] *Good Practice Guide on Urban Freight Transport*, commissioned by the European Union in 2007, the following measures may be implemented to manage urban freight:

- *designated routes for freight vehicles;*
- *limits on the size and weight of delivery vehicles;*
- *limits on access times to town centres;*
- *restrictions on emissions and noise produced by delivery vehicles; and*
- *"nearby delivery areas" – areas of street space dedicated to goods vehicles for the loading and unloading of goods destined for nearby shops.*

[Allen, Thorne & Browne 2007]

Such measures obviously have to be designed and introduced in collaboration with local business, and special attention should to be paid to the transport of hazardous goods.

Traffic flow management

Looking at the current traffic management of the town, it is clear that the existing control infrastructure may be utilised to far better effect. Traffic control must be appropriate and signals should be coordinated.

Long-term answers

Steps such as those described above may bring about significant changes to the flow of traffic through or around the town in the short to medium term.

However, the long-term, sustainable solution to congestion lies in a greater use of public transport – by commuters in particular – and in encouraging non-motorised transport for shorter trips. The aim must be to move more people with fewer vehicles, which would address both congestion and road safety in direct and lasting ways.

As a university town, Stellenbosch has tremendous scope for introducing and celebrating cycling as a major form of local transport. For this to happen, however, dedicated and safe cycle lanes must be made a priority by the local council. The present level of cycling collisions in the town indicates that current provision for cyclists is unsatisfactory, and that without the necessary commitment and resources, the potential for cycling will remain unrealised. The scale and amenities of the town also create an opportunity to make more provision for pedestrians. In terms of encouraging NMT, it is imperative that the town prioritise the provision of pedestrian and cycle spaces that are safe and attractive and do justice to the historic and cultural elements of the town.

Ultimately, the town must also consider whether there is merit in disincentivising vehicular traffic through the town, as is being done elsewhere by means of congestion charging and higher parking fees (see the 'Congestion charging' case study at the end of the chapter). However, this cannot happen until suitable alternatives, in the form of both efficient public transport and opportunities for NMT, are achieved.

Addressing road safety issues

When considering the vulnerability of pedestrian and other forms of non-motorised transport, it is commonplace to blame the engineering profession for a lack of protective elements. Indeed, the lack of safe public transport options certainly stands out as a key missing piece in the transport plan for Stellenbosch and is undoubtedly one area of improvement that the engineering industry should bring to fruition. There are clearly areas in town that accommodate large numbers of pedestrians on a daily basis, where safer pedestrian shelters and crossing points, and even barriers to pedestrian movement, could directly reduce pedestrians' exposure to vehicular traffic. However, a recent survey of pedestrian behaviour along Bird Street (the main pedestrian arterial through the town), indicated a marked lack of use of the crossings provided, with pedestrians motivated more by immediacy of access than safety (Roux, AW 2010). High-risk crossing behaviour was recorded on a frighteningly regular basis. While the study indicated that physical solutions may be found for some of the areas indentified

---

5   Best Urban Freight Solutions.

in it, any long-term answer to the pedestrian problems of Stellenbosch must involve re-educating pedestrians about not only their right to safety, but also their responsibilities.

A similar lack of compliance to traffic laws was noted by Liebenberg (2010), in his study of the yielding behaviour of cyclists at a four-way stop in Stellenbosch. Liebenberg found that cyclists exhibited little regard for traffic laws and operated inconsistently, sometimes functioning as vehicles and at other times behaving similarly to pedestrians. If the town is to achieve higher levels of cycling as a form of NMT, it will be essential to offer education and guidance to the town's cyclists in order to ensure more consistent and safer cycling behaviour.

Alcohol has been shown to play a causal role in injury collisions in South Arica in general; in terms of the Western Cape, the 2007 MRC/UNISA study on fatal injuries, based largely on autopsy results, found that 56.2% of drivers and 64.7% of pedestrians showed elevated blood alcohol levels (Donson 2008). This is certainly a challenging problem to overcome, but concerted efforts must be made to re-educate the public about the dangers of alcohol to all road users, and alternative means of transport must be prioritised in order to ensure that drunk driving may be avoided and inebriated pedestrians' opportunities for conflicts with vehicles are minimised.

Conclusion

Congestion and road safety already affect the daily lives of Stellenbosch residents, and constitute the core transport challenges to be addressed in developing a new and sustainable future. By identifying the common issues that contribute to these problems, namely the lack of public transport, the poor provision for non-motorised transport, and the challenges pertaining to traffic management, we are better placed to identify the key components of a future transport system for the town.

Stellenbosch needs an integrated, intermodal transport system that makes adequate provision for all the people of the town and effectively addresses their need for mobility, accessibility and safety. This must be attained through responsible planning to provide a system that is economically efficient, socially sound and environmentally sustainable. Such a system must be founded on a greater emphasis on public and non-motorised transport, yet without neglecting the requirements of vehicles that are necessary for essential services and economic activities.

Stellenbosch has an extraordinary opportunity to rethink its future as far as traffic flow, road-user safety and transport efficiency are concerned, and, through reprioritising the human users of our roads, to develop a new vision of transport that improves both environmental and human quality of life.

REFERENCES

Allen J, Thorne G & Browne M. 2007. *BESTUFS Good Practice Guide on Urban Freight Transport*. Manual commissioned by the European Union. University of Westminster: School of Architecture and the Built Environment.

Arup Transport Planning. 2007. *Stellenbosch Town Transport Master Plan (STTMP)*. Draft project report prepared for Stellenbosch Municipality. Cape Town: Arup Transport Planning.

Carslaw DC, Ropkins K & Bell MC. 2006. Change-point detection of gaseous and particulate traffic-related pollutants at a roadside location. *Environment Science and Technology*, 40(22):6912-6918.

Donson H (ed). 2008. *A profile of fatal injuries in South Africa, 2007*. Medical Research Council of South Africa/University of South Africa Crime, Violence and Injury Lead Programme. [Accessed 8 November 2010] http://www.doh.gov.za/docs/reports/2007/sec1a.pdf

Peden M, Scurfield R, Sleet D, Mohan D, Hyder AH, Jarawan E *et al*. 2004. *World report on road traffic injury prevention*. Geneva: World Health Organisation.

Pienaar WJ. 1981. Car operating costs under various traffic conditions in South African cities. Unpublished MA thesis. Stellenbosch: Stellenbosch University.

RTMC (Road Traffic Management Corporation of South Africa). 2008. *Road Traffic Report*, March. [Accessed 3 March 2010] http://www.rtmc.co.za/RTMC/Files/Traffic_Reports/Financial%20Year/Mar%202008%20Report.pdf

Roux AW. 2010. Analysis of pedestrian flows along Bird Street. Final-year project. Stellenbosch University: Department of Civil Engineering. Stellenbosch: Stellenbosch University.

Roux GB. 2010. Speed distributions: the effect of speed cameras at intersections. Final-year project. Stellenbosch University: Department of Civil Engineering. Stellenbosch: Stellenbosch University.

SSI Engineers & Environmental Consultants. 2009. *Stellenbosch Non-Motorised Transport Network Plan.* Draft project report prepared for Stellenbosch Municipality. Stellenbosch: Stellenbosch Municipality.

Transportation Research Board. 2000. *Highway Capacity Manual.* Washington, DC: Transportation Research Board.

Vela VKE. 2010. Moving Stellenbosch. Project report for Stellenbosch University. Stellenbosch: Stellenbosch University.

Verlinde KJS. 2010. Vehicle speeds through stop-controlled intersections. Final-year project. Stellenbosch University: Department of Civil Engineering. Stellenbosch: Stellenbosch University.

Vuchic VR. 1999. *Transportation for liveable cities.* Rutgers, New Brunswick: CUPR Press.

WCPG RNIS (Western Cape Provincial Government Road Network Information System). 2010. *Road Network Information Reports.* [Accessed 7 March 2010] http://rnis.pgwc.gov.za/rnis/rnis_web_reports.main

WHO (World Health Organisation). 2002. *A 5-year WHO strategy for road traffic injury prevention.* [Accessed 13 March 2010] http://www.who.int/violence_injury_prevention/publications/road_traffic/5yearstrat/en/index.html

WHO. 2009. *Global status report on road safety.* [Accessed 13 March 2010] http://www.whqlibdoc.who.int/publications/2009/9789241563840_eng.pdf

---

CASE STUDIES

Multimodality

There are many examples of towns and cities across Europe where transportation policy has increasingly reflected the significance of ecological considerations, and where changes have been made to successfully reduce emissions and improve quality of life, while maintaining economic efficiency. **Freiburg**, in Germany, is one such example. While it benefitted from an early introduction of a tram system to its urban infrastructure, Freiburg has committed itself to reshaping its transport infrastructure to achieve the following objectives:

- increasing non-motorised mobility;
- increasing use of public transport;
- reducing car mobility;
- reducing car parks; and
- reducing energy consumption.

Over the past thirty years, Freiburg has concentrated its efforts on introducing an extensive network of 30 km/h zones, promoting bicycle use, and improving the public transport sector. This includes improvements to the size and synchronisation of the tram and bus systems and the introduction of a single regional fee card for public transport, which allows commuters to use multimodal transport opportunities without duplicate charges.

This policy of enhancing opportunities for public transport and making it affordable and accessible to residents has had very positive results: between 1976 and 1996, car use declined from 60% to 43%. Moreover, 90% of the town's university students use public transport or bicycles as their sole source of transport within the town.

Congestion charging

London is running the biggest congestion charging experiment in the world. As such, the city is widely used as an example when considering the potential benefits that congestion charging may bring.

In 2003, congestion charging was implemented in the central London boroughs. It had an immediate impact, reducing the amount of traffic in the heart of the capital by about 15%. Transport for London (TfL), who manages the scheme, claims that the total quantity of traffic in the city fell by 21% between 2002 and 2006, with public transport absorbing the bulk of the difference. In addition, bicycles have increased by 49%, taxis by 13%, and buses and coaches by 25%.

Overall, the scheme appears to have resulted in a significant reduction in congestion levels, and the surplus income generated has been reinvested back into improved public transport. In spite of these achievements, a high level of public opposition to theses charges remains and there are claims that small businesses have suffered. Such claims are, however, disputed by TfL.

Other cities, including San Francisco, are currently developing similar schemes to the London model, with congestion charging to be used to generate funds for the improvement of public and non-motorised transport facilities.

Minibuses as a component of public transport

Hong Kong transport officials have embraced small buses as a complement to their larger passenger buses for those city streets that are too narrow for conventional buses and for routes where the numbers of potential passengers are too low to support conventional buses. This system has also allowed the city to legalise the minibus taxi sector, much of which was previously operating without any formal management.

Minibuses, or 'light' buses, offer advantages over conventional buses in that they are typically faster and more efficient and can be deployed more frequently along routes at lower cost. Like their South African counterparts, the minibus taxis of Hong Kong carry a maximum of 16 seated passengers, with no standing room. The city has chosen to adopt formal routing and scheduling for the majority of these buses (known as 'green-line' buses), while allowing other minibuses to operate informally on demand ('red-line' buses).

In 2002, the government offered subsidies to convert minibuses from diesel to liquefied petroleum (LP) gas; it is now looking at repeating the scheme to facilitate a transfer to hybrid, or even electric, vehicles.

The Hong Kong example demonstrates how an existing transport sector may be used more efficiently and tied into a more synchronised system of multimodal transport opportunities to enhance passenger convenience.

RESPONSE TO
TRANSPORT

MATTHEW MOODY

The chapter by Sinclair, Bester and Van Dyk provides an overview of transport in Stellenbosch and the measures required to achieve a more sustainable system. For the authors, the core transport challenges facing the town are congestion and traffic safety. As remedies to these problems, they suggest enhancing alternative modes of transport, improving traffic management, and educating cyclists, pedestrians and the broader public about road safety issues.

In response, I would argue that the transport problems faced by Stellenbosch reflect the global dominance of a powerful "system of automobility" (Urry 2004:26), which Henderson describes as follows:

> ... automobility derives from a system calculated to coerce individuals into driving, that subordinates all other modes of transport and ways of dwelling, that requires enormous state subsidy and regimentation of urban space for maximum throughput and speed, and requires a centralized state-backed capitalist oligopoly of oil, highway, automotive manufacturing and real estate control over transportation policy.
>
> [Henderson 2006:295]

The use of the private car continues to grow around the world and many cities have been – and continue to be – shaped primarily in a way that facilitates car mobility (Martin 2009). This reflects the extraordinary power of the automobile/oil industrial complex, as well as that of the global middle class in ensuring the generation of urban forms that primarily support their continued economic reproduction. This 'system of automobility' expresses itself in widespread car use among all those who can afford it, low-density sprawling urban forms and car-oriented urban design. Such an environment necessarily –

- undermines sustainable modes of transport;
- depends on diminishing supplies of oil (thus threatening future urban stability);
- threatens local and global ecologies;
- damages human health and well-being;
- degrades the quality of urban life and public space; and
- (most importantly, from the perspective of a developing country) severely constrains the ability of the poor to effectively participate in society.

[Martin 2009; Newman & Kenworthy 1999; Newman 2008; Vasconcellos 2001]

Many cities have been – and continue to be – shaped in such a way that car use is necessary, even essential, to negotiate daily life. As a result, life for those without a car is severely constrained. Stellenbosch is not immune from these global trends, as the following statement from the town's interim transport plan makes clear:

> Extrapolating on recent trends, private vehicle-induced sprawl, traffic congestion, reduced safety and environmental deterioration will continue to worsen.
>
> [Arup Transport Planning 2007:3]

In a highly inequitable, developing country context like that of Stellenbosch, the most critical transport issue is not congestion or, indeed, traffic safety. Rather, it is the relationship between transport, inequality and sustainability. Car-dominated transport systems facilitate the resource-intense mobility of the middle and upper classes. This directly undermines the mobility and access of the poor majority who either use poor quality public transport, walk or cycle. As car use grows, so cities sprawl and streets and public spaces are reshaped to facilitate car mobility (Gehl & Gemzøe 2006). In such a context, walking and cycling become unfeasible and unsafe, and the poor increasingly rely on low-quality public transport that struggles to provide an effective service in a sprawling, car-dominated urban environment (Peñalosa 2006).

Therefore, in a developing country context, transport interventions should primarily focus on enhancing the accessibility of the poor, not resolving the congestion of the already highly mobile car-owning minority (Baeten 2000; Peñalosa 2006). This is not to say that congestion should be ignored. Growing congestion may very well undermine the urban economy, but this is a result of the excessively dominant role played by the private car in the daily life of the economic elite. As Sinclair and her co-authors suggest, this must change. Fortunately, the measures required to resolve the mobility and access problems of the poor and the congestion problems of the elite may be mutually reinforcing – as I will now explore.

The primary focus of a sustainable transport approach is to reduce the dominance of the car by "reclaiming urban spaces from automobiles" (Henderson 2006:294). According to the literature, this may be achieved by:

- *creating a slower, more human urban environment, which is conducive to non-motorised travel;*
- *creating shorter distances between locations by encouraging walkable, dense, mixed-use urban villages connected by high-quality transit to form a polycentric urban network;*
- *enhancing the quality of public and non-motorised transport;*
- *ensuring that the design, availability and distribution of scarce public space, including streets, encourage travel by sustainable modes of transport.*

[Banister 2008; Kenworthy 2006; Newman & Kenworthy 1999]

The ultimate objective of sustainable transport interventions is to "design cities of such quality and at a suitable scale that people would not need to have a car" (Banister 2008:74).

There are cities and towns around the world that have embraced a more sustainable transport approach, including Vancouver, Bogotá, Portland, Paris, Copenhagen, New York City, Curitiba and Freiburg. These cities have collectively embraced the principles of sustainable transport mentioned above and have sought to create more sustainable, equitable, high-quality urban environments that support effective and safe alternatives to the car, cause less environmental damage and facilitate equitable accessibility to urban opportunities. What is common to all of these places is the recognition that sustainable urban development cannot rely on the car as both the dominant means of urban transportation and the major force in shaping urban form and design.

The challenge for Stellenbosch is to embrace a more sustainable transport approach in a global context where the dominant economic system and classes are reliant on the continued expansion of car use, car-related industries and car-dominated urban forms (Baeten 2000; Paterson 2008). The priority should be to retain and enhance the ability of the poor majority to access urban amenities and employment opportunities through the use of viable public and non-motorised modes of transport and, crucially, to ensure urban forms that are primarily supportive of green modes of transport, rather than maximum car mobility. It is important to bear in mind the following argument by Peñalosa:

> *Do we dare create a transport system giving priority to the needs of the poor majority rather than the automobile-owning minority? Are we trying to find the most efficient, economical way to move a city's population, as cleanly and as comfortably as possible? Or are we just trying to minimise the upper class's traffic jams?*
>
> [Peñalosa 2006:4]

REFERENCES

Arup Transport Planning. 2007. *Stellenbosch Town Transport Master Plan (STTMP)*. Draft project report prepared for Stellenbosch Municipality. Cape Town: Arup Transport Planning.

Baeten G. 2000. The tragedy of the highway: empowerment, disempowerment and the politics of sustainability discourses and practices. *European Planning Studies*, 8(1):69-86.

Banister D. 2008. The sustainable mobility paradigm. *Transport Policy*, 15:73-80.

Gehl J & Gemzøe L. 2006. *New City Spaces*. Copenhagen: The Danish University Press.

Henderson J. 2006. Secessionist Automobility: Racism, Anti-Urbanism, and the Politics of Automobility in Atlanta, Georgia. *International Journal of Urban and Regional Research*, 30:293-307.

Kenworthy JR. 2006. The eco-city: ten key transport and planning interventions for sustainable city development. *Environment and Urbanization*, 18:67-85.

Martin G. 2009. The Global Intensification of Motorisation and its Impacts on Urban Social Ecologies. In: J Conley & AT McLaren (eds). *Car Trouble: Critical Studies of Automobility and Auto-Mobility*. Farnham: Ashgate.

Newman P. 2007. Beyond Peak Oil: Will Our Cities Collapse? *Journal of Urban Technology*, 14(2):15-30.

Newman P & Kenworthy J. 1999. *Sustainability and Cities: Overcoming Automobile Dependence*. Washington: Island Press.

Paterson M. 2007. *Automobile Politics: Ecology and Cultural Political Economy*. Cambridge: Cambridge University Press.

Peñalosa E. 2006. *The role of transport in urban development policy*. GTZ (Deutsche Gesellschaft für Internationale Zusammenarbeit) Sourcebooks. Germany: GTZ.

Urry J. 2004. The 'System' of Automobility. *Theory, Culture and Society*, 21(4):25-39.

Vasconcellos E. 2001. *Urban Transport, Environment and Equity*. London: Earthscan.

Section 3

Ecosystem services

Section Introduction

Ecosystem services

Ecosystems and the services they render are indispensable to human socio-economic systems and thus to our quality of life. Stellenbosch's unique and biologically diverse ecosystems have formed the backbone of the area's rich agricultural heritage. This section reports on the state of these ecosystems and examines some of the key factors that threaten them.

Most of the threats to Stellenbosch's ecosystems are interconnected. Intensive agriculture and unsustainable farming practices are causing a decline in soil health and the capacity of the ecosystem to supply ecological services. It is also increasingly vulnerable to climate change, peak oil, the financial crisis, and other shocks, because of its dependence on high levels of expensive, imported fossil fuel-based inputs. In terms of water resources, groundwater is being pumped faster than recharge rates, and surface water withdrawals are threatening water supply, as well as habitat uses. Agriculture accounts for more than half of all water consumption in the Western Cape. Furthermore, declining water quality from human activity negatively impacts agriculture and endangers the area's indigenous vegetation. Almost half of Stellenbosch's natural vegetation is critically endangered due to habitat loss and fragmentation, invasive alien plant species, unnatural fire patterns, and climate change.

Some of these issues are addressed by the municipality's Spatial Development Framework, but overall there has been ineffective integration of environmental needs into planning. A new perspective is needed that views society and nature as deeply interconnected. The chapters in this section propose the following as future solutions:

- Intensive 'high external input' agriculture should be phased out in favour of a more resilient agro-ecological approach, which recognises the importance of soil health and water quality.
- Agriculture must be restorative if it is to remain viable. Merely doing less harm is no longer an option.
- A new management and implementation model for water services is required, one that provides the requisite degree of long-term certainty that households and businesses need in order to invest in a future sustainable Stellenbosch.
- The environment requires a powerful voice within the municipality and the support of relevant interest groups to ensure that its needs are incorporated during the planning process.

Chapter 12

SOIL HEALTH
Sustaining Stellenbosch's roots

JOHANN LANZ

Introduction

Soil, underfoot and out of sight, is frequently neglected in our thinking about the world and its future. Yet, soil is the foundational resource of all agricultural production; without it, we cannot eat. But soil is also much more than that. It is the recycling works of the earth's biomass and the storehouse of the raw material from which all life on earth, including humans, is formed. Part of the purpose of this chapter is to communicate something of the importance of soil and to provide some scientific appreciation of it in the hope that it might take its rightful place at the forefront of sustainability thinking.

The chapter aims to identify and explain, in an accessible way, the sustainability challenges of agricultural soil use in the Stellenbosch area. In order to do so, some background information is given to create some understanding of the different soil types and the issue of soil health is discussed, with an emphasis on identifying the kind of thinking that is required to ensure that soil health is maintained into future.

Much of the information provided here about the soils of the Stellenbosch area is based on the author's own experience, gained through soil surveys and assessments performed in the area over many years. The perspective is an agro-ecological one, which recognises the critical importance of the ecological function of soil, both for agriculture in particular and environmental quality in general.

The title of this chapter alludes not only to 'historical roots', in the sense of Stellenbosch's long agricultural history, but also to the underlying 'physical roots' – that which sustains what we see above the surface. When we admire a beautiful tree, how aware are we of the roots that sustain it? Likewise, when we admire the totality of the agricultural and natural landscape surrounding Stellenbosch, how aware are we of what sustains it from beneath the surface? As the *Soil Atlas of Europe* so aptly observes: "We should never forget that soil is the beating heart of the ecosystem. Remove the soil and life within that ecosystem will collapse" (Jones, Montanarella & Jones 2005:13).

Stellenbosch soils: What do we have?

In order to understand what Stellenbosch has in the way of soil resources, it is useful to begin by briefly considering what soil is. Soil is not simply weathered rock. Almost all definitions of soil recognise that it is made up of four essential components, namely mineral grains derived from the weathering of rock, soil moisture from precipitation (containing dissolved salts), air within pore spaces, and organic matter (both living and dead). The functioning of soil is dependent on the interactions between these four components, and material that does not exhibit these interactions is not soil (Gobat, Aragno & Matthey 2004).

The amount of life contained in soil is mind-boggling. In terrestrial ecosystems, there is usually more biodiversity – and sometimes greater biomass as well – below the surface of the soil than above it (*ibid*.). The agro-ecological view of soil emphasises the importance of this living component of soil and actually defines soil as a living body.

The role of soil in agriculture is not simply that of a physical support structure that holds up plants. Rather, soil is a system of complex, interacting ecosystem processes that play a role in all aspects of plant growth, from nutrient supply to disease suppression. Soil is also important beyond its agricultural role for performing several essential ecosystem services. These may be listed as follows (adapted from Seybold, Mausbach, Karlen & Rogers 1998):

- sustaining biological diversity, activity and productivity;
- regulating and partitioning water and solute[1] flow;
- filtering and buffering, degrading, immobilising and detoxifying organic and inorganic materials, including industrial by-products and atmospheric deposition;
- storing and cycling nutrients and carbon within the earth's biosphere; and
- providing physical stability and support for plants or socio-economic structures, or protecting archaeological artifacts associated with human habitation.

Stellenbosch has a range of different soil types with varying characteristics and agricultural capabilities. Differences between soil types are the result of the differential effects of five soil formation factors, namely geological parent material, climate, topography, biota[2] and time. The soils in the Stellenbosch area are derived predominantly from granite, which underlies most of the agricultural land. A small proportion of soils is derived from (or influenced by) Malmesbury shale or Table Mountain sandstone that have been eroded from the higher lying mountains.

Soil type is also strongly influenced by position in the landscape. Thus, similar soils tend to occur in similar landscape positions across the greater Stellenbosch area and differ from those in different landscape positions. Examples of landscape positions include crests, the mid- and lower slopes of mountains and hills, and valley floors.

The scientific classification of soils into different soil types is based on an understanding of the processes that formed a particular soil. In the South African soil classification system, soils are divided into soil forms (based on the particular sequence of soil horizons that occurs), which in turn are sub-divided into families (Soil Classification Working Group 1991). All soil forms are given a South African place name. Common soil forms in the Stellenbosch area are thus named Clovelly, Oakleaf, Tukulu, Glenrosa, Klapmuts and Kroonstad.

Soil classification is extremely inaccessible to non-experts, with the result that the general understanding of soil variation and soil types is very limited. A new basis for grouping soil types at a higher level than the soil form was recently developed by Fey (2010). It provides a more accessible way of describing generalised soil types and will be used here to describe the soils of the Stellenbosch area (see Table 12.1).

There is a very high degree of soil variation in the Western Cape – so much so that soil types may vary over a distance of only tens of metres. Small-scale soil maps, such as the land type ones that have been produced on a national scale, are of limited usefulness for 'on-the-ground' soil use decisions, which usually require detailed soil surveys. Table 12.1 provides a general overview of the soil types in the Stellenbosch area.

---

1 Any substance that is dissolved in a liquid solvent to create a solution.

2 The living organisms of a region.

Table 12.1 Generalised summary description of the prominent soils of the Stellenbosch area

Soil group	Characteristics	Parent geology	Landscape position	Agricultural suitability
Oxidic	Deep, well-drained, uniformly red- to yellow-brown-coloured, loamy soils with good root development potential and high water holding capacity	Granite and associated with relic, deeply weathered surfaces	Slopes of higher lying ground	Very high
Lithic	Depth-limited, reasonably drained, usually brown, loamy, residual soils on weathering rock with reasonable root development potential above the rock, but water holding capacity frequently limited by high stone content	Shales	Convex high-lying crests and slopes	From medium to high, depending on depth
Cumulic (alluvial sands)	Deep, variably drained, either light- or dark-coloured, sandy soils	Alluvial sediment	Valley bottoms	From medium to high, depending on drainage and colour
Cumulic (other)	Moderately deep, reasonably drained, brown, loamy soils, usually with high stone content	Colluvial sediment	Concave higher lying slopes	Medium to high
Gley	Depth-limited, very poorly drained, grey, sandy soils underlain by clay	Granite/shale	Lower lying and flat	Low
Duplex	Depth-limited, reasonably drained, brown to grey, light-textured, usually gravelly soils underlain by clay	Granite/shale	Lower lying slopes	Medium
Plinthic	Variable depth, moderately drained, mottled brown- to grey-coloured, light-textured, very gravelly soils with hard, cemented horizons	Granite/shale	Lower lying slopes	Low to medium-high, depending on depth and drainage

An understanding of how 'good' the Stellenbosch soils are, is essential to any consideration of the sustainability challenges related to them. The concept of agricultural soil capability is used to assess how good a soil is (Schoeman, Van der Walt, Monnik, Thackrah, Malherbe & Le Roux 2002). It comprises an evaluation of the soil's inherent physical and chemical fertility, based largely on the presence or absence of soil limitations that limit plant growth.

There is a tendency among non-experts to associate soil fertility purely with the chemical characteristics of the soil. In fact, as far as soil capability in the Stellenbosch area is concerned, physical factors are usually more important, as nutrient limitations may be fairly easily overcome by means of agricultural inputs. Important physical factors of soil capability are root development potential, water holding capacity, drainage, supply of plant nutrients and pH suitability. Soil limitations that are common around Stellenbosch include depth limitations due to clay or rock in the subsoil; poor drainage due to landscape position or less permeable layers within the soil profile; low nutrient holding capacity; and in sandy soils, low water holding capacity. From an agricultural use perspective, it is more important to understand a soil's limitations than its classification.

The better soils around Stellenbosch generally occur in higher lying positions, particularly along the higher cultivable slopes of the sandstone-topped mountains – the Simonsberg, Jonkershoek and Stellenbosch mountains, and the Helderberg in particular – and the upper slopes of the Bottelary and neighbouring hills. This distribution is illustrated in Figure 12.1. The most sought-after vineyard soils occur on relics of a prehistoric land surface that was deeply weathered in a climate different from the present one (Oberholzer & Schloms 2010). Such relic land surfaces are fairly common around Stellenbosch. Soils of lower agricultural capability generally occur on the lower slopes and valley bottoms.

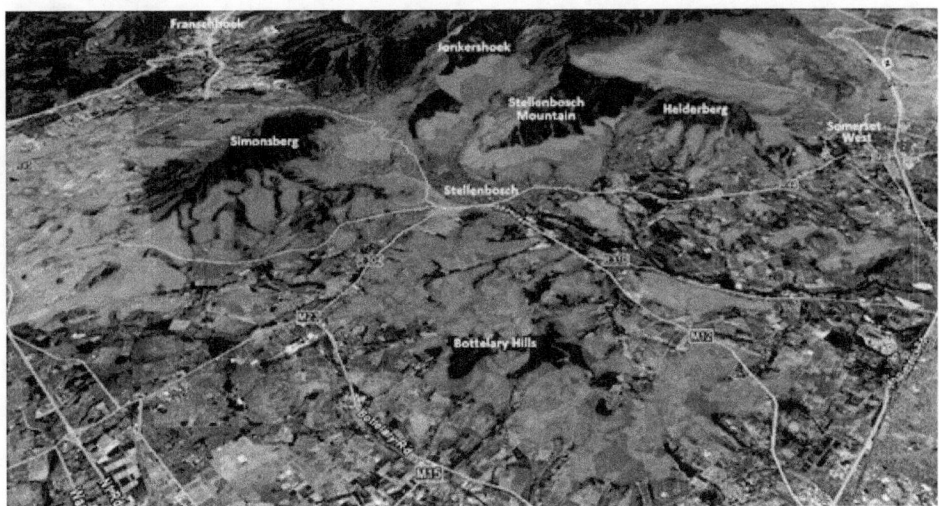

Figure 12.1   Distribution of high-potential soils in the Stellenbosch area [Adapted from Oberholzer & Schloms 2010 © AfriGIS Pty (Ltd); Google earth: 33°48'23.18"S   18°46'29.59"E]

The quality of Stellenbosch soils may be contextualised within the broader South African soil profile, where arable soils constitute only 13% of the total land surface, with only 3% considered as having high potential (Republic of South Africa Department of Agriculture, Forestry and Fisheries (RSA DAFF) 2008). In comparison, Stellenbosch has a far higher proportion of high-potential and high-value agricultural land, as shown in Table 12.2.

Table 12.2   Proportions of the different soil potential categories for the cultivation of wine grapes of soils in the Stellenbosch area[3] [Generated from Oberholzer & Schloms 2010][4]

Soil potential	Potential description	Area (ha)	% of total area
Low	Not recommended	38 106	35
Medium	Conditionally recommended	42 981	40
Medium-high	Recommended	4 058	4
High	Recommended	22 892	21
Total		108 039	100

There are several reasons why the proportion of high-potential soils is higher in Stellenbosch than nationally. Soil resources are mostly evaluated in combination with climate – because there is little value in having good soils if the climate cannot support agricultural production – and the agricultural potential of much of South Africa is largely limited by climate. Fortunately, Stellenbosch is blessed with a favourable climate for a range of high-value crops, particularly wine grapes;[5] importantly, it also has irrigation water available.

The quality of Stellenbosch soils should also be contextualised on a global scale. Although the better Stellenbosch soils are rated as high potential in the South African context, they are still of significantly lower agricultural capability than many soils of other agricultural regions. All Stellenbosch soils are limited in terms of nutrient reserves. They have a low storage capacity for nutrients, with low levels of the essential macronutrient phosphorus in particular, but frequently also of potassium and other essential trace elements, particularly zinc and boron. Moreover, they are naturally acidic and therefore require amelioration with lime to make

---

3   Excluding urban and mountainous areas.

4   Excluding the Durbanville and Cape Flats areas included in the source map.

5   Wine grapes do not really require high-fertility soils. Stellenbosch has a range of different soil qualities and wine may be successfully produced from grapes grown on most of them.

the pH suitable for agricultural production. Probably the most limiting fertility factor of South African soils in comparison to soils in other parts of the world is their very low levels of organic carbon – a result of South African climatic conditions. Low carbon content limits natural nitrogen supply, biological activity, and favourable physical characteristics such as friability and water-holding capacity.

An important point to consider in assessing soil quality and sustainable soil use is that South African soils are inherently very susceptible to degradation (Mills & Fey 2003). This is due in part to the inherent erodibility of the soil and parent material (Le Roux, Newby & Sumner 2007), but also to the very low natural levels of organic carbon in South African soils (De Villiers, Pretorius, Barnard, Van Zyl & LeClus 2002).

Soil degradation is a serious issue in South Africa and high levels of degradation have occurred (Musvoto 2008; Bai & Dent 2007; Gibson, Paterson & Newby 2005; Hoffman & Todd 2000), with erosion (predominantly by water, but also by wind) being perceived as its principal cause. Other recognised forms of soil degradation include compaction, crusting, acidification, nutrient depletion, salinisation, pollution, water logging, and the loss of high-value agricultural land due to urban expansion.

However, in the Stellenbosch area, levels of these overt forms of soil degradation (with the exception of the last one) are minimal. Lanz (2007) proposed several reasons for this, one of which being the challenges posed by local soil conditions, which have resulted in technologically advanced agriculture. High-income, long-term crops requiring high investment predominate, which tends to encourage good agricultural practices. Good agricultural practices are further encouraged by market forces and their requirements for producer accreditation with systems such as EurepGAP and the Integrated Production of Wine (IPW). Other reasons include effective agricultural extension services, farmer study groups, government-subsidised soil conservation works, and the strict application of agricultural legislation.

The agricultural soils of the Stellenbosch area are used mainly for wine production, mainly because of the suitability of the climate. Hence, although future land use may change, a present-day consideration of soil sustainability must focus primarily on the growing of wine grapes. The wine industry works proactively to promote the idea of sustainability and there are several industry initiatives with a sustainability focus,[6] such as Sustainable Wine South Africa, IPW, and the Biodiversity and Wine Initiative (BWI). Moreover, there is a strong market drive for proactiveness around sustainability, particularly from the European market.

For a non-expert who requires an understanding of the soil resources of Stellenbosch, the above discussion may be summarised as follows: soils play a fundamental role, not only in agriculture, but also in environmental quality in general. There is a high degree of soil variation in Stellenbosch and soils differ in their inherent characteristics, agricultural capability, and management requirements. Although Stellenbosch has good soils compared to the rest of South Africa, they still have substantial limitations and significantly lower capability than many soils of other regions. To conclude this section, it may be said that Stellenbosch has good soils supporting a vibrant agriculture; however, because of the national scarcity of high-potential agricultural land and the susceptibility of our soils to degradation, it is imperative that the quality of Stellenbosch's soil resources be maintained in order to ensure sustainability.

Declining soil health: A threat to sustainable agriculture?

Given that Stellenbosch has a high proportion of good agricultural land and low levels of soil degradation, it may appear as though there are no problems regarding the sustainability of soil resources in the area. However, the argument will be made here that declining soil health poses a potential sustainability challenge and that it should be appropriately and timeously addressed.

The concept of soil health

Soil health is a relatively new concept in agriculture that has been developed over the last two decades in response to increasing awareness of environmental concerns and sustainability issues. Although there are slightly different understandings of soil health in the literature, the one advocated in this chapter is as follows:

---

6 An additional contextual factor in considerations of soil and sustainability in Stellenbosch is the trend of decreasing profitability in Stellenbosch wine farming (Van Wyk & Le Roux 2010).

soil health is the capacity of soil to function ecologically and, in so doing, supply a range of ecosystem services, including those that support agricultural production (Kibblewhite, Ritz & Swift 2008; Gruver & Weil 2007; Doran, Sarrantonio & Liebig 1996). Soil health is not about preserving soils in their natural state, but rather about optimising their biological function for the delivery of ecosystem services. Examples of ecosystem services that support agricultural production include the maintenance of soil structure, nutrient cycling and disease suppression.

The soil health concept is primarily a response to concerns about subtle forms of soil degradation that are not easily detectable in the short term, but which may lead to longer term loss of soil function, and consequently the loss of the ecosystem services supported by soil function. Soil health recognises soil biology as being central to the functioning of soil and acknowledges that biological properties are interlinked with chemical and physical ones (Haynes & Graham 2004). It is a holistic and integrative approach to understanding soils (Herrick, Brown, Tugel, Shaver & Havstad 2002), based on ecological and systems thinking. Unlike the concept of soil capability discussed above, soil health focuses on those aspects of soil condition that may be changed over fairly short periods of time through appropriate agricultural management. Declining soil health is associated with intensive agricultural use, with the depletion of soil organic matter and the reduction of biological function in soils presenting particular areas of concern (Karlen, Andrews & Wienhold 2004; Weil & Magdoff 2004; Pankhurst, Doube & Gupta 1997).

Many farmers relate readily to the concept of soil health, often referring to it as 'life in the soil'. They distinguish between healthy soils, which have life in them, and unhealthy (or 'dead') ones; soils that are in the process of being regenerated from dead soils are said to be 'getting the life back into them'.

Despite the attention paid internationally to the concept of soil health in agriculture and scientific literature, much about it remains scientifically unknown and uncertain. There are different perceptions of exactly what constitutes soil health and how important it is, particularly in relation to crop production.

Difficulties in measuring soil health

A large number of different biological, chemical and physical indicators of soil health have been used and evaluated with the aim of assessing and measuring soil health. Commonly used indicators include organic matter content, active carbon content, aggregate stability, bulk density, water holding capacity, soil depth, field penetration resistance, microbial biomass, soil respiration, potential nitrogen mineralisation, and standard chemical composition (see Gugino, Idowu, Schindelbeck, Van Es, Wolfe, Moebius, Thies & Abawi 2007; Pankhurst *et al.* 1997; Doran *et al.* 1996).

Any assessment of soil health requires the use of several different indicators. A minimum data set for soil health is the smallest set of soil properties that may be used to characterise or measure soil health, and this varies based on the intended land use, soil type and climate. A single minimum data set of indicators will probably always remain undefined, because of the inherent variability among soils. Thus, in summing up the state of soil quality assessment indicators, one may concur with Karlen *et al.* (2004) when they say that there is no conclusive agreement in the international scientific community on the relative importance of the different indicators or their measurement methods.

The local context of soil health

In South Africa in particular, the soil health concept is neither well understood nor widely accepted. There is a lack of scientific expertise in those aspects of soil science that pertain to soil health (Haynes & Graham 2004) and its importance for crop production and agricultural sustainability remains contested in South African agricultural science. As a result, there is almost no local data available about soil health trends, and methods and capacity to monitor soil health are largely non-existent. In contrast, most European countries have established scientific programmes to monitor soil health (Bloem 2006).

As for Stellenbosch, the concept of soil health is also not mainstreamed in the wine industry. For various reasons (which were explained earlier in this chapter), there is minimal overt soil degradation in the Stellenbosch area. However, these reasons do not address issues of soil health. For example, IPW requirements protect the soil resources in basic ways, but do not address soil health or the biological aspects of soil in any direct way. Thus, although there may not be obvious soil degradation in Stellenbosch, there may well be declining soil health. Moreover, the research capability to measure whether soil health is declining or not is inadequate at this stage. Given this situation and the resultant lack of hard data regarding soil health decline, there is a real threat that soil health may well be declining. The possibility alone demands that something be done about it; at the very least, a concerted effort should be made to achieve increased certainty and understanding of this issue.

A challenge to the established paradigm

The failure to address soil health results not only from a lack of knowledge, but also from a constraint in the thinking that currently informs agriculture. The current agricultural scientific paradigm is based on a need for conclusive evidence. The understanding of soil ecology has increased immensely in recent years, but because of its complexity and the large knowledge gaps that still exist, science is not yet at a point where it is able to provide conclusive evidence for relationships between soil biological function, crop production and agricultural sustainability. Furthermore, soil is a complex system and conclusive scientific evidence is only obtainable through a reductionist analytical approach, which is inherently ineffective for understanding complex systems. Therefore, current agricultural scientific thinking cannot effectively address the issue of soil health. Science tends to focus on things that may be measured to the exclusion of those that cannot, which may lead to a refutation of the importance of those things that cannot be measured. This is also the case with soil health, with a fairly significant voice within the Stellenbosch agricultural community claiming that it is of no consequence.

However, it is the contention here that the danger of declining soil health represents a major sustainability challenge in terms of soil resources in Stellenbosch. Declining soil health means a loss of soil resources, which will have negative consequences for the already-challenged profitability of viniculture, as a decline in soil health results in a reduction in the quantity and/or quality of production, or a need for increased inputs to achieve the same levels of production. Such inputs would include increased fertilisation, increased physical soil manipulation and increased pest control. Peak oil (and potentially peak phosphorus as well) are likely to impact even further on the future costs of these inputs.

If the key challenge for the sustainability of soil resources is to ensure the maintenance of soil health and current thinking presents a constraint to doing so, then the question must be asked: what kind of thinking will make it possible to address this challenge?

How might we think differently?

The kind of thinking required to address the challenge of potentially declining soil health must be holistic and integrative. It must be rigorous, but cannot be reductionist. It must be a thinking that is appropriate to understanding living systems, which, as Ulanowicz (2009) has argued, differs radically from the thinking required to understand non-living systems. In this regard, it is interesting to note that the thinking on which soil science is based has always been directed at understanding soil as a non-living system. Perhaps it is time for this to change.

Agro-ecology offers a way of thinking that meets the abovementioned requirements. Agro-ecology may be described as the application of ecological principles and approaches to agricultural ecosystems, and involves designing the strengths of natural ecosystems – such as efficiency, diversity, self-sufficiency, self-regulation and resilience – into the agro-ecosystem (Magdoff 2007). Agro-ecology fits well with an integrative understanding of soil health (as described above and by Kibblewhite *et al.* 2008), which emphasises ecological function and recognises the capacity of soil to maintain its own health through feedback.

Justification for agro-ecological thinking is strengthened by the similarity it bears to the complexity thinking advocated by Cilliers (1998). Complexity thinking argues that the behaviour of complex systems exhibits important differences from the behaviour of non-complex systems and it highlights the limitations of reductionism in understanding the behaviour of complex systems. Soils display all the characteristics of complex systems as described by Cilliers (*ibid.* and 2000) and must therefore be considered as such. However, acknowledging a system as complex carries certain implications.

The implications that may be derived from a complexity perspective and applied to soils suggest the following: in order to develop an effective understanding of a complex system, it is necessary to focus on relationships and organisation within the system, rather than simply on the analysis of its components. Soil fertility (and soil health) may be understood as emergent properties of the whole soil system, and they result from and are influenced by all the interactions within the system. For this reason, an understanding of soil behaviour cannot be based on a separation between the biotic and abiotic components of soil, as has been the case historically in the discipline of soil science. The implications of a complexity perspective demand that the importance of soil biology be acknowledged, and that the ecological functioning and biodiversity of soil be recognised as important components of soil fertility.

Agro-ecology and the concept of soil health represent complexity perspectives on soil. Such thinking requires a paradigm shift from the way in which soils have historically been understood in agricultural science. There are farmers in the Stellenbosch area who have successfully made this shift and conversations with them reveal some of the characteristics of this mind shift and its impact on their farming practices.[7]

Following a different paradigm

The primary characteristic of these farmers' thinking is that it is holistic and approaches the whole farm as a system. Management interventions are not aimed at a single outcome, but are performed primarily with a view to balancing the system as a whole. This tends to shift the focus from what is *most efficient* to what is *most sustainable*. Although not expressed in such terms, soil is seen as a functioning ecosystem and fertility as an ecosystem service, not something that comes from a bag of fertiliser; and soil health is seen as the basis of more sustainable farming. These farmers display a respect for life and a belief in the restorative capacity of natural processes. Their motivations for adopting such an approach are twofold: market opportunities and an environmental ethic that encourages them to minimise environmental impact through the agricultural practices they choose.

These farmers evaluate their management practices primarily by assessing crop performance in the field, which requires close observation and an intimacy with what is going on in the vineyard. They express a need to be in the vineyard, not the office, and rely on a well-developed intuitive capacity for assessing the condition and reaction of their vineyards in order to monitor and make decisions about their management practices. They place less emphasis on control and display a greater willingness to let nature run its course, for example by using weeds to provide groundcover and harbour beneficial organisms, rather than controlling them. They also make little use of analytical measurement methods such as soil analysis, because their experience is that a reliance on analytical results retards the development of their intuitive capabilities.

The farmers attribute improvements in their soil to more environmentally friendly agricultural practices. Common observations include an increase in the sponginess of the soil underfoot; a strengthening of the soil structure (which may be seen in the tendency to cling to a pulled-up root); a darker soil colour; and a compost-like smell in the soil (indicating biological activity).

These agro-ecological farmers acknowledge that it is not an easy management approach and that there are numerous challenges and practical problems to overcome. However, they express a passion for facing up to these challenges and a belief in the possibility of overcoming them. They also declare increased enjoyment of the farming experience and pleasure at the greater engagement, creative response and ongoing learning that it requires.

The approach of these farmers may be described as a holistic one that includes the concept of 'self'. Such an approach draws into question the efficacy of a purely objective approach, which relies exclusively on rationality, for dealing with complex systems. In describing his approach to dealing with a complex world, Kauffman (2008) states:

---

7   These conversations, which serve as the basis for the section that follows, took the form of a series of interviews conducted by the author as part of a soil health project (see Lanz 2007).

In confronting a complex world, we confront the reality of the need to understand our full humanity. Reason alone is an insufficient guide to our actions. We are called on to re-integrate our entire humanity in the living of our lives.

Kaufmann's words resonate with Cilliers's (1998) idea that action with respect to a complex system requires an ethical choice, because an objectively correct choice can never be absolutely determined. This approach suggests that we need to move beyond the requirement of conclusive proof before being able to take action. In terms of this approach, making soil management decisions requires ethical choices, which must be made within the wider context of an interdependent global ecology and for which the individual must assume responsibility. This raises the question: are complex agricultural systems best managed through measurement and control or through the development of intuition? The answer is that, in future, managing these systems will probably require both.

The thinking required to maintain soil health must be able to make sense of the complexity of the challenge and allow for the possibility of acting differently to overcome them. In order to address the sustainability challenges presented by declining soil health, the kind of thinking described above will not only have to be adopted by farmers, but will also have to form part of the science that will inform decisions around soil resource use and management into the future. Such thinking is a direct challenge to some of the assumptions of the current agricultural and scientific paradigm.

Conclusion: Embracing a healthy future, from the ground up

What are the potential outcomes of the kind of thinking advocated above? What might such thinking make possible?

Thinking more holistically and allowing for different possibilities stimulates creative solutions to local problems. This is important, because a given combination of soil, climate and other conditions creates unique circumstances. For this reason, a simple transfer of solutions that proved effective in other contexts is often not successful, and locally applicable solutions, based on indigenous knowledge, must be developed. The recognition that conscious action is required despite the lack of an objectively determined correct choice for guidance stimulates a culture of experimentation and a more responsive and flexible approach to agriculture that is likely to adapt more effectively to changing future conditions, including climate change. Furthermore, it provides internal motivation for achieving sustainability and encourages land managers to make responsible choices, rather than to simply comply with externally imposed regulations, which is a less effective way of bringing about positive change.

Taking soil health seriously also provides opportunities in terms of the green economy and investments in future soil productivity. Although there are several obstacles to agricultural land managers receiving carbon market funding to support and incentivise soil health improvements, it remains a possibility. Other ways of attributing financial value to soil health as natural capital may also be developed. Another benefit that may accrue as a result is the use of natural soil fertility as a buffer against the cost increases associated with future oil peak (and potentially phosphorus peak as well).

The thinking advocated in this chapter offers an important opportunity to ensure a healthy foundation for our agro-ecosystem into the future by maintaining soil health. If, however, we continue to pursue current thinking and so largely ignore the issue of soil health, the failure to address it may over time lead to the irreversible loss of our valuable soil resources and threaten the sustainability of agricultural production in Stellenbosch and its surrounding areas.

REFERENCES

Bai ZG & Dent DL. 2007. *Land degradation and improvement in South Africa. 1. Identification by remote sensing*. Report 2007/03. Wageningen: International Soil Reference and Information Centre (ISRIC) World Soil Information.

Bloem J. 2006. Monitoring and evaluating soil quality. In: J Bloem, DW Hopkins & A Benedetti (eds). *Microbiological methods for assessing soil quality*. Wallingford: CABI Publishing.

Cilliers P. 1998. *Complexity and Postmodernism: understanding complex systems*. London: Routledge.

Cilliers P. 2000. What can we learn from a theory of complexity? *Emergence*, 2(1):23-33.

De Villiers MC, Pretorius DJ, Barnard RO, Van Zyl AJ & LeClus CF. 2002. *Land degradation assessment in dryland areas: South Africa*. Paper prepared for the United Nations Food and Agriculture Organisation (FAO) Land Degradation Assessment in Drylands project. [Accessed 19 October 2010] http://www.lada.virtualcentre.org/eims/download.asp?pub_id=97320&app=0

Doran JW, Sarrantonio M & Liebig MA. 1996. Soil health and sustainability. *Advances in Agronomy*, 56:1-54.

Fey MV. 2010. *Soils of South Africa*. Cape Town: Cambridge University Press.

Gibson D, Paterson G & Newby T. 2005. *Land: Background Research*. Paper prepared for the South African Environment Outlook (SAEO) report on behalf of the Department of Environmental Affairs and Tourism (DEAT). [Accessed 19 October 2010] http://www.soer.deat.gov.za/dm_documents/Land_-_Background_Paper_Z6OqR.pdf

Gobat J-M, Aragno M & Matthey W. 2004. *The living soil: fundamentals of soil science and soil biology*. Enfield, NH: Science Publishers.

Gruver JB & Weil RR. 2007. Farmer perceptions of soil quality and their relationship to management-sensitive soil parameters. *Renewable agriculture and food systems*, 22(4):271-281.

Gugino BK, Idowu OJ, Schindelbeck RR, Van Es HM, Wolfe DW, Moebius BN, Thies JE & Abawi GS. 2007. *Cornell Soil Health Assessment Training Manual*. New York: Cornell University Press.

Haynes RJ & Graham MH. 2004. Soil biology and biochemistry – a new direction for South African soil science? *South African Journal of Plant and Soil*, 21(5):330-344.

Herrick JH, Brown JR, Tugel AJ, Shaver PL & Havstad KM. 2002. Application of soil quality to monitoring and management: paradigms from rangeland ecology. *Agronomy Journal*, 94:3-11.

Hoffman MT & Todd S. 2000. A national review of land degradation in South Africa: the influence of biophysical and socio-economic factors. *Journal of Southern African Studies*, 26(4):743-758.

Jones A, Montanarella L & Jones R (eds). 2005. *Soil Atlas of Europe*. Luxembourg: European Commission.

Karlen DL, Andrews SS & Wienhold BJ. 2004. Soil quality, fertility and health – historical context, status and perspectives. In: P Schjønning, S Elmholt & BT Christensen (eds). *Managing soil quality: challenges in modern agriculture*. Wallingford: CABI Publishing.

Kauffman S. 2008. *Reinventing the Sacred: A New View of Science, Reason and Religion*. New York: Basic Books.

Kibblewhite MG, Ritz K & Swift MJ. 2008. Soil health in agricultural systems. *Philosophical Transactions of the Royal Society*, 363:685-701.

Lanz J. 2007. Land and soil study as part of SEA for the Cape Winelands District Municipality. Unpublished report submitted to the Council for Scientific and Industrial Research (CSIR). Stellenbosch: CSIR.

Le Roux JJ, Newby TS & Sumner PD. 2007. Monitoring soil erosion in South Africa at a regional scale: review and recommendations. *South African Journal of Science*, 103:329-335.

Magdoff F. 2007. Ecological agriculture: principles, practices and constraints. *Renewable Agriculture and Food Systems*, 22(2):109-117.

Mills AJ & Fey MV. 2003. Declining soil quality in South Africa: effects of land use on soil organic matter and surface crusting. *South African Journal of Science*, 99: 429-436.

Musvoto C. 2008. *Emerging Issues Paper: Land Degradation*. Paper prepared for the Department of Environmental Affairs and Tourism (DEAT). [Accessed 19 October 2010] http://www.soer.deat.gov.za/Land_XizdF.pdf.file

Oberholzer B & Schloms H. 2010. *Catena: Soil Associations of Stellenbosch*. Paarl: Winetech.

Pankhurst CE, Doube BM & Gupta VVSR. 1997. Biological indicators of soil health: synthesis. In: CE Pankhurst, BM Doube, & VVSR Gupta (eds). *Biological Indicators of Soil Health*. New York: CAB International.

RSA DAFF (Republic of South Africa. Department of Agriculture, Forestry and Fisheries). 2008. *Abstract of Agricultural Statistics*. [Accessed 17 September 2010] http://www.agis.agric.za

Schoeman JL, Van der Walt M, Monnik KA, Thackrah A, Malherbe J & Le Roux RE. 2002. *Development and application of a land capability classification system for South Africa*. Final report to the Department of Agriculture: Directorate of Agricultural Statistics, GW/A/2000/57. [Accessed 15 April 2005] http://www.agis.agric.za

Seybold CA, Mausbach MJ, Karlen DL & Rogers HH. 1998. Quantification of soil quality. In: R Lal, JM Kimble, RF Follett & BA Stewart (eds). *Soil Processes and the Carbon Cycle*. Boca Raton, FL: CRC Press.

Soil Classification Working Group. 1991. *Soil Classification – a taxonomic system for South Africa. Memoirs on the Agricultural Natural Resources of South Africa No 15*. Pretoria: Department of Agricultural Development.

Ulanowicz RE. 2009. *A third window: natural life beyond Newton and Darwin*. West Conshohocken: Templeton Foundation Press.

Van Wyk G & Le Roux F. 2010. VinPro-opname – die produksieplan, die koste van druiweproduksie en produsentewinsgewendheid. *Wineland*, May 2010:81-86.

Weil RR & Magdoff F. 2004. Significance of soil organic matter to soil quality and health. In: F Magdoff & RR Weil (eds). *Soil organic matter in sustainable agriculture*. Boca Raton, FL: CRC Press.

AGRICULTURE
From vulnerability to viability

GARETH HAYSOM & LUKE METELERKAMP

Introduction

In 2007, the then head of the International Food Policy Research Institute (IFPRI), Joachim von Braun, stated that the world food situation was being rapidly redefined by new driving forces and that changes in food availability, rising commodity prices, and new producer-consumer linkages held crucial implications for the livelihoods of poor and food-insecure people. While this statement refers specifically to food security, the underlying challenge applies to agriculture and the agricultural economy as a whole. The point had (earlier) been articulated slightly differently by Uphoff (2002), who asserted that the agricultural technologies that had been developed and extended over the previous four decades had contributed to unprecedented growth in world food production; for example, it had resulted in a doubling of global grain output between 1965 and 1990 – a remarkable achievement drawing on the skills and innovations of thousands of scientists, extentionists and farmers[1] and backed by the supportive decisions of policy makers. However, the same author cautioned that "there are now growing concerns that this strategy of agricultural development may not be the best, or the only way, to promote agriculture in the future as, amongst other things, it has costs as well as benefits" (Uphoff 2002:3).

Uphoff's warning requires further attention and understanding when considering the future of agriculture in the Stellenbosch region. Can Stellenbosch's agriculture continue – as it has over the past years – to innovate and adapt to changes in what is arguably a reactionary manner, or is an alternative approach required? What are the likely consequences of continuing with a business-as-usual approach, and how are farmers and those involved in the agricultural sector introducing changes that are radically different? The question is further complicated by the range of issues confronting farmers today – a situation that makes planning for change over the next five years a considerable challenge, and renders any strategies beyond that and for the next twenty years even more complex and challenging.

This chapter is not intended to be read as a detailed empirical analysis of the existing Stellenbosch agricultural system; nor does it purport to offer a look into a metaphorical crystal ball to map out what agriculture should resemble twenty years from now. Instead, it offers an overview of trends and nodes of innovation, informed by case studies and the stories of farmers active in the agricultural sector in the broader region. This approach is informed by the core contextual reference point of the chapter, namely that current and emerging issues must be considered and the impact of these issues on the specific sector under review, now and into the future, must be better understood before any attempt can be made to imagine future scenarios. For this reason, the 'polycrisis' concept (to be discussed in greater detail further in the chapter) will serve as a basis from which to argue for an alternative perspective in considering the future of agriculture, and particularly that of agriculture in the Stellenbosch region.

---

1 Here and elsewhere in this chapter, 'farmers' refer to both men and women involved in the production and beneficiation of a variety of agricultural typologies in a wide variety of contexts. The term is used generally and refers to all farming, unless specifically stated otherwise.

This chapter does not purport to provide a definitive view of the future of agriculture; rather, it is intended to be read as a 'think piece', aimed at stimulating discussion and debate. In order to facilitate this dialogue, the chapter will provide a brief overview of the general agricultural landscape in South Africa before focussing more narrowly on that of Stellenbosch.

Two sets of case studies will be presented to develop an argument for an alternative agricultural future. The studies were specifically conducted in order to gain an understanding of and document the shifts taking place within certain agricultural communities, and will serve as a basis to argue that shifts are indeed already taking place, but have not as yet entered into the agricultural mainstream, and that although such shifts remain on the periphery, they are nonetheless significant indicators of change.

Finally, the chapter is also informed by the overarching principles of sustainable development and is therefore oriented towards a more sustainable approach to agriculture. Such an orientation often presents a challenge, as discussions pertinent to sustainable agriculture are often boxed into the organic agriculture paradigm. Regrettably, this tends to confine debate to simplistic views either advocating for or against this type of agriculture. The authors are of the opinion that the term 'organic agriculture' does not necessarily denote sustainable agriculture; we therefore prefer to use the term 'agro-ecology' to describe the envisioned sustainable farming systems. Agro-ecology is described by Miguel Altieri as the discipline providing the basic ecological principles for the study, design and management of agro-ecosystems that are both productive and conserve natural resources, while at the same time being culturally sensitive, socially just and economically viable (Altieri 1995).

The relationship between sustainable development and new perspectives on agriculture was demonstrated by Reijntjes, Haverkort and Waters-Bayer:

> ... today, the question of agricultural production has evolved from a purely technical one to a more complex one characterised by social, cultural, political and economic dimensions. The concept of sustainability, although controversial and diffuse due to existing conflicting definitions and interpretations of its meaning, is useful, because it captures a set of concerns about agriculture which is conceived as the result of the co-evolution of socio-economic and natural systems.
>
> [Reijntjes *et al.* 1992:2]

Our discussion of the agricultural landscape of Stellenbosch will therefore be informed by the complementary concepts of sustainability and agro-ecology.

Sustainability, the polycrisis and agriculture

Swilling and Annecke refer to the convergence of the current global stresses of climate change, peak oil, ecosystem degradation, increasing inequality, rapid urbanisation (particularly in sub-Saharan Africa), increasing food and water scarcity, hegemonic shifts and an ongoing series of financial crises as a "global polycrisis" (2012:xxii). In doing so they join a growing number of individuals and institutions highlighting the role that agriculture has played in precipitating this polycrisis (Swilling and Annecke 2012; Bates & Hemenway 2010; Lal 2010; the United Nations Food and Agriculture Organisation (FAO) 2009; the Intergovernmental Panel on Climate Change (IPCC) 2007a; Magdoff 2007; the *Millennium Ecosystem Assessment* (MEA) 2005).

However, as Swilling and others also demonstrate, agriculture is not only a significant driver of the polycrisis globally; it is also being adversely affected by it. This suggests that in order to achieve long-term sustainability, agriculture will have to adapt to the impacts of the polycrisis by seeking not only to mitigate, but reverse the negative trends it helped to create. A reduction in non-renewable inputs, facilitated by increased resource efficiency and a shift towards increasingly (agro-ecological) self-produced inputs, is an important criterion for the restoration and adaptation of agriculture in the context of the global polycrisis (Lal 2010; Pretty 2006; Scherr 2000; Altieri 1999).

The South African agricultural landscape

Vink and Van Rooyen describe South African agriculture as follows:

> [It is a sector] highly exposed to global markets, as farmers receive few subsidies; international trade (imports and exports) makes up a large proportion of total production; and trade at the country's borders has been substantially liberalised. Farmers' incomes are therefore highly dependent on movements in the exchange rate and on global economic conditions [...] There has been a significant increase in the concentration of farm holdings within the commercial agricultural sector. In 1996, there were 60 000 farming units, but by 2007 this had declined to fewer than 40 000 units, suggesting a consolidation of landholding into larger units of ownership and production.
>
> [Vink & Van Rooyen 2009:5,13]

This consolidation is expected to continue and some commentators place the total farming units at present at less than 35 000 (Pettersson 2011). These figures are, however, misleading, as they do not take into consideration the vast number of subsistence and small-scale farms in South Africa. Citing data from a Presidency Fifteen Year Review Project, Vink and Van Rooyen (2009) reported[2] that of the estimated 8 million households living in the non-metropolitan areas of South Africa at that time, 17% (or 1.3 million households) had access to land for farming purposes. Most of these households (97%) engaged in some farming activity, mostly on relatively small plots of land.

Geographically, 35% of the land surface of South Africa receives sufficient rain for dryland crop production, but only 13% (or 14 million hectares) is suitable arable land. In addition, most of the available 14 million hectares is marginal land, with only 3% of the land considered as high-potential land. If one were to apply the international norm of 0.4 hectares of arable land to feed one person, South Africa's 14 million hectares would feed at most 35 million people. Hence, if we only followed the international norm, the net result would be overexploitation as we exceed the carrying capacity of our soils (Swilling 2008). This points to an agricultural system that is at, or beyond, its limits. The reliance on an ever-shrinking group of farmers to produce the bulk of South Africa's food; a large but largely unappreciated group of smaller scale farmers left on the margins; and ecological stresses that are being compounded by the converging challenges of the polycrisis all point to the fact that South African agriculture is a high-risk endeavour and is increasingly vulnerable to a variety of challenges. In addition, there is a number of key focus areas illustrating the need for great urgency in redesigning the agricultural sector in South Africa so that it may respond to these challenges, which include declining soils and water scarcity; unresolved land reform and dualistic nature of South African agriculture; innovation focused only on large-scale industrial agriculture; and the fact that people are removed from food sources.

However, discussions on the national level, while indicative of the broader issues, must take local contexts into consideration. There is a variety of different 'agricultures' in South Africa, Stellenbosch being a case in point.

The Stellenbosch agricultural system

From an agricultural perspective, Stellenbosch is often considered to be successful, with financially viable farms offering premium products that are sought after on both local and international markets. In fact, this is very far from reality, with many farms operating at the margins. The Stellenbosch agricultural sector, dominated as it is by wine, experiences many of the issues raised by Vink and Van Rooyen, specifically in terms of their exposure to exchange control fluctuations and international market trends.

However, not all challenges associated with South African agriculture are directly applicable to the Stellenbosch agricultural sector. For years Stellenbosch farmers have added value to the regional economy by being more than just farmers and applying strategies that have contributed greatly to the increased viability of Stellenbosch agriculture. Arguably the two most important such strategies are integration into the value chain, where farms beneficiate agricultural produce into viable and value-added retail items (grapes into wine); and diversification, as evidenced by the ever-growing agritourism industry, which have seen farms transformed into tourist

---

2   Vink and Van Rooyen drew this data from: Tregurtha N, Vink N & Kirsten J. 2008. Presidency Fifteen Year Review Project: Review of agricultural policies and support instruments 1994-2007. Pretoria, Unpublished.

destinations. While these strategies provide for increased viability, they also potentially mask the real threats related the sustainability of production methodologies.

Research compiled by Kelly and Schulschenk (forthcoming) used existing statistical information to present a general overview of the current status of Stellenbosch agricultural production, distribution and consumption in order to determine key vulnerabilities, as well as opportunities to strengthen resilience. This research identified wine and stone fruit farming as the dominant agricultural activities (both in terms of land use and rand value), with vegetable and essential oil production occurring on a smaller scale. Deciduous fruits (including wine grapes) contribute 87.5% to gross farm income, vegetables 9.9%, and other horticultural products 2.6%. Wine grapes occupy the largest surface area (71.5%), followed by peaches (9.6%). Both the wine grape and fruit markets are export-oriented. Commercial vegetable production (5 211 tonnes per year) consists mainly of cabbage, tomatoes, onions and green beans. The contribution of Stellenbosch agriculture to overall South African exports is significant: 27% of viticultural products, 29% of deciduous fruit, and 17% of table grapes exported in 2006. In 2010, wine exports amounted to 54 million litres, or 49% of the total harvest (Matoti 2011; Louw 2009; Statistics South Africa 2007; Statistics South Africa 2006 in Kelly and Schulschenk forthcoming).

Although agriculture is not in itself a large employer in Stellenbosch, it indirectly supports a number of other sectors (Stellenbosch Municipality 2010), in particular manufacturing (see Table 13.1). Manufacturing is the largest employer in Stellenbosch, followed by wholesale, retail, and community and personal services, in that order. This underscores the importance of the agricultural sector in terms of providing livelihoods and indirectly contributing to wage income for people employed in the region. The *Stellenbosch Municipality IDP 2010* presents a picture of an economy still immersed in the rural nature of the region with "manufacturing strongly linked to the agricultural activities of the region" (*ibid.*:19). In fact, an estimated 20% of the Stellenbosch economy is linked directly to agriculture. The agricultural landscape further adds significant value to the tourism and leisure sector. If these mutually supportive values are considered collectively, agriculture contribution could be as high as 30%.

Table 13.1   Percentage of workers employed in the various economic sectors (2007) with author estimates for indirect contributions added [Adapted from Stellenbosch Municipality 2010]

Percentage of workers employed in the various economic sectors in 2007		
Sector	%	Established agriculture[*] %
Agriculture, forestry and fishing	6.9	6.9
Mining	0.9	0
Manufacturing	20.1	6.0
Electricity and water	0.5	0
Construction	7.7	1.0
Wholesale and retail trade; catering and accommodation	16.2	8.0
Transport and communication	2.0	0.5
Financial and business services	8.3	2.0
Community, social and other personal services	15.9	3.5
Undetermined	13.8	1.0
Unreported in IDP**	7.7	
Total	100%	19.9%

* All figures excluding listed agriculture are estimates calculated at approximately 25% of specific sector for all sectors excluding tourism where estimated at 50%.
** This item was added and calculated by the authors due to the shortfall in the *Stellenbosch Municipality IDP 2010* data.

A strategic response to the global polycrisis within the Stellenbosch agricultural system would need to focus on building resilience to future shocks (Kelly & Schulschenk forthcoming) through further diversification; a transition to more diversified agricultural production; and a reduction in the exposure to external constraints such as increased input costs, market threats and increasing ecological threats. These ecological and economic threats are most evident in the scenarios projected by ongoing climate change work conducted by a collection of South African climate scientists, led by the School of Bioresources Engineering and Environmental Hydrology at the University of KwaZulu-Natal. The research, which entails overlaying a number of different climate models, projects an increase in evaporation of between 5 and 10% by 2050, and a further increase of between 15 and 25% by 2090. Perhaps even more significantly for the Stellenbosch agricultural system, with its emphasis on fruit production, is the finding that there will be a marked reduction in Positive Chill Units (PCUs): these units are critical to the formation and development of fruit and are expected to decrease by about 60% by 2090, with a smaller but still significant reduction between now and 2030. Climate change is also already resulting in drastic changes in pest behaviour, with, for example, increases in the mean annual mating hours of the African sugar cane borer (*Eldana saccharina*; Lepidoptera: Pyralidae)[3] (Schulze 2011).

These climate-related threats and other components of the polycrisis present real challenges for current production typologies and call for alternative responses and innovation in agriculture. Agriculture has been, and remains, one of the core drivers of climate change; but agriculture, through different management approaches, is also one of the few industries that has the potential to play a critical role in mitigating its effects. In order to understand how farmers might and should respond to these threats, valuable lessons may be learned from those farmers already engaged in transformative activities.

The agricultural challenge

Shifts in production approaches in the Global North tend to be driven by a growing public demand for food that is both healthier and less harmful to the environmental and social systems in which it is produced and consumed (Pretty 2006; Taylor, Madrick & Collin 2005; Halweil 2004). The trend is different in the South, where it is increasingly argued that agriculture should be founded on the principles of resilience and self-reliance (FAO 2008; International Assessment of Agricultural Science and Technology for Development (IAASTD) 2008b; Pretty 2006). It is believed that this may be achieved through the use of localised, low external input (LEI) forms of agriculture, which work in much closer partnership with natural systems than modern forms of agriculture have thus far done (Holt-Gimenez & Patel 2009; FAO 2008; Pretty 2006; Altieri 1999). These systems place the power of production in the hands of farmers and local communities, encouraging them to form restorative partnerships with soils, animals and other living systems in order to reduce their dependence on credit providers, agrochemical multinationals and international trade policies (Holt-Gimenez & Patel 2009; Altieri 1999). As Pretty (2006) shows, the debate at an academic level appears to be polarised between those who agree with the change towards LEI systems and those in favour of maintaining and expanding high-input Green Revolution (GR)[4] methods, with very little meaningful dialogue between the two camps.

As we shall see, the reasons for the changes in approach of the farmers in the Swartland case study are more pragmatic. The cost of agricultural inputs such as fertilisers, pesticides, diesel and farm machinery have risen steeply over the past decade – especially during the last five years (Grain SA 2009; National Agricultural Marketing Council (NAMC) 2009) – with a significant knock-on effect on production costs. There are a number of reasons for this trend, including the close connection between the price of oil and the cost of synthetic chemical inputs (and transport in general), the corporate monopolisation of the input market, and supply shortages (FAO 2009; Holt-Gimenez & Patel 2009; Hopkins & Holden 2007). In fact, the impact of continually rising oil prices on a broad range of agriculture-related fields is a key component of the argument for decreasing the dependence of farming systems on oil and other external inputs (Swilling & Annecke 2011; Hopkins & Holden 2007; Heinberg 2003).

From an ecological perspective, declining soil health also plays a major role in the cost of producing food, as degraded soils are less productive relative to the inputs applied to them (Intergovernmental Panel on Climate Change (IPCC) 2007a; Pretty 2006). Put another way, degraded soils require more inputs (which translates into

---

3 While of little consequence for Stellenbosch agriculture, *Eldana saccharina* is a real threat in sugar and maize growing regions. It is mentioned here to demonstrate the relationship between pests and climate change.

4 'Green Revolution' refers to farming practices that make use of synthetic petrochemical-based inputs and fertilisers, and often incorporates proprietary hybrid seeds, irrigation systems and access to credit of some kind to fund farming operations.

higher costs) compared to healthy ones in order to deliver equal yields, and ultimately they may fail to deliver altogether (Lal 2006; Scherr 1999). As a result of mismanagement by both traditional and green revolution farmers, one-third of all the world's agricultural land is either moderately or severely degraded (IAASTD 2008a). Conway (1997) notes that the progressive loss of on-farm biodiversity and concomitant increase in the use of pesticides are resulting in growing pest problems, because pests adapt to external control measures and natural pest control mechanisms are inadvertently removed from the system.

Considering that declining soil health leads not only to impoverished ecosystems, but also to less competitive and productive farms (and that up to 30% of employment in Stellenbosch derives from farming – see Table 13.1), protecting and improving Stellenbosch's soils should be key to the development strategy of the region. Such a drive towards enhancing soil health should not be seen as an environmental quest to restore soils to their *natural* state, but should instead be based on the principle that soil constitutes an expandable ecological asset that has the potential to underpin sustainable social growth in the region.

Such cost and soil health considerations have resulted in a resurgence of LEI agro-ecological farming techniques, which are less dependent on non-renewable inputs, able to restore degraded soils and water systems, and offer farmers a far greater degree of independence (Pretty 2006; Altieri 1999). Indeed, the benefits of LEI techniques make them far better suited to ensuring long-term production without necessitating profit trade-offs (Magdoff 2007; Pretty 2006; Altieri 1999).

Farmer case studies

This section describes two separate case study projects involving seven farms in the Swartland region (Metelerkamp 2011) and nine farms in the Stellenbosch area (Landman, Kate and Haysom, 2010)[5], respectively. Both sets of farmers have incorporated significant changes into their farming practices, whether by necessity (in the case of the Swartland farmers) or choice (the Stellenbosch farmers). While the two regions represent different farming systems,[6] the Swartland case study is included here both as indicative of the changes happening in agriculture and as bearing directly on the Stellenbosch scenario in the sense that the two farming communities face similar challenges, even if these challenges appear more acute in the case of the Swartland.[7]

Firstly, although not predominantly a grape growing region,[8] the Swartland area has many climatic, agronomic and economic features in common with the Stellenbosch region. As such, the climatic changes experienced by farmers in the Swartland, such as changes in rainfall patterns and the length of hot, dry spells, may already be affecting Stellenbosch farms as well, but may be masked by the fact that most Stellenbosch farmers have access to pressurised irrigation from the Theewaterskloof system (potentially reducing their vulnerability to some of the initial climatic shifts), whereas Swartland farmers rely primarily on rain-fed irrigation. Secondly, Swartland farmers do not beneficiate their production in the manner that Stellenbosch wine farmers do and issues of viability and vulnerability are therefore more immediate in their case. Lastly, both sets of farmers are confronted by the ever-rising costs of agricultural inputs: whereas wine farmers may not rely as heavily on fertiliser inputs as their wheat growing Swartland counterparts do, the costs associated with pest control and other comparable inputs are certainly rising.

Swartland case studies

The aim of the research was to determine whether or not examples exist of commercial grain farmers in the Swartland region moving away from high external input agricultural production systems towards production systems based instead on ecologically restorative partnerships with soils and other natural systems. Where such farmers were identified, the researchers sought to understand why they were changing their approach to farming and investigated the specific alternative technologies and practices implemented by these farmers.

---

5 This report was commissioned as part of a background study within a broader Stellenbosch food security initiative research project, the Stellenbosch Food System Strategy Project, coordinated by Haysom.

6 When the dominant agricultural practice is considered at a regional level.

7 A notable exception is the effect of annual crop rotation and regular tillage, which was a stress point for the Swartland case study farmers.

8 There is viticulture in the Swartland, but it does not constitute the dominant regional agricultural practice.

Table 13.2   Swartland case study farms [Metelerkamp 2011]

Location	Farm	Farmer	Farm size
Malmesbury	Silvermyn	Peter Steyn	1100ha
Bo-Hermon	Elandsberg	Mike Gregor	6500ha
Philadelphia	Uitkyk (1)	Junior Heroldt	1800ha
Moorreesburg	Uitkyk (2)	Cobus Bester	2250ha
Piketberg	Partyskraal	François Ekstien	2486ha
Pools	n/a	Aubrie Rigter	1000ha
Malmesbury	Elim	Dirk Lesch	395ha

It emerged from the reviews of the abovementioned farms that economic pressures, risk minimisation and the desire for intergenerational sustainability (motivated by the tradition of the 'family farm'), as well as the logical incentive to shift towards a system that appears to be more profitable and resilient, were the four main drivers of change, with most farmers citing economic pressure as the primary driver for change in their farming practices. Predictably however, further questioning revealed that economic pressure was not an event in itself, but rather the culmination of a range of other factors, such as changes in local and international trade policies, rising input prices, rising input requirements, rising machinery costs, high land prices, crop failures, variable weather, theft and produce price fluctuation (Statistics South Africa 2007; Agricultural Research Centre (ARC) 2010; Metelerkamp 2011).

As a result of these pressures, the farmers have engaged in significant and ground-breaking research and practice to revert to more traditional farming approaches, such as increasing crop and livestock diversity, using organic fertilisation and leaving crop residues on the land, in combination with innovations such a minimum tillage and precision agriculture. As shown above, these interventions were initiated largely for financial reasons and were not motivated by environmental concerns. However, as the farmers started working *with* nature (as opposed to the industrial input-based practice of controlling and dominating nature), they progressively developed a deep appreciation for the natural environment and have come to appreciate it as an ally in their farming enterprises, based on the understanding that if nature is in balance, it supports their farms.

The polycrisis served as a conceptual reference point in the evaluation of the changes in farming practice and the drivers of these changes to allow for an agricultural analysis that went beyond the farm gates to take cognisance of the broader social-ecological issues that affect and are affected by agriculture. Figure 13.1 shows some of the connections that were established in this way. The primary driver denotes the main challenge identified by the farmers (i.e. economic pressure), while the secondary drivers are the reasons given by the farmers for the existence of the primary one. Links were then drawn between these secondary drivers and the polycrisis.

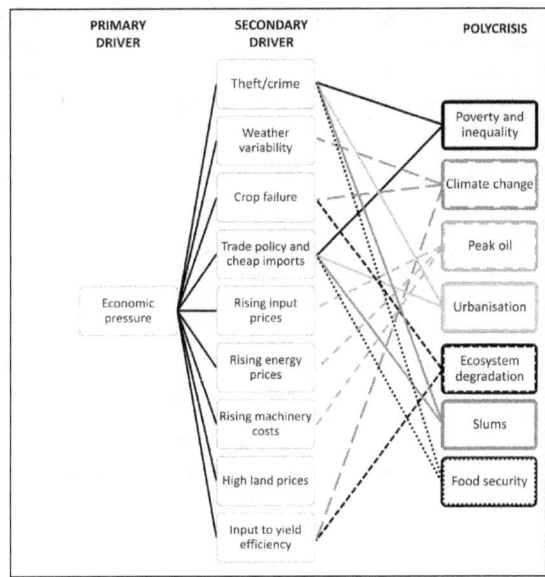

Figure 13.1   Indirect links between the polycrisis and economic pressures on Swartland farms [Metelerkamp 2011]

On-site observations and interviews with the farmers revealed that the degree to which they were altering their practices varied significantly. However, four key technologies and practices were identified to be of greatest significance and common to all seven farmers. These were:

- increased use of legume rotations;
- a reduction in tillage intensity;
- new styles of planters; and
- increased farm sizes.

Thirty other recently adopted technologies and practices were also identified, most of which indicate a shift towards LEI systems, including increased use of organic fertilisers, the retention of crop stubble on fields, and the diversification of crops and livestock.

Table 13.3   New Swartland farm management practices [Adapted from Metelerkamp 2011]

Newly adopted technology or practice			
New farming practice	Agro-ecologically aligned	New farming practice	Agro-ecologically aligned
Albrecht system of soil balancing	√	More sheep and cattle per hectare	√
Animal rotation	√	New cultivars	√
Bigger farms		New planters	√
Bigger machinery		New poisons	
Compost making	√	One pass planting	√
Compost tea	√	Organic fertiliser alternatives	√
Contour bunding	√	Owl boxes	√
Reduced tillage	√	Precision planting	√
Crop rotation	√	Reduced fungicide use	√
Developing own marketing channels	√	Reduced pesticide use	√
Foliar sprays over pelletised fertiliser		Reforestation	√
GPS technologies	√	Stopped burning crop residue	√
Increased herbicide use		Trace element fertiliser application	√
Leaving organic material on soil surface (mulching)	√	Use of cow manure	√
Legume rotation	√	Use of pelletised chicken manure	
Reduced labour		Virgin land conversion to farmland	
Microbial stimulants	√	Virgin land preservation	√

When aggregated for the seven farms included in the research, the overall effect of these changes has been an improvement in soil fertility and increased drought resistance. In addition, there has been an absolute decoupling of a number of non-renewable farm inputs (such as diesel and synthetic nitrogen) from production levels and a significant reduction in the financial risk profile of these farms.

While these LEI systems may produce slightly less wheat, the farmers reported dramatic increases in other outputs such as sheep and cattle. The potential improvement in price stability stem predominantly from increased efficiency in input usage and the utilisation of natural fertility and pest management practices, which are less susceptible to monopolistic input sales structures, international shortages and increasing fossil fuel costs. Secondary data also strongly suggest that wheat grown by these methods may be produced at significantly lower cost than that cultivated elsewhere in the region by means of traditional methods.

The findings of this research show that although commercial farmers in the Swartland may not have conceptualised the challenges of sustainable development in terms of Swilling and Annecke's polycrisis, it is indeed clearly reflected in many of the challenges they face. The case studies revealed a growing body of knowledge and practice developing in the commercial agricultural sector in the Swartland in response to these challenges. This body of knowledge addresses some of the key sustainability issues raised, including peak oil, ecosystem degradation, climate change and food insecurity. Moreover, the insights gained in terms of how more sustainable and restorative agricultural practices may be applied to larger scale farm systems is likely to become an increasingly valuable asset in addressing various elements of the polycrisis.

While there is little direct evidence of a similar paradigm shift in farming practices in the Stellenbosch area, especially in the competitive wine industry that dominates agriculture in the region, there are a few farms where some changes are evident, mostly due to international customer demands for reduced carbon footprints or other such prerequisites to market access. Backsberg, Reyneke Wines and Spier are examples of farms that have started implementing change, with farmers, winemakers and others involved in farming operations seeing themselves as custodians of nature instead of trying to control it.

Stellenbosch case studies

It would seem that Stellenbosch agriculture, and specifically the wine industry, is making some effort to respond to the challenges of sustainability, as is evident from two initiatives launched during the past 15-odd years, namely the Integrated Production of Wine (IPW) initiative and the Biodiversity and Wine Initiative (BWI).

The IPW was established by the broad South African wine industry in 1998 and is a voluntary environmental sustainability scheme stipulating a set of compliance criteria for sustainable viti- and viniculture in terms of production, processing and packaging of products.[9] While internationally recognised and aligned to international standards, it has lately been criticised by some for being embedded in a sustainability approach of 'doing less harm', rather than adopting newer trends that are altering the fundamentals of agriculture. Another challenge presented by the IPW is how to ensure that its stringent compliance measures do not limit sustainability-oriented innovations, such as water recycling (Liebenberg 2010).

The BWI, which came into being in 2004, is aimed at more comprehensive biodiversity conservation and sustainable wine farming through a partnership between the country's wine industry and conservation sector "to protect and conserve [the] unique natural heritage within the Cape Winelands – an outstanding place with iconic species – whilst maintaining living, productive landscapes". To date, over 126 000 hectares of natural area have been conserved by BWI producers, meaning that the South African wine industry's conservation footprint is well in excess of its current vineyard footprint of 102 000 hectares (WWF South Africa 2012). The BWI calls for stewardship and greater care of nature, with its primary focus being to prevent farmers from utilising more land for production. While doubtlessly a critical intervention, as in the case of the IPW, one may again argue that only encouraging farmers to do less harm does not set Stellenbosch up for increased social and ecological prosperity in the long run.

Such concerns notwithstanding, both these initiatives are to be applauded as key to the development of sustainability-oriented farming practices in the region by not only providing farmers with the tools to shift to sustainable agricultural practices, but also refocusing the industry on broader production issues.

In 2010, Landman, Kate and Haysom conducted a study to identify and document agro-ecological activities and innovations on farms in Stellenbosch and the surrounding areas in order to lay the groundwork for similar future initiatives. As in the case of the Swartland case studies, the research was also informed by the various polycrisis issues described earlier, but with a special emphasis on soils. Table 13.4 provides a list of the farms reviewed.

---

9   See http://www.ipw.co.za for more information

Table 13.4 Farms in the Stellenbosch region employing agro-ecological farming methods[10] [Landman et al. 2010]

Farm	Farmer	Farm size	Production type	Agro-ecological mode/type
De Zalze	Francois Malan	30ha	Grapes, lemons, clementines, vegetables, eggs, milk, sheep, chickens and ducks	Organic and biodynamic
Farm 502	Eric Swartz	10ha	Vegetables	Organic
Spier BD Farm	Angus Macintosh	300ha	Beef, chickens, eggs, grapes, vegetables, olives, medicinal herbs and gogi berries	Organic and biodynamic
Uitzicht	Johan Reyneke	45ha	Grapes	Biodynamic
Villeria 2	Erick Zanzele	1.5ha	Vegetables, chickens and eggs	Organic
Gumtree	Skye Fehlmann	9ha	Fruit and vegetables	Organic
Bloublommetjies-kloof	Wendy Lilje	38ha	Natural biodynamic soaps, herbs, cheese and buttermilk	Biodynamic
Foxenburg Estate	Marianne Hemmes	120ha	Mushrooms, goat milk cheese, yogurt and olives	Organic
Waterkloof Farm	Tomke Heeren	30ha	Medicinal plants	Biodynamic
Tulbagh Mountain Vineyards	Rebecca Tanner and Kazik Nicholls	180ha	Wine grapes, vegetables and sheep	Organic

At the time of the research (2010) all these farms displayed significant agro-ecological innovations; however, the farmers were not sharing their knowledge or the lessons learned. Such a network has subsequently developed specific to certain agricultural practices. This is led by the local office of the Biodynamic Agricultural Association of South Africa (BDAASA) local Stellenbosch office, offering lectures, training and a space for sharing of information and collaboration on marketing and sales initiatives. There is, however, scope for greater participation in this process, particularly from conventional farmers in the early stages of transition to alternative, more agroecological farming approaches.

While the Stellenbosch farmers' reasons for adopting agro-ecological farming methods may not have been primarily financial, they are in fact reaping financial benefits from it; indeed, when discussing their interventions with other farmers, financial gains tend to dominate discussions. So, for example, some farmers experienced higher total profits and significantly lower exposure to the current downturn in the wine industry thanks to agro-ecological specialisation, while others reported regularly being offered land in the vicinity of his farm by conventional farmers forced to move from the land as their farms become progressively less viable (Landman et al. 2010). Yet, the farmers generally agree that the premium they are currently receiving for their produce as a result of the agro-ecological farming methods is only a short-term benefit that allows them to accelerate the restorative processes on their farms (Landman et al. 2010).

The Stellenbosch case study farms reflected the common features of continuous on-farm innovation by trial and error, with farmers applying various interventions and identifying the most appropriate intervention for each context through a process of elimination. Most of the farmers reviewed utilised multiple resources to build their knowledge base, and all identified individual specialists as being instrumental in their respective learning processes. And finally, most of the farmers asserted that, in spite of the ongoing economic downturn and various market-based challenges, their farms were more viable at the time of the study than they had been five years previously.

Conclusion

While the impacts of the polycrisis may not yet be evident to the broader farming community in the Stellenbosch agricultural system, their implications require urgent attention. Stellenbosch farmers' adherence to business-as-usual approaches will serve to continually reduce the resilience of the system, which will not only diminish

---

10 Italicised terms indicate that the farmer is either in the process of converting to, or is farming according to, a specific agro-ecological method, but chooses not to certify.

agricultural returns in the region, but also reduce the ability of agriculture to provide employment and assume the role of environmental custodian. As the varied and interconnected components of the polycrisis show, a dependence on external technical inputs alone will not provide the necessary insulation to ensure a viable farming sector. New approaches are required: approaches focussed on responding more consciously to the polycrisis by proactively taking its challenges as clues about how to evolve – as opposed to only reacting as and when its manifestations reach critical levels. Encouragingly enough, the Stellenbosch case study shows that a growing number of farmers in the region appear to be responding to these challenges in innovative ways.

The case study of farmers in the Swartland region showed that they have developed context-specific technologies that are relevant to their needs, such as working in partnership with natural systems to reduce their reliance on petrochemical inputs. Economic pressure was the primary driver for these changes, and while the farmers who were included in the study reported marked increases in returns, they also affirmed that supporting ecosystem services and facilitating their functioning resulted in more resilient farming systems. This result was mirrored on the farms included in the Stellenbosch case study, with the farmers there also reporting increased economic resilience as a result of the shifts made towards agro-ecological approaches (even though these farmers appear to have been acting from a more ideological or pre-emptive position). Considering Kelly and Schulschenk's (2011) argument that building resilience to future shocks should form the core focus area for reform of the Stellenbosch food system, both these case studies offer encouraging prospects for future food security in the broader region.

Transforming the Stellenbosch farming system will require complex strategies; for example, reducing the predominance of grape production may result in alternative agricultural typologies that require more water, which may not be possible given the context of water scarcity in the region (which will probably become progressively worse under the influence of climate change). Responses to the various challenges facing the farming system will therefore require small but radical steps to adapt to the looming (and arguably already present) threats. Existing region-wide sustainability-oriented programmes, while laudable, have not brought any fundamental changes to the ways in which agriculture is practiced in the region; as such, they should be seen as initial and valuable components of a process of change, but not as an end in themselves. The same applies to many of the organic certification schemes, which, while rewarding changes in agricultural practices and facilitating more sustainable forms of agriculture, do not automatically drive the change necessitated by the polycrisis – for example, a large lettuce farming operation, whether organic or not, remains a monocrop farm, with the same challenges of pest infestation and other related industrial farming issues.

Farmers take advice from researchers, extension officers and, to a large extent, so-called *gifsmouse*.[11] However, the presence of Stellenbosch University and other institutions, such as Winetech and the ARC, provides an ideal opportunity for the research community to play a leading role in the development of agro-ecological farming in the region. While it is appreciated that funding for research into these new types of agriculture falls outside the dominant funding sources of the past few decades, this research is of critical importance. Moreover, the experiences of the farmers reviewed in the two case studies presented reveal a changing agricultural landscape and responsible research institutions and researchers should orient their work accordingly, regardless of where funding seeks to direct research.

In terms of developing a broad knowledge base, the Swartland case studies in particular demonstrated that farmers are often very proactive about learning from one another, a situation that may be exploited to assist with the transfer of appropriate technologies and skills to respond to the current crisis by employing these innovative and agro-ecologically-oriented farmers in the region as farmer educators. Indeed, finding ways to facilitate such exchanges will be critical to the future of agriculture in the region.

There have been some visible changes in agricultural practices in the Stellenbosch region in recent years, for example, the reappearance of cattle and other livestock on the landscape; intercropping and the use of groundcover crops in vineyards offer further evidence of innovation. However, these changes remain isolated and, considering the scale of the polycrisis, are arguably tokenistic. The first step on the road to a drastic overhaul of agricultural practices in the Stellenbosch region would be to ensure that the farmers whose

---

11 *Gifsmous* (literally 'poison pedlar') is a term used colloquially for pesticide (and sometimes also fertiliser) sales representatives/staff, who are often the first point of call for many farmers and play a key role in offering extension advice to farmers.

innovative agro-ecological farming methods were recorded in the case studies form the vanguard of a broader movement that will rapidly come to represent the status quo.

Drastic changes in farming practices and, even more importantly, in the farming ethos, are required if farms are to remain viable. Simply doing less harm is no longer a tenable proposition; agriculture must be restorative if it is to remain feasible. Agriculture is one of only a few industries that have the potential, through changed practice, to insulate society from the challenges associated with the polycrisis by building the necessary resilience to future shocks and providing for an industry that is critical to the sustainability of Stellenbosch and the region into the future.

REFERENCES

Altieri MA. 1995. *Agroecology: The science of sustainable agriculture*. Boulder, CO: Westview Press.

Altieri MA. 1999. *Agroecology: principles and strategies for designing sustainable farming systems. Agroecology in Action*. Berkeley: University of California.

ARC (Agricultural Research Council). 2010. *Guidelines: Production of small grains in the winter rainfall region*. Bethlehem Agricultural Research Council: Small Grain Institute.

Bates A & Hemenway T. 2010. From Agriculture to Permaculture. In: L Starke & L Mastny (eds). *State of the World: 2010. Transforming Cultures: From Consumerism to Sustainability*. New York: Earthscan. pp. 47-54.

Conway G. 1997. The Doubly Green Revolution. In: J Pretty (ed). 2005. *Sustainable Agriculture*. London: Earthscan.

FAO (Food and Agriculture Organisation). 2008. *The State of Food Insecurity in the World: 2008*. Rome: FAO.

FAO. 2009. *The State of Food and Agriculture: 2009*. Rome: FAO.

Genis A. 2008. Minimumbewerking en wisselbou is wenresep. *Landbou Weekblad*, 5 September:46 48.

Grain SA. 2009. *Price indices/Prysindekse*. [Accessed 20 June 2010] http://www.grainsa.co.za/documents/prysindekse_1.xls

Halweil B. 2004. *Eat Here*. New York: WW Norton & Company.

Hardy M. 1998. Towards sustainable crop production in the Swartland: A short review. *Elsenburg Journal*, 2:42-45.

Heinberg R. 2003. *The Party's Over: Oil, War and the Fate of Industrial Societies*. Vancouver: New Society Publishers.

Holt-Gimenez E & Patel R. 2009. *Food Rebellions*. Cape Town: UCT Press.

Hopkins R & Holden P. 2007. *One Planet Agriculture: Preparing for a post-peak oil food and farming future*. Bristol: Soil Association.

IAASTD (International Assessment of Agricultural Science and Technology for Development). 2008a. *Global Summary for Decision Makers: International Assessment of Agricultural Knowledge, Science and Technology for Development*. Washington, DC: Island Press.

IAASTD. 2008b. *Synthesis Report: International Assessment of Agricultural Knowledge, Science and Technology for Development*. Washington, DC: Island Press.

IPCC (Intergovernmental Panel on Climate Change). 2007a. *Climate Change 2007*. Geneva: UNEP (United Nations Environment Programme).

IPCC. 2007b. *Climate Change 2007: Synthesis Report, Summary for Policymakers*. [Accessed 12 June 2008] http://www.ipcc.ch/pdf/assessment-report/ar4/syr/ar4_syr_spm.pdf

Kelly C & Schulschenk J. (Forthcoming). Assessing the vulnerability of Stellenbosch's food system and possibilities for a local food economy. *Development Southern Africa, Special Edition on Food Security in Southern Africa*. Midrand: DBSA (Development Bank of Southern Africa).

Lal R. 2006. Perspective: Managing soils for feeding a global population of 10 billion. *Journal of the Science of Food and Agriculture*, 86:2273-2284.

Lal R. 2010. Managing soils for a warming earth in a food-insecure and energy-starved world. *Journal of Plant Nutrition and Soil Science*, 173:4-15.

Landman A, Kate T & Haysom G. 2010. Agroecological Farms in and around Stellenbosch Report 2010. Unpublished research report. Stellenbosch: Sustainability Institute/Stellenbosch University.

Liebenberg A. 2010. Personal interview. Stellenbosch: 14 November.

Louw D. 2009. *Stellenbosch Agricultural Sector Overview*. Cape Town: CNdV Africa.

Magdoff F. 2007. Balancing food, environmental and resource needs. *Renewable Agriculture and Food Systems*, 22(2):77-79.

Matoti B. 2011. Impact of Investment in Agriculture in the Western Cape. Presentation to the OMEGA investment conference on food security prepared on behalf of the Western Cape Department of Agriculture. V&A Waterfront, Cape Town: 11 May.

MEA (Millennium Ecosystem Assessment). 2005. *Ecosystems and Human Wellbeing: Synthesis. A Report of the Millennium Ecosystem Assessment*. Washington DC: Island Press.

Metelerkamp L. 2011. Commercial agriculture in the Swartland: Investigating emerging trends towards more sustainable food production. MPhil thesis. Stellenbosch: Stellenbosch University.

NAMC (National Agricultural Marketing Council). 2009. *Update: Trends in Selected Agricultural Input Prices.* Pretoria: NAMC.

Pettersson TJ. 2011. Response to question 977 for written reply: National Assembly, Second Session, Fourth Parliament. Mr JH van der Merwe (Inkhata Freedom Party) to ask the Minister of Agriculture, Forestry and Fisheries. [Accessed 8 June 2011] http://www.politicsweb.co.za/politicsweb/view/politicsweb/en/page72308?oid=239919&sn=Detail&pid=72308

Pretty J. 2006. *Regenerating Agriculture*. London: Earthscan.

Reijntjes C, Haverkort B & Waters-Bayer A. 1992. *Farming for the future*. London: MacMillan Press Ltd.

Scherr SJ. 1999. *Soil Degradation: A Threat to Developing-Country Food Security by 2020?* Washington: IFPRI (International Food Policy Research Institute).

Scherr SJ. 2000. A downward spiral? Research evidence on the relationship between poverty and natural resource degradation. *Food Policy*, 25(4):479-498.

Schulze R. 2011. Adapting to Climate Change in the South African Agricultural Sector: What's facing us and are we up to the challenge? Presentation to the OMEGA investment conference on food security. V&A Waterfront, Cape Town: 10 May.

Statistics South Africa. 2006. *Report on the survey of large- and small-scale agriculture*. Pretoria: Statistics South Africa.

Statistics South Africa. 2007. *Census of Commercial Agriculture 2007 (Preliminary)*. Pretoria: Statistics South Africa.

Stellenbosch Municipality. 2010. *Stellenbosch Municipality IDP 2010*. [Accessed 13 March 2011] http://www.stellenbosch.gov.za/jsp/util/document.jsp?id=3264

Swilling M. 2008. Sustainability and infrastructure planning in South Africa: a Cape Town case study. *Environment and Urbanisation*, 18(1):23-51.

Swilling M & Annecke E. 2012. *Just Transitions: Explorations of Sustainability in an Unfair World*. Cape Town: UCT Press.

Taylor J, Madrick M & Collin S. 2005. *Trading Places: The local economic impact of street produce and farmers' markets*. London: New Economics Foundation.

Uphoff N (ed). 2002. *Agroecological Innovations: Increasing Food Production with Participatory Development*. London: Earthscan.

Vink N & Van Rooyen J. 2009. *The economic performance of agriculture in South Africa since 1994: Implications for food security*. Development Planning Division Working Paper Series No 17. Midrand: DBSA.

Von Braun J. 2007. The World Food Situation: New Driving Forces and Required Actions. *Food Policy Report*:1-27. Washington: IFPRI.

World Commission on Environment and Development. 1987. *Our Common Future*. Oxford: Oxford University Press.

WWF (World Wildlife Foundation) South Africa. 2012. *Biodiversity and Wine Initiative*. [Accessed 29 August 2012] http://www.wwf.org.za/what_we_do/outstanding_places/fynbos/biodiversity___wine_initiative

http://www.stellenbosch.org.za/jsp/util/document.jsp?id=3264

Chapter 14

WATER

Supply and quality

JO BARNES

We already have the statistics for the future: the growth percentages of pollution, overpopulation, desertification. The future is already in place.

[Günter Grass 1990:37]

Introduction

Water is a renewable resource, provided that it is used sustainably – quite a profound statement, seeing that very few communities over the millennia have yet managed to achieve this for any length of time. In order for the greater Stellenbosch area to reach this dream, there are many challenges to be faced. This chapter looks at water supply, consumption and quality, as well as some of the most urgent problems that need to be overcome to achieve the goal of renewable water use.

It is important at the outset to note that there is no single concept called 'water quality'. What constitutes water of an acceptable quality depends on its intended use. There is thus a range of definitions of acceptable water quality or minimum standards of water, depending on whether the water is intended for human consumption, agricultural applications (for example, irrigation of edible crops or livestock watering) or the many industrial processes requiring substantial volumes of water. A further essential factor that often only receives lip service is the amount and quality of water needed by the environment itself, especially our rivers and streams, in order to function as ecologically healthy natural systems.

Water resources

South Africa is an arid country, with only 8.6% of the rainfall available as surface water (Walmsley, Walmsley & Silberbauer 1999). The available freshwater resources in South Africa are already almost fully utilised or even under stress (De Villiers & De Wit 2010). At the projected population growth, economic development rates and the rate of urbanisation, it is unlikely that the future demand for water will be sustainable if it continues to be used in the same way. If there is no system change to massively increase efficiencies and reduce negative environmental impacts, water supply will become a major restriction to future socio-economic development.

Even though nature provides water for 'free', complicated infrastructure is needed to deliver water of an acceptable quality to where it is needed. The infrastructure investment for the proper management of water resources is vast and extremely costly; therefore the utmost care should be exercised in planning future systems of water utilisation.

The interrelated nature of the water processes on earth and the cyclic nature of the supply of water are often referred to as the water cycle, as presented in Figure 14.1. The water cycle is a continuously spinning wheel – it does not matter where one enters the cycle to assess the process, although a convenient point is usually rainfall, which is why this element of the cycle is always depicted at the top of the wheel. However, it is very clear that all the elements of the cycle impact either directly or indirectly on all the other elements.

Surface water

The Stellenbosch region has a Mediterranean climate with winter rainfall and relatively hot, dry summers. Due to the limited storage capacity of the region's natural surface water bodies, a particularly high number of reservoirs and dams have been constructed. Their primary function is the storage of water for irrigation, livestock watering and urban water supply (Department of Agriculture Western Cape (DAWC) 2007). The main reservoirs providing water to the Stellenbosch area include the Theewaterskloof, Steenbras and Berg River dams. The first two dams are outside the greater Stellenbosch area and represent water transfers from other, similarly water-stressed regions. The Berg River Dam was designed to increase the yield of the Western Cape water system by 81 million m³ (18%) to 523 million m³ per year (Dalgliesh, Steytler & Breetzke 2004). In addition to large dams, there are over 40 000 farm dams in the Western Cape, storing in excess of 100 million m³ of water. When compared to the rest of South Africa, the concentration of dams in the Western Cape is by far the highest. However, the proliferation of these dams is not without sustainability challenges. Apart from the high capital costs, dams are often responsible for a build-up of contaminated water and silt, giving rise to the accumulation of pollutants in the environment. Examples of such pollutants are heavy metals and toxins excreted by cyanobacteria.

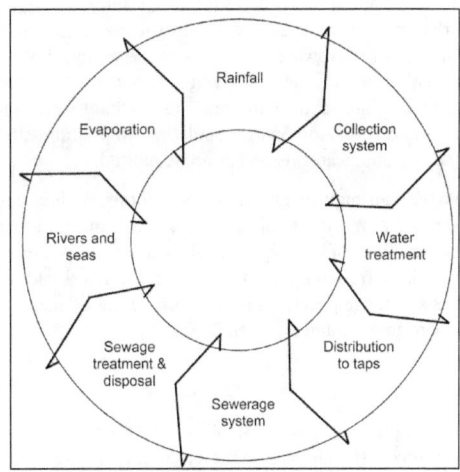

Figure 14.1   The water cycle [Barnes 2003]

The two major catchments in the greater Stellenbosch area are the Berg River (which flows through the Franschhoek area) and the Eerste River (which flows through the Stellenbosch area). Both these rivers have numerous tributaries, but many of them are not perennial. Were it not for controlled releases of water from the Theewaterskloof and Berg River dams during the summer months, the lower reaches of the Berg River would also run dry during the summer, as was observed by the author after 2010's low-rainfall winter. The Plankenbrug River (a tributary of the Eerste River) is an exception; its lower reaches have become perennial over the summer months because of run-off and discharges from urban Stellenbosch. The Eerste River suffers from over-abstraction in its upper reaches and would have stopped flowing in the summer months, were it not for the Plankenbrug joining it just below the town of Stellenbosch. It also receives a large volume of water from the Stellenbosch waste water treatment works just after the confluence. Both of these two sources of water augmenting the Eerste River are at times heavily polluted with sewage. In fact, the Eerste River is reputed to consist of 80% effluent and storm water drainage in the summer months, though no formal study has been published.

Mean annual potential evaporation – the amount of water that could evaporate if sufficient water were available – far exceeds mean annual precipitation in every part of South Africa (De Villiers & De Wit 2010). Evaporation in the Western Cape is significantly greater than rainfall, meaning that less water finds its way into rivers, dams and wetlands (Dalgliesh *et al.* 2004). Rainfall moreover tends to be highly seasonal and also erratic. Unpredictable periods of drought are therefore an ever-present risk. Added to this natural variation in rainfall, droughts may become more frequent as a result of climate change (Midgley, Chapman, Hewitson, Johnston, De Wit, Ziervogel, & Mukheibir *et al.* 2005).

According to the *Africa Water Atlas* (United Nations Environmental Programme (UNEP) 2010), the rainfall in Southern Africa has decreased on average by an overall 15% since 1980 and is projected to diminish even further by 2050. The data provided in the atlas indicate that most of Southern Africa, including South Africa, will reach a state of water deficit around 2025. This is supported by local data for the Western Cape (Midgley *et al.* 2005; Dalgliesh *et al.* 2004). Available water sources in the region are already fully committed in most parts and there is little scope for further development of water resources.

A high proportion of the South African population resides in urban areas. By 2009, 61% of the population already lived in urban areas (Central Intelligence Agency (CIA) n.d.). In terms of the Western Cape, this large urban population is augmented by the high influx of people from other provinces, which puts pressure on the already-stressed water resources in the region (Dalgliesh *et al.* 2004). Most of the new arrivals settle in peri-urban informal settlements, where they overwhelm the ability of the local authorities to provide services such as clean water and sanitation. The resultant water management challenges are daunting. In order to supply water to this burgeoning population, impoundments such as reservoirs and dams are needed or more and more groundwater resources are exploited.

Water consumption in certain areas of the province already exceeds water supply (especially when an adequate ecological reserve is factored in), which means that competition for scarce water resources will inevitably become part of life for the inhabitants of the Western Cape. Water restrictions are also likely for Stellenbosch and the surrounding towns. Because water is already a limiting factor in the area, any projected climate change has serious implications for the competing demands of socio-economic development and the interests of the environment (Midgley *et al.* 2005).

Groundwater

In comparison to the rest of South Africa, the Western Cape is relatively well-endowed with groundwater resources. The mountainous areas of the south-western Cape receive some of the highest rainfall in the country (>2 500 mm per year). This water recharge compensates to some extent for the lack of storage space in aquifers due to the host formations being predominantly geologically old rocks, the pore spaces of which have been destroyed by compaction.

Brief overview of water use

United Nations Water (2011) reports that worldwide about 8% of extracted water is used for domestic purposes, about 70% for irrigation and about 22% for industry.

Domestic water use

Water used in and around the home falls into three major use categories: human consumption (mainly used for drinking, food preparation and personal hygiene), sanitation purposes (cleaning, home hygiene, waterborne sewage disposal), and outside uses (watering of gardens, washing of vehicles, etc.). These use categories also represent different health risk classes. Water for human consumption should be of the highest quality, containing no disease-causing organisms or harmful chemicals. Water used for cleaning and toilet flushing can be slightly less pure, but care should be taken to instruct the occupants about the limitations of using a lower class of water for sanitation purposes. If this is not done, and subsequently careless use of second-class water leads to disease, then the water savings are not worth the threat to human health. The same goes for the risks associated with the use of recycled water or domestic grey water for irrigation and the washing of paved areas, etc.[1]

Water services managed by water services authorities (mostly municipalities) are mainly utilised for domestic purposes and industrial activities in towns. The Department of Water Affairs and Forestry (DWAF)[2] (RSA DWAFa n.d.) estimated that this segment uses less than 15% of the available freshwater resources, although it is an essential water sector.

---

1   This does not mean that all water for human consumption should originate from purification works. The reuse of water may be introduced successfully through education about the higher risks involved and the proper ways to reduce those risks. Supervision should also be provided to ensure that regulations are adhered to. It is important to note that not all advice on the reuse of grey water cited by so-called 'green' recycling companies is suitable for South African conditions and may constitute unacceptable risks to people and the environment.

2   In 2009, the forestry responsibility was transferred to the newly formed Department of Agriculture, Forestry and Fisheries and the name of this department changed to the Department of Water Affairs (DWA).

Agricultural use

From the 1960s onwards, the introduction of large-scale irrigation systems was part of the so-called 'Green Revolution' in many developing countries, greatly enhancing their ability to produce enough food to feed their populations. In some areas of the world irrigation is essential in order to grow any crops at all, while in other areas it may enable the growing of more profitable crops or enhance crop yields. Increasing competition for water, inefficient irrigation practices and crops that are not water-wise or are unsuitable for the climatic conditions will all be important constraints to future sustainable water use in the agricultural sector (International Assessment of Agricultural Knowledge, Science and Technology (IAASTD) 2008).

Irrigation accounts for 50% of the total water use in South Africa (RSA DWAFb n.d.). According to O'Keeffe, van Ginkel, Hughes, Hill & Ashton (1996) the largest single group of users of fresh water in South Africa at that time was farmers, who accounted for 67% of directly used water and more than half of all water used. Furthermore, half of the water used by farming operations was used for irrigation and a further 1.5% went into livestock watering. A few years later, the reported irrigation usage had decreased to about 62% of the total water requirements (Juana, Kirsten & Strzepek 2006); yet, at the time, agriculture accounted for only about 4% of the gross national product and employed about 11% of working people in South Africa.

Food production is a strategic activity that cannot be limited. Rather, water use in the sector should be optimised. In addition, the large quantity of water used for food production means that, should irrigation water be polluted, waterborne pollution could enter the food chain. This is another reason why poor sanitation systems cannot be allowed to contaminate natural watercourses.

Agriculture accounts for nearly 55% of all water consumption in the Western Cape (DAWC 2007). The water sources available for agriculture are already nearly optimally exploited, and any expansion of wine, fruit or vegetable production is limited by the availability of irrigation water. Any future expansion will only be possible through savings achieved in other current areas of water use.

Industrial and technological use

Probably every manufactured product on earth requires water during some part of the production process. Industrial uses include fabricating, washing, processing, diluting, the cooling or transport of a product (e.g. flume water) and sanitation. Some industries are large consumers of water, notably the food, paper, chemicals, heavy metals and refined petroleum industries. These industries also have the potential to pollute water sources with vast amounts of mainly chemical pollutants.

The United Nations Educational, Scientific and Cultural Organisation (UNESCO) and the United Nations Industrial Development Organisation (UNIDO) (n.d.) estimate that industries worldwide account for 22% of water use. However, this figure hides a large divergence: industries in high-income countries use 59% of the total water used, while low-income countries generally use only about 8% of total extracted water for industrial purposes. Water use has been growing at more than twice the rate of population increase over the last century, clearly pointing to a looming crisis.

Globally, industry and energy together account for 20% of water demand (United Nations (UN) 2009). Comparable data for South Africa are difficult to access or interpret, since official statistics sometimes report separately on mining and other industries, while at other times they are added together. The DWA reports on urban water use by including urban industries under this heading. The *Water Conservation and Water Demand Management Strategy for Industry, Mining and Commercial Water Use for South Africa* stated that the industry, mining, commercial and power generation water use sector accounted for more than 10% of the total of 20×10^9 m^3 of water used annually in South Africa at the time (RSA DWAF 2003).

Environmental (habitat) needs

As water demands from agriculture, industry and human habitation increase, there is less water left to sustain the environment, especially the rivers and aquifers. This portion of water is referred to as the ecological reserve. The loss of ecological reserve water comes at a high environmental cost, including stagnant or poorly functioning river environments and the loss of biodiversity, wetlands and the ability to recover from damage inflicted by pollution. It is estimated that half of the world's wetlands have disappeared during the last century and that more than 20% of the world's freshwater fish species are now endangered or extinct (World Business Council for Sustainable Development (WBCSD) 2009). Wetlands are converted to agricultural land and desiccation will further degrade those that remain. Well-functioning wetlands do not only perform an important natural purification function, but also provide natural flood control.

One important consequence of the damage inflicted on the environment as a result of the reduction of the ecological reserve is that it impairs the ability of natural systems to clean up pollution. This is a significant loss, since in many places humans are polluting water sources faster than nature can process the damaging pollutants. In addition, reduced water volumes in rivers will also have an impact on estuaries (Midgley *et al.* 2005).

The *National Water Act* (Republic of South Africa (RSA) 1998) greatly improved the prioritisation of water for the purpose of maintaining ecological systems. Accordingly, primary human consumption (water for drinking, washing and cooking) and ecological needs (known collectively as the 'Reserve') enjoy priority with regard to water allocation. In consequence, the volume of water available for secondary human use (gardens, industries and agriculture) will have to be drastically reduced and consumption curtailed in certain areas in order to address Reserve needs.

Recreational use

Though important to people's well-being and 'sense of place', recreational water use is usually a small and non-consumptive component of general water use. The ability to pursue water sports or enjoy the natural beauty of rivers and streams for outdoor pursuits such as picnics and play are important privileges that may teach people respect for nature and enhance their lives. The fact that one cannot attach a price to such activities does not mean that they are any less important.

Due to environmental concerns about using water for non-essential purposes, the irrigation of sports fields is a grey area. In some countries, such water use is classified as agricultural use. The fertiliser run-off from sports fields may also contribute to overall water pollution, but probably to a considerably smaller degree than fertiliser run-off from agriculture. Thus, looking at the overall water resources in a country, the water used for sports fields may have a very small impact on total water supply; however, it may reduce the availability of water for other purposes at specific times in a given place.

Water quality

Measuring water quality

The complexity of the concept of water quality is reflected in the many types of water quality indicators used to measure the fitness of water for different uses.

There are a few simple measurements that may be carried out at the riverside, but these do not tell us much about the risk such water may pose. More complex determinations must be made in a laboratory, which requires the collection and proper transport of a water sample for analysis at another location. Such determinations can be expensive; moreover, South Africa has a shortage of properly equipped water analysis laboratories and experienced technical personnel. Due to the expense and effort involved, most monitoring programmes are carried out by government institutions. However, these programmes[3] have very few sampling points and do not cover any river or area well.

- *Physical assessment*

The physical attributes of water are important for many water uses. Cloudy or turbid water (containing many silt particles or organisms such as algae) may clog filters and is expensive to purify. Suspended particles also provide shelter for harmful organisms, which means that cloudy water poses an increased risk of being contaminated with organisms that cause disease. In fact, turbid water should always be regarded with suspicion and not used for drinking purposes unless absolutely necessary. Muddy, silty rivers – resulting either from the

---

3   The DWA's National Microbial Monitoring Programme is one such programme. Unfortunately, its data is not available to the public.

transfer of turbid water from the Theewaterskloof Dam or from earthmoving activities in drainage areas – are of particular concern in the Stellenbosch area. The existing legislation (*National Water Act* (RSA 1998)) prohibits the pollution of natural water sources by run-off from earthmoving activities; however, in practice, such incidents are seldom followed up and hardly ever acted upon.

Surface water in and around human habitations may also experience temperature disturbances, called thermal pollution. Thermal pollution is an increase or decrease in the temperature of a natural body of water as a consequence of human actions. Heated water originating from cooling or industrial manufacturing plants is a common cause of this type of pollution is. Increased water temperature depletes oxygen in water, which may kill river plants and fish. Warm water may also affect ecosystems by allowing species that are better adapted to higher temperatures to replace the indigenous species in the water body, thus upsetting the ecological balance. Run-off from urban areas may also raise the temperature of surface waters.

- *Chemical assessment*

The chemical attributes of the water in a given river may affect its toxicity and determine whether or not it is safe to use, and are therefore important indicators of water quality. As such, chemical water quality is an important indicator of the health risk to humans and the plants and animals that live in and around a river.

The assessment of chemical water quality includes measurements of numerous chemical compounds dissolved or suspended in water. Commonly measured chemical parameters include pH, alkalinity, hardness, nitrates and nitrites, ammonia, orthophosphates and total phosphates, dissolved oxygen, and biochemical oxygen demand. Electrical conductivity and turbidity are also used as indicators of water quality.

Agriculture has long used only chemical parameters to judge the fitness of water for irrigation and this practice is still very much in evidence today. However, there has been a rapid rise in the microbiological pollution (especially by sewage) of the surface waters available to most farms for irrigation and farmers who ignore microbiological assessments when judging the fitness of water for irrigating edible crops increase the health risk to consumers.

- *Microbiological assessment*

Microbiological attributes refer to the number and types of organisms that inhabit a waterway and are important indicators of water quality. The poorer the water quality, the fewer the number and types of beneficial organisms that are able to live in it and the higher the number of harmful or disease-causing organisms (pathogens) that are present.

Microbiological examination provides the most sensitive (although not the most rapid) indication of the pollution of drinking water supplies (National Health and Medical Research Council (NHMRC) & Agricultural and Resource Management Council of Australia and New Zealand (ARMCANZ) 1996/2001; RSA DWAF 1996). Unlike chemical or physical analysis, it involves a search for small numbers of viable organisms (unless the water is heavily polluted). Due to the fact that the growth medium and conditions of incubation may influence the species isolated and the count of these organisms, microbiological examinations may have variable accuracy (RSA DWAF 1996). This means that the standardisation of methods and laboratory procedures are of great importance if criteria for microbiological water quality are to be uniform across different laboratories (Niemi, Heikkila, Lahti, Kalso & Niemela 2001; Grasso, Sammarco, Ripabelli & Fanelli 2000; NHMRC & ARMCANZ, 1996//2001; RSA DWAF 1996; Nixon, Rees, Gendebien & Ashley 1996).

For technical and financial reasons, it is not practical to assess the safety of water by testing for the numerous pathogens that might be present (NHMRC & ARMCANZ 1996/2001; RSA DWAF 1996). Therefore, so-called 'indicator organisms', which point to the presence of harmful organisms, are generally used for routine monitoring of the potential presence of pathogens in water.

There are many indicator organisms available, all with their distinct advantages and disadvantages, but no single one meets all requirements. Neither is there a universal combination that ensures good predictive ability in all cases. Each situation has to be considered individually. Considerations in the choice of indicator organism(s) depend (among other things) on the intended water use, risk of infection, potential sources of pollution, financial resources and the availability of laboratory facilities and expertise.

During the 1980s, *Escherichia coli* (*E. coli*) was chosen as the biological indicator for water treatment safety (NHMRC & ARMCANZ 1996/2001). *E. coli* is an organism that occurs in the gut of warm-blooded animals (including humans) and is a reliable indicator of the presence of poorly treated or untreated sewage in water. Due to methodological deficiencies, *E. coli* surrogate tests (such as the faecal coliform and total coliform tests) were developed and became part of water quality testing guidelines. With the advent of modern testing techniques, *E. coli* count was reintroduced into drinking water regulations and it is now the indicator organism of choice whenever water quality screening tests indicate a possible problem.

Pollution: a serious threat to water quality

Whether water is polluted in the technical sense depends on whether it is physically suitable for its intended use (Olson 1992). Water may be unacceptable for drinking, but still suitable for irrigation or industrial cooling purposes. Water pollution may originate from human sources (such as sewage discharge, urban storm water drainage, refuse disposal, agriculture, and mining and industrial activities) or from natural causes (such as siltation in rivers during times of flood.)

Water pollution is generally categorised as coming from point sources or diffuse sources. Abatement efforts have largely concentrated on point sources (because of their ready accessibility) and on the symptoms of pollution, rather than its causes (*ibid*.). In fact, because of their easy identification and accessibility, there has been a tendency in the past to classify water pollution as originating from point sources (even when a closer look would have revealed this to be false). As a result, diffuse sources were ignored and all abatement efforts concentrated on the point source, which led to poor results and the eventual abandonment of efforts to stop the pollution.

▣ *Point and diffuse sources of pollution*

The major point sources of water pollution in the Stellenbosch area are:

- poor provision of sanitation, especially in low-cost housing areas;
- poorly functioning sewerage systems, poor inspection and maintenance, and delayed unblocking of drains;
- old waste water treatment works that are operating far beyond their capacity; and
- winery and food processing effluents.

Diffuse pollution comes from numerous sources that are not easy to pinpoint. In the Stellenbosch area, these sources may include urban run-off (especially contaminated storm water), farming activities, rural dwellings, etc. Urban storm water is a particularly underestimated source of sewage pollution. During any rain episode, storm water enters the sewerage systems of the town (whether by design or not), causing sewage drain overflows in most low-lying areas. The additional water reaching the sewage treatment works also results in major overflows into the nearest receiving water body, which greatly exacerbates the pollution of local waterways by sewage.

Although agricultural run-off is most often mentioned as a serious source of pollution, this was not borne out by the observations made by the author over a period of six years of study of the Plankenbrug River (Barnes 2003). Neither the microbiological nor chemical water quality of the river showed any gross pollution in the farming areas over that period (regardless of the season), but both deteriorated sharply once the river flowed past the urban edge of Stellenbosch town proper.

▣ *Types of water pollution*

Generally speaking, microbial pollutants from sewage often result in infectious diseases that negatively affect human and animal health and aquatic life. This may lead to increased illness and even mortality within a given environment. However, it is difficult to substantiate the direct relationship between water pollution in the Stellenbosch area and related illnesses due to the complete absence of reliable and detailed water pollution data and only very rudimentary waterborne disease data for the area. Nonetheless, it is obvious that high levels of microbiological pollution, which indicate the presence of sewage in the water, is highly undesirable and that such water carries an increased risk of disease.

Heavy metals and toxic compounds from industrial processes and industrial waste may accumulate in water bodies. These substances are toxic to marine life such as fish and shellfish and may also affect the rest of the food chain. Some toxins affect the reproductive systems of marine life and may therefore disrupt the species mix within a given aquatic environment. This means that entire animal communities may be badly affected by this type of pollutant. There are very few efforts made at monitoring such pollutants in the Stellenbosch area.

Organic matter and nutrients cause an increase in aerobic algae and depletes oxygen from the water body. This process is known as eutrophication and it causes fish and other aquatic organisms to suffocate. The toxins produced by these so-called 'blue-green' algae (of which microcystins are the most important) are also harmful to human health and are closely related to liver damage and certain types of cancers.

Suspended particles such as silt and mud reduce the amount of sunlight penetrating water, which disrupts the growth of photosynthetic plants and micro-organisms and prevents the ultraviolet rays of the sun from reducing the number of pathogens in the water by natural means.

Conclusion: Water strategies for a sustainable future

The scope of this chapter prevents a full treatment of all the present needs, but the suggestions below should help to direct further debate.

Strategies for immediate implementation

- *Reuse, reuse, reuse*

One of the most feasible ways to dramatically improve water resources management is to add a water loop in the flow from 'raindrop to crop'. This means first using water for human consumption (mainly in urban areas) and then reusing it for irrigation (subsequent to appropriate treatment to make it safe for agricultural use) (Barnes 2003). This is an especially viable option for smallholdings surrounding urban areas that produce vegetables and other cash crops, as the large volumes of waste water produced in towns need only be piped over relatively short distances in most cases. Water reuse has been proposed as the only viable long-term solution to the water shortage in areas such as Stellenbosch. However, given the current quality of sanitation and efforts to prevent water pollution in the Stellenbosch area, this solution is untenable at present – yet another reason for taking drastic action to quite literally 'clean up our act'.

It is, however, prudent to introduce a word of caution at this point regarding the essential safeguards that should be in place to allow waste water to be reused safely. The shorter the loop between the production of waste water and its point of reuse (especially without any intervening treatment), the higher the risk of contamination and resultant health problems. For instance, waste water produced in a domestic setting should only be reused directly on the same premises if an intermediate treatment step is introduced; and it should never be used on edible crops or on areas such as lawns where children play, because they are most susceptible to infections from the pathogens in household waste water.

- *Improve sanitation systems*

These improvements include increasing the availability of sanitation in low-income areas and seriously improving the reticulation systems in all areas of Stellenbosch. Treatment capacity and quality at the water purification works and waste water treatment works are of the highest priority. Almost all the waste water purification works in the greater Stellenbosch area are running beyond capacity and on outdated technology and there is a serious shortfall in the funds and efforts spent on maintenance of the systems already in place. The primary responsibility of local political systems to allocate money for such essential services is proving to be a serious obstacle to preventing the problems from escalating, let alone making improvements.

The provision of sanitation in all informal settlements in the area should be reviewed and strenuous efforts made to engage residents in decision making regarding all new toilet and tap facilities in order to encourage responsible use.

- *Provide education about hygiene and sanitation*

Many households may benefit from guidance on hygiene and sanitation, especially in terms of the constraints of particular households. This does not only apply to poor households (where the need is apparent), but also to higher income households, where inefficient water use, water-wasting gardens, littering and inappropriate diversion of storm water from properties into sewers occur far too regularly. In a study in Kayamandi (a township right outside of Stellenbosch), door-to-door visits by trained community health workers proved very effective over the short term (Barnes 2003).

- *Improve stormwater management*

As in the case of many other municipalities, storm water management in Stellenbosch Municipality is the responsibility of the Roads Department, which in effect means that concerns about water volumes and possible flooding override any concern about water contamination. This allows large volumes of contaminated storm water to reach the nearest water course unhindered. Moreover, the Roads Department is not equipped or trained to deal with contaminated water and storm water should therefore rather be managed by a combined team of road and sanitation engineering staff. The argument that combining departments in this way will inevitably result in friction is not a valid objection.

- *Prevent litter*

Any sustained anti-litter campaign will help to reduce the number of blockages in drains and storm water systems and thereby reduce the amount of pollution reaching the rivers. Sporadic anti-litter efforts – which appear to be little more than window dressing – do more harm than good. A sustained anti-litter campaign involves much more than simply inviting a few concerned citizens to pick up plastic bags while being photographed with a local dignitary.

- *Conserve water*

When water restrictions have to be implemented, the local authority (mainly the municipality) usually concentrates on household users, policing the restrictions with varying amounts of zeal. Unfortunately, no one polices the local authority itself, where in certain cases vast amounts of water may be lost due to poor reticulating systems, burst pipes, poor water management, etc. In fact, the volume of water lost in this way may exceed the savings achieved by household users by several orders of magnitude. It makes more sense for authorities to first consider large-scale users (and itself) and implement a sustained programme of consultation to reduce water demand at all times before issuing periodic draconian measures for household users, only to lift them when the next wet season produces enough water for the foreseeable future. A more holistic approach with a fairer spread of the burden of water saving will gain much better acceptance than the present system.

Strategies for the medium term

Water strategies for the medium term should pursue several courses of action. Any further destruction of any wetland should be prevented and more sustainable and eco-friendly farming practices encouraged. Better control of sediment pollution must be enforced at all sites where excavation or soil moving occurs. In addition, all levels of law enforcement must be greatly improved to prevent the large-scale pollution that is taking place at present, including policing the local authorities and not only businesses and private citizens. Monitoring of water courses must be improved by using a scientific system of water analysis, so that management decisions may be based on objective and valid data, rather than opinions. Finally, as many citizens as possible should be introduced to safe alternatives to waterborne sanitation and these alternative systems should be properly overseen to ensure their proper use.

Strategies for the longer term

In order to achieve sustainable water use in the longer term, there is an urgent need for a new model of management and implementation of basic services that may be protected from parochial local politics. The present politicised municipal system, with its inherent risk of staff instability and uncertain financial planning, should be replaced with a non-partisan structure that may ensure service delivery in a representative, transparent and responsible manner. As water resources diminish, competition for this scarce resource will escalate, and so will the conflicts surrounding its availability. There is no doubt that the long-term economic and social survival of the Stellenbosch area will depend on the creation of a non-partisan body to oversee all aspects of water use and the prevention of water pollution.

The Stellenbosch area is fortunate to have many experts and facilities available locally: Stellenbosch University, a range of research centres, numerous agricultural institutes and the provincial headquarters of the Department of Agriculture are all located within its area boundaries. These resources should be utilised in a much more coherent and effective manner. A cooperative body, allowing much more concrete participation by technical specialists and professionals, should be set up to provide expertise and support to the present local authority structures. At present, this resource is not even recognised. Again, the effective functioning of such a body will depend on the cooperation and political will of local leaders.

If narrow partisan interests are allowed to prevail, the future of greater Stellenbosch's water resources – and the heritage-rich area it supports – looks bleak indeed.

REFERENCES

Barnes JM. 2003. The impact of water pollution from formal and informal urban developments along the Plankenbrug River on water quality and health risk. PhD thesis. Stellenbosch: Stellenbosch University.

CIA (Central Intelligence Agency). 2009. Urbanization Rates of Countries. In: J Cutler (ed). *Encyclopaedia of Earth*. Cleveland, Washington, DC: Environmental Information Coalition, National Council for Science and the Environment.

Dalgliesh C, Steytler N & Breetzke B (eds). 2004. *Western Cape State of the Environment Overview Report*. Report prepared for the DEA&DP (Western Cape Department of Environmental Affairs and Development Planning). Report No 329585/1. Cape Town: DEA&DP.

DAWC (Department of Agriculture Western Cape). 2007. *Resource Utilisation: Natural resources of the Western Cape*. [Accessed 22 January 2011] http://www.elsenburg.com/trd/resource/natres.html

De Villiers S & De Wit M. 2010. *H2O-CO2-energy equations for South Africa: Present status, future scenarios and proposed solutions*. Africa Earth Observatory Network (AEON) Report Series No 2. Cape Town: University of Cape Town.

Grass G. 1990. Interview in *New Statesmen & Society*, 22 June 1990. [Accessed 3 September 2012] http://quotes.dictionary.com/we_already_have_the_statistics_for_the_future

Grasso GM, Sammarco ML, Ripabelli G & Fanelli I. 2000. Enumeration of *Escherichia coli* and coliforms in surface water by multiple tube fermentation and membrane filter methods. *Microbios*, 103(405):119-125.

IAASTD (International Assessment of Agricultural Knowledge, Science and Technology). 2008. *Executive Summary of the Synthesis*. [Accessed 3 June 2008] http://www.agassessment.org/docs/iaastd_exec_summary_jan_2008.pdf

Juana JS, Kirsten JF & Strzepek KM. 2006. Inter-sectoral water use in South Africa: efficiency vs equity. Paper presented at the 26th International Association of Agricultural Economist Conference. Gold Coast, Australia: 12-18 August. [Accessed 3 June 2008] http://ageconsearch.umn.edu/bitstream/25486/1/cp060412.pdf

Midgley GF, Chapman RA, Hewitson B, Johnston P, De Wit M, Ziervogel G, Mukheibir P, Van Niekerk L, Tadross M, Van Wilgen BW, Kgope B, Morant PD, Theron A, Scholes RJ & Forsyth GG. 2005. *A status quo, vulnerability and adaptation assessment of the physical and socio-economic effects of climate change in the Western Cape*. CSIR Report No ENV-S-C 2005-073. Report to the Western Cape Government. Stellenbosch: CSIR.

NHMRC (National Health and Medical Research Council) & ARMCANZ (Agricultural and Resource Management Council of Australia and New Zealand). 1996 (as amended 2001). *Australian Drinking Water Guidelines*. Canberra: NHMRC & ARMCANZ.

Niemi RM, Heikkila MP, Lahti K, Kalso S & Niemela SL. 2001. Comparison of methods for determining the numbers and species distribution of coliform bacteria in well water samples. *Journal of Applied Microbiology*, 90(6):850-858.

Nixon SC, Rees YJ, Gendebien A & Ashley SJ. 1996. *Requirements for water monitoring*. Topic Report No 1/1996. Copenhagen: EEA (European Environment Agency).

O'Keeffe JH, Van Ginkel CE, Hughes DA, Hill TR & Ashton PJ. 1996. *A simulation analysis of water quality in the catchment of the Buffalo River, Eastern Cape, with special emphasis on the impacts of low cost, high-density urban development on water quality*. Volume I. Report No 405/0/96. Pretoria: Water Research Commission.

Olson BH. 1992. Environmental Water Pollution. In: WN Rom (ed). *Environmental and Occupational Medicine*. 2nd Edition. Boston, MA: Little, Brown & Co.

RSA (Republic of South Africa). 1998. *National Water Act (Act No 36 of 1998, as amended)*. *Government Gazette*, 19182. Pretoria: Government Printing Works.

RSA DWAF (Republic of South Africa. Department of Water Affairs and Forestry). 1996. *South African Water Quality Guidelines*. 2nd Edition. Vol. 1. Domestic Use. Pretoria: Council for Scientific and Industrial Research (CSIR) Environmental Services.

RSA DWAF. 2003. *The Water Conservation and Water Demand Management Strategy for Industry, Mining and Commercial Water Use Sector*. [Accessed 20 May 2012] http://www.dwaf.gov.za/WaterConservation/Programs_IMP.htm

RSA DWAFa: Directorate Water Conservation. n.d. *Local Authorities & Water Services Institutions*. [Accessed 23 January 2011] http://www.dwaf.gov.za/WaterConservation/Programs_Arc.htm

RSA DWAFb: Directorate Water Conservation. n.d. *Water Conservation and Demand Management in the Agricultural Sector*. [Accessed 23 January 2011] http://www.dwaf.gov.za/WaterConservation/Programs_Arc.htm

UN (United Nations). 2009. *World Water Development Report 3: Water in a Changing World*. [Accessed 3 September 2012] http://www.unesco.org/new/en/natural-sciences/environment/water/wwap/wwdr/wwdr3-2009/downloads-wwdr3/

UNEP (United Nations Environmental Programme). 2010. *Africa Water Atlas. Summary for Decision Makers*. [Accessed 21 January 2011] http://www.unep.org/pdf/Africa_Water_Atlas_Executive_Summary.pdf

UNESCO (United Nations Educational, Scientific and Cultural Organisation) & UNIDO (United Nations Industrial Development Organisation). n.d. *World Water Assessment Programme – Water and Industry*. [Accessed 26 January 2011] http://webworld.unesco.org/water/wwap/facts_figures/water_industry.shtml

United Nations Water. 2011. *Statistics: Graphs & Maps: Water use*. [Accessed 26 January 2011] http://www.unwater.org/statistics_use.html

Walmsley RD, Walmsley JJ & Silberbauer M. 1999. *National State of the Environment Report. Freshwater systems and resources*. [Accessed 25 January 2011] http://www.ngo.grida.no/soesa/nsoer/issues/water/index.htm

WBCSD (World Business Council for Sustainable Development). 2009 (updated). *Facts and trends – water*. Version 2. Project Director: J Griffiths. Geneva, Switzerland: WBCSD.

Chapter 15

ECOSYSTEM SERVICES
Protecting Stellenbosch's natural systems

BLAKE ROBINSON

Introduction

Stellenbosch is known around the world for its rich agricultural heritage, in particular its wine industry. This has been made possible by the goods and services provided by its unique ecosystems, namely fresh water, fertile soils, pollination, and suitable climatic conditions. As markets for Stellenbosch's produce have expanded internationally and farming practices have become more industrialised, demand for land in the area has seen human interests conflicting with those of the natural systems that assist food production. Unless action is taken to halt and reverse the damage inflicted on life-supporting natural systems, the region runs the risk of losing the natural assets that make it so special.

Healthy ecosystems provide the goods and services necessary for human quality of life. The *Millennium Ecosystem Assessment* (United Nations Environmental Programme (UNEP) 2005) provided the first detailed analysis of this relationship, categorising the ecosystem services on which socioeconomic systems depend as follows:

- provisioning services: oxygen, water, food, fibres, natural medicines, pharmaceuticals, genetic resources and biochemicals;
- regulating services: air quality, water regulation, water purification and waste treatment, pollination, erosion regulation, CO_2 absorption, climate regulation, disease regulation, pest regulation and natural hazard regulation;
- cultural services: spiritual and religious values, aesthetic values, ecotourism and recreation; and
- supporting services: soil formation, nutrient cycling and primary production.

Despite the crucially important role these goods and services play in supporting life, the relationship between ecosystem health and human well-being is often misunderstood or tends to be overlooked. In an effort to translate the value of ecosystems into measures that are more readily accessible to decision makers, attempts have been made to ascribe a financial value to the goods and services they provide. In 2009, Cape Town researchers valued the city's ecosystem goods and services (EGSs) at R2 billion to R6 billion per year, and found that R8.30 to R13.50 worth of EGSs are generated for every rand invested into the city's natural environment. Based on these returns, municipal spend on the environment is 1.2 to 2 times more worthwhile in terms of generating public value than expenditure on the economy (De Wit, Van Zyl, Crookes, Blignaut, Jayiya, Goiset & Mahumani 2009). There is therefore a strong economic rationale for promoting ecosystem health and municipalities would be ill-advised to make environmental trade-offs to advance short-term human interests.

While maintaining ecosystem health is vitally important worldwide, the issue is perhaps even more pressing in Stellenbosch due to the uniqueness and vulnerability of its local ecosystems. This chapter starts with an overview of the importance of Stellenbosch's endemic ecosystems, building an appreciation for the richness of the area's biodiversity. This is followed by an account of the current threats to ecosystem health in Stellenbosch and the impacts they have and are likely to have in the future. Turning to solutions, a number of initiatives that attempt to address these challenges are reviewed. The chapter concludes by looking at the extent to which ecological issues are currently embraced by Stellenbosch Municipality and identifies enhanced in-house capacity as a key requirement for effectively integrating ecological concerns into development planning.

Stellenbosch's ecological assets

The Western Cape is home to the Cape Floral Kingdom, the smallest of the world's six floral kingdoms. Covering a mere 0.06% of the earth's surface, the area contains over 9 000 plant species and its exceptional pattern of species richness makes it a treasure trove of biodiversity.[1] Considering that 69% of these species are found nowhere else in the world and that many of them are threatened, its conservation is important from both a national and global perspective (Republic of South Africa Department of Environmental Affairs and Tourism (RSA DEAT) 2007).

The Cape Floral Kingdom is one of nine priority conservation areas for biodiversity in South Africa and the Stellenbosch municipal area occupies a significant portion of it. Stellenbosch also has the highest density of endemic species in South Africa, with 0.05 unique species per square kilometre (Gerber 2005). As the map below shows, the vegetation is characterised by a combination of mountain fynbos, West Coast Renosterveld and sand plain fynbos. The term 'fynbos' refers to vegetation that is specifically adapted to the Cape's Mediterranean climate and periodic fires. It is constituted by the iconic Proteaceae, reed-like Restionaceae, heaths and geophytes. These plants grow predominantly in acidic, coarse sandy soils with little nutrient value, but some are found in the richer alluvial soils of the Renosterveld and alkaline marine sands to the south. They are of particular interest to the global scientific community due to their relationship with fire, their use of ants and termites to disperse seeds, and the high level of pollination by insects, birds and mammals (UNEP 2008).

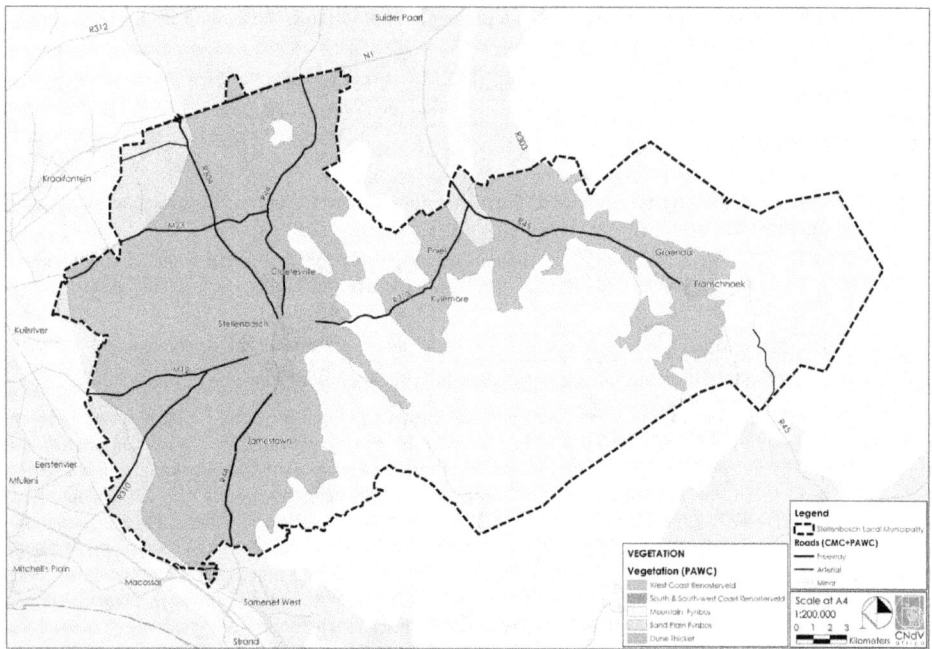

Figure 15.1   Distribution of vegetation types across the Stellenbosch municipal area [CNdV Africa 2010a]

---

1   The term 'biodiversity' is used to refer to "the richness and variety of life forms in a particular area and the interconnections that link them in a mutually beneficial web of life" (Local Action for Biodiversity (LAB) 2010:8).

Although not as spectacular or diverse as the flora, many endemic species of mammals, birds, reptiles, amphibians, fish and insects also occur in the region. These species play an important role in the functioning of ecosystems by maintaining a balance between species, aerating and feeding the soil, dispersing seeds, and facilitating pollination. They include a number of rare species, such as the endangered Table Mountain ghost frog (*Heleophryne rosei*), micro frog (*Microbatrachella capensis*), Cape clawed toad (*Xenopus gilli*), velvet worm (*Peripatopsis leonine*) and geometric tortoise (*Psammobates geometricus*) (UNEP 2008).

The 2004 *National Spatial Biodiversity Assessment* (NSBA) classified around half of the indigenous vegetation in the Stellenbosch municipal area as 'Critically Endangered'. According to the definition, this means that so much of the original habitat in a given area (or areas) has been destroyed that ecosystem functioning has been impaired, and that a significant portion of the species once associated with the ecosystem has been lost. Due to the interconnected nature of ecological systems, the extinction of one species may have unforeseen knock-on effects on other species, which may eventually result in ecosystem collapse. The main areas of concern are those to the west of the Stellenbosch-Klapmuts axis, the Dwars River Valley and the Franschhoek Valley. Another quarter of the region's indigenous vegetation is considered 'Endangered', with the areas along the higher lying mountain ranges to the east being 'Least Threatened' (CNdV Africa 2010a). This is illustrated on the map in Figure 15.2.

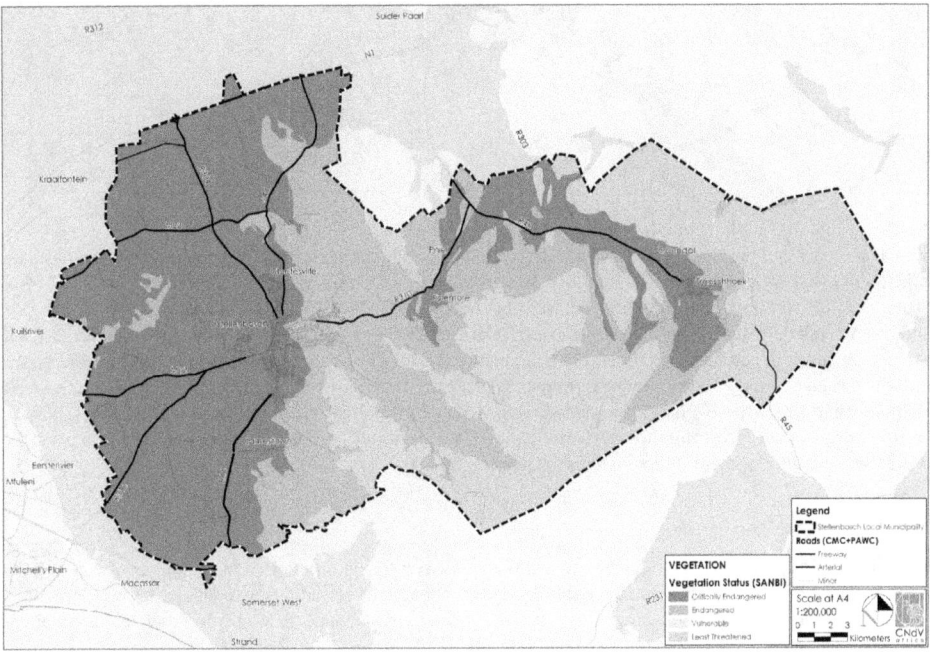

Figure 15.2   Ecosystem status across the Stellenbosch municipal area [CNdV Africa 2010a]

A comparison between this map (Figure 15.2) and the location of conservation areas (Figure 15.3) reveals that very little of the Critically Endangered and Endangered areas in the Stellenbosch Municipal Area have been given formal conservation status, hence they are under severe threat of being damaged by human activities.

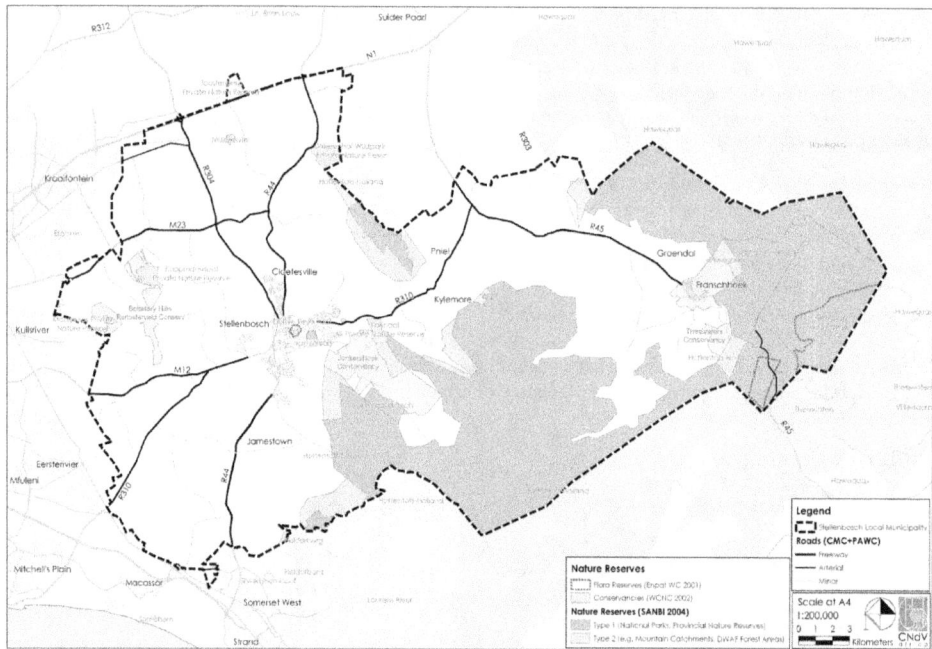

Figure 15.3   Reserves and protected areas within the Stellenbosch municipal area [CNdV Africa 2010a]

Spatial planning and the rapid growth of human settlements in the Stellenbosch region in recent decades have contributed to a situation where human interests are directly threatening those ecosystems most in need of protection. The National Environmental Management Act (Act No 107 of 1998), Biodiversity Act (No 10 of 2004), and the Amended Protected Areas Act (No 31 of 2004) specify that planning for biodiversity in South Africa is the responsibility of local municipalities. Their mandate includes incorporating environmental considerations into the design of Spatial Development Frameworks (SDFs) in the interest of achieving integrated development planning that meets the needs of people, as well as that of the environment (Gerber 2005). The following section introduces some of the threats to Stellenbosch's ecosystems and provides some insight into what the protection and development of ecosystems in the area entails.

Threats to Stellenbosch's ecological assets

The provision of goods and services by Stellenbosch's ecosystems is threatened in a number of ways by the encroachment of human activities. Over the years, the impacts of settlement expansion, farming practices, pollution, waste management and other activities have manifested in the interconnected problems discussed below.

Loss and fragmentation of natural habitats

When natural habitats are disturbed or broken up it becomes difficult for organisms to maintain the interconnections between them that allow ecosystems to function. While conservation areas are helpful in protecting specific sections of natural habitats, different species require different minimum areas in order to survive. Conservation areas that are too small or isolated are thus limited in their ability to effectively prevent biodiversity loss (LAB 2010).

The main cause of habitat loss and fragmentation of ecosystems in the Stellenbosch region is industrial agriculture. The replacement of vast areas of indigenous vegetation with monocultures that are fertilised and protected with chemical products wreaks havoc on natural flora and fauna and impairs the ability of organisms to move across this land. Of the indigenous flora found in the area, Renosterveld vegetation has been the most severely affected by human activity, because it typically grows in fertile soils that are well-suited to agriculture. As demand for new types of wine cultivars grows, the expansion of vineyards up mountainsides is also placing strain on mountain fynbos vegetation (*ibid.*).

The clearing of indigenous vegetation to make way for cultivation on private farms is difficult to monitor and police and is one of the top challenges for conserving biodiversity and maintaining healthy ecosystems in Stellenbosch. Urban sprawl also contributes significantly to habitat loss, not only in terms of the vegetation cleared for construction and fire breaks, but also in terms of the impact of growing populations on natural resource extraction and waste sinks (*ibid.*).

Invasive alien plant species

Invasive alien plants threaten biodiversity and water security in the Western Cape, particularly in the mountain catchments and riparian zones. It is estimated that up to 70% of natural fynbos in the Western Cape has been invaded to some degree, and 2.5% is severely invaded. Away from their natural predators, invasive species are able to thrive and in many cases take over from their indigenous predecessors. They are also often better able to utilise available resources and are thus capable of growing at a faster rate to achieve dominance (LAB 2010).

Alien species may disrupt ecological processes and change the relationships between ecosystem elements in a manner that impairs the delivery of ecosystem services. When they displace indigenous plants, they may reduce the amount of fresh water available by extracting more than their fair share or blocking water courses. In addition, some species sterilise the surrounding soil by burning at high temperatures, while others release toxins into the ground that prevent indigenous seeds from growing (*ibid.*).

Invasive alien species may also be found in watercourses, particularly in slow-moving, nutrient-rich waters, such as those polluted by agricultural or human waste. Species like water hyacinth grow very fast and form dense strands that may cover whole sections of river, in some cases restricting the flow and causing blockages. They may also smother indigenous species and prevent fish, birds and insects from accessing open waters (*ibid.*).

While it is possible to remove these plants and rehabilitate invaded ecosystems, the process is arduous and resource-intensive. Large-scale interventions to remove alien vegetation may also result in erosion, the sedimentation of downstream areas, or the germination of alien seedlings. Removal of waterborne species is typically done by machine, which may destabilise river banks and damage aquatic ecosystems. Moreover, the process is never 100% effective, so that regular maintenance of the area is required to prevent these plants from re-establishing themselves (*ibid.*).

Unnatural fire patterns

Stellenbosch's indigenous vegetation is prone to fires; in fact, some species are dependent on periodic fires (every fifteen years or so) for reproduction (LAB 2010). However, if the periods between fires are too long or too short, growth may be adversely affected. Human activities in fynbos regions and the encroachment of highly flammable alien species are resulting in more frequent fires than are necessary. Excessively frequent fires may reduce biodiversity and impair ecosystem functioning by not allowing indigenous plants enough time to mature and reproduce and by destroying organic matter in the soil that helps to retain moisture and reduce erosion (*ibid.*).

The maintenance of appropriate fire regimes is an important part of maintaining Stellenbosch's biodiversity, as it allows for the benefits of fire to be achieved without harm to human settlements and ecosystems. In addition to fire suppression and planned burning, this also requires efforts to control invasive alien species and improve public awareness (*ibid.*).

Deteriorating water quality

As a result of pollution from human activities, the quality of river water in the Stellenbosch area is poor. Pollutants include *E. coli* (*Escherichia coli*) from human waste, phosphates and nitrates from farms and gardens, and heavy metals and carbons from road run-off (CNdV Africa 2010b). Existing waste water treatment works are unable to deal with the high levels of non-toxic pollutants that cause eutrophication and may lead to system collapse. According to the *Green Drop Report 2009: South African Waste Water Quality Management Performance*, the quality of waste water treatment provided by Stellenbosch Municipality's six treatment works was "less than satisfactory", with the average performance score ranging between 47% and 59% (RSA DWA 2010).

According to the South African National Biodiversity Institute (SANBI), most Western Cape rivers have a critically endangered conservation status, and even those in the mountains are considered vulnerable to contamination from human activities. Damaged rivers are unable to filter and clean water to safe levels, with negative knock-on effects on the ecosystems and communities that depend on them. In addition to the effect on local ecology and the well-being of Stellenbosch communities, there is mounting concern that, should water quality fall below European Union standards, the fruit and wine export industry may suffer in the near future (CNdV Africa 2010b).

Climate change

A very long period of climatic stability played an important role in allowing a rich diversity of species to establish themselves in the Cape Floral Kingdom (UNEP 2005). The effects of relatively rapid climate change on the Western Cape's ecosystems may be slow to manifest, but will be severe in impact (Freeth, Bomhard & Midgley 2007). Compounding the challenges mentioned above, climate change is likely to bring rising temperatures, reduced and erratic rainfall, and more frequent fires. Drier weather will exacerbate water shortages, and increasing competition for water resources will place additional pressure on rivers, estuaries and wetlands. The drying of these ecosystems will compromise their ability to provide goods and services, with negative repercussions for humans and other dependent species. A rise in sea level will result in more incidents of flooding in Stellenbosch's low-lying areas, along with erosion and the intrusion of salt into groundwater reserves. Higher levels of CO_2 in the atmosphere will aid the spread of woody alien species, further adding to water scarcity and fire risks (*ibid*.).

Biodiversity plays an important role in adaptation to climate change and ecosystems with a complex interplay of species are more likely to survive than those found on industrial farms where only a few species predominate. That said, it is estimated that the Western Cape's biodiversity hotspots could lose up to 30% of their species as a result of climate change. The actual impact is difficult to predict due to the interrelated nature of a number of influencing factors, but the ability of ecosystems to adapt and survive will be determined by the pace and extent of climatic changes (*ibid*.).

Like humans, plant and animal species can only live within certain tolerances, and as their surroundings change, they need to relocate or adapt in order to survive. In addition to efforts to slow climate change, maintaining the rich diversity of the Cape Floral Kingdom's many species requires that ecosystems be given space to adapt and migrate to more suitable conditions.

Interventions to protect Stellenbosch's ecological assets

The ecological challenges mentioned above have not gone unnoticed and a number of different groups have undertaken efforts to address them. Some of the more noteworthy endeavours of particular relevance to the Stellenbosch region are discussed below.

Stellenbosch Municipality's 2010 Spatial Development Framework

Stellenbosch Municipality has incorporated a number of strategies into the 2010 revision of its *Spatial Development Framework* (SDF) to address some of the concerns raised in the previous section. Although their influence is largely limited to the approval of new development plans and the zoning of land, these strategies provide insight into how the municipality plans to address conflicts between man and the environment. They include:

- *encouraging landowners outside formal conservation areas to conserve Endangered and Critically Endangered vegetation types, and to link with existing conservancies;*
- *supporting projects to eradicate alien vegetation in non-agricultural areas;*

- *protecting conservation areas as a means of ensuring water quality and quantity, with a focus on the upper reaches of river systems;*
- *implementing river conservation zones of between 10 m and 30 m in width (depending on the width and maturity of the river) on each bank to protect riverside ecosystems from all human activities save passive recreational pursuits;*
- *upgrading waste water treatment works to achieve minimum DWA water quality standards;*
- *focusing development in low-density areas, infill, and brownfield land before considering greenfield sites; and*
- *encouraging forms of tourism that reinforce Stellenbosch's unique sense of place.*

[CNdV 2010b]

While these strategies indicate that the municipality is aware of the conflict between built environments and natural areas, they do not appear to regard the region's ecological assets (except perhaps those that are endangered) as being of any greater importance than farming activities and do little to prevent trade-offs from being made between agriculture and biodiversity.

The municipality's ability to identify and manage conservation areas by means of the SDF is inhibited by a lack of fine-scale mapping of its ecological assets. The most recent *Environmental Management Framework* (EMF) for Stellenbosch was produced in 2005 and applications for financial resources to update it in line with the 2010 revisions to EMF regulations have thus far been unsuccessful. Maps of agricultural land, which would make a valuable contribution to the planning of biodiversity corridors and water resources management, are likewise unavailable (De la Bat 2011).

The Cape Winelands Biosphere Reserve

In 2007, UNESCO designated over 300 000 ha of Western Cape land as the Cape Winelands Biosphere Reserve under the Man and Biosphere (MAB) Programme, thereby incorporating it into a global network of 550 protected biospheres (see Figure 15.4). The core area covers a part of the Cape Floral Kingdom with the highest floristic diversity and greatest concentration of endemic and threatened species. The reserve extends from the Kogelberg Biosphere Reserve and Theewaterskloof Dam in the south to the Voëlvlei Dam in the north and includes the towns of Stellenbosch, Paarl, Wellington, Franschhoek, Villiersdorp and Klapmuts.

The Biosphere Reserve aims to ensure that human development coexists sustainably with the natural environment, with the goal of becoming "[an] area of excellence and good practice for people, culture and nature", in accordance with relevant legislation and policy (RSA DEAT 2007:1). It is managed by a Section 21 company and receives technical assistance from partner municipalities.

The intention behind the Biosphere Reserve is to add value to municipal planning processes by monitoring and coordinating SDFs and EMFs across municipalities (Balie 2011). At the time of writing, an EMF is being prepared for the eastern portion of the Cape Winelands District Municipality, but this excludes the Stellenbosch Municipality. Moreover, in the absence of a fine-scale EMF or any form of legislative power, the Biosphere Reserve is limited in its ability to influence spatial planning in Stellenbosch.

The Cape Winelands District Municipality has aligned its business plans with the functions of the Biosphere Reserve and coordinates several practical projects that create employment while furthering the interests of biodiversity in the region. These projects include the clearing of alien vegetation (in association with Cape Nature and Working for Water), organising an 'Environmental Week', and coordinating the 'Youth and Environment Programme', which introduces teachers and students from underprivileged areas to ecological issues (*ibid.*). Such initiatives not only help to address some of the main threats to local ecosystems, but also foster an appreciation for the environment, which may in turn engender behavioural change.

Cape Action for People and the Environment (C.A.P.E.)

C.A.P.E. is a partnership between government and civil society and aims to conserve and restore biodiversity in the Cape Floral Kingdom while at the same time benefitting the people of the region. Its twenty-three signatory

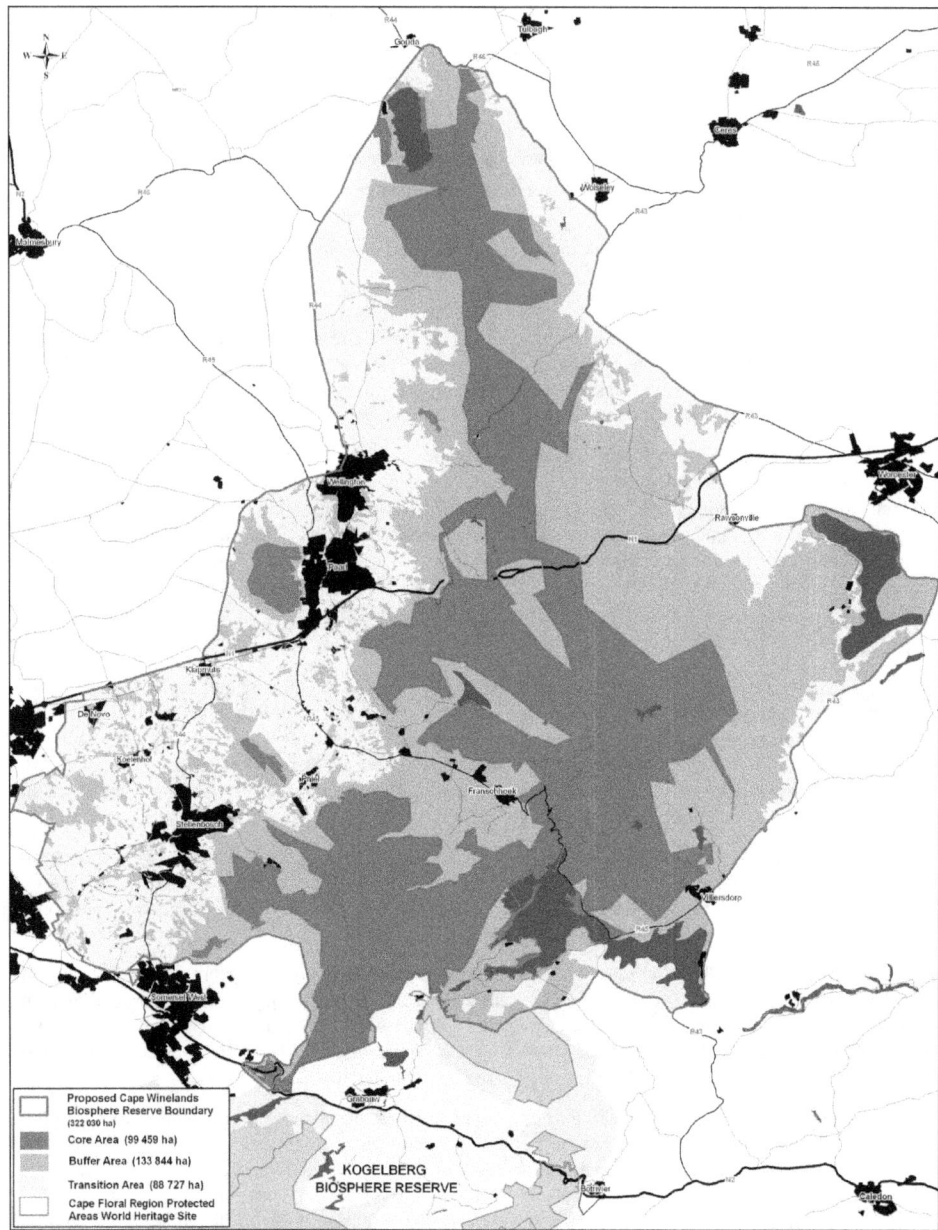

Figure 15.4   The Cape Winelands Biosphere Reserve plan [RSA DEAT 2007]

partners represent government, NGOs and community-based organisations and are coordinated by the C.A.P.E. Coordination Unit, hosted by SANBI at the Kirstenbosch Botanical Gardens (C.A.P.E. 2011).

The first phase of the programme used donor funding to develop new approaches to conservation that are relevant to the Western Cape context. Task teams were assembled to work on an array of interventions, including the following:

- providing conservation education;
- building capacity within institutions;
- providing fine-scale biodiversity plans;
- stimulating economic development that is compatible with biodiversity protection;
- facilitating conservation stewardship outside of state-owned protected areas;
- creating nature reserves and corridors;
- protecting water resources;
- developing new fire management systems; and
- coordinating strategies for the removal of invasive alien species.

From 2004 to 2009, C.A.P.E. was involved in a project called 'Integrating Biodiversity into Land Use Decision Making', in association with SANBI and the Western Cape Department of Environmental Affairs and Development Planning (DEA&DP). The project was designed to improve the integration of biodiversity in land use planning and decision making by providing useful information and support to decision makers, raising awareness of biodiversity issues and supporting cooperative governance (*ibid.*).

C.A.P.E. is also involved in a project headed by CapeNature to collate information about, among others, ecosystems, water resources, vegetation types and protected areas on a shared website, the BiodiversityGIS website.[2] The site provides free maps, research, databases and tools to aid the dissemination of expertise developed by C.A.P.E., CapeNature, SANBI and other institutions. The data may be disaggregated by municipality and provide a wealth of information, from wetland and forest inventories to information on soil and threatened species (*ibid.*).

Working for Water (WfW)

The Working for Water programme was launched in 1995 by the then Department of Water Affairs and Forestry (DWAF).[3] The programme aims to reduce the numerous threats posed by invasive alien plant species to indigenous ecosystems and water resources and employs local communities to clear these plants. As of 2012, it is running over 300 projects across South Africa and has created employment for approximately 20 000 people from marginalised communities in the clearing of more than 1 million ha of invasive aliens (RSA DWA 2012). The programme is supported by government, private companies and research foundations and has been internationally recognised for its outstanding contribution to environmental conservation and job creation.

The Biodiversity and Wine Initiative

In the centuries following the colonisation of the Cape, a significant portion of Stellenbosch's ecologically rich land has been privatised, cleared and planted with rows of imported grapevines. Almost 95% of South Africa's wine is grown in the Cape Floral Kingdom and there is a strong link between the wine industry and the fragmentation and deterioration of this ecosystem. The Biodiversity and Wine Initiative was formed in 2004 as a collaborative effort between local wine producers, the Botanical Society of South Africa, Conservation International and the Green Trust to protect natural habitats on wine estates and encourage sustainable grape farming practices (World Wide Fund for Nature (WWF) 2011).

---

2   See http://www.bgis.sanbi.org

3   The Department of Water Affairs and Forestry was split in 2009 and the forestry responsibility transferred to the Department of Agriculture, Forestry and Fisheries. The Department of Water Affairs (DWA) falls under the responsibility of the Minister of Water and Environmental Affairs.

Since its inception, the initiative has seen over 127 000 ha of natural area conserved by its 167 members, making the South African wine industry's conservation footprint larger than its current vineyard cover of 102 000 ha. The size of the conservation area is growing daily, with every additional hectare converted to vineyard being matched by a commitment from members to conserve another hectare of natural vegetation. In return, a distinctive sugarbird logo is used by member winemakers as a marketing tool to communicate their commitment to environmental responsibility to discerning buyers (*ibid.*).

Organic and biodynamic farming practices

Organic farming is based on the recognition that soil is alive and that healthy plants need healthy soil to thrive. Farms are viewed as 'closed-loop' systems, with organic waste products (leaves, stalks, grass clippings, spoiled produce, animal waste, etc.) being used on site to enhance the fertility of the soil (instead of chemical fertilisers, which have a sterilising effect). Pests and weeds are viewed as indicators of natural imbalances and are controlled by mixing plant species, encouraging natural predators such as birds and ladybirds, and using plant-based insecticides instead of chemical products. Biodynamic farming is similar, but is based specifically on the insights of Dr Rudolf Steiner and goes beyond organic principles to incorporate more subtle influences (for example, the effect of solar, lunar and earth rhythms on plant growth) to optimise fertility, vitality and nutrition wherever possible.

Considering the time it takes for plants to regain their natural resilience and the short-term effect this has on the productivity of certain crops, weaning a farm from chemical fertilisers and pesticides may be a risky undertaking. Nonetheless, the owners or management of a number of well-known farms in the Stellenbosch region are convinced of the advantages of more ecologically sensitive approaches to farming and are starting to benefit from a growing demand for organic products both locally and internationally, for example:

- Spier is adopting biodynamic approaches in the cultivation of its grapes, beef, chicken, eggs and vegetables, with over 300 ha of its land being farmed in this manner (Spier 2011).
- Laibach Wines started a gradual process of converting to organic practices in 2000. Vineyard management principles include, among other things, building the soil with cover crops, controlling weeds by physical rather than chemical means, and encouraging natural predators of pests by creating breeding grounds for them around the vineyards (Laibach Wines 2011).
- Over the past several years, Reyneke Wines has transitioned from high external input farming to biodynamic practices, allowing its vines to re-establish their natural defences. In addition, its cellar practices are conducted in accordance with natural and cosmic systems in an effort to ensure that its wines most accurately reflect the uniqueness of the soil and climate (Reyneke Wines 2011).

Reconciling planning and ecology

Considering the large proportion of Stellenbosch land classified as critically endangered and endangered (Figure 15.2) and the poor correlation between the areas that are threatened and those that are being actively conserved (Figure 15.3), it appears that the municipality's efforts to safeguard its ecosystems have thus far been inadequate. As the previous section shows, there are a number of overlapping initiatives focusing on improving the relationship between man and the environment, which translates into a growing body of local expertise on these issues. This raises the question of how this knowledge might more effectively influence decision making in Stellenbosch Municipality.

Ineffective integration of environmental needs and priorities into planning lies at the heart of Stellenbosch's ecological problems. Environmental concerns tend to be seen as an optional extra or a luxury in the face of pressing social and economic issues, resulting in trade-offs being made, with long-term negative repercussions. However, with a degree of innovation, human development does not need to conflict with nature, and it is possible to build human settlements in ways that may positively contribute to the environment (see Birkeland 2008). This requires a shift in perspective to viewing the region as an integrated social-ecological system in which humans and nature interact in a mutually beneficial manner (Resilience Alliance 2006).

Although Stellenbosch's most recent SDF shows signs of a willingness to conserve the region's ecosystems and water resources, the tough decisions necessary to advance environmental interests above those of property developers in the pursuit of long-term sustainability goals are not being made. According to the municipality, this is an issue of capacity. Responsibility for the environment currently falls within the Planning and Development and Parks and Gardens portfolios, but is not the core focus of either. While these portfolios have attempted to introduce environmental thinking into the municipality and have worked on a handful of projects in Stellenbosch – as and when time and budgets have allowed them to do so – the resources needed to effectively tackle ecological issues in the region are lacking.

The municipality needs experts on board to represent the interests of the environment, particularly when development plans are approved. The appointment of such individuals would help to integrate ecological interests into the planning of human activities and improve the municipality's functioning as a social-ecological system. Job descriptions have been compiled for a Senior Environmental Planner and an Air Quality Control Officer, but as of March 2012 these roles had not yet been approved as priority appointments (De la Bat 2012).

Conclusion

Stellenbosch is an area renowned for its rich ecological resources, but these are under threat from human activities. Environmental custodianship is largely confined to small and disconnected conservation areas and is poorly integrated into the functioning of the municipality. As the pollution of water courses, clearing of indigenous vegetation, extermination of species, and emissions of greenhouse gases continue, the region's unique ecology is at risk of being destroyed, and the negative implications of this for its communities and agriculturally-oriented economy are likely to become increasingly apparent in the coming years.

Although a number of groupings and organisations are actively engaged in trying to help Stellenbosch to grow in a manner that would allow for mutually beneficial human development and ecological preservation, the resources to meet this goal, both within and outside governmental structures, are limited. Unless action is taken to incorporate a strong and powerful voice for the environment into the region's planning processes and the way in which development is conceived, the risk of losing many of the ecological assets that make Stellenbosch attractive to visitors, residents and investors will continue to rise.

REFERENCES

Balie Q. 2011. Personal interview. Stellenbosch: 3 March.

Birkeland J. 2008. *Positive development – from vicious circles to virtuous cycles through built environment design*. London: Earthscan.

Bolton P. 2011. Personal interview. Stellenbosch: 22 March.

C.A.P.E. (Cape Action for People and the Environment). 2011. *About us*. [Accessed 13 April 2011] http://www.capeaction.org.za/index.php?C=about

CNdV Africa. 2010a. *Stellenbosch Municipal Spatial Development Framework – draft status quo report*. Stellenbosch: Stellenbosch Municipality.

CNdV Africa. 2010b. *Stellenbosch Municipal Spatial Development Framework – draft strategies report*. Stellenbosch: Stellenbosch Municipality.

De la Bat B. 2011. Personal interview. Stellenbosch: 22 March.

De la Bat B. 2012. Personal interview. Stellenbosch: 14 March.

De Wit M, Van Zyl H, Crookes D, Blignaut J, Jayiya T, Goiset V & Mahumani B. 2009. *Investing in natural assets: a business case for the environment in the City of Cape Town*. Cape Town: City of Cape Town.

Freeth R, Bomhard B & Midgley G. 2007. *Adapting to climate change in the Cape Floristic Region: Building resilience of people and plants in protected areas*. Report prepared for the International Union for Conservation of Nature (IUCN). South African National Biodiversity Institute (SANBI), Climate Change Research Group. Cape Town: Kirstenbosch Research Centre.

Gerber L. 2005. Biodiversity risk assessment of South Africa's municipalities. MSc thesis. Stellenbosch: Stellenbosch University.

LAB (Local Action for Biodiversity). 2010. *Cape Winelands Biodiversity Report*. Report prepared for the Cape Winelands District Municipality. South Africa: Cape Winelands.

Laibach Wines. 2011. *Organic Farming*. [Accessed 19 March 2012] http://www.laibachwines.com/index.php/contents/view/2

Resilience Alliance. 2006. *Glossary*. [Accessed 20 April 2011] http://www.resalliance.org/index.php/glossary

Reyneke Wines. 2011. *Biodynamics*. [Accessed 5 April 2011] http://www.reynekewines.co.za/biodynamics.htm

RSA DEAT (Republic of South Africa. Department of Environmental Affairs and Tourism). 2007. *Application for nomination of Cape Winelands Biosphere Reserve.* Stellenbosch: Dennis Moss Partnership.

RSA DWA (Republic of South Africa. Department of Water Affairs). 2010. *Green Drop Report 2009: South African Waste Water Quality Management Performance.* [Accessed 13 April 2011] http://www.dwaf.gov.za/Documents/GreenDropReport.pdf

RSA DWA. 2011. *Welcome to the Working for Water webpage.* [Accessed 19 March 2012] http://www.dwaf.gov.za/wfw

Spier. 2011. *Farm Visits.* [Accessed 5 April 2011] http://www.spier.co.za/what_to_do_at_spier/farm_visits

UNEP (United Nations Environment Programme). 2005. *Millennium Ecosystem Assessment.* Washington: World Resources Institute.

UNEP. 2008. *Cape Floral Region protected areas.* [Accessed 12 April 2011] http://www.unep-wcmc.org/medialibrary/2011/06/29/2823bc8a/Cape%20Floral.pdf

WWF (World Wide Fund for Nature). 2011. *Biodiversity and Wine Initiative.* [Accessed 5 April 2011] http://www.wwf.org.za/what_we_do/outstanding_places/fynbos/biodiversity___wine_initiative

RESPONSE TO
SECTION ON ECOSYSTEM SERVICES

KAREN J ESLER

That people have been in a reciprocal relationship with the Stellenbosch environment for millennia is no exaggeration. Where I live, it is not uncommon to come across a hand axe turned over in a newly ploughed vineyard. More than 300 000 years ago, Stone Age people would have looked across the same landscape, teeming with game and abundant with natural resources. The stone tools they left behind are referred to by archaeologists as the Stellenbosch Complex. In fact, the oldest archaeological evidence for modern humans comes from the Cape (from between 200 000 and 150 000 years ago). Clearly this was a great place to live.

Fast forward to 2012. Our reliance on the natural environment has not changed; and it is still a great place to live. In the valleys, soil and water support a now predominantly agricultural landscape that in turn supports a community of temporary and permanent farm workers and many thriving agricultural concerns (vineyards, orchards, vegetables, grazing). There are families who have lived for generations in a single valley. It is the environment, the magnificent backdrop of the Cape Fold Mountains, the Winelands, and the historical presence of people that draws tourists to this area. It is no accident that among Stellenbosch University's core and leading research competencies is a focus on the natural environment. Our reciprocal relationship with the natural environment is both multifaceted and complex. Healthy soils, clean air, clean water and beautiful landscapes are central to this relationship.

Great advances have been made in our understanding of the interacting biological, physical and social aspects that influence the state of our natural environment and the services that we obtain from it. For example, we have unequivocal evidence that intact wetlands and riverside (riparian) vegetation help to retain or store water and sediments. This is a critical 'service' that the natural environment provides, resulting in a suite of benefits to those living in, or deriving a livelihood from, the catchment. These so called 'ecosystem services' include benefits related to the moderation of flood events; increasing river flows in the dry summer season; recharging groundwater aquifers; and improving water quality by assimilating nutrient and waste products. It is not difficult to imagine that healthy wetlands provide tourism and recreation benefits, but how many of us realise that the suite of benefits critically extend to our ability to produce food; to provide a constant flow of clean, uncontaminated fresh water for domestic and agricultural use; and to minimise the need for disaster management? We need to make these links obvious and non-negotiable. Our local economy and our very existence rely on it.

Sadly, these services provided by the natural environment (for free) are under enormous stress, as the foregoing chapters describe. Development without a 'big picture' plan has carved up our landscape, consuming its beauty. A mountain of garbage greets visitors to our town; and now we have run out of space to dump more waste. Water supply is already at, or close to, exploitation limits. Diminished by water-hungry, non-native trees that invade many catchments and most riverbanks, what is left passes though developed areas that bleed their excess into the water. Little or no protective riparian vegetation or wetlands remain to provide their critical filtration and bank stabilisation function. How tragic that the water of the Eerste River now poses a health hazard to any downstream water user; how tragic that our children can no longer dabble their toes in search of tadpoles. Regionally, the loss of water quality and quantity is a pending crisis likely to be exacerbated in a climatically uncertain future. The future is not what it used to be.

Paradise, however, is not lost. I believe that it is still possible to engineer our future into one where a clean, safe and abundant environment is a given; a right that is not only enshrined in our constitution, but a reality. But brave, bold steps are needed by institutions, individuals and communities.

There is no doubt that we need to better harness the knowledge and talent that exist right on our own doorstep. Stellenbosch has a remarkable collection of academic talent from which to draw on. Through Stellenbosch University, the Council for Scientific and Industrial Research (CSIR) and other local institutions, much of the management, engineering, agricultural, technological and ecological knowledge already exists to appropriately manage our natural resources and even to restore that which needs fixing. New, exciting possibilities and creative innovations can and are being developed. The future is bright with endless possibility. However, a massive challenge is how to unlock this knowledge, which in many cases remains inaccessible to farmers, managers and policy makers. Not all academic knowledge is generated for application, and tragically much of it, even if applicable, is locked up in publishing houses whose business is to sell that information, often at huge cost if you are not in the academic/scientific system. Stellenbosch University is a leader in Open Access Publishing (for example, anyone can now obtain university dissertations with ease), but even more egalitarian and creative thought is needed. For example, we urgently require systems and people in place to translate academic solutions into actionable strategies and demystify the environmental jargon (whatever happened to our agricultural extension officers?); we need this information to be readily at hand (for example, where would I go to obtain information on water quality or the state of the soils on my farm?); we also need greater investment in application-driven, locally relevant research. The latter is not necessarily popular with academics and publishers who garner more visibility and recognition from basic research that is international in scope (see Esler, Prozesky, Sharma & McGeoch 2010), but creative incentives for an expanded variety of knowledge products would go some way to achieving this. Finally, although there is a vast reservoir of untapped knowledge out there, there are still yawning gaps. We might have some knowledge of soil type and its subtle link to *terroir*, but what of the links between management systems, soil health and future productivity? What are the long-term consequences of extensive use of pesticides and herbicides – both to the natural environment and human health? What does an uncertain climatic future mean for agriculture? Can we build resilience into our natural and agricultural systems, so that these systems are able to deal with ongoing change and future shocks; and if so how?

As individuals, we must take responsibility for our own role in the erosion of our natural resources, many of which are no longer available in abundance. In an environment of scarcity, we now need to deal with our excess, the malaise of the twenty-first century. It is those of us who use the most resources who need to learn how to live with less. Consider for example, the role you may play in driving the demand for that out of season fruit or vegetable. An agricultural landscape converted to a sterile sea of plastic may be what it takes to get that product on the shelf. Reducing that which we consume (water, electricity); walking rather than driving; reusing that which we can (shopping bags, tyres – in fact everything possible); and recycling instead of discarding (oil, organic matter, glass, paper, plastic, metal) are all simple actions at the household level. It is not only the environment that wins. There are social, economic and ecological benefits to doing these things. We need to change our mind set into a simpler one where people with ecological consciousness exist in a local economy where we are fair and kind to each other. We need to give back – to our family, community, society and environment, both natural and built. Above all, we need to be mindful of the decisions we make. Don't write off those small, seemingly insignificant, pro-environment actions, because they can have a profound collective impact when coordinated.

Coordination may occur at many levels, and already does to some extent. Schools, community groups, churches and the university are all playing a role, and there are some wonderful, generous initiatives already taking place. But more could happen. Regular and frequent communication about sustainable options would add value to our local newspaper (what happens to my rubbish when it leaves my home; who deals with it, and where?). Community involvement requires being informed; and if information is insufficient to guide action, having an appropriate range of incentives (for example, rates or tax rebates) and disincentives (for example, sliding scale costing for water, electricity and rubbish removal) in place would further help. Collective decision making and debate around budget allocations may help direct our rates and taxes towards real solutions at the district and local municipal levels. Municipalities need to include both built and ecological (green) infrastructure in their audits. Acknowledging the important water purification and waste treatment function of wetlands, and investing in their restoration, would minimise the need for costly new waste treatment plants, freeing up money for delivering important public services. Providing a functional sewage and sanitation infrastructure in our towns would go a long way to reducing unwanted river inputs. With newly promulgated Spatial Development Plans, municipal leaders now have the opportunity to steer development towards more sustainable trajectories, to plan our towns and landscapes in ways that are multifunctional and that welcome, rather than shun, the natural environment. Models already exist for alternative, environmentally friendly and organic agricultural practices; what is more, these are viable, going concerns that are also socially responsible. Indeed, a wide range of pro-environment options already exist for all sectors (e.g. Working for Water programme, Biodiversity and Wine Initiative and C.A.P.E government/public partnership). Strong local leaders with long-term vision have an opportunity to gather these options together and to coordinate them so that their combined impact is a significant one.

If common sense and a sense of community exist, I believe that the challenges that lie ahead may be turned into blessings. It is possible to restore and maintain healthy environments and healthy communities. All we need is collective action, long-term vision, budgets and champions to drive the process. The natural environment is the very reason why we have existed in this landscape since the dawn of humanity. We urgently need to reconnect with that which made us unique.

Acknowledgements

This commentary benefitted from a range of conversations with students, friends and colleagues (in particular Nadia Sitas, Alanna Rebelo, Helen de Klerk, Richard Cowling, Myke Scott, and Christy and Jan Momberg).

REFERENCES

Esler KJ, Prozesky H, Sharma GP & McGeoch M. 2010.
How wide is the "knowing-doing" gap in invasion biology?
Biological Invasions, 12:4065-4075.

Section 4

Social dynamics

Section Introduction

Social dynamics

Stellenbosch is considered as one of the most unequal places in the world and there is little "cross-border" interconnection between people of different backgrounds. Despite government efforts to reform health, housing, education and sport, poverty and inequality have actually increased since the mid-1990s. Widespread unemployment, lack of housing and unstable political leadership have severely impacted on the town's social dynamics, stunting the development of a socially cohesive society.

Heritage and culture: As the second oldest (white) settlement after Cape Town, Stellenbosch is considered a historical town and is characterised by colonial Dutch, Georgian and Victorian architecture. Although the town is renowned for its long historical legacy, several questions arise. Whose history is associated with the architecture, public spaces, gardens and artefacts? What economic interests have contributed to the preservation of what is now regarded as heritage Stellenbosch? What memories are being preserved, and with whom in mind?

Although Stellenbosch today consists of modern as well as historical elements, tourists come to see a selected past, and African narratives and images appear foreign in the avenues that make up the centres and suburbs of Stellenbosch. How many times have we heard tourists, friends and family exclaim: "This feels so much like Europe!" The narratives of brown and black heritage remain either under-represented or absent, despite efforts such as the writing up of the histories of people who lived in the centre of town before apartheid relocations. As a keyword search of Stellenbosch across social networks will reveal, the growing flow of (mainly youthful) tourists into townships such as Kayamandi has started to create memories and experiences of contemporary black urban life in juxtaposition with the manicured images of the Cape Dutch and Huguenot heritages. Will these diverse heritages ever merge into a shared memory of the future? Can we find an authentic mix that becomes really attractive or must we always contrive to preserve the postcard image of Stellenbosch?

Public space and art: Public spaces are not just physical spaces constructed by planners and engineers; they also exist as cultural explorations of the mind and imagination. This is why we need to critically reflect on the implications of the public art that is commissioned, produced and generated within the spaces traversed every day by everyone, regardless of who they are, where they live or how much they earn. Experiencing public art, like experiencing public transport or sport, becomes a meeting point of minds, where diversity is consumed consciously or unconsciously. But it is never exempt from power. When we see public spaces being encroached upon by corporations and government who seek to control and benefit financially from the public sphere without thought for the impact on society or culture, we need to reflect, comment, engage, counteract and protest. Three engagements with public art are described in the chapter by Marthie Kaden. They include the heritage and cultural practices of artist Dylan Lewis' 'Predators and Prey' urban sculpture trail (which was installed in and around Stellenbosch during 2008 to 2010); the Solms-Delta historic farm in the Drakenstein region near Simondium; and the collaborative counter-practice of Lara Kruger and Safe House residents' 'Cosy for a rhino' (2009). In Kaden's own words, these three different approaches may be …

> [...] viewed as public contact zones for diverse groups to encounter the stories, expressions and aspirations of the different communities involved. Each of these three cultural practices mentioned above opens up possibilities for diverse interpretations and meanings to flow from the interactions and confrontations within and between the different groups, communities and heritages involved. These practices may differ or conform to the dominant imagination of Stellenbosch, from where members of the community will either accept or reject it. This will shift and change, or assert and replicate the dominant imagination. These practices, as do others, engage in the making of Stellenbosch as a place.

Poverty and inequality: About one third of the total population of Stellenbosch is classified as poor – and this excludes farm workers. Indicators of inequality show that Stellenbosch has a higher level of inequality than the provincial average, but lower than that of the country as a whole. Unemployment estimates are unreliable, but official figures from 2007 estimated that approximately 40% of economically active people were unemployed at that time, with most of the unemployed falling into the 15 to 34 age group.

Social cohesion: Social cohesion is the glue that holds communities together and allows people to support one another despite differences or adversity. It is nurtured by good political leadership, dialogue (inclusive rather than extractive), institutions and joint action to heal divisions and form a stronger community. Due to the high levels of inequality between its inhabitants, unemployment, lack of housing and unstable political leadership, Stellenbosch is a ticking time bomb of potential unrest. Reducing inequality and linguistic fractionalisation is key to building cohesion. By focusing on shared place and blood, the right to hospitality for all and the desire for sustainability, the town may develop a sense of 'we' to infuse the emergence of a social cohesion movement that crafts visions and pathways to a very different sustainable future.

Business strategies: Businesses today face the challenge of balancing financial viability with environmental sensitivity and social responsiveness. Key is the realisation that businesses are not separate from the social-ecological system, but rather embedded in and dependent on it for their survival. Therefore, instead of simply reducing undesirable practices, businesses should be strengthening the sustainability of the social-ecological system. This requires access to new knowledge and broad innovations (not only technological) that support change. As such, Stellenbosch is an ideal location for businesses due to the potential for collaboration with local learning and research institutions, which may help businesses to adapt and so ensure their survival.

Health: Unsafe sex, interpersonal violence, road accidents and alcohol harm are the greatest risks contributing to premature death in Stellenbosch, followed by unsafe water, poor sanitation, inadequate hygiene and poor nutrition. While all these factors should be of concern to Stellenbosch Municipality and the wider stakeholder community, it is significant that, as the authority responsible for water, sanitation, waste collection, aspects of public health and traffic and related safety issues, the municipality is in a position to considerably reduce the greatest risk factors contributing to premature death.

Behaviour change: In trying to drive the transition to sustainability, it is essential to have an understanding of how human behaviour may be changed. Changing minds, rather than just behaviour, is a more challenging goal, yet a necessary one if we are committed to a more sustainable future. For Stellenbosch University in particular, issues of identity and exclusion must be addressed in order to build relationships of mutuality and trust before staff and students may assume their proper role as brokers of sustainable development.

Education for sustainability: While many believe that education creates awareness and is therefore key to more sustainable living, current education could in fact be part of the problem by serving to replicate a failing system. Teacher-dominated pedagogies, highly structured curricula, inappropriate teacher-learner ratios, rigidity in the schooling system, and cosmetically tagging on environmental education in curricula present barriers to education for sustainability. A socially critical education that investigates local problems, is less structured, has weaker boundaries between schools, classrooms and communities, and whose teachers act as co-investigators is necessary to create a sustainable Stellenbosch. School-university partnerships that focus on building relationships in which informal education may occur would also be beneficial.

Sport: Historically, sport in Stellenbosch was plagued by inequality and racism and helped to perpetuate a lack of social cohesion. It has, above all, been the arena of intense struggle as a diversity of race and class interests harnessed the organisation of sporting activities to achieve their respective political aims. Nevertheless, the town is endowed with an exceptional sporting infrastructure that has benefitted from many decades of investment by the university and other public agencies (and is reinforced by an extraordinary natural setting for sporting activities). Since 1994, these agencies have collaborated to open these facilities up for everyone to use. This collaboration includes proactive programmes that go beyond mere non-racialism (which is an achievement in itself) to ensure that the poorer sections of the community may gain access. However, an integrated, long-term sports plan for the town that is aligned with provincial and national sports plans must still evolve from the collaborative efforts promoted by the Rector-Mayor Forum.[1] Anyone who strolls around the university soccer fields on a Saturday will realise that sport probably plays a more important role than any other public activity in creating a sense of social cohesion and common identity.

---

1 The Rector-Mayor Forum was established in 2005 to facilitate monthly meetings between Stellenbosch University and Stellenbosch Municipality to discuss the challenges facing the town and find ways of combining forces to solve them.

Children: As stated in the opening line of this introduction, Stellenbosch is considered as one of the most unequal places in the world and there is little "cross-border" interconnection between children of different circumstances. However, that all children should be given the opportunity to develop to their full potential is self-evident. The town should become a place where children are cherished at every age, not only through better health and education systems, but also by providing them with safe public spaces, access to nature, and the opportunity to contribute meaningfully to sustainable development. Collaborative leadership from relevant local authorities and organisations is vital if Stellenbosch is to build a vibrant culture of participative citizenship for a sustainable future.

The authors of this book are almost unanimous in suggesting that social sustainability in Stellenbosch requires good, collaborative political leadership and inclusive, rather than extractive, institutions. Policy makers, in partnership with NGOs, businesses and residents, should promote sustainable lifestyles and help to build a vibrant culture of participative citizenship. Addressing the twin challenges of poverty and inequality should infuse all actions, from formulating public policy to social activism and allocating private investments; in short, poverty must cease to be an abstract category and become a lived reality that affects everyone. If children can be made the central focus of public and family life, future generations may inherit social cohesion.

How we imagine and practice our heritage, public art, sport, education, businesses and health will be what translates the broader aspiration to build socially cohesive communities into the dynamics of everyday life. But changing behaviour will not be easy. Behaviour is both conditioned by its context and reshapes it as new desires emerge in response to creative modes of messaging the future.

Chapter 16

HERITAGE

Contextualising heritage production

ALBERT GRUNDLINGH & DORA SCOTT

Introduction

One of the pitfalls of being an historian is the tendency to take the past for granted. It is not always appreciated that such an interest is actually a curious preoccupation tolerated by society which does not necessarily have access to or is under no obligation to underwrite the arcane workings of a profession which often prides itself in wrapping its findings in dense and impervious prose, buttressed by numerous footnotes imparting a sense of authority. In contrast to this, public history – particularly in the form of heritage – is much more open and free ranging, there for everybody to observe. In the words of one observer, heritage is the "continuous nourishing tradition" (source unknown, quoted in Lowenthal 1998:5).

In a way, this opens up an avenue for descriptions of heritage to abound – as they tend to do in Stellenbosch – without interrogating the underlying dynamics of heritage production. The aim of this chapter is to outline some of the more salient features of this process and take a tentative look into the future of heritage in the town.

Stellenbosch and self-representation: Normalising heritage?

Established in 1679, Stellenbosch is the second oldest (white) settlement after Cape Town. As such, the town can legitimately lay claim to a right to exude 'pastness' as an element of its self–representation. Orbasli and Woodward note that "the way in which historical towns are presented to tourists and the way in which tourists experience them are [...] very different from other types of 'designated' heritage attractions" (2009:318). On the surface, such towns may assume a certain authenticity that stands in contrast to more obviously 'manufactured' heritage sites, such as historical theme parks. However, it is precisely this projection of normality in historical towns that may and should be probed.

Stellenbosch has a varied architectural legacy, with numerous colonial Dutch, Georgian and Victorian buildings (many of which are protected as historical monuments) bordering tree-lined streets (Bulpin 2001:4). The setting consistently invites comparison with European places, which serve as a type of benchmark, as may be seen from the following excerpt from a travel guide-cum-public relations text:

> *For Stellenbosch has indeed charm and beauty. Not the muted grey beauty of Burgos; not the brick-faced charm of Lübeck; nor the terracotta loveliness of Taormina, but the charm of lime-washed gables, of green oaks and blue mountains. And Stellenbosch has beautiful lines: the lines of undulating old walls, of the curves and hollows of its white Cape Dutch gables, of its pediments and turrets, of drip mouldings across Georgian facades, of the cast iron lacework decorating Victorian verandahs, and of the plump convexity of the barrel vault on the old Kruithuis.*
>
> [Van Huyssteen 1993:1]

The core historical part of the town mainly comprises the area circumscribed by Dorp, Plein, Drostdy, Piet Retief and Van Ryneveld Streets – a quarter conspicuous for its old buildings. This zone is especially aimed at tourists and is filled with restaurants, bistros, curio shops, galleries and other businesses that rely on tourist expenditure. Ironically, according to various shop owners and people who work along Church Street, the hub of tourism activity, most tourists only visit the town *en route* to the many wine farms surrounding it (Visagie 2008; Koopman 2008; Barnard 2008; Strydom 2008). Thus, the town is not necessarily a destination in its own right, but rather part of an itinerary that may also include places such as neighbouring Franschhoek. Nevertheless, tourism still accounts for almost 25% of the town's income (*Eikestadnuus*, 10 September 2010 in Retief 2011).

It would not be inappropriate to describe this touristic part of town as an "open-air museum of itself" (Kirshenblatt-Gimblett 1988:151), specifically preserved to present the town's history and heritage. Visitors are encouraged to "discover the rich historical and cultural heritage of this area, where splendid examples of Cape Dutch architecture have been preserved to honour our ancestors" (Stellenbosch Tourism 2008a). The uses these buildings are put to generally do not correspond to their original function, as most of the houses in this area were used for residential purposes in the past. As the town expanded and businesses took over the town centre, it became increasingly unpopular as a place to live and residences were adapted to accommodate businesses. The reverse of this process has since occurred, with many apartments above shops having been revamped and made available for rental. Nonetheless, these buildings remain shells of history, filled with the physicality of the past and incorporating the present on their own terms. This has resulted in the town becoming a mimesis, or an 'impression (or imitation) of itself'. Kirshenblatt-Gimblett comments on this occurrence and notes that once sites, buildings, objects, technologies, or ways of life can no longer sustain themselves as they formerly did, they "survive" (by being made economically viable) as representations of themselves (1998:151). Locations become museums of themselves within the tourism economy.

The historical part of town is supposed to represent the 'essence' of Stellenbosch – a historical, placid and leisurely place. Visitors are promised that they will "experience the culture, tradition and true romance of Stellenbosch by strolling through the old town centre" (Stellenbosch Tourism 2008b). The term 'experience', which has become ubiquitous in tourism and museum marketing, signals a self-conscious shift in orientation away from artefacts and modern museums toward thee visitors themselves. It indexes an engagement of the senses, emotions and imagination (Kirshenblatt-Gimblett 1998).

Clearly, Stellenbosch is 'aware' of itself as a tourist attraction that may be 'experienced' – and intentionally presents itself in that way. The experiences to be had in the town are of a particular nature in that it is controlled and manufactured – a simulated environment. This environment is self-consciously constructed from diverse elements from the past (different architectural styles) to form a new, contemporary architectural location. The sum of the different parts of the town is thus greater than the individual styles and this creates a new setting, where the present may be consumed through the refraction of the past. Metaphorically speaking, Stellenbosch is a patchwork quilt of history, with an almost seamless stitching together of various pieces of history – even if some pieces, such as the town's slave history, are not readily apparent. Equally, this very seamlessness tends to obscure the more multifaceted character of the town. As one commentator recently explained in response to the question of what Stellenbosch stands for:

> For some, a picturesque and historic student town in the Wineland district of the Western Cape; for others, a sharp race and class frontier; at times it is a signifier of modernity and its limits; but also a home and a destiny.
>
> [Du Plessis 2010:53]

Such contradictions, which are inherent to 'the town-as-a-tourist-attraction', are evident in Stellenbosch. Presented as the preserved heritage of the town, the town centre is the only part of the town (except for the university campus) that benefits from diligent preservation efforts. Therefore, there is an argument to be made for the town centre as a 'heritage synecdoche' – a small part is made to stand for the whole, with the greater part of the town being neither as polished nor as geared towards tourism as its historical core.

Ashworth and Turnbridge (1990) suggest that the reason why people travel to historic towns is that they have a desire to experience the past – to see how people used to live and to encounter buildings of a symbolic or historical nature. In the case of Stellenbosch, one may say that visitors desire to see 'old Europe' as it was transposed to the 'uncivilised' Cape Colony. When they stroll through the historical tourist area, they experience a part of their own history filtered back at them and the familiar being presented in an unfamiliar setting. In this way, they get to see themselves and their own heritage through the eyes of the past and the lens of Africa, however distorted that lens may be at the southernmost tip of the continent. Hall and Zeppel refer to this phenomenon as "romanticising the past" (1992:50) and it perhaps especially true of Dutch tourists, who pick up more than just a whiff of the glory days of the Dutch East India Company when they visit this corner of the old Cape Colony.

Stellenbosch is not a colonial town anymore, but an assimilated urban space with modern and older elements existing side by side. But, as far as tourists are concerned, the modern may be easily consumed elsewhere and is often all too pervasive in their own day-to-day lives; they find the selected past in a tourist setting more appealing. The nature of that selected past is determined by those who have the power to direct the tourist gaze and the cultural goods they have at their disposal or are able to manufacture. European heritage generally dominates the tourist routes of the Western Cape. Very little mention is made of Africa or the 'African' experience in Stellenbosch. The presence of at least four African curio shops in Church Street alone and the Africa-themed Sosati restaurant impart a distinctly commodified dimension to the notion of 'Africa', where Africa is 'produced', rather than 'found' (compare Kirshenblatt-Gimblett 1998). Likewise, in close proximity to the town on the Spier wine estate, the upmarket Moyo restaurant – replete with ethnic accoutrements such as (presumably) Xhosa face painting and tribal dancing – seeks to manufacture an African experience under controlled touristic conditions; and for the slightly more intrepid visitor, the AmaZinc restaurant in neighbouring Kayamandi offers an 'authentic' black township experience.

It is a case of 'Stellenbosch-in-Africa' and 'Africa 'light'-in-Stellenbosch'. On one level, it may even appear as though 'Africa' is the foreigner in town, not the tourists from around the globe. Stringent influx control during the apartheid years and the dictates of tourism are only two of the many factors that have conspired to produce this particular heritage set. Yet, it is a heritage that by and large succeeds in the market place – which in turn raises further questions about the developments that underpinned this process.

Material and discursive underpinnings

It is relatively easy to discern the hand of apartheid in the segregated nature of the town; it is somewhat more difficult to trace the dynamics of the less obvious developments that shaped the dynamics of conservation. Although the timeframe of historical preservation may be spread over several decades, the 1960s and early 1970s stand out as periods of accelerated heritage production.

These junctures are not incidental. In a broader analysis, they coincided with the rapid rise of an Afrikaner middle class during a period of unprecedented economic growth (at an average of 6%) in the 1960s. In general, there was a trend away from unskilled or semi-skilled labour to better remunerated positions with stable career prospects, aided by government policies that promoted Afrikaner education in a variety of ways. As a result of the financial upswing and other facilitating factors, a new class of industrial and commercial capitalists emerged at the upper echelons to make their presence felt alongside their English-speaking counterparts. Along with the seemingly unassailable power of the National Party during the 1960s, a newfound and assertive confidence permeated Afrikaner ranks (Grundlingh 2008). It was a period when they could look back at their past with a certain equanimity and serene satisfaction; for a while, their days of struggle to gain what they considered a respectful place in the country appeared to be over, inviting a seemingly prosperous future with a less pressing need than before for a robust past.

The British historian EP Thompson has accurately observed that there is "no such thing as economic growth which is not, at the same time, growth or change of a culture" (1967:97). The trajectory of that change, however, is not always predictable. In terms of the cultural heritage of Stellenbosch, two conflicting trends emerged. The first reflected the outlook of the newly emergent wealthy Afrikaners who embraced modernising impulses with almost unthinking vigour. The past, having served its purpose, could now be safely jettisoned, and what they regarded as antiquated buildings in Stellenbosch had to be replaced with new and shiny contemporary structures. The second trend, which stood in opposition to rampant modernisation, was embodied in people such as Erika Theron, the academic and former Mayoress of Stellenbosch who proved particularly vigilant about protecting the town's "image as a site of natural beauty, high culture, learning, and historic significance" (Tayler 2010:187). In later years Theron observed that, if unreflective modernising inclinations had been allowed to progress unchecked, the town would have been a treeless zone of small skyscrapers (Theron 1983). Likewise, the local newspaper expressed concern that the greater good of the town was being put at risk by the greed of rapacious individuals (Anonymous 1965b).

The conservationist lobby, though also benefitting from the exceptional economic growth, represented a more thoughtful and considered position, imbued with a finely honed sense of aesthetics and the potential of historical heritage. This grouping – in which the industrialist Anton Rupert of Rembrandt tobacco fame and certain university staff members loomed large – sought to resolve tensions by preserving the old in a way that was also financially viable and sustainable over the long term.

Rupert not only reflected the innovative entrepreneurial spirit of his generation, which came to full fruition during the 1960s; he also had a special feel for historical towns such as Stellenbosch (and was also actively involved in heritage conservation in the Karoo town of his birth, Graaff-Reinet) and called for ...

> [... an] alliance between capital and culture: in order to help preserve that which has been bequeathed to us, but also to provide favourable circumstances for those who follow us to create cultural goods for successive generations. History provides proof that material progress does not have to be in opposition to cultural growth.
>
> [Quoted in Dommisse 2005:454 – authors' translation]

Similarly, the historian and Rector of Stellenbosch University, HB Thom, contended in 1964 that Afrikaners were now in the privileged position that there could be a healthy interaction between economics and culture, and that Stellenbosch might demonstrate that symbiosis (Thom 1964). In line with this thinking, Anton Rupert took the lead in 1966 by establishing a company called Historical Homes with 136 founding members, including 36 public companies. Restoration went hand in hand with business principles, with buildings initially rented out as homes and later as offices, flats, restaurants, shops and antique shops. The first financial year showed a profit of R13 143 (a considerable amount at the time) and the company consistently maintained this performance over the ensuing years, posting reserves of more than R10 million by 2002 (Dommisse 2005).

Rupert was also acutely aware of how tourism might contribute to the economic well-being of the town, arguing in 1975 that more leisure time and higher incomes with more money available for discretionary spending would fuel tourism, with potential benefit to historical towns (Rupert 1975). The assumption was not unwarranted. By the late 1960s tourists were already flocking to Stellenbosch during the holiday season, as reported in the town newspaper:

> [... an] unequalled number of tourists in the town proved that Stellenbosch's beautiful historical buildings and other attractions had become more popular. Everywhere during the holidays one could see groups of people with cameras capturing buildings and scenes of nature on film.
>
> [Anonymous 1969]

The present-day historical town centre and its touristic appeal are therefore built squarely on capitalist foundations, which may thus be seen as central to the town's heritage construction. The same process occurred elsewhere. In Europe and the United Kingdom, heritage tourism was also closely linked to the rise of the new middle class, both as shapers and consumers of what was on offer. As in the case of Stellenbosch, cultural goods and services came to constitute capitalist assets and were integrated as market commodities (Richards 1996).

However, capitalism was not the only shaping force in heritage production in Stellenbosch. Heritage was also embedded in a specific discursive formation, bound up with notions of ownership and a deeply rooted past. The historic and cultural buildings of the town were intended to convey to visitors that the inhabitants of this beautifully preserved town were "an established 'volk' with a rich heritage" of more than three centuries (Anonymous 1965a). In a similar vein, from 1965 until 2000, Stellenbosch's colonial past was proudly and publicly celebrated with considerable pomp and pageantry in a parade and historical re-enactments during an annual festival in honour of Simon van der Stel, the town's founder (Hanekom 2010).

Stellenbosch in the 1960s was of course closely associated with the Afrikaner nationalist movement, which to all outward appearances reached its peak during this time. Its alignment was of a special kind though. While Rupert, as a far-sighted businessman, was anything but a knee-jerk Afrikaner nationalist supporter, he did subscribe to elements of white nationalism coupled to a strong sense of place, and such considerations informed his conservation drive:

> We must preserve, because buildings from the past are proof that we have already been here a long time and belong here. The historical buildings are the title deeds to the land which we love.
>
> [Rupert 1975:18 – authors' translation]

As we have noted earlier, a more muted and less stridently heroic interpretation of the past than that which was prevalent during the nationalist march to power began to surface slowly in the 1960s. Yet, even if economic growth and sufficient self-confidence was starting to dampen the fire and brimstone nationalist versions, there was still an underlying need for historical legitimisation. Heritage in Stellenbosch represented a softer and milder form of nationalistic expression, as opposed to the febrile nationalism exuded by the overpowering structure of the Voortrekker Monument in Pretoria. The apparent ambiguity in nationalism's linking of progressive modernisation to aspects of the past is not uncommon, as Mitchell (2001:212) aptly demonstrates:

> One of the odd things about the arrival of the era of the modern nation-state was that for a state to prove that it was modern, it helped if it could also prove it was ancient. A nation that wanted to show that it was up to date and deserved a place among the company of modern states needed, among other things, to produce a past.

In the case of Stellenbosch, the modern and the ancient were woven almost imperceptibly into one. Indeed, the heritage of the town reach extends beyond its formal boundaries to the adjoining countryside with its wine farms located in dramatic natural surroundings, which add greatly to the already pleasing milieu by imparting an air of gentle sophistication, high culture, serenity, wealth and timelessness. Unlike the harsher environments of the Free State and the former Transvaal,[1] which were ravaged by cataclysmic events such as the South African War of 1899 to 1902 and (especially in the case of the Transvaal) marred by a process of unrelenting industrialisation and mining extraction, the countryside of the Boland[2] is characterised by stability and continuity. This sense of rootedness is reflected in Van Huyssteen's description of a 'Bolander':[3]

> [A Bolander is someone who] looks at the world from the vantage point of a long past and many generations of experience [...] His admiration is for the old families with their farms and old money. Farms and houses are not sold easily; a man keeps what has been bestowed on him; he cherishes it [...] If an old family farm is sold, it is almost as if a whole community is saddened.
>
> [Van Huyssteen 1983:20 – authors' translation]

The linkages between town and countryside were consolidated by the development of wine tourism and the establishment of a wine route in the early 1970s, which coincided with the almost simultaneous heritage drive in town. The Stellenbosch Wine Route sought to combine tradition and landscape, a connection that ensured the wider appeal of the region as a whole (see Randle 2004). At the same, it may be argued that, in real terms, this linking of tradition and landscape largely precluded other development options (and perhaps even contributed to the current evolutionary dead end as far as the landscape is concerned).

Towards a 'new' heritage

It is a commonplace that the heritage projection of Stellenbosch does not do justice to all inhabitants of the town, particularly as brown and especially black heritage is either woefully under-represented or entirely absent. This is not to deny the purposeful attempts to develop a heritage awareness through, for example, the current historicising of Luckhoff School (which was rezoned under the notorious Group Areas Act during the 1960s) and the publication of the memoires of people who were relocated under the Act, as well as

---

1. In 1994, the new constitution subdivided the Transvaal Province into Limpopo, Mpumalanga, North West and Gauteng.

2. The names 'Boland' (literally 'top country' or 'land above') and 'Cape Winelands' are often used interchangeably. The Boland region has no defined boundaries, but is roughly situated to the northeast of Cape Town, in the middle and upper courses of the Berg and Breede rivers, around the mountains of the central Cape Fold Belt.

3. Someone who was born and bred in the Boland.

deliberations to promote Ida's Valley as a cultural landscape of the Winelands (see Biscombe 2006 and Pistorius & Todeschini 2010). Nevertheless, despite these valiant efforts, harmonising the dissonant character of Stellenbosch's heritage should perhaps not be seen as simply a process of adding a few new choir members under the baton of a new conductor. More intriguing is the question of whether there is not a song, long forgotten, that needs to be rediscovered and retuned – one which sang of a past, albeit in different tones, of a common sense of place and destiny.

While the Group Areas Act and forced removals in Stellenbosch gave rise to much bitterness (Biscombe 2006), it did not succeed in obliterating a dual awareness that, in Stellenbosch, brown people have a joint claim to that which has been projected as white heritage, and that the town's history cannot be that easily disentangled. On a practical level, the historic core of the town owes much to slaves, brown artisans and master builders (Giliomee 2007). Even if only subliminal – and notwithstanding the aberrations and ravages of apartheid – there was, in terms of identity formation, a persistent strain that could not be wholly suppressed: brown and white people had much in common on the basis of language, religion and cultural aspirations, though the trajectories of such interests had diverged in the past. This line of thinking is evident in what a brown teacher told the well-known journalist, Piet Cillie, in 1978 when the teacher was quizzed on the 'group history' of the brown community:

> *Sir, my history is your history, my traditions are your traditions – and I won't be surprised if my forefathers were also your forefathers!*
>
> [Cillie 1980:22]

Even more telling is the remark that a Mr AA Cupido of Stellenbosch made before the Group Areas Board in 1956:

> *"We have always been here in Stellenbosch. Here we were born and here we grew up and worked, and helped to establish what is here today; and I cannot see how it can be expected to say farewell to all of this."*
>
> [Quoted in Giliomee 2007: front page]

Turning such historical ideas into heritage would be a daunting task, as heritage production tends to privilege those aspects of the past that are relatively easy to project and commodify. Intangible dimensions are more difficult to showcase and invest with a presence in order to convey a message that is subtle rather than strident. Nevertheless, the challenge of future heritage production will not only be to foreground and make good the arrears that have accumulated in certain disadvantaged geographical areas of Stellenbosch, but also to explore joint strands of history that have been deeply submerged by the debris of apartheid. Conceptually this calls for alertness to the notion of hybridity rather than purity. Inevitably, such developments will over time also have the effect of recalibrating that which has up to now passed for white heritage.

Conclusion

This chapter has, in an anthropological sense, sought to make Stellenbosch appear 'strange'. It has sought to present the historical core as virtually 'a museum of itself' and to demonstrate the processes through which such a view is kept intact. The process of heritage production *per se* was also explored by situating it within the materialist and discursive formations that prevailed during a period of accelerated growth in the 1960s and early 1970s in particular, a time during which heritage in the town was related to wider Afrikaner nationalism, but in a nuanced fashion, underpinned by influential businessmen such as Anton Rupert, who placed the enterprise on a sound financial footing.

The trajectory of future heritage in the town remains to be debated and raises the possibility of exploring hitherto grey areas.

History never reaches an end. A range of class and cultural interests persist, transform and recombine, and new ones emerge from time to time in bursts of energy that both challenge and thrive on representations that are frozen in the aesthetics of built form. While developers move into heritage areas to reinforce the café culture of an urban elite looking for high-density living that is located safely away from the more threatening environments of the big cities, the homeless appropriate a place like Enkanini (a short walk away from Stellenbosch CBD), where 6 000 people now live without services, but ever hopeful that Stellenbosch might also be defined one day as a place that regards them as full citizens. The gulf between the chattering classes in their cafés and these largely illiterate, recently arrived shack dwellers sets the stage for a new set of contested outcomes that are impossible to predict.

Acknowledgements

We are indebted to Linde Dietrich, Santi Basson, Bill Nasson, Sandra Swart and Mark Swilling for constructive critical comments and locating material.

REFERENCES

Alen G & Brennan F. 2004. *Tourism in the New South Africa. Social Responsibility and the Tourist Experience*. London: IB Tauris.

AlSayyad N (ed). 2001. *Consuming tradition, manufacturing heritage: global norms and urban forms in the age of tourism*. London: Routledge.

Anonymous. 1965a. *Eikestadnuus*, 7 May.

Anonymous. 1965b. *Eikestadnuus*, 2 July.

Anonymous. 1969. *Eikestadnuus*, 17 January.

Ashworth GJ & Turnbridge JE. 1990. *The tourist-historic city*. London: Belhaven.

Barnard L. 2008. Personal interview conducted by Dora Scott. Stellenbosch: 2 October.

Biscombe H (ed). 2006. *In ons bloed*. Stellenbosch: Sun Press.

Bulpin TV. 2001. *Discovering Southern Africa*. Cape Town: Tafelberg Publishers.

Cape Tours. 2008. Personal interview conducted by Dora Scott. Stellenbosch: 2 October.

Chaney D. 2004. The power of metaphors in tourism theory. In: S Coleman & M Crang (eds). *Tourism. Between Place and Performance*. New York: Bergham Books.

Cheong SM & Miller ML. 2000. Power in tourism: a Foucauldian observation. *Annals of Tourism Research*, 27(2):371-390.

Cillie P. 1980. *Eet jou rape eerste: 'n bonteboek uit die geskrifte van PJ Cillie*. Cape Town: Tafelberg Publishers.

Cohen E. 1988. Authenticity and commoditization in tourism. *Annals of Tourism Research*, 15:371-386.

Coleman S & Crang M (eds). 2002. *Tourism: Between Place and Performance*. London: Berghahn Books.

Dommisse E. 2005. *Anton Rupert: 'n lewensverhaal*. Cape Town: Tafelberg Publishers.

Du Plessis I. 2010. The exploded view and beyond. *South African Review of Sociology*, 41(1):51-55.

Garrod B & Fyall A. 2000. Managing heritage in tourism. *Annals of Tourism Research*, 27(3):682-708.

George R (ed). 2007. *Managing Tourism in South Africa*. Cape Town: Oxford University Press.

Giliomee H. 2007. *Nog altyd hier gewees: die storie van 'n Stellenbosse gemeenskap*. Cape Town: Tafelberg Publishers.

Goeldner CR. 2006. *Tourism: principles, practices, philosophies*. New York: Wiley.

Grundlingh A. 2008. Are we Afrikaners becoming too rich? Cornucopia and change in Afrikanerdom. *Journal of Historical Sociology*, 21(8-9):119-135.

Hall CM. 1999. *The geography of tourism and recreation: environment, place and space*. London: Routledge.

Hanekom W. 2010. The Simon van der Stel Festival: Constructing heritage and the politics of pageantry. Unpublished BA (Hons) long essay. Stellenbosch: Stellenbosch University.

Harrison D. 2001. *Tourism and the less developed world: issues and case studies*. Wallingford, UK: CABI Publishing.

Hollinshead K. 1998. Tourism, hybridity and ambiguity: the relevance of Bhabha's 'third space' cultures. *Journal of Leisure Research*, 30(1):121-150.

Karp I & Lavine SD. 1991. *Exhibiting cultures – the poetics and politics of museum display*. Washington: Smithsonian Institution Press.

Kirshenblatt-Gimblett B. 1998. *Destination culture: tourism, museums and heritage*. Berkeley, CA: University of California Press.

Koopman J. 2008. Personal interview conducted by Dora Scott. Stellenbosch: 29 September.

Leask A & Yeoman I (eds). 2002. *Heritage visitor attractions: an operations management perspective*. London: Continuum.

Lowenthal D. 1998. *The heritage crusade and the spoils of history*. Cambridge, UK: Cambridge University Press.

MacCannell D. 1973. Staged authenticity: arrangement of social space in tourist settings. *American Journal of Sociology*, 78(3):43-60.

Macey D. 2001. *Dictionary of Critical Theory*. London: Penguin.

Millar S. 2002. An overview of the sector. In: A Leask & I Yeoman (eds). *Heritage visitor attractions: an operations management perspective*. London: Continuum.

Mitchell T. 2001. Making the nation: the politics of heritage in Egypt. In: N AlSayyad (ed). *Consuming tradition, manufacturing heritage: global norms and urban forms in the age of tourism*. London: Routledge.

Moore K. 2002. The discursive tourist. In: GMS Dann (ed). *The tourist as a metaphor of the social world*. Wallingford, UK: CABI Publishing.

Nederveen Pieterse J & Prekh B (eds). 1995. *The decolonization of the imagination: culture, knowledge and power*. London: Zed Books.

Orbasli A & Woodward S. 2009. Tourism and heritage conservation. In: T Jamal & M Robinson (eds). *The Sage handbook of tourism studies*. London: Sage Press.

Pearce DG & Butler RW (eds). 1993. *Tourism research: critiques and challenges*. London: Routledge.

Pistorius P & Todeschini F. 2010. *Idas Valley as an example of the cultural landscape of the Cape Winelands*. [Accessed 4 August 2010] http://www.international.icomos.org/studies/viticoles/viticole8.pdf

Richards G (ed). 2007. *Cultural tourism: global and local perspectives*. New York: The Haworth Hospitality Press.

Randle T. 2004. Grappling with grapes: wine tourism of the Western Cape. Unpublished MA thesis. Cape Town: University of Cape Town.

Retief P. 2011. Stellenbosch 360. Unpublished paper. Woordfees, 2011. Stellenbosch.

Richards G. 1996. Production and consumption of European cultural tourism. *Annals of Tourism Research*, 23(2):261-283.

Richards G (ed). 2007. *Cultural tourism: global and local perspectives*. New York: The Haworth Hospitality Press.

Robinson M, Evans N, Long P, Sharpley R & Swartbrooke J. 2000. *Tourism and heritage relationships: global, national and local perspectives*. London: Centre for Travel and Tourism in association with Business Education Publishers.

Rupert A. 1975. *Is die bewaring van ons erfenis ekonomies te regverdig?* Pretoria: University of Pretoria.

Schouten F. 2007. Cultural tourism: between authenticity and globalization. In: G Richards (ed). *Cultural tourism: global and local perspectives*. New York: The Haworth Hospitality Press.

Smith MK. 2003. *Issues in Cultural Tourism Studies*. London: Routledge.

Smuts F (ed). 1979. *Stellenbosch Drie Eeue*. Stellenbosch: Stellenbosch City Council.

Stellenbosch Tourism. 2008a. *A town for all seasons*. Stellenbosch: Stellenbosch Tourism.

Stellenbosch Tourism. 2008b. *Stellenbosch and its Wine Routes 2009*. Stellenbosch: Stellenbosch Tourism.

Stellenbosch Museum. 2008. *Stellenbosch Village Museum*. Stellenbosch: Stellenbosch Museum.

Strydom U. 2008. Personal interview conducted by Dora Scott. Stellenbosch: 8 October.

Tayler J. 2010. With her shoulder to the wheel: the public life of Erika Theron (1907-1990). Unpublished DLitt et Phil thesis. Pretoria: University of South Africa.

Theron E. 1983. *Sonder hoed of handskoen*. Cape Town: Tafelberg Publishers.

Thom HB. 1964. *Eikestadnuus*, 9 October.

Thompson EP. 1967. Time, work discipline and industrial capitalism. *Past and Present*, 38:32-57.

Urry J. 1992. *The tourist gaze: leisure and travel in contemporary societies*. London: Sage Publications.

Van Huyssteen T. 1983. *Hart van die Boland*. Cape Town: Tafelberg Publishers.

Van Huyssteen T. 1993. *Footloose in Stellenbosch*. Cape Town: Tafelberg Publishers.

Visagie H. 2008. Personal interview conducted by Dora Scott. Stellenbosch: 8 October.

Warren S. 1998. Cultural contestation at Disneyland Paris. In: D Crouch (ed). *Leisure/Tourism geographies: Practices and geographical knowledge*. London: Routledge.

Zeppel H & Hall CM. 1992. Arts and heritage tourism. In: B Weiler & CM Hall (eds). *Special interest tourism*. London: Routledge.

Chapter 17

SHARED SPACE

Power, heritage, play

MARTHIE KADEN

The constitution of public spaces of action and political discourse depends [...] upon the creation of numerous spheres of appearance in which individuals can disclose their identities and establish relations of reciprocity and solidarity.

[D'Entrèves 1992:146]

On the horizon, then, at the furthest edge of the possible, it is a matter of producing the space of the human species – the collective (generic) work of the species – on the model of what used to be called "art"; indeed, it is still so called, but art no longer has any meaning at the level of an "object" isolated by and for the individual.

[Lefebvre 1991:422]

I'm not like you
how should I call you
so at last we may
fairly
place signals
between us

[Marcil-Lacoste 1992:141]

Introduction: The fragile common

Heritage, culture and social memory insist on the fundamental principle of a shared space: that everything – from language to politics; from art to nature; from technology to the earth; from the past to our hopes for the future – belongs irreducibly to all of us. This shared space or public realm, as it has come to exist, is increasingly under threat as every activity and value of social and cultural life is converted into a commodity for purposes of profit and a display of powerful financial allegiances and private ownership. More and more under neoliberal capitalism, public space is manipulated, owned, disowned, privatised and gated – to the exclusion of certain individuals and groups of people belonging to an equally free and general public (Mouffe 1992; Sennet 1978).

Thinking about (and through) our shared cultural heritage as a collective space may prove to be a valuable way – as well as a resource – for generating new and alternative forms of challenging neoliberal and imperial hegemonies and to invent practices for a common and public space that will benefit every member of a society. Based on Chantall Mouffe's (1992; 2002) perspectives on a radical and plural democracy, one may take the position that cultural heritage, as a similarly constructed and coherent practice as that of a capitalist economy, may then be employed as a symbolic resource to counter and critique that very tradition. Her call to "extend the principles of equality and liberty to an increasing number of social relations" opens the field for heritage and cultural production to be considered as a form of "democratic politics", which is closely linked to the "existence of a public sphere where people [can] act as citizens" (1992:3,9).

This immediately raises questions about how one would *equally* accommodate the many different social memories and heritages that prevail in a multicultural society without excluding or privileging the one over the other, or simply following the now-unproductive idea of a 'rainbow nation'. Mouffe's (1992; 2002) pluralistic view of a democratic society allows a range of different ethnic and cultural identities to coexist and acknowledges that we may hold many different views of citizenship at the same time. Her agonistic pluralism is a critique of essentialism and acknowledges that homogeneity on a national level cannot continue to form the basis of a democratic society anymore. Instead, it is suggested that a complex equality, based on the differences produced in and through an ensemble of social relations, may provide a set of oppositions from within which certain agreements or correspondences in different forms of life may be reached (Mouffe 1992; Marcil-Lacoste 1992).

To achieve a radical form of democratic society or space one must first recover and unify the community around the idea of the 'common good'. As a set of values, the common good allows one to discriminate between differences that should or should not exist, and to seek some kind of identification, connection and agreement in shared forms of life (such as heritage and cultural production) in the midst of the many tensions and antagonisms present in a multicultural society (Marcil-Lacoste 1992; D'Entrèves 1992; Mouffe 1992).

This chapter seeks to reflect on the possible uses of heritage and social memory in a multicultural society as having purpose and meaning for the present, as well as for the future. Within the context of this book, it weighs some of the possibilities that current constructions of the past and symbolic actions and representations, as practiced by different individuals and communities in and around Stellenbosch, may contribute towards its sense of place or *habitus*.[1] It will also show how heritage practices and its resulting cultural objects, spaces and interactions may facilitate or prohibit change. It also considers what power structures are at play in these cultural and social discourses and to what extent it is possible for democratic will and agency to turn these discourses into productive sites for diverse people to negotiate their multiple identities, citizenship and a sense of belonging to a place and a community.

The heritage and cultural practices of artist Dylan Lewis's *Predators and Prey Urban Sculpture Tour* (which was installed in and around Stellenbosch from 2008 to 2010), the historic Solms-Delta farm in the Drakenstein region near Simondium, and the collaborative counter-practice of Lara Kruger and Safe House residents' *Cosy for a rhino* project (2009) present three different approaches to cultural and heritage work. Playing out in different locations and spaces in and around Stellenbosch, these practices may be viewed as public contact zones for diverse groups to encounter the stories, expressions and aspirations of the different communities involved. Each of these three cultural practices mentioned above opens up possibilities for diverse interpretations and meanings to flow from the interactions and confrontations within and between the different groups, communities and heritages involved. These practices may differ or conform to the dominant imagination of Stellenbosch, from where members of the community will either accept or reject it. This will shift and change, or assert and replicate the dominant imagination. These practices, as do others, engage in the making of Stellenbosch as a place.

Power, display, hegemony: Dylan Lewis's *Predators and Prey Urban Sculpture Tour*

In 2008, bronze sculptures by internationally-acclaimed artist Dylan Lewis appeared in various locations in Stellenbosch, his hometown. Lewis installed nineteen monumental animal sculptures in the historic heart of Stellenbosch, while four sculptures were installed in the 'outlying areas' of Ida's Valley, Jamestown, Kayamandi and Tennantville (Stellenbosch Sculpture Tour n.d.). Lewis's objective with the *Predators and Prey Sculpture Tour* was to "pay homage [to] extinct creatures" and to place his sculptures against "architectural urban markers of civilization" (*ibid*.). The 'Predators and prey urban sculpture trail' further aimed to create awareness of "nature [and] conservation". A cell phone platform, from which profits were donated to the World Wide Fund for Nature, allowed the public to obtain information about the artist and each individual sculpture on the route. A walking map showing the locations of the sculptures in the historic centre of the town (Figure 17.1)

---

1   *Habitus* refers to sociologist Pierre Bourdieu's concept of place or context as a "generative machine" based on the "active and creative relation between people and their worlds" (Hiller & Rooksby 2005:7,11).

was also available, as well as a website[2] with information about the artist, the tour, a coffee table book, and catalogue of the artist's work. The site also invited the public to post photographs and comments about their encounters with the bronze animal sculptures. In addition, the Rupert Museum ran Lewis's *Shapeshifting* exhibition concurrently with the sculpture trail.

The *Predators and Prey* sculptures were strategically placed in front of historic, public, commercial, or institutional buildings of civic status, with careful consideration of the formal relations and spaces that ensued between each sculpture, building and immediate street environment. Attention was also given to the specific coupling of animal species, overall physiognomy and the selected site in order to set subtle narratives and a range of associative and possible meanings into motion: the animal fragments were aptly installed at the Sasol Museum and the group's lone lion held firm its territory in front of the Ou Hoofgebou (literally 'old Main Building', the oldest building on Stellenbosch University campus), while the inert and endangered rhinoceros found itself on the edge of the town common, Die Braak. Although the official exhibition period for this expansive installation was October 2008 to July 2010, most of these sculptures still found themselves coupled with their architectural partners until mid-2011. By that time, these pairings were familiar fixtures in the social memory and psyche of the Stellenbosch public.

Lewis's sculpture trail caused general excitement and curiosity in and around Stellenbosch and was also a popular tourist attraction. The monumental form of the sculptures intensified their presence and visual impact, as did the dramatic contrast between the dark bronze, baroque-like forms and traditional white-washed architecture and the repetitive occurrence of Lewis's signature style appearing all over the same place. With this expansive outdoor exhibition, Lewis turned the historic centre of Stellenbosch and the other locations into a vast *Wunderkammer* or chamber of curiosities,[3] but turned it out onto the street blocks, pavements and entrance gates to historic buildings, reaching as far as the outlying suburbs of the town. In short, he turned Stellenbosch into a personal museum for his sculptures.

Lewis's *Predators and Prey Sculpture Tour* corresponded with some of the practices of modern museums, one of which is to select a central site in a town or city because of its general associations with "civic status" and "authority" (Forgan 2005:579). By displaying the bulk of his bronzes in the central district of Stellenbosch, Lewis seemed to appropriate this central position of power to authorise the installation of four sculptures in the outlying areas as a benevolent recognition of the 'workers'. Lewis's use of prestigious architectural sites for the display of his bronzes in Stellenbosch also reflected the modern museum's use of the physical features and visual language of buildings to "encode knowledge in material forms" (*ibid*.:572). His use of bronze (an expensive material generally used for important civic statues); the colonial and imperial ideologies encoded in the buildings on the sculpture trail; and the alliance with the status of the commercial and public institutions occupying these buildings were all constitutive of a set of relations that may be described as belonging to an exclusive, intellectual, and affluent class. The institutions and corporate concerns housed in these buildings require, in different ways, a certain qualification, facility, or position for individuals to enjoy affiliation or membership – an exclusive association that Lewis and his sophisticated bronzes claimed for themselves. These institutions and buildings all occupy prime sites in central Stellenbosch that are also in walking distance from one another to complete the circle – except for the four sculptures in the outlying areas.

When Lewis installed his twenty-three monumental sculptures across Stellenbosch, "allowing the animals to claim back the space they once roamed" (Stellenbosch Sculpture Tour n.d.), he ironically appropriated animal behaviour and several tons of bronze to mark a substantial territorial space. Staking out his site with weighty and unyielding animals also held a second irony: it mirrored the very hegemonic and colonial practices that factored the extinction of many animal species and for others to become endangered. By staking out a central site of prominence and authority in the heart of historic Stellenbosch, the trail configured a geography of

---

2 http://www.stellenboschsculpturetour.co.za

3 *Wunderkammer* (German for 'rooms of wonders') or 'cabinet of curiosities' refers to an encyclopedic collection of objects belonging to history, geology, ethnography, archeology, art, etc. The baroque *Wunderkammers* were precursors of our modern museums. (OED n.d., s.v. 'Wunderkammer'). The *Wunderkammer* was regarded as a microcosm or theater of the world, and was a symbol of a patron's control of the world through its microscopic indoor reproduction. For the development of the *Wunderkammer*, see Oliver Impey and Arthur MacGregor, *The origins of museums: the cabinet of curiosities in sixteenth- and seventeenth-century Europe* (2001).

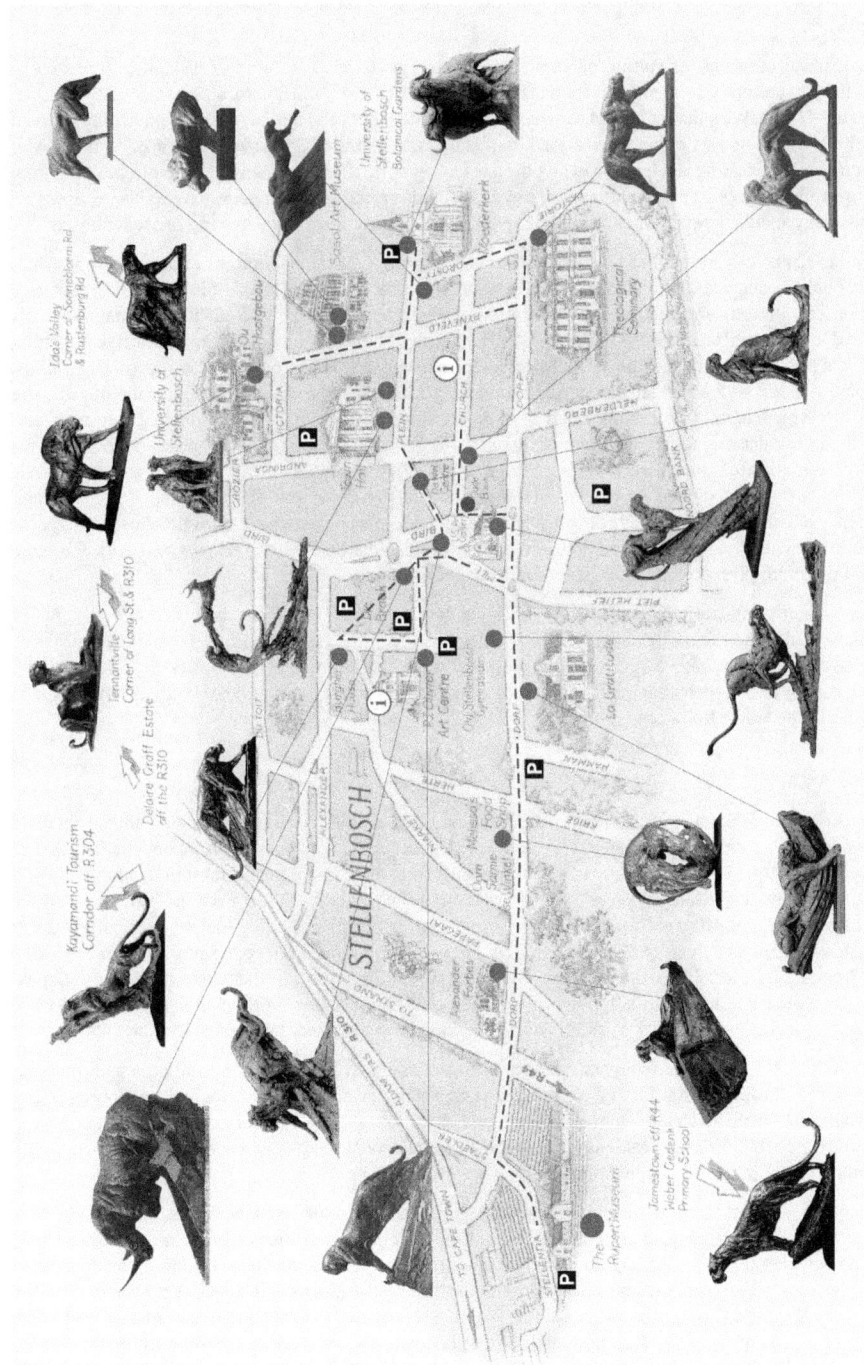

Figure 17.1   *Predators and Prey Sculpture Tour* by Dylan Lewis, an installation of monumental work by Dylan Lewis, Oude Bank, Corner of Church and Bird Street, Stellenbosch, South Africa. Enquiries: +27 (0) 21 880 0054 or info@dylanart.co.za [Stellenbosch Sculpture Tour n.d., reproduction by kind permission of Pardus Art]

privilege and power, banishing others to the outskirts of the town where they were kept in place by four heavy predators.

Landscape is never innocent. As culture performed on nature, landscape is a complex discourse. It is a social construct and textual practice signifying humans' imagined relationship with nature (Schama 1995; Mitchell 1994; Cosgrove 1984). With his sculptural overwriting of the Stellenbosch landscape, Lewis physically mapped a substantive territory as part of his spatial domain. If landscape, as Denis Cosgrove (1985) argues, is a way of seeing nature and structuring the world, Lewis's bronze cartography and accompanying map offered him, from a high centre above the rest, a commanding bird's eye view of this magnificent landscape. With his ever-present bronzes all over the area, Lewis demonstrated his symbolic appropriation and ownership of Stellenbosch.

Ernesto Laclau (2002:47) argues that "representation is constitutive of the hegemonic relation". When Lewis installed his bronze animals on a colonial and post-apartheid site, rendering visible his network of collector clients, sponsors, patrons and institutional aspirations, he established a specific egalitarian imaginary that reflects the dominant official and corporate powers. The placing of the baroque-like bronze animals in relation to the buildings did not displace the underlying colonial and imperial discourses in these configurations, but occupied the space in a similar manner, perpetuating colonial, imperial and neoliberal ideologies. The *Predators and Prey Sculpture Tour* conformed to the prevailing and dominant community of the town as a whole, without any discrimination or acknowledgement of plurality, taking up what Laclau (2002:38) terms as the "representation of the universality". There was no difference or dialectic between the hegemonic dimensions of the *Predators and Prey Sculpture Tour* and the particular and diverse body of the working-class Stellenbosch citizen it represented. In Laclau's terms, one may assert that the Stellenbosch citizen or community was by itself an unmediated and universal entity captivated in the *Predators and Prey Sculpture Tour*, which therefore presented itself as hegemonic imagery.

Social space is not democratic if external interferences have control over others, and a certain asymmetry between universality and the particularity of the social agent is required for civil society and a public sphere to emerge. One may therefore also accept that if representation is "constitutive of democracy" (Laclau 2002:48), its function cannot be purely passive, but should open up new forms and ways of democratic community – as we might see in the Solms-Delta case.

Heritage, place, community: The heritage practice of Solms-Delta

The heritage practice of Solms-Delta, a historic farm dating back to 1687, is an encompassing and sustainable project involving an extended community living and working on the farm and contributors beyond its physical borders. It is a story that embraces the lives of active agents from all levels of the social sphere who are taking part in the significant change and wholesale transformation of their *habitus*. This started in 2001, when owner Mark Solms, a world-renowned neuroscientist, addressed the 300-year legacy of colonial history of Zandvliet (as the farm was then called) to excavate a complex history of multiple intersecting stories dating back to the Early Stone Age. He established a museum on the estate, Museum van de Caab ('Museum of the Cape'), to explore the farm's history "from all points of view", specifically focusing on the "personal experiences of dispossessed Khoe-San [sic], pioneer settlers, slaves, and the current resident labourers" (Randle 2008:3).

In order to redress the unfortunate past of the South African wine industry, Solms also initiated a partnership with British philanthropist Richard Astor to set up a trust for disadvantaged residents and employees of the estate. Based on a "realistic wealth-sharing model", the shareholding grant to tenant workers living on the estate makes it possible for farm residents to now enjoy "comfortable homes and social programmes" and a general improvement in quality of life (Solms-Delta Wine Estate n.d.).

The Solms-Delta network includes a plurality of individuals and groups (archaeologists, academics, farm residents, etc.) who are related in an integrated socio-cultural system to contribute, from different dispositions, towards the complex processes of tacit knowledge production. Here, the Dik Delta Fynbos Culinary Garden with its indigenous edible plants rescues many flora under threat of extinction and revives the culinary tastes and uses of the Khoi and Cape's first settlers in the Fyndraai restaurant. A rural Cape music project, Music van de Caab, centres on the Winelands community to develop their own resources and supports fieldwork for formal academic research into the origins of traditional Cape music (*ibid.*).

On the level of agricultural enterprise and the daily struggles of workers, Solms-Delta has generated a *habitus* similar to sociologist Pierre Bourdieu's concept of an "active and creative relation between people and their worlds" (Hiller & Rooksby 2005:7). As an all encompassing project that integrates the lives of the community, the environment and the produce of daily work, Solms-Delta is a total practice that may be described as a result of the actions and interactions shaped concurrently by the *habitus* and capital of agents, as well as the dynamics of shared participation in a common endeavour.

Solms-Delta presents the idea of a *Gesamtkunstwerk*[4] as a picture of radical democracy. As a total work of art, the *Gesamtkunstwerk* incorporates the full spectrum of all people's actions, which come into being by means of its collective efforts. It also presents the public sphere as a space where people act as citizens and that requires the active participation of everyone. In a *Gesamtkunstwerk*, everyone is an artist or social agent within the collective; however, the social agent or artist does not act as a unitary subject, but takes on an ensemble of different subject positions in relation to the other actors and within the specific discourses at play. In terms of the value of the *Gesamtkunstwerk*, John Cage describes it thusly:

> *Art instead of being an object made by one person is a process set in motion by a group of people. Art's socialised. It isn't someone saying something, but people doing things, giving everyone (including those involved) the opportunity to have experiences he would not otherwise have had.*
>
> [Cage in Heimbecker 2008:477]

The model of the *Gesamtkunstwerk* presents, as in radical democracy, a pluralistic framework where everyone participates in different positions in a heterogeneous collective towards the common good of all. As an ecological project, the Solms-Delta *Gesamtkunstwerk* resembles an intertwined and dialectical structure between nature and people,[5] emphasising the inevitable designation between land and the life worlds of its inhabitants. The idea of the Solms-Delta community seems to grow from an intimate link with the soil, the landscape, and their heritage and traditions – a genuine and symbolic relationship that integrates community life with the environment. According to Cosgrove (1984:19), for the insider, there is no clear separation between self and the prospect, and he describes such a landscape composition as "much more integrated and inclusive with the diurnal course of life's events – with birth, death, festival, tragedy – all the occurrences that lock together human time and place".

In imagining a new idea of landscape, Solms-Delta simultaneously uncovers and exposes the cartographic palimpsest of a colonial paradigm written over the land that went before. Landscape here is a process through which people render the past visible as a basis from which to produce new imaginations for a future landscape. The interaction between members of the Solms-Delta community; the excavation of previous life stories and activities; the renewed care of the land; and the opening out of original relationships also make it possible for fresh folklore to develop from these tensions from the past. Landscape, as the intertwining of the "lived-body and the entity-complex", enacts a certain 'geographicity' (Backhaus 2009:16,25), a gestural language that changes the land, as well as implicates how we view and treat the environment. The Solms-Delta landscape is a complex image allowing multiple views to coexist.

At Solms-Delta, ancient and recent past, as well as the present and near future, intersect with memories of the land, agriculture, and multiple social and cultural affiliations. These representations stand in a dialectic and dialogical relationship to each other, as well as to the wider geographical region. The narratives that emerge from and between the configurations in the Museum van die Caab, Dik Delta Fynbos Culinary Gardens, and Music van de Caab, connect and collaborate with daily life and several other projects and events on the estate

---

4   After Richard Wagner's use of the term, *Gesamtkunstwerk*, translating into a 'total work of art'. Wagner stresses the unification of the different arts into a complete new art form for the future (Millington n.d., s.v. 'Gesamtkunstwerk'). Broadly speaking, the eclectic mix and collective nature of a *Gesamtkunstwerk* may also refer to involving several senses in an aesthetic encounter (Jewanski n.d., s.v. 'Synaesthesia'). The Wagnerian concept of the *Gesamtkunstwerk* stresses the collective nature of a 'total work of art' and the *Kunstwerk der Zukunft* (art of the future). It may also be viewed as a "critique of capitalism as a corrupt and decadent ideology of modernity" (Andreas Huyssen in Goebel 2007:494).

5   David Pepper terms such a structure as eco-socialism. Eco-socialism acknowledges and promotes a democratic, pluralist society; radical revolutionary system-changing activities; ecological change from a social justice point of view; the communal nature of general will and dignity of labour; etc. (Pepper 1992).

to constellate a life story shared by the farm community, specialists, academics, neighbours, and visitors. Solms-Delta is a hybrid memory site or an alternative and new arena that makes it possible for multiple voices, democratic strategies, indigenous knowledges, and cross-cultural translations to emerge.

Representation at Solms-Delta allows a range of modes and particular voices to emerge from their everyday life, interactions and experiences, rendering a co-produced, textured and plural democracy. In Lara Kruger's project with Safe House residents, questions about appearance and the visibility of symbolic actions and interactions come to the fore as it is played out and enacted in public space. Here, participation in artifice or symbolic acts in public presents the idea that 'making art on the commons' is an enabling democratic act towards citizenship.

Play, appearance, agency: The *Cosy for a rhino* project

For Hannah Arendt, citizenship is constituted in public space through visible actions of civic engagement and collective deliberation. This requires 'appearance' and can be constituted only if we "share a common world of humanly created artefacts [...] and settings". The public sphere or the world that we hold in common is activated when we participate in its artifice through our own collective and creative actions, "human artefact, [...] fabrication of human hands, as well as [...] the affairs which go on among those who inhabit the man-made world together". Appearance and the "presence of others who see what we see and hear what we hear assures us of the reality of the world and of ourselves" (Arendt in d'Entrèves 1992:146-148).

Art worker and teacher Lara Kruger's knitting project in collaboration with residents from Safe House, a home for abused women and children in Stellenbosch, affirms Arendt's views on the importance of appearance and participation in public space through artefact. The collective action and creative participation of the Safe House residents in a public intervention provided them with a platform for dialogue and discussion from which a sense of self-worth, agency and citizenship could emerge.

Kruger's work in the field of social sculpture[6] has led her to collaborate with the Safe House residents. As social sculptor Kruger assumes the role of facilitator and "responsible participant" (Kruger 2010) in managing and steering participation and collective interactions within social groups and processes. Kruger's first collaboration with Safe House residents was to knit squares for the social justice activist group, Code Pink, who called on knitters worldwide to contribute to their *Knitters for Peace* project. These squares were to be made into a cosy to cover the gates of the White House in Washington to celebrate Mother's Day in 2009. Taking part in something constructive excited the Safe House participatory group, so the women decided to knit a cosy from recycled plastic bags to cover something in Stellenbosch to coincide with the Washington event. Through dialogue and conversation, which also included a walk to the village green, a democratic decision was reached among the Safe House collective to knit a cosy to cover the life-sized Dylan Lewis bronze rhinoceros installed on Die Braak in Stellenbosch – hence the name of the project, *Cosy for a rhino*.

The knitting of the cosy was done at Safe House and during two days on site on Die Braak (see Figure 17.2). Throughout the process, the Safe House collective engaged in conversation. Kruger relates that discussions centred on the symbolic value of knitting and knit work, teamwork, and solving the problem of fitting the cosy onto the bronze sculpture. According to Kruger, the following ideas flowed from the collaborative process:

- *knit work / knitted cosy = gift, warmth, protection = mother = 'safe house';*
- *abused women and children = vulnerable; in need of protection = rhinoceros: endangered animal, abused, value, beauty, strength;*
- *act of knitting with plastic bags = transformation, regeneration, recycling, environmental issues = 'safe house': regeneration, protection, transformation.*

[Kruger 2010]

---

6   Social sculpture refers to a "socially engaged art practice" aimed at the "radical transformation of the world" (Social Sculpture Research Unit n.d.). It is as such in direct contrast to the idea of the artist as individual genius working for and in the gallery space of a capitalist-driven art market. Social sculpture is based on Joseph Beuys' idea that everyone is an artist and that the world is shaped (or sculpted) by thinking, spoken and material forms and processes. It may be broadly defined as an "evolutionary process through which we shape the world in which we live" (Social sculpture n.d.). See Jeff Barnum's article, Social sculpture: enabling society to change itself (2010), and Laurie Rojas article, Beuys' concept of social sculpture and relational art practices today (2010).

Figure 17.2 *Cosy for a rhino* project on Die Braak, Stellenbosch, 2009 [Photographs by Lara Kruger]

Dialogue and interaction were not only restricted to members of the Safe House collective. The small knitting ensemble on Die Braak attracted the attention and interest of members of the public who also engaged in conversation with them. Kruger recounts that the public's positive response towards their work "contributed to a deep sense of pride in all the participants involved". When asked about the purpose of the project, they countered that they "just felt the need to do something for the Stellenbosch community" (Kruger 2010). Yet, the good intentions of the collective knitters were met with some hostility too, and Kruger recalls an incident where a passing street teenager abused the women verbally. He also threatened to damage the group's work on Die Braak. This generated social problem-solving processes among the Safe House collective, who used the incident as an opportunity to build relations with the car guards in the area and quickly established a 24-hour watch over the cosy.

Kruger also tells of a member of the Safe House collective who now knits for a community project for the prevention of sexual abuse of children. She also believes knitting in the exposed and 'unprotected' public space of Die Braak facilitated interaction with other community members, which greatly contributed towards the women taking ownership of the project.

It is ironic that women who seek shelter from domestic abuse should find self-worth and agency in participation in a public intervention. One may only assume that the collective effort of the residents of Safe House contributed to their solidarity, and that their public act of knitting – of creating artifice – brought a sense of purpose and direction to their actions. Seyla Benhabib (in D'Entrèves 1992:162) points out that in Arendt's participatory conception, public sentiment is not encouraged towards "reconciliation and harmony, but rather *political agency and efficacy"* and a sense that what we do in life can indeed make a difference. In the Arendtian sense, knitting a cosy for a statue holds the possibility of reactivating the circumstances for a democratic and active citizenship that is based on self-determination.

The shared activities and participatory processes of the Safe House knitters created a sense of collective identity within the group. The discussions and practical chatter among themselves and debate with bystanders about the legitimacy of their project strengthened their solidarity, as did the confrontation with the youth who threatened to vandalise their work. Their collective sense of purpose and specific positions within the collective allowed the knitting group to act constructively even in the face of adversity. Such a concept of citizenship is what Nancy Fraser (in D'Entrèves 1992:158) calls "the standpoint of the *collective concrete other*", referring to the specific circumstances, needs, vocabularies and resources available in encountering one another. Unique or culture-specific identities are of less importance here; instead, a new group identity emerged from the middle ground of shared experiences and the collective and common skills in ordinary women's handiwork. In the case of the Safe House collective, Fraser's view would be that they were able to act because the stress was on "solidarity rather than on [...] compassion, on respect rather [than] on love or sympathy, and on autonomy" (D'Entrèves 1992:159).

For the Safe House collective, the public space of Die Braak was integral to the knitting of the cosy and provided the mediating link between symbolic action (knitting) and agency. Knitting the cosy in the closed and private spaces of the Safe House premises would not have engendered the same sense of autonomy and worth in the women. The public space of Die Braak provided a stage for democratic conditions, which were opened up by their active performance and knitting of a cosy for a sculpture. These actions are much more far-reaching than the women's own self-determination. Their appearance and knitting also activated Die Braak as a public space and place where experiences may be shared among different members of the community. Public space, even on Die Braak, has no meaning or value until it is activated with our actions and life stories. Here, citizenship constitutes a public space of action and interaction, of showing up, of revealing and playing out our identities and differences, of a culture, and of our common world.

The knitted cosy or artefact could have been anything (any colour or material); it merely presents the women's collective powers and efforts in similar terms. It is purely symbolic action, a play or even comic inspiration. It is the artefact that emerges from ordinary women's handiwork: to knit a simple cosy or comforter as a token or gift for where it may be most emblematically needed – a shared public space. What these women have to offer in the form of handiwork mediates all their hopes, actions and relations to others and to the world.

In a society where public space and culture are dominated and regulated by neoliberal and institutional structures, the Safe House women's actions might have translated into a form of civil disobedience. Knitting a plastic cosy for the colossal and unyielding bronze sculpture of an internationally-acclaimed artist is a symbolic and subversive act of self-assertion against dominant power relations. Ironically, it was this same hegemonic landscape and terrain that allowed the Safe House women to renegotiate agency and democratic will through their own self-initiated actions.

Conclusion

We have suggested that public space is threatened by corporate and government appropriations, sanctions and control. As a zone of possible contact between groups of people and communities, cultural heritage – as artifice and as productive symbolic action – may provide a viable option to revive and reclaim the public sphere as a radically democratic and shared space. This requires that we invent and facilitate innovative and imaginative practices to perform in public that will be of general advantage to the common good of all members of the community. This is in opposition to neoliberal and capitalistic developments aimed at benefitting investors, with little or no concern for the social dimension of local cultures.

Representation is central to our understanding of the world. As a construct and as a human artifice, our experiences of the world and others, as well as our expectations for the future, are directly shaped by it. A picture of a radically democratic Stellenbosch becomes possible if we can imagine every community's active and public participation in a range of interlinking but diverse cultural and symbolic productions. Such a landscape is not dominated by the continued homogeny of colonial and imperial images, power developments and corporate advertisements, but allows appearance and play among and between the many locally textured voices. Pluralistic cultural practices open up the possibility for new forms of identification and a complex agonistic equality to emerge from a range of interactions between individuals and communities. Such connective cultural practices allow us to envision hopeful scenarios for a sustainable future Stellenbosch.

Acknowledgement

I am grateful for permission from Pardus Art to reprint the *Predators and Prey Urban Sculpture Tour* map, as well as to Lara Kruger for permission to reproduce the photographs of the Safe House collective's *Cosy for a rhino* on Die Braak.

REFERENCES

Backhaus G. 2009. Introduction: The problematic of grounding the significance of symbolic landscapes. In: G Backhaus & J Murungi. *Symbolic landscapes.* Houten: Springer.

Backhaus G & Murungi J (eds). 2009. *Symbolic landscapes.* Houten: Springer.

Barnum J. 2010. Social sculpture: enabling society to change itself. [Accessed 20 September 2012] http://www.reospartners.com/node/341

Bennett T. 2006. Exhibition, difference and the logic of culture. In: I Karp, CA Kratz, L Szwaja & T Ybarra-Frausto (eds). *Museum frictions: public cultures/global transformations.* Durham: Duke University Press.

Bourdieu P. 2005. Habitus. In: J Hiller & E Rooksby (eds). *Habitus: a sense of place.* 2nd Edition. Aldershot: Ashgate.

Cosgrove D. 1984. *Social formation and symbolic landscape.* London: Croom Helm.

Cosgrove D. 1985. Prospect, perspective and the evolution of the landscape idea. *Transactions of the Institute of British Geographers,* New Series 10(1):45-62.

D'Entrèves M. 1992. Hannah Arendt and the idea of citizenship. In: C Mouffe (ed). *Dimensions of radical democracy: pluralism, citizenship, community.* London: Verso.

Forgan S. 2005. Building the museum: Knowledge, conflict, and the power of place. *Isis,* 96(4):572-585.

Goebel RJ. 2007. Gesamtkunstwerk Dresden: Official urban discourse in Durs Grünbein's poetic critique. *The German Quarterly,* Fall:492-510.

Heimbecker S. 2008. HPSCHD, Gesamtkunstwerk and Utopia. *American Music,* 26(4):474-479.

Hiller J & Rooksby E. 2002. *Habitus: a sense of place.* 1st Edition. Aldershot: Ashgate.

Hiller J & Rooksby E. 2005. *Habitus: a sense of place.* 2nd Edition. Aldershot: Ashgate.

Impey OR & MacGregor A. 2001. *The origins of museums: the cabinet of curiosities in sixteenth- and seventeenth-century Europe*. London: House of Stratus.

Jewanski J. "Synaesthesia." Grove Music Online. [Accessed 1 February 2011] http://www.oxfordmusiconline.com/subscriber/article/grove/music/48564

Karp I, Kratz CA, Szwaja L & Ybarra-Frausto T (eds). 2006. *Museum frictions: public cultures/global transformations*. Durham: Duke University Press.

Kruger L. 2010. Personal conversation. Stellenbosch: 3 October.

Laclau E. 2002. Democracy and the question of power. In: J Hiller & E Rooksby (eds). *Habitus: a sense of place*. 1st Edition. Aldershot: Ashgate.

Leach N. 2005. Belonging: towards a theory of identification with space. In: J Hiller & E Rooksby (eds). *Habitus: a sense of place*. 2nd Edition. Aldershot: Ashgate.

Lefebvre H. 1991. *The production of space*. Oxford: Blackwell.

Marcil-Lacoste L. 1992. The paradoxes of pluralism. In: C Mouffe (ed). *Dimensions of radical democracy: pluralism, citizenship, community*. London: Verso.

Millington B. "Gesamtkunstwerk." The New Grove Dictionary of Opera, edited by Stanley Sadie. [Accessed 1 February 2011] http://www.oxfordmusiconline.com/subscriber/article/grove/music/O011027

Mitchell WJT. 1994. *Landscape and power*. Chicago: University of Chicago Press.

Mouffe C (ed). 1992. *Dimensions of radical democracy: pluralism, citizenship, community*. London: Verso.

Mouffe C. 2002. Which kind of space for a democratic habitus? In: J Hiller & E Rooksby (eds). *Habitus: a sense of place*. 1st Edition. Aldershot: Ashgate.

Olick JK & Robbins J. 1998. Social memory studies: from 'collective memory' to the historical sociology of mnemonic practices. *Annual Reviews: Sociology*, 24:105-140.

Pepper D. 1993. *Eco-socialism: From deep ecology to social justice*. London: Routledge.

Randle T. 2008. Solms-Delta: a story van de Caab. Reprint from *Capensis: Quarterly Journal of the Western Cape Branch of the Genealogical Society of South Africa*, 1:3-16.

Rojas L. 2010. Beuys' concept of social sculpture and relational art practices today. [Accessed 20 September 2012] http://www.chicagoartmagazine.com/2010/11/beuys'-concept-of-social-sculpture-and-relational-art-practices-today

Schama S. 1995. *Landscape and memory*. London: Fontana.

Sennett R. 1978. *The fall of public man: on the social psychology of capitalism*. New York: Vintage Books.

Social Sculpture. n.d. [Accessed 2 February 2011] http://www.walkerart.org/archive/F/A44369B1F42E32026178.htm

Solms-Delta Wine Estate. [Accessed 28 January 2011] http://www.solms-delta.co.za

Stellenbosch Sculpture Tour. [Accessed 29 January 2011] http://www.stellenboschsculpturetour.co.za

Waterton E & Watson S (eds). 2011. *Heritage and community engagement: collaboration or contestation?* London: Routledge.

Wunderkammer. OED. Oxford University Press. [Accessed 25 September 2012] http://www.oed.com/view/Entry/243535?redirectedFrom=wunderkammer

Chapter 18

POVERTY AND INEQUALITY
Stocktaking of the social landscape of Stellenbosch

JOACHIM EWERT

Introduction

Poverty and inequality in society is deemed problematic for a number of reasons. Firstly, in most moral frameworks,[1] whether rooted in religion or in secular social philosophy, poverty and inequality (especially in its extreme form) is regarded as inhumane, undesirable and deserving of charity. Secondly, poverty is 'bad' for the economy, because it points to a small or limited market – simply put, poor people do not have a lot of purchasing power. Thirdly, poverty and inequality may become socially and politically destabilising, giving rise to conflict, undermining economic development, and so forth.

This is no different in South Africa, exhibiting as it does the typical profile of a developing country – albeit a 'middle-income' one – or even that of an 'emerging market'. Eighteen years after the transition to democracy, more than 40% of the population is still living in poverty.[2] Moreover, measured by the Gini coefficient, the gap between rich and poor has widened since the twilight years of apartheid. A lack of education and skills and slim chances on the labour market constitute some of the main reasons behind unemployment and widespread, chronic poverty. On the other hand, those in possession of scarce skills and qualifications have benefited from the democratic transition and South Africa's integration into the world economy, deepening the divide between 'winners' and 'losers'.

In many ways, Stellenbosch[3] is a microcosm of South African society, with inequality perhaps slightly more pronounced than in most municipal areas, but with a lower percentage of people living in extreme poverty. Even so, it is a situation that most of those involved in 'development' would want to change and not sustain. In other words, in the years to come they would want to see greater Stellenbosch with fewer people living in poverty and less inequality than at present – for one or more of the reasons mentioned in the opening paragraph.

This raises the central question of this chapter, namely: how may poverty and inequality best be reduced? Over the last sixty years, a lot has been written about 'development' and 'development strategies' as a way out of poverty. Although schools of thought have come and gone – and 'big theory' has been declared dead (Sachs 2005) – there has emerged a kind of minimalist consensus among those that still 'believe' in development

---

1 But by no means all. For example, extreme neo-liberals would argue that people's lives are their own responsibility; if they end up in poverty that is their 'deserved station' in life.

2 Some would put it closer to 50% (May 2010).

3 'Stellenbosch' includes the towns of Stellenbosch, Jamestown, Raithby, Klapmuts, Kylemore, Pniel and Franschhoek, plus the farm land in between. Henceforth in this chapter 'Stellenbosch', 'greater Stellenbosch' and 'Stellenbosch municipal area' will be used interchangeably.

and think that the United Nations' Millennium Development Goals are worth pursuing. In line with this view, it will be argued that poverty can be effectively reduced if three things are put in place. The first is economic growth, because growth is necessary (if not sufficient in itself) for the reduction of poverty (and, conversely, inequality slows growth); the second is good governance, which includes functioning institutions that are not undermined by corruption; and the third is healthcare and education systems that deliver people who are 'fit' for the labour market of a modern, open economy.

Although there are important differences (for example, the emphasis on growth), this argument is not entirely unrelated to the view that defines 'development' as 'human development' – a broad approach that puts people first and has the building of 'human capabilities' as its defining rationale. The 'capabilities' approach emphasises functional capabilities or 'substantive freedoms', such as the ability to live to old age, engage in economic transactions, or participate in political activities, instead of utility (happiness, desire fulfilment) or access to resources (income, commodities, assets). Consequently, poverty is understood as capability deprivation (Sen 1999). In the process of helping people build capabilities and get out of poverty, health and education are seen as two key areas.

In line with this guiding perspective, this chapter devotes less space to conceptual issues such as definitions of poverty or yardsticks to measure it by, and more to an analysis of the local economy and governance, as well as the concrete, empirical situation with regard to education and healthcare in Stellenbosch and the impact this has on life chances, social mobility and, ultimately, the reduction of poverty and inequality.

Poverty measurements and trends in South Africa

Although it is generally accepted that poverty is a complex and multidimensional phenomenon, the methods that have been employed to get a sense of its extent and impact have been mostly quantitative, and more often than not monetary ('money-metric'). For instance: some of the earliest studies done in Cape Town in the late 1950s used a so-called 'poverty datum line' (PDL); during the 1980s, various institutions used the 'minimum living level' (MLL) or the 'household subsistence level' (HSL) to measure the extent of deprivation (Wilson & Ramphele 1989); the first nationally representative survey (undertaken in 1993) used a 'national poverty line' of R322 per person per month (May 2010); and the Office of the President used R250 (or USD30) per person per month as its yardstick in 2004. By 2010 the 'poverty line' for a 'household of one' had shifted up to R1 315 per month (South African Institute of Race Relations (SAIRR) 2011). The United Nations Development Programme (UNDP), on the other hand continues to use USD1 as its yardstick for those living in 'extreme poverty', with those surviving on USD1-2 per day being considered to live in 'moderate poverty' (Sachs 2005).

A more complex way of looking at poverty is the Human Development Index (HDI), developed by the UNDP. The HDI is a composite statistic used to rank countries, regions or groups by level of 'human development'. The statistic is composed from data on life expectancy, education and *per capita* gross domestic product (as an indicator of standard of living). The lower the score on a scale from 0 to 1, the worse the human development or 'poverty'.

As we shall see below, Stellenbosch Municipality uses none of these benchmarks when trying to determine the number of 'indigent' people living within its boundaries. Instead, households with an income of R2 800 or less per month are considered 'indigent' or poor and may apply for subsidised basic services. It is a simple monetary measure that is relatively easy to apply and appears to 'work' under the circumstances.

As far as poverty trends in South Africa are concerned, some data collected between 2000 and 2008 (May 2010) suggest that poverty has declined. If the number of people living on less than USD2 per day is taken as the yardstick for absolute poverty, then data produced by the South African Institute of Race Relations (2011) would suggest that only 5% of the total population is poor. This is 58.5% less than in 1996. However, it is safe to say that this figure would be considerably higher if it were not for the government's social grants, of which there are over 16 million recipients in 2012 (SAIRR 2011). Many of these have little or no independent income.

Seen from the perspective of the HDI, it would appear that South Africa is one of the few countries whose HDI declined between the early 1990s and the late 2000s (from 93rd in 1992 to 175th position in the world in 2008), with high adult mortality and lowered life expectancy being the main reason (May 2010).

Far less contested than quantitative measures of poverty are indices of inequality, with the Gini coefficient being almost universally accepted today. Also in this regard the trend for South Africa is not positive. The country's coefficient rose from 0.66 in 1993 to 0.68 in 2000 and 0.70 in 2008,[4] with inequality rising most markedly within the African community (May 2010).

The government has implemented a number of policy measures to respond to these trends. These include higher government expenditure, with the lion's share going to social services. Of this allocation, education has received the largest share, followed by health, social security and housing. The government now spends more than R80 billion per year on social grants, of which there are a projected 16 222 516 beneficiaries in 2012 (which is considerably more than the 9 219 000 people in formal employment) (SAIRR 2011). High spending (including the primary school feeding program) and a greater focus on education have led to an increase in enrolment figures and an improvement in functional literacy. In 2008, the enrolment ratio stood at 78.8% of eligible children and the adult literacy rate at 87.6% (May 2010). However, as pointed out below, high spending and enrolment figures can be misleading and often conceal the poor performance of many schools and children, also in Stellenbosch.

Another important shift has been the decentralisation of the responsibility to 'do' social development. The provision of a large part of the so-called 'social wage' (for example basic services, not including housing) falls mainly within the mandate of the sub-national spheres of government, that is, provincial government and municipalities. At the same time, only 14% of municipal budgets are derived from provincial and national transfers (May 2010).

That notwithstanding, Stellenbosch Municipality, like provincial government, has been going about its poverty reduction and 'development' task with considerable efficiency and effectiveness, to the point where the basic needs of the 'indigent' are provided for. From here on, upward mobility and permanent escape from poverty now largely depends on parents' and children's will to succeed and the concerted efforts of communities to take more responsibility for their own 'development'.

A close-up look at Stellenbosch

Population and demographic profile

The size of the population in the Stellenbosch municipal area is estimated to be somewhere between 175 000 and 270 000.[5] Of this, approximately 56% are coloured people, 22% whites, 20% Africans and 0.22% Asians; 75% of the population is Afrikaans-speaking (Econex 2010).

Regardless of which total figure one subscribes to, there is no question that the area's population has grown over the last fifteen years or so, a period during which the African population has shown the biggest increase (estimated at over 40%), due largely to in-migration, mostly from the Eastern Cape. As a result, Kayamandi alone now has an estimated 40 000 inhabitants.

On average, the Stellenbosch population is older than the national average, with a far smaller cohort of individuals below the age of 20 (Econex 2010). Nevertheless, the child dependency ratio is fairly high, standing at 36.9 in 2007. In that same year, 48.7% of the total population was male, while females accounted for 51.3% (Western Cape Provincial Treasury (WCPT) 2010).

---

4   The Gini coefficient measures inequality among the values of a frequency distribution, with 0.0 expressing perfect equality and 1.0 expressing maximal inequality.

5   Estimates range widely between 175 000 and 270 000. For instance, the Western Cape Provincial Treasury estimates the total population at 270 000 (2010). On the other hand, the Director of the municipality's Community Services uses a figure of roughly 175 000 (Linde 2011). A senior manager concerned with economic development uses the figure of 200 000, even going up to 230 000 (Moses 2011).

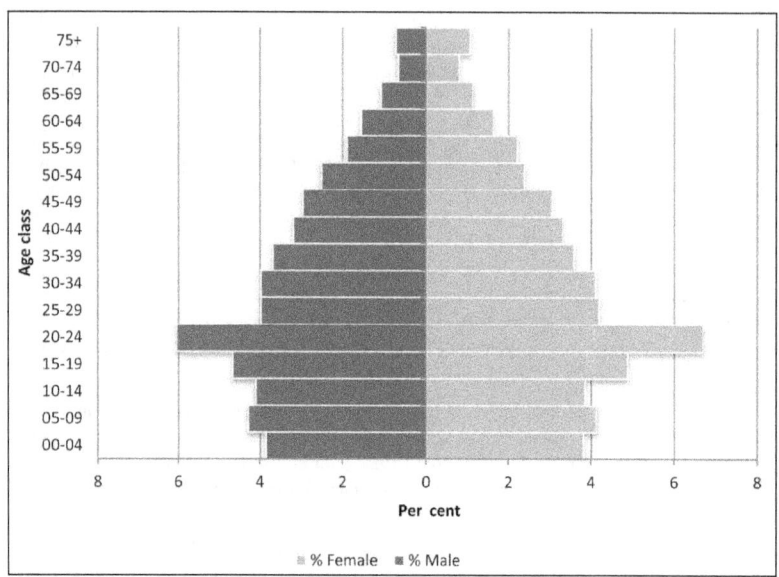

Figure 18.1   Population pyramid for Stellenbosch, 2009 [IHS 2010]

Economy and employment

Stellenbosch Municipality accounted for 24.4% of the Cape Winelands District's economy in 2009, making it the second largest economy in the region. Stellenbosch's regional Gross Value Added figure (GVA-R) increased from R3.834 billion in 2001 to R5.234 billion in 2009, at an average annual rate of 4.0% compared to the 3.2% of the Cape Winelands District over the same period. However, in the wake of the global financial crisis, growth in the municipality slumped to 0.6% in 2009 (WCPT 2010).

Given that economic growth is one of the vital ingredients of job creation, development and the reduction of poverty, this is not good news. Even if the area returns to a sustained 4% growth in the near future, it does not necessarily mean that this will reduce the high level of unemployment, due to the apparent mismatch of qualifications and labour market demand – which again underlines the crucial importance of education and the school system (about which we have more to say below).

Community services, finance, manufacturing and trade are the most important sectors of the local economy in terms of their contribution to GVA (27.1%, 25.1%, 23.5% and 10.1% respectively). Agriculture, a traditional employer of uneducated and unskilled labour, only comes in at number five, contributing a mere 5.4% in 2009 (WCPT 2010).

According to the Western Cape Provincial Treasury, the 'potentially' economically active population in the municipal area stood at 141 179 people in 2007.[6] In its estimate, those employed increased from 43 516 people in 2001 to 75 021 in 2007, meaning that 31 505 more people found employment in the region during that period (*ibid.*).

Manufacturing, community services, wholesale and retail trade, and agriculture[7] are the four most important sectors in terms of employment, providing 20.2%, 17.4%, 16.3% and 12.4% of all jobs, respectively. According to the Treasury, of all those employed, almost 25% are 'low-skilled'[8] (*ibid.*).

---

6   Based on the Treasury's 2007 estimate of the total population of 200 518 (WCPT 2010).

7   Interestingly, while employment has decreased, the Value Added created by agriculture over the period 1996-2009 has gone up, pointing to an increase in productivity.

8   The report does not define 'low-skilled'.

Despite the fact that new jobs were created at a rate of 9.5% per year during the period 2001 to 2007, unemployment remains high, ranging between 17.1% and 40%.[9] Among the unemployed, the majority are female, coloured and young. Youths aged between 20 and 29 represent over 46% of all unemployed, while those between 15 and 34 years of age account for 70.1% of all unemployed (*ibid.*). Unfortunately, the Treasury's report contains no data about the qualifications and skills profile of this category; however, one suspects that a lack of qualifications and skills is the main reason for not gaining entry into the labour market (see Education below).

Income, poverty and inequality

As is the case in most other South African communities, income distribution in the greater Stellenbosch area is markedly uneven. According to the Western Cape Treasury, in 2009, 30.4% of households earned an income of between R0 and R42 000 per year; 36.1% earned between R42 000 and R132 000; 28.5% between R132 000 and R600 000; and 4.9% earned above R600 000 (WCPT 2010). In other words, two-thirds earned between R0 and R132 000 per year and the remaining third R132 000 and more.

Not surprisingly given the history, the distribution of income is also racially skewed, with, for example, most whites earning an income between R192 000 and R360 000 per year and Africans clustered in the R18 000 to R30 000 per year income bracket (Figure 18.2).

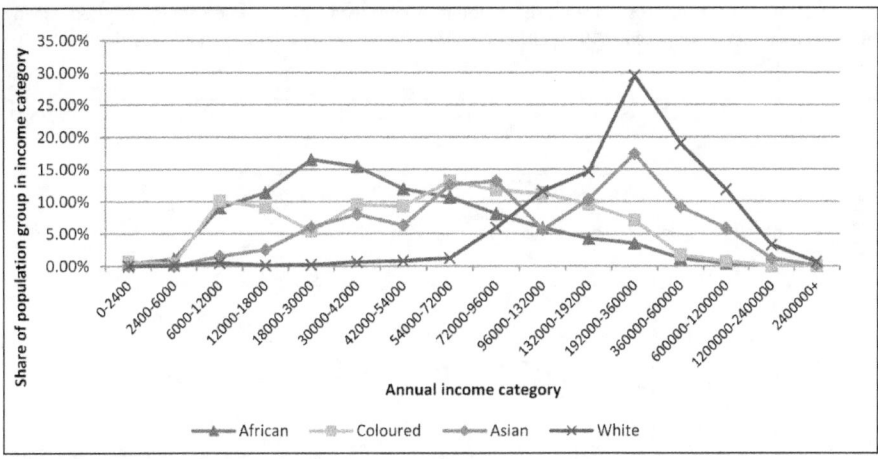

Figure 18.2   Income distribution in Stellenbosch, 2009 [IHS 2010]

Inequality in the Stellenbosch population as a whole amounted to a Gini coefficient of 0.61 in 2009, which was higher (that is, more unequal) than the provincial figure of 0.59, but lower than inequality in the country as a whole (0.65).[10] Between 1996 and 2009, inequality increased within all groups, except for whites (Figure 18.3).

---

9   The Treasury cites a figure of 17.1% for 2007, but a simple calculation of its own figures (potentially economically active minus employed), shows that it stood closer to 40% in 2007. This would be in line with the 'wider' measure of unemployment at national level.

10 Incidentally, this means that inequality at the national level has hardly been reduced since the late 1980s, when it stood at 0.66 (Wilson & Ramphele 1989).

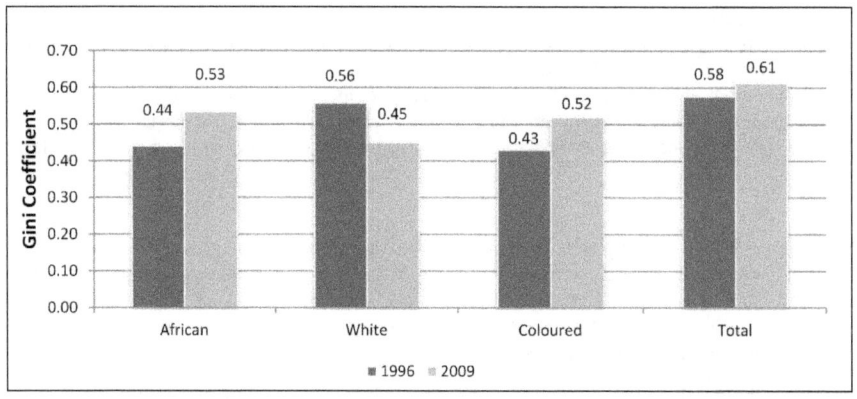

Figure 18.3   Inequality in Stellenbosch, 1996 and 2009 [IHS 2010]

According to research undertaken by IHS, 27%[11] of all inhabitants of Stellenbosch municipal area were living on less than USD1 a day in 2009 (that is, in 'extreme' poverty). Some 42% of all Africans, 37% of coloureds, 12% of Asians and 2% of whites tried to survive on this amount. Except for whites, the percentage of people living in poverty increased in all groups since 1996, although it was higher around 2002 than in 2009 (Figure 18.4).

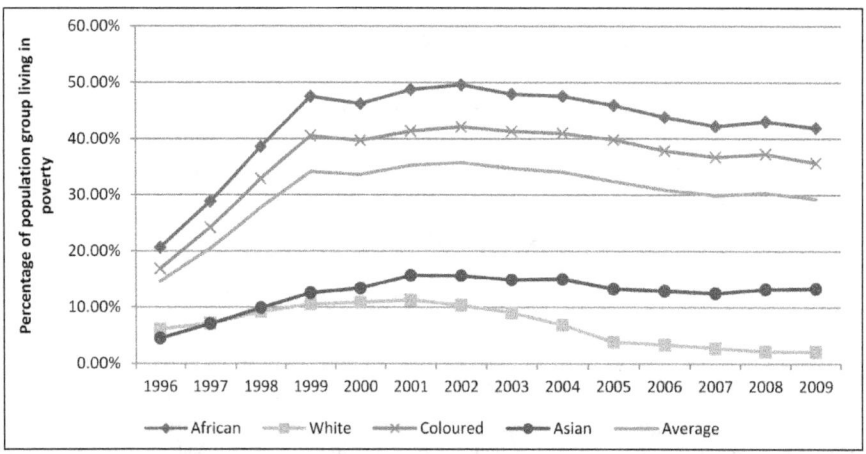

Figure 18.4   Percentage of population groups living in poverty in Stellenbosch, 1996 to 2009 [IHS 2010]

As revealing as these figures may be, Stellenbosch Municipality does not use the USD1 a day indicator when identifying the 'poor'. In order to register for 'indigent' assistance, a household may not have a combined income of more than R2 800 per month (Linde 2011).[12] Proof to this effect must be furnished when applying for registration. Officials at the municipality try to verify the correctness of the information supplied, but according to a manager at the Indigent Division, these checks "are not always what they should be" (Vosloo 2011).

At the time of writing, approximately 10 390 households were on the municipality's 'indigent' list (Linde 2011). Most of them live in the informal settlements of Kayamandi and Langrug (Franschhoek). These are also the areas that have experienced a high influx of migrants, mostly from the Eastern Cape. (As is shown below, it is also in these same areas where some of the schools with the poorest performance are to be found).

---

11   Although sizable, this was considerably lower than the national figure of 42%.

12   This is the current amount. It is adjusted upward from time to time.

Depending on the total population figure used, the above gives us some idea of the extent of poverty in the area. For instance, if one assumes (conservatively) that each household consists of six people, 10 390 households would translate into 62 340 'poor' individuals. Calculated as a percentage of 200 000, this amounts to 31% of the total population – a considerable burden on the state, and coming on top of welfare grants as it were, of which there were 28 342 beneficiaries in Stellenbosch in 2007 (WCPT 2010).

High as it is, the number of 'poor' would increase significantly if farm workers, backyard dwellers and tenants of private property (for example, a house), were to be included. Although counted as inhabitants of the Stellenbosch municipal area, they do not qualify for most of the subsidies for 'indigents'.[13] If registered on the 'indigent' list, poor households qualify for 'free basic services' in the form of credit for electricity and/or water and refuse removal. However, there are a number of qualifications: firstly, the applicant has to be a South African citizen; and secondly, he or she has to own a dwelling (valued at not more than R60 000), rent it from the municipality or live in an informal settlement where legal tenure has been granted. Moreover, 'free' does not mean open-ended; the subsidy is limited. For instance, only the first 50 kWh of electricity and the first 6 kl of water are free. (In the case of informal settlements, the situation is slightly different; there, water is supplied in bulk and is completely free). The cost of graves and the rental of the municipal hall are also subsidised.

As a result of the municipality's 'indigent' policy, almost all households in the area have access to basic services. For instance, in 2007, 97.9% had access to electricity; 95.9% to flush toilets; 98.9% to piped water; and 88.4% to refuse removal (WCPT 2010). This is in line with the superior performance of the Western Cape as a whole compared to the other eight provinces (Sapa & Thomas 2011).

Most of this is paid for by means of 'redistribution'. Although the bulk of the funds spent on 'indigent' services are covered by transfers from the National Treasury, ultimately it is the middle classes and the corporate sector that bear the major burden of 'redistribution'. Apart from VAT and some other minor forms of tax, it is approximately 6 million tax payers and the biggest 200 companies in South Africa that represent the tax base from which this has to be funded (Bishop 2011) – rather small in the context of a total population of approximately 51 million. In addition to income and company tax, local property owners contribute to the municipality's coffers in the form of property tax, which in the case of Stellenbosch, increased significantly in 2009.[14]

The small tax base is not the only challenge. An area such as Stellenbosch – which is relatively prosperous, enjoys modest economic growth, and is relatively well-managed in terms of service delivery – acts as a magnet for a constant flow of migrants. In this sense, 'indigent' subsidies may become a bottomless pit, which would be neither viable nor fair on the small number of tax payers. But that is not all. More important is the fact that free basic services and welfare payments can never be more than poverty relief. They help people survive, but they do not build human capabilities. In that sense they are not a long term solution.

The answer, therefore, to achieving a less poor and more equal Stellenbosch does not lie in more 'redistribution', as some would have it, but in better human capabilities. That is why a health care and educational system that 'works' is of crucial importance for the area's future.

That much needs to be done in this regard becomes obvious from a quick glance at the HDI for the area. In 2009, the HDI score for whites, coloureds and Africans was 0.87, 0.57 and 0.53 respectively (Econex 2010). Not only are the last two scores low by international standards; what makes it worse is the fact that they have improved only marginally (for coloureds) or not at all (for Africans) since 1996.

Part of the reason for this is that building human capabilities is much harder to do than delivering services to address basic needs. While service delivery is largely a question of good planning, logistics and dedicated, honest administration, improving people's health and education requires the right values and attitudes, sacrifice, and exemplary leadership – qualities that are sorely lacking in some parts of the community.

---

13 So far, farm workers are being subsidised when renting a municipal hall or when purchasing a grave site.

14 Unlike in the past, when property tax was pegged at 'market-related' prices, in 2009 they were linked to the price the property was likely to fetch at current prices based on recent sales in a particular area (in the case of residential property, for instance). In some parts of Stellenbosch, this meant a 25% increase in market price, while in others it involved increases of up to 700%.

Healthcare

Most poor people in South Africa are uninsured[15] against health problems and illness (that is, they are not members of a medical fund), which means that private health care is unaffordable for most people[16] and they have to make use of public health care services.

Mindful of this situation, the post-apartheid government placed great emphasis on better access to health care,[17] setting the tone with the introduction of free medical care for pregnant mothers and children under the age of six and expanding the number of clinics at the community level. As a result, primary health care (for the uninsured) is practically free.

The first point of contact is the clinic, which is nurse-driven and doctor-supported, with doctors working at the Stellenbosch district hospital providing one day per week of service at one or more of the eight clinics in the district (WCPT 2010) to examine 'concerned' (i.e. serious) cases. The emphasis is on prevention; it is programme-driven and focuses on breast and cervical cancer, diabetes, HIV/AIDS[18] and tuberculosis. The percentage of the Stellenbosch population with HIV/AIDS has increased from approximately 1% in 1996 to 5.71% in 2009, which represents an increase of 595% (Econex 2010).

Should patients require more specialised attention after examination at the clinic, they are first referred to the Stellenbosch district hospital and then to the Paarl regional hospital or one of the central hospitals (Tygerberg, Grootte Schuur or Red Cross), depending on their illness and the treatment required. Patients admitted to the hospital are classified into four income categories, ranging from H0 to H3. The poorest patients (that is, falling into H0 – pregnant women, children under six, pensioners) pay nothing, regardless of the treatment required. Patients falling into categories H1 to H3 pay a standard fee, but one that is affordable.[19] While this arrangement is beneficial for the patient, it translates into low revenue for the hospital. In the case of Stellenbosch district hospital, for instance, patient fees only contribute R1.3 million to a total budget of R90 million to R100 million (Davids 2010).

Dental care is provided by dentists at the Ida's Valley clinic at minimal costs for patients. Optical services are also provided at the clinic level (where patients pay only R90 for a pair of glasses, for example). Transport by ambulance is provided by the Metro ambulance services and is free of charge.

According to the CEO: Health Care of the Stellenbosch Sub-District, access to health care in the Western Cape has improved significantly over the last 10 years or so, not least because of good planning[20] and "very good management" by the people heading health care in the province, including the current Member of the Executive Council (MEC).[21] A new minister at the national level has also "made a difference" (Davids 2010).

---

15 According to the CEO: Healthcare, Stellenbosch Sub-District, Dr R Davids, only about 20% of the Stellenbosch population is insured (Davids 2010). This means that, based on the assumption of a total population of 200 000 people, the public health system in the district has to provide care for 160 000 people.

16 For instance, a normal consultation at a private practise in Stellenbosch now costs R340, excluding medication. To put this into perspective, the gross pay of a farm worker earning the minimum wage now stands at approximately R347.10 per week.

17 Leaving the disastrous policy on HIV/AIDS aside for a moment.

18 The aim is to test 800 patients per quarter for HIV/AIDS. According to the CEO: Healthcare, Stellenbosch Sub-District, very few pregnant women refuse to be tested. The problem is that many women test too late, when transmission of the virus has already occurred. On average, 16% of all pregnant women test positive and 60% of all tuberculosis patients are also HIV positive. The HIV/tuberculosis hotspots are Kayamandi, Groendal (in Franschhoek) and Klapmuts (Davids 2010).

19 H1, H2 and H3 patients pay R20, R65 and R108 respectively. H1 patients pay nothing for x-rays, a sonar scan or medication, but H2 and H3 pay a subsidised fee. The different categories are: H1 Singles: R0-R3 000 per month; Married persons: R0-R4 166 p.m.; H2: Singles: R3 001-R6 000 p.m.; Married: R4 167-R8 333 p.m.; H3: Singles: R6 001-R8 003 p.m.; Married: R8 334 p.m. and higher. Patients have to furnish proof of their income.

20 In 2003, the Province approved the *2010 Strategic Health Care Plan* (currently under revision), from which flowed the *Comprehensive Service Plan*, approved in 2007. The latter spells out the details and is really the nuts and bolts of the health care system, for example the number of clinics, nurses, doctors, etc.

21 Theuns Botha of the DA.

However, the system still faces a number of important challenges:

- *There is a shortage of doctors. In 2010 there were eight full-time doctors employed at Stellenbosch hospital, one of whom is a dedicated HIV/AIDS doctor. All doctors do duty at the various clinics, as explained above. Ideally, the CEO would like to appoint three to five more doctors in order to provide more dedicated care to Kayamandi, Cloetesville and the Franschhoek communities. However, more funding is needed to achieve this. Doctors work up to sixty hours a week, including paid overtime. Generally speaking, they manage – except over weekends, when only three are on duty and the casualty cases are higher than normal.*
- *Waiting times at the clinics and emergency unit "are still too long". However, a "fast line" has been created for family planning, children, and those with chronic illnesses.*
- *Patients at the clinics in and around Stellenbosch town have an attitude of "entitlement". They insist on seeing a doctor and are not satisfied with being treated by nurses. Clearly, there is still a lot of ignorance regarding the idea of primary health care and the 'nurse-driven, doctor-supported' concept.*
- *The "quality of service at the hospital needs to improve". This includes better hygiene, security, and drug supply.*
- *The Stellenbosch hospital requires a new casualty ward, as part of a "revitalisation plan".*
- *Infant mortality has increased over the last ten years, partly due to the HIV/AIDS pandemic. A problem in this regard is the habit of many pregnant women in the African community of reporting too late for tests at the clinics. Fearing stigmatisation in the wake of a miscarriage, they do not 'announce' their pregnancy before the fifth month or so. But by then it may be too late to prevent mother to child transmission, as it may already start at fourteen weeks. In other words, the challenge is to have women report to the clinic much earlier.*
- *The district wants to introduce the novel concept of a 'farm health post'. Under this scheme, farm workers are trained as 'carers' to supplement the services supplied by mobile clinics (currently five for the whole district).*

[Davids 2010; Doctors' Focus Group 2011]

On the whole, it would appear that health care for the poorest in the community is both accessible and affordable, although patients have to put up with queues and not particularly helpful staff at some of the clinics.

Something that is rather unfortunate is the almost complete breakdown in interaction between the public and private health sectors in Stellenbosch. Up until a few years ago, Stellenbosch hospital enjoyed the support of the local community and a number of private doctors provided service free of charge to the hospital. For instance, some of the equipment still in use was donated by private benefactors over the years. Likewise, if not for private donations a year or five ago, the hospital would not have a new roof. Unfortunately, it would appear that donors and private doctors alike have been alienated by the hospital. As a result, private doctors now refuse to help out at the hospital.[22] Although the reasons for the breakdown in the relationship are not entirely clear, the community is certainly worse off for lack of this potentially productive and fruitful cooperation.

Education

Unequal and inferior education for black South Africans was one of the defining features of apartheid, seriously impacting on people's life chances. It was therefore to be expected that the equalisation of education would be one of the top priorities of the post-apartheid government.

Although there have been some grave errors at the policy level,[23] there has also been significant progress in the direction of providing equal education for all, specifically with regard to state funding. For instance, the government now spends more *per capita* on a black child than a white one. It has also introduced 'no (school)

---

22 In consequence, among other things, general practitioners at the hospital have to administer anaesthetics during operations.

23 One of these, so-called 'Outcomes-Based Education' (OBE), has now been officially abandoned.

fee' schools, exemption from school fees and primary school feeding schemes, and bus transport is available for most learners, including those living on farms. Four years ago, the government abandoned the idea that children should attend 'the school closest to their home', for fear of reinforcing and perpetuating the old apartheid divisions on a residential basis. Parents now have the freedom to choose which school they want their children to attend.

At first glance, the education situation in the greater Stellenbosch area is not a particular cause for concern. In 2007, roughly 80% of the population was literate[24] and learner enrolment increased from 24 744 in 2007 to 26 489 in 2010 (WCPT 2010).

However, figures such as these do not tell us much and leave a number of important questions unanswered. For instance, what is the dropout rate between Grades 1 and 9, or Grades 8 and 12? What subject combinations do learners take in their matric year? Specifically, how many take subjects such as mathematics, physical science or accounting? How many Grade 12s do actually pass? And how many of these with a university exemption? With regard to all these questions, how do different schools compare?

Important as they are, many of these statistics are not (freely) available.[25] However, even a cursory investigation reveals that, despite the steps taken by government to create more equality of opportunity, there are big differences between schools as far as educational performance and attainment is concerned. The clear dividing line is between former Model C schools on the one hand, and former House of Representatives (HOR; that is, coloureds only) or Department of Education and Training (DET; that is, black Africans only) schools on the other. As is explained below, these differences relate to income (class), values, parental and community involvement, and learner motivation, as well as leadership, resolve and dedication on the part of teachers.

One has only to compare the 2010 matric results of the eleven high schools in greater Stellenbosch for these differences to become evident, as Table 18.1 illustrates.

Table 18.1   2010 matric results for high schools in the Stellenbosch municipal area [Maliwa 2011]

School	Number of candidates who completed matric exam	Number of candidates who passed	Pass rate (%)	University exemption
Bloemhof High School	140	140	100	134
Cloetesville High School	148	94	63,5	17
Franschhoek High School	17	17	100	10
Groendal Secondary School	158	100	63,2	10
Kayamandi Secondary School	176	122	69,3	26
Kylemore Secondary School	129	91	70,5	23
Lückhoff Secondary School	142	107	75,4	24
Paul Roos Gymnasium	243	241	99,2	205
Rhenish Girls' High School	122	122	100	118
Stellenbosch High School	133	131	98,5	107
Stellenzicht Secondary School	70	37	52,8	9

Of the schools on the list, Bloemhof, Paul Roos, Rhenish, Stellenbosch High and Franschhoek High are former Model C schools. Three have a 100% pass rate (with Bloemhof and Rhenish – both girls' schools –among the ten best performing schools in the province) and the other two fall short of a perfect pass rate by less than 2%. No less than 80% of all candidates at four of the schools obtained university exemption.

At present, on average 70% of the learners at these schools are white, with coloured and black learners constituting the remaining 30%. The latter originate from a mix of working- and middle-class families, with the former possibly dominating. However, most have come through racially mixed, former Model C primary

---

24   'Literate' is defined as having completed seven years of schooling.

25   Requests for some of these statistics were put to the Western Cape Education Department, but without success.

feeder schools. Clearly, the aspirations of parents account for their choice of these schools.[26] Their children gain entry to these high-performing schools after a selection process, where academic performance is one, but by no means the only criterion.[27] Once their child has been admitted to the school, 'poor' parents may apply for exemption from school fees, which may be as high as R17 600 per year, but this is granted only in desperate cases. Even if the parents are professional, middle-class people, school fees of this magnitude require tight budgeting. In fact, the lengths parents are prepared to go to for a good education for their children are illustrated by the example of some of the pupils at Rhenish Girls' High, who commute from as far as Blouberg to attend it.

The schools, in turn, use the revenue from school fees to appoint extra teachers, finance the maintenance of buildings and grounds, or pay for new facilities and educational tours. Not only are many of the parents at these schools middle-class earners; they are also educated, travelled and possess a wide frame of reference. The value of achievement is a normal part of their home culture, as are books and access to the Internet.

This is a world apart from the standard situation that prevails at some of the 'underperforming'[28] schools and their learners' homes. In many ways, the opposite is the case: low income; not much more than a primary school education; a restricted world view; both parents working or absent; low aspiration; little interest in the child's school life; alcohol and drug abuse; and dysfunctional families. Books in the home are an exception, not to say access to the Internet – not a very supportive home situation.

To be sure, this is the other extreme, and not all children at underperforming schools originate from such homes. Even so, a random survey of the homes of Kayamandi, Groendal or Stellenzicht Secondary school learners would probably show that it applies to more than a handful of families or single-parent homes.

Where this is the case, children receive little or no inspiration from their parents and do not bring much motivation or confidence to the class room. Under the circumstances, it is extra important that the school not only be a place of refuge, but more importantly, a place of inspiration, learning, self-worth, discipline and joy.

▣ *'Underperforming' schools: a case study*

Unfortunately, even a superficial investigation at one of the underperforming schools in the area shows that it is some way off the ideal painted above. To begin with, the school entrance does not look very inviting, sports fields hardly sport a blade of grass, and school buildings are in need of maintenance. Merely judging by the physical appearance, it is difficult to see how learners can develop any sense of pride in their school.

Beyond appearance, the academic performance of the school is the worst in the municipal area, with only 52% of Grade 12s passing matric in 2010 (although this is up from 44% in 2009) and nine of the seventy candidates[29] obtaining university exemption. The seventy-four matrics represented 60% of all those who started in Grade 8. In 2011, not one of the matrics took mathematics, accounting or physical science. In fact, the school stopped offering physical science a few years ago, because there "was not enough interest from the learners' side".[30]

---

26 Their perception that their children will get a better education at former Model C schools is supported by facts: 88% of all black pupils enrolled at these schools passed matric in 2009, compared to 55% at all other schools (Personal communication, school principals 2011).

27 African children and their parents try to gain access to these schools despite the fact that Xhosa is not offered as a home language, with the result that the two languages they offer for matric are actually their second and third languages. Clearly, the perception of obtaining a better education outweighs the consideration of mother tongue education. In fact, learners and their parents may purposefully make this choice in order to become proficient in English – something that will count in their favour as they pursue their education or careers.

28 The Western Cape Education Department defines an 'underperforming school' as a school where less than 60% of matrics obtain a pass. In order to pass, a learner must obtain a 30% mark in three of his/her subjects and 40% in the other three. The latter must include the home language. In order to qualify for a university exemption, a learner must obtain a mark of 50% in four of the subjects.

29 There were seventy-four matrics but only seventy wrote the exams.

30 According to the former principal the lack of interest on the part of the pupils did not justify a science teacher. This means that none of the learners, even if they obtain university exemption, can go on to study science, engineering or medicine, all of which require physical science (telephonic interview, Stellenbosch, 15 May 2007). For ethical reasons the principal shall remain anonymous.

The weak academic performance of the school could be behind the dwindling enrolment numbers: from 800 learners in 2005 to 607 in 2011.[31]

Eighty-five per cent (85%) of all current learners are farm workers' children, coming from as far afield as the outskirts of Kuils River. Although they get to school by bus, many still have to walk some distance to the pickup spot. While most farm worker homes sport a radio or television these days, many of these children still grow up in a limited intellectual world, 'inherit' their parents' low self-esteem, and have little belief in their own abilities. To get to matric is a huge achievement, but in order to get there they minimise the risk of failing by avoiding mathematics and science, because these subjects are deemed to be too difficult.[32]

It would also appear that once these children get to school, there is little to inspire them. Poor leadership, devoid of any vision, over the years and lamentable discipline make matters worse. Parents' attendance at parent-teacher meetings is 30% at most,[33] nor, it would seem, is there much enthusiastic involvement by the middle-class community right next door, some of whose children must surely be attending the school. The situation as a whole raises questions over the school's governing body.

The fact that this school is a 'non-fee paying' one is not an unqualified blessing. The governing body requests a donation of R150 per year from parents, but cannot enforce it. Having been defined as a 'no fee' school, it receives a state grant of R840 per child per year. This translates into a budget of R434 000 for 2011, for instance, 50% of which was spent on text books; 24% on municipal services (electricity, water, etc.); and 20% on 'local purchases' (for example printing paper). This left only 4% for the maintenance of buildings and grounds[34] – admittedly, this is not much; however, one cannot escape the impression that there is no concerted effort or desire on the part of teachers or parents to improve the situation. It would appear that the manual skills that many of the latter must surely possess are not being mobilised and put to good use.[35]

When reflecting on ways of how to turn this into a better school, senior teachers at the school tended to emphasise more finances and sport. There were not many ideas forthcoming as to how to address the issues of discipline, children's motivation, or parental involvement.

The very least that this school – and perhaps other underperforming schools in the area – need, is new leadership with vision. The new powers of the Minister, which includes the power to remove principals and take away the powers of governing boards at schools that consistently underperform, could be employed to good effect in such a case.

Another idea that has been mooted is to turn the school into a double-medium school (as Paul Roos, for instance), in the hope of attracting a different kind of learner who, at 'critical mass', could go some way towards changing the culture of the school. A shift to this direction could be facilitated by actually moving the school back into the centre of town, where it was before the early 1990s.

A third, and more drastic measure, would be to close the school altogether and integrate the learners into the existing high schools in Stellenbosch. By spreading them over so many different schools, they would be exposed to a completely different school culture and set of values, and the self-perpetuating psychology of poverty might be broken. While this may seem extreme, the situation at the school is desperate and calls for decisive interventions.

---

31 Telephonic interview with Acting Principal, Stellenbosch, 22 January 2011. For ethical reasons the school shall remain anonymous.

32 In various interviews and focus group sessions with farm children conducted by the author over the last 10 years, they frequently said that these subjects are 'too hard' ('te swaar').

33 Telephonic interview with the Acting Principal of the school (as of January 2011), Stellenbosch, 22 January 2011.

34 *Ibid*.

35 A senior official in the school circuit management commented that the parents and the communities linked to some of the underperforming schools, "do not look after them".

Governance and community involvement

When thinking about the quality of governance as a vital component of any effort to reduce poverty at the local level, the municipality is the key player. It is not only 'good' governance that matters; a medium- to long-term development plan and the ability to pursue it in a focused and resolute manner are also required. In such endeavours, comfortable majorities at the council level make a real difference. Unfortunately, this is not the case in Stellenbosch. It is one of twenty-two municipalities in the Western Cape that are governed by a coalition of parties, often with a paper thin majority.[36]

Since 2007, control of the council has been see-sawing between the ANC and a DA-led coalition. At the moment, non-ANC councillors outnumber the ANC by five, but whether the five non-DA councillors align themselves with the latter depends on the issue at the hand. Over the last year or two, the tail has been wagging the dog as it were, because the one independent member has been crossing the floor between the two blocks depending on whether a particular decision would advantage her particular community or not – in the process throwing the IDP off course to a greater or lesser extent. However, floor crossing was outlawed recently, which may bring greater stability to the governing of the area.

According to the then Director Community Services, the chopping and changing between a coalition and an ANC-dominated council does not significantly disrupt the normal day-to-day management and administration of the municipality. However, depending on who is in control, it does affect the motivation of some 20% of the officials, while 60% have a professional ethos that entails carrying on doing their jobs (Linde 2011).[37]

The apparent passivity and non-involvement of some of the resident communities seem to be much more of a problem. The municipality views its work as that of a tripartite partnership between the council, the officials and the 'community'. According to some top officials some communities are not active partners at all, but simply passive recipients of municipal services and welfare payments, and do not look after the facilities that have been provided by the council. An example is the Kayamandi administration hall, which was vandalised to the extent that its upgrading cost the municipality R700 000 over the last few years. In order to prevent more of this happening, the municipality spends R140 000 per month on protecting its halls and sports fields (*ibid.*). Clearly the communities concerned are not taking ownership of what has been handed to them and are not playing their part in what is meant to be a symbiotic relationship.

Conclusion

Both poverty and inequality in the Stellenbosch municipal area have increased since the mid-1990s. Except for whites, inequality has also increased within each of the other racial groups.

The in-migration of fairly large numbers of lowly educated and low-skilled people is mainly, but not exclusively, responsible for the changing picture. These in-migrants form a large part of that third of the population who is defined as 'indigent' – and farm workers are not even included in that category.

Although a third of the population is defined as 'poor' or 'indigent', this does not necessarily mean that they all live in grinding, chronic poverty – more than 90% live in (often free) formal dwellings and close to a 100% enjoy free basic services. Health care is practically free, albeit somewhat inconvenient, and so is schooling (in the sense of 'no fee' schools). In addition, many receive state welfare grants, especially the child support grant. State spending (including the 'local state') and 'redistribution' of this magnitude helps people survive and prevents them from sinking into extreme poverty. But these are not long-term solutions; in fact, there is evidence to suggest that they are counterproductive.

It is only when people's capabilities are improved – when they are equipped with an education and skills that put them in a position to earn a decent living in the labour market – that they may permanently escape from poverty, become economically independent and no longer be a burden on the state.

---

36 At only eight municipalities does the governing party have a majority.

37 As of 2011, the municipality has had a staff of 1 028.

Unfortunately, the situation in this regard does not look particularly encouraging. As one important indicator, the matric pass rate at the schools situated in the 'pockets of poverty',[38] is low (despite the fact that matric passing standards are arguably not particularly high). Although data is lacking, it is probable that the dropout rate is relatively high and that many of those who do make it to matric offer a portfolio of subjects that does not count for much in the labour market. How else does one explain the fact that 70% of all unemployed in the Stellenbosch area fall into the 15 to 34 age group?

It is here where parents, children, teachers and the community must take more responsibility. With their basic needs provided for, they must develop the will to get ahead and turn their schools into places of learning and pride, platforms for upward mobility.

Certain schools in close proximity to Stellenbosch (e.g. Kayelitsha, Kraaifontein) serve as examples of what could be achieved. Here leadership, combined with assistance from provincial government and support from parents, have turned things around. Local underperforming schools should learn from them – like beacons of light, they point the way to the future.

While education in the poor communities is in need of urgent attention, significant progress has been made in the field of healthcare and the satisfaction of basic needs. Economic growth, like healthcare and education, are greatly dependent on national and/or provincial policies and there are limits to what the municipality can do in this regard. However, it could play a facilitating role by, for instance, setting up an integrated HDI forum, where local officials, business people, educators and healthcare professionals could engage with policy makers at the national and provincial level to discuss ideas on how best to address local development issues and, ultimately, the reduction of poverty and inequality.

---

38 Stellenzicht, Kayamandi, Cloetesville; and Groendal and Langrug in Franschhoek.

REFERENCES

Bishop A. 2011. Civil servants wages eat into ability to build crucial capacity. *Cape Times*, 25 January:14.

Davids R. 2010. CEO: Healthcare, Stellenbosch Sub-District. Personal communication. Stellenbosch: 13 December.

Doctors' Focus Group. 2011. Personal communication. Stellenbosch Hospital: 1 February.

Econex. 2010. *A Comparative Economic Analysis of Stellenbosch*. Stellenbosch: Econex.

IHS Global Insight: Regional Explorer. 2010. Advanced country analysis and forecast (Rex Version 2.3 F). [Accessed 5 July 2011] http://www.ihs.com/products/global-insight/country-analysis/advanced-forecasting.aspx

Linde H. 2011. Director Community Services, Stellenbosch Municipality. Interview, Stellenbosch: 25 January.

May J. 2010. Poverty Eradication: the South African experience. Unpublished paper prepared for the Department of Economic and Social Affairs: Division for Social Policy and Development, Economic Commission for Africa: Economic Development and NEPAD Division Expert Group Meeting on Poverty Eradication, Addis Ababa, Ethiopia: 15-17 September.

Maliwa MW. 2011. Circuit Team Manager (Circuit 1), Cape Winelands Education District. Interview, Stellenbosch: 10 January.

Moses W. 2011. Department of Planning and Economic Development, Stellenbosch Municipality. Interview, Stellenbosch: 25 January.

Sachs J. 2005. *The End of Poverty*. London: Penguin.

Sachs W (ed). 2005. *The Development Dictionary*. Johannesburg: Witwatersrand University Press.

SAIRR (South African Institute of Race Relations). 2011. South Africa Survey 2010/2011. Johannesburg: SAIRR.

Sapa & Thomas D. 2011. Province streets ahead in terms of basic services. *Cape Times*, 25 January:1,6.

School principals of former Model C schools. 2011. Telephonic interviews, Stellenbosch: 25 & 26 January.

Sen A. 1999. *Development as Freedom*. Oxford: Oxford University Press.

Vosloo S. 2011. Deputy Director Community Services, Stellenbosch Municipality. Interview, Stellenbosch: 25 January.

WCPT (Western Cape Provincial Treasury). 2010. Regional Development Profile Cape Winelands District. [Accessed 5 July 2011] http://www.westerncape.gov.za/other/2010/11/dc02_cape_winelands_sep_profile_2010_final_19_dec_10.pdf

Wilson F & Ramphele M. 1989. *Uprooting poverty: the South African challenge*. Cape Town: David Philip.

Chapter 19

SOCIAL COHESION
Pipe dream or possibility?

JEROME SLAMAT, THUMAKELE GOSA & CHRIS SPIES

Introduction

In this chapter, we shall focus on the issue of social cohesion (and especially the lack thereof) in the Stellenbosch area. A short exploration of social cohesion in recent literature provides an anchor to describe why the current lack of social cohesion renders the town unsustainable, and on what basis and along which routes social cohesion could be built in Stellenbosch for a sustainable future.

The authors agree with Wilkinson and Pickett's notion that inequality "destroys relationships between individuals born in the same society but into different classes; and its function as a driver of consumption depletes the planet's resources" (2009:n.p.). We argue that an investment in human capital is an urgent priority if Stellenbosch is to respond creatively to the inequalities and divisions that still plague its population. Such an investment requires the coming together of people to reach consensus on a credible and inclusive process and structure for generative dialogue and visionary problem solving (*ibid*.). Stellenbosch, known for its cultural heritage and architecture, has the potential to replace its legacy as the birthplace of the apartheid ideology with that of the community that successfully designed and institutionalised a peace and development infrastructure that transforms destructive conflict into springboards for cohesion and development.

When the only spaces for interaction to address key issues are contested and confrontational spaces, our town is in trouble. Building social cohesion would require the skilful crafting and vigorous protection of safe spaces, where the intention is not only to convince, but to analyse, understand and act together. The symbiotic and complementary coexistence of contested and safe spaces is a necessary building block of such an infrastructure for peace and development.

Social what?

The term 'social cohesion', a popular concept in the field of economic and social sciences, originated in Canada, spread to Europe and is now widely used in literature and policy papers across the globe (Jenson 1998).

In scientific terms, cohesion is an attraction between molecules of the same substance, whereas adhesion is an attraction between molecules of different substances. 'Social cohesion' conveys the idea of the "glue that holds us together" (Koelble 2003:176) and describes the outcome of processes whereby people hold on to one another despite differences, hardships or adverse circumstances. Cohesive societies are internally connected by shared universal human values.

The common key concept that underpins most definitions of social cohesion is the one of a shared, integrated, stable and more equal society that faces shared challenges through respectful collaboration in safe spaces. Such a society builds on the resilience of its people and institutions to unlock the potential of each and every one of its members. It provides the climate for change and a space "where all those living there feel at home.

It respects everyone's dignity and human rights while providing every individual with equal opportunity. It is tolerant. It respects diversity" (Club of Madrid 2011). It also develops strong and positive relationships between people from different backgrounds and circumstances in the workplace, in schools and in neighbourhoods (Institute of Community Cohesion (iCoCo) 2005).

Cohesive societies do not happen by accident. People nurture good (political) leadership so as to build defences against attempts to "expose and exacerbate social fault lines" and "harness the potential residing in their societal diversity" (Easterly 2006:5). Dimensions of social cohesion include belonging, inclusion, participation and recognition, as well as legitimacy (Fairbairn 2006), social connectedness, inequality and the cultural environment (AHURI 2005).

The Council of Europe (2004:3) using a democratic rights-based approach, defines social cohesion as follows:

> [...] the capacity of a society to ensure the welfare of all its members, minimising disparities and avoiding polarisation. A cohesive society is a mutually supportive community of free individuals pursuing these common goals by democratic means.

The problem with this definition, says Turzi (2008:131), is that it presupposes the existence of a well-functioning welfare state with "the notion of social citizenship". While it is true that social cohesion is linked to the efficacy of established social inclusion mechanisms, the behaviour and values of members of society play an equally important role.

These reflections on cohesion help us to understand that social cohesion is both a desired outcome as a result of inclusive, value-based processes to overcome common threats, and a dynamic, ongoing joint action to harness potential to build resilience from within.

Social cohesion and *ubuntu*

People in our part of Africa are more familiar with the term '*ubuntu*', which is understood to refer to one of three things:

- 'humanness' (viewed as something an '*umuntu*' (person) may possess);
- an identity view, as expressed by the proverb '*umuntu ngumuntu ngabantu*' (most often translated as 'a person is a person through other people'); and
- a broad cluster of beliefs about 'humanness', according to which 'humanness' must be understood within the context of the identity view (Gade 2009).

Ubuntu appeals to South Africans because it resonates with core human values. Archbishop Desmond Tutu has said that "a person with *ubuntu* is open and available to others, affirming of others, does not feel threatened that others are able and good, for he or she has a proper self-assurance that comes from knowing that he or she belongs in a greater whole and is diminished when others are humiliated, when others are tortured or oppressed" (1999:34).

Former President Thabo Mbeki advocated the notion that *ubuntu* is a "timeless core value system" and "conviction in a truly universal bond of sharing that connects all humanity", combined with an "ethos which bind communities together". His notion is that these values and ethos drive community members to act in solidarity with the weak and the poor and help them to behave in particular ways for the common good. In practice, however, there is little evidence to support the notion that *ubuntu*-driven social cohesion is indeed happening and, as Mbeki says, is showing itself through "effectiveness, reconciliation, truth sharing and loving criticism, the honouring of feelings and a celebration of differences. It faces the future with hope and confidence" (Mbeki 2004:n.p.).

Despite Mbeki's references to social cohesion – and more recently, President Jacob Zuma's frequent use of the term and the subsequent national Social Cohesion Summit – it is our experience that the term 'social cohesion' remains for many a foreign concept that stubbornly resists consumer-friendly translation into our context. It is unfamiliar and exotically strange to the vast majority of the population and people generally find it confusing. The term still lacks clarity and struggles to break through conceptual vagueness and murkiness. "I can't find the diamond in the concept," a Stellenbosch resident complained in a meeting; while for some, it is a "naïve suggestion that socially cohesive societies are always harmonious, devoid of political conflict or dissent" (Easterly 2006:3). Others are sceptical of the notion that bringing people together to engage in dialogue is the right approach, and prefer to do things together [rather] than to sit in talk shops without any action taken as a result of the talks. Much energy is therefore spent on trying to explain what cohesion is and why it is important. So far the concept has proven unattractive to many and it would appear that what is needed is a more basic definition, stated in simpler language, such as that of Koelble's metaphor of the "glue that holds us together" (2003:176), mentioned earlier.

Social cowesion

South Africans will have to find their own words to describe what they hope to see in the world. Although we realize that it will probably not resolve the confusion attached to the notion of 'cohesion', the authors have considered proposing a new concept, namely social 'cowesion', to connote:

- the extent to which people unite and include others to constructively satisfy fundamental human needs and rights for everyone;
- a sense of unity and purpose to design and implement societies, systems and institutions that are just, fair and empowering for everyone;
- the quality of relationships between people who value, build and respond to constructive conflict; and
- a shared and integrated society that values 'we-ness instead of focusing on 'otherness' and that which divides us.

The wounds of the past

No existing definition of social cohesion will fully satisfy people who have been shaped by shared pain, "multiply woundedness" (Cabrera 2002:n.p.) forced separation, racism, exclusion, loss, internalised oppression, violence, injustice and inequality. Cabrera, a Nicaraguan psychologist, helped traumatised communities to draw up "inventories of woundedness", because

> [...] multiply wounded societies run the risk of becoming societies with inter-generational traumas [...] When people begin to talk about their history, assume it and reflect on it, a fundamental process takes place: people find meaning and significance in what they have lived through.

And professor Pumla Gobodo-Madikizela argues that

> [...] suggesting that people must 'forget the past' creates great conceptual mischief [...] Some people were victims, some perpetrators, and others bystanders or beneficiaries of apartheid privilege.
>
> [2010:n.p.]

Gobodo-Madikizela therefore calls for dialogue about the past "in a spirit of shared pain about the past [...] to remember it in order to transcend it."

What is clear is that we will have to use our creative imagination to enter into new and extraordinary experiences, which include dialogue and joint action, to discover the meaning of social cohesion for us here in Stellenbosch.

During a recent dialogue session on community healing, organised by the Institute for Healing of Memories, it became clear that one of the most severe wounds inflicted on all South Africans was the loss of 'we-ness' as a result of a system that assigned man-made identities to every citizen, causing us to fracture along artificial fault lines that subverted the preconditions for *ubuntu*.

Keeping two worlds alive

In Stellenbosch and South Africa today we are living in a post-conflict society. The conflict we experienced during the past more than three centuries was on the basis of race. Race – in the form of institutionalised racism called 'apartheid' after 1948 – determined the distribution privileges and life chances amongst South Africans.

White South Africans were the privileged with access to all the wealth and opportunities of the country, while other groups were generally excluded; and the descendants of past white generations are still benefitting from their inherited wealth. Yet, according to the Institute for Justice and Reconciliation (IJR) barometer, today race is no longer the biggest dividing factor; inequality is (IJR 2010). The minority black nouveau riche is fast accumulating wealth for themselves through the current government policy of black economic empowerment (BEE), while the majority of the population still lives in poverty and happens to be black.[1] South Africa is now officially the most unequal country in the world, and Stellenbosch must probably count as one of the most unequal towns in the country, thus making it quite possibly one of the most unequal places in the world.

The historical separate 'group areas', with their stark contrasts, are still intact and clearly visible. Absolute opulence and abject poverty, two different worlds along the fault lines of class and race, exist on opposite sides of the same road or railway line. For the most part, they exist separately from each other, but regularly collide in various ways.

At well as being a post-conflict society with an inherited inequality based on race, Stellenbosch and South Africa is also a post-Cold War society with a strong emphasis on individualism and self-interest. Our thinking about society or communities is informed by the dominant view of persons and society, or 'social ontology', to use Gould's term (1998). According to this view, individuals are rational isolated egos, driven by self-interest and the quest for ultimate freedom, which in this context is understood as the absence of external constraint (negative liberty). People exist independently from each other, related only in external ways, thus leaving their basic nature unchanged even in contact with other persons. Against this background, there is little room for community or social cohesion, yet it continues to dominate – and we allow it to dominate. Hence, we are reduced to a disconnected aggregation of atomistic individuals, each seeking their own interests, but in many cases secretly yearning and longing for 'community'. This atomism gives rise to destructive behaviours that render our communities fragmented and torn.

Eighteen years after liberation, no great strides had been made to increase contact between the groups that had been separated by apartheid. In fact, a recent IJR report affirmed that 50% of South Africans stated that they have no contact with persons of a different racial group on any given day (IJR 2010). This means that historical racial divisions are to a large extent perpetuated in the new South Africa and that pressures are building up. If these pressures are not addressed, they are bound to be explosively released at one time or another.

Among others, the following features are constitutive of present-day Stellenbosch:

- **Inequality:** In 2008, research showed that the richest 10% of households in South Africa earned almost forty times more than the poorest 50% and nearly one hundred and fifty times more than the poorest 10%. It further found that inequality is interrelated with other intractable social problems, such as poverty, unemployment, social exclusion and marginalisation (Centre for the Study of Violence and Reconciliation (CSVR) 2008). As stated earlier, Stellenbosch must be one of the most unequal towns in the country, as the following statement by Paul Hendler attests:

    > I have recently been involved in assisting the municipalities of George, Hermanus, Kimberley, Knysna, Paarl, Saldanha and Stellenbosch to draw up sustainable human settlement plans (HSPs), a statutory requirement. What struck me is the extent to which mainly poorer black residents of these places still live in segregated townships, many in the squalor of shanty structures, while the richer minority continue to live on the other side of the proverbial railway track, close to the centre of business and amenities.

    [Hendler 2010:9]

- **Unemployment:** By the end of the first decade of the new millennium, the unemployment rate in the Western Cape stood at 19.7% compared to more than 23% countrywide. In 2006 and 2010, unemployment in Stellenbosch was 12% and 13% respectively, with almost one in four (22%) of the African population being jobless (Statistics South Africa 2010; Stellenbosch Municipality 2011).

---

1   According to Van der Berg (2010), 39.6% of blacks, 15.5% of coloureds, 5.8% of Indians and only 0.9% of whites lived below the poverty line of R3 532 per capita per year in 2000.

- **Inadequate housing:** While there appears always to be room for more luxury estates in Stellenbosch, lower income housing projects are fifteen to twenty years behind schedule. In terms of access to housing, the development of affordable housing (below R500 000) is very rare and development trends are largely biased to the higher income sector, ranging from R550 000 to more than R1.65 million per house (Sustainability Institute n.d.). The estimated housing backlog in Stellenbosch is more than 20 000 and is projected to reach a total of 44 361 by 2030. This total includes 14 361 housing units in informal settlements and 30 000 in the already overcrowded formal ones (Stellenbosch Municipality 2009).
- **Unstable political leadership:** Stellenbosch has been unable to put a stable local government in place during the last five years. Across the board there is consensus that stable local government is crucial for sustainable local development. However, divisive party politics continue to destabilise the town. What is needed is leadership that will put Stellenbosch first and rise above divisive party politics in the interest of long-term, sustainable development for the town. This can only happen if the two main political parties adopt and internalise the notion of a 'loyal opposition'. If electoral defeat continues to translate into actions that undermine governance in order to discredit the party in power, then social cohesion and development will remain impossible.

Unequal societies lacking social cohesion are unsustainable

If growth is valued above equality and inequality is not tackled as a national priority, social cohesion will remain a chimera. Inequality undermines cohesion and stifles the potential for survival. As epidemiologists Richard Wilkinson and Kate Pickett put it:

> [...] the life-diminishing results of valuing growth above equality in rich societies can be seen all around us. Inequality causes shorter, unhealthier and unhappier lives; it increases the rate of teenage pregnancy, violence, obesity, imprisonment and addiction; it destroys relationships between individuals born in the same society but into different classes; and its function as a driver of consumption depletes the planet's resources."
>
> [Wilkinson & Pickett 2009:n.p.]

Reducing inequality is not just a slogan. It is absolutely essential for our survival and the well-being of every citizen.

There are at least two other compelling arguments why the status quo has to change. Firstly, from a governance and economic perspective, research confirms the vital link between the role of the state in building social cohesion and high-quality institutions that serve all citizens fairly and justly. However, ethnic divisions hinder the development of social cohesion, which is necessary to build good institutions; therefore, the "lowering of economic (and other) divisions has been, and remains, a vital task for countries wrestling with development" (Easterly 2006:12). Put simply, if people are divided, institutions are weak and the country suffers. Conversely, more social cohesion leads to better institutions, and better institutions in turn lead to higher growth. In Easterly's view, this is true regardless of how we measure institutions.

Easterly's research has found that two factors in particular pose serious challenges to cohesion building: initial inequality and linguistic fractionalisation. The chances that societies with lower initial inequality and more linguistic homogeneity will successfully build cohesion and strong institutions that lead to higher economic growth are far greater than those of societies such as Stellenbosch, with its high levels of initial inequality and linguistic fractionalisation. Language, and especially the lack of learning (and teaching) each other's language, further reinforces exclusion in Stellenbosch. It follows that investment in Stellenbosch will have to be much more than only economic in nature: it has to be investment in human relations and social capital in order to help us to overcome these challenges.

Secondly, from an emotional and psychological point of view, the inventory of 'wounds' (see Cabrera 2002) has for many citizens not shrunken, but rather grown longer. We are sitting on a time bomb in Stellenbosch, as indeed in South Africa as a whole. Many South Africans still do not grasp and appreciate the miracle of South Africa's peaceful revolution. With the new South Africa came expectations and promises of a better life for all South Africans. More than a decade down the line that better life has still not materialised for the majority of black South Africans, especially insofar as racial polarisation and white privilege are concerned. Moreover, the problem is now shifting towards intergroup inequality, or what is more commonly referred to as class-based inequalities.

Whereas the overall levels of poverty are on the decline, intergroup inequality has risen dramatically (Van der Berg 2010). A few 'black diamonds' (African business tycoons) seem to have benefited from the BEE policy, while the majority of the black population is still marginalised. Pressures are building because of this experienced inequality and are vented in the form of service delivery protests and social unrest. Disenchantment and disillusionment are setting in, with even the populist and supposedly pro-poor Zuma government coming increasingly under attack from various quarters, including trade unions, for failing to live up to its promises and not being attentive to the needs of the poor.

The contact between the haves and have-nots in South Africa (and Stellenbosch) is mainly utilitarian in nature, with the have-nots who work as gardeners, security guards and domestic servants for the haves and witnessing on a daily basis how the rich live.[2]

After economic inequality, politicians are the second most dividing factor (IJR 2010). We take pride in *ubuntu*, but don't see how it operates in practice. We are told that we are free, but find ourselves caught in the crossfire between criminals on the one hand, and on the other, squabbling politicians whose vocation to serve others appear to be increasingly lost in clouds of allegations of corruption and self enrichment. We are afraid, humiliated and hopeful at the same time. So who are we in Stellenbosch? Who is a *Stellenbosser* (a resident of Stellenbosch)? Who are outsiders and *inkommers* (literally 'incomers')? Who are we when we are disappointed, angry and proud at the same time? And what do we do with these conflicting emotions?

Dominique Moïsi argues convincingly that emotions matter. Governments that fail to respond constructively to people's emotions risk the ignition of chaos. Moïsi affirms that it is the task of governments "to study the emotions" and "diagnose the emotional state of the population"; to "capitalize on them if they are positive and to try to reverse or contain them if they are negative" (2009:29). He argues that the emotions of fear, humiliation and hope – always present in variable proportions – are shaping the geopolitics of the world. All three of these emotions are directly related to and reflect the level of confidence that people have, which in turn determines their ability to rebound after a crisis and to respond or adjust to challenging circumstances. Where humiliation and fear reign, hope cannot grow and prosper. In such a climate, sustainable development is not possible. And with no safety net in place, things are bound to break and fall apart:

> [...] *deliberate humiliation without hope is destructive, and too much fear, too much humiliation, and not enough hope constitute the most dangerous of all possible social combinations, the one that leads to the greatest instability and tension.*
>
> [Moïsi 2009:15]

On the other hand, emotions may be changed. Fear may give way to hope, says Moïsi. We need to put measures in place to foster hope, so that our confidence as a nation may grow. It is important to grasp that what we do to secure the future of the poorest in our society stands in direct relation to the stability and sustainability of our town and country.

Are there some tentative bases for an "us" in Stellenbosch?

As we have indicated earlier, the impact and consequences of having identities assigned to us by the draconian apartheid system will continue follow us like a shadow for a long time to come. According to Manfred Max-Neef (1991), identity is a fundamental human need. The frustration of our need for identity, freedom, participation, understanding, creation, leisure, security and acceptance causes poverties of those needs. The effects of denying us the freedom to define ourselves, especially our identity, and causing us to still live separately, unequally and deprived of healthy and free interaction across divisions, continue to haunt us to this today.

---

2   One of the authors experienced the case of an elderly lady in Langa (Cape Town) who started a township restaurant after seeing a restaurant bill of her employer (for whom she worked as a domestic servant) that amounted to more than what she earned in a month.

In a way it may be said that South Africans are struggling to come to grips with the question: who are we? Our country has changed (in some respects); our borders are not really borders anymore; many of our former heroes disappointed us; we have become roaming citizens of the world in one respect, yet in another we attack foreigners and call them *makwerekwere*.[3] What is it that binds 'us' (as South Africans) together? In order to be sustainable, that which binds us together must be something more than a person or an icon, such as our dear Madiba; he will someday die and not be with us anymore. What binds us together must be more than national symbols, such as our beautiful and colourful national flag, which was waved with tremendous pride during the 2010 FIFA World Cup month. The 'glue that holds us together' must be more profound and more enduring than personalities or symbols.

Who are 'we' and what binds 'us' together in Stellenbosch? And is there a sustainable basis for an 'us' in Stellenbosch? What is there that tells us that we belong together? These are difficult questions, given the division and lack of social cohesion in Stellenbosch and South Africa today. After the negotiated political settlement that brought an end to institutional apartheid, we were careful to emphasise difference and diversity and to ensure that every community has the right to freely celebrate its own heritage and culture – the motto on our national coat of arms reads 'Unity in Diversity' in the Khoisan language. What we have neglected to do, in retrospect, is to emphasise what binds us together as South Africans (and as residents of Stellenbosch).

The authors are of the opinion that there are some possible bases for an 'us' in Stellenbosch and that we have a good chance to negotiate some kind of 'us'.

Ties of place and blood: geographical, biological, historical and social bases

It is a not well-known fact that the founder of Stellenbosch, Governor Simon van der Stel, was a man of mixed blood (South African History Online n.d.). This simple historical fact was suppressed by successive white governments, because it was an uncomfortable truth for a regime that attached paramount importance to racial purity in its scheme of white racial supremacy. Governor Van der Stel was portrayed as a pure, heroic and larger than life Dutchman in the official school history books; his mixed ancestry was conveniently omitted. So, already in the veins of the founder of Stellenbosch flowed blood from different groups; an 'us' was already embodied in his person. Moreover, Van der Stel was by no means an exception; in fact, Hans Heese (1984) and Hermann Giliomee (2007) both give interesting accounts of the composition and nature of the population of the Cape Colony during the early colonial period.

Giliomee (2007) identifies the lack of eligible European women as one of the most important social factors that shaped the early history of the settlement at the Cape. According to Giliomee's research, in 1690 the ratio of men to women was 2,8:1 in the Stellenbosch district, which means that many settlers were unable to marry women of European descent. Between 1657 and 1687, approximately one-fifth of all marriages were contracted between settlers and former slave women, one of the children born of such a marriage was an ancestor of 19th century presidents Paul Kruger and MT Steyn. Numerous children of mixed blood were also born from illicit liaisons between settlers and slaves (and it was not uncommon at the time for a European master to father children with his slave women, whether by consent or not). Moreover, slave women often served as wet nurses and surrogate mothers to settler children.

Hence, what we have in Stellenbosch today are the descendants of people who were bound by **ties of place**. They lived in close proximity, even sharing living spaces in many instances. The lives and experiences of slaves and settlers were heavily intertwined and in some cases the two groups were even bound to each other by **ties of blood**, as in many other colonial settings. The amazing reality is that one might find people of different backgrounds and life experiences in present-day Stellenbosch who are in fact 'family' in the narrowest sense of the word.

In terms of the relationships between different groups in Stellenbosch in more recent times, before the introduction of the Group Areas Act, it is not hard to find personal accounts bearing witness to the **sense of community** that prevailed in 'mixed areas' (where people from different backgrounds and racial groups used to live together) in Stellenbosch. Former neighbours from different backgrounds and racial groups tell of the positive community spirit that pervaded their neighbourhoods (Biscombe 2006). They also talk with heartache

---

3   A pejorative term for black foreigners.

and yearning about how this spirit was destroyed and their neighbourhoods broken up by the proclamation of the Group Areas Act. For those who were forcibly removed to racially assigned other 'group areas', the emotional loss was aggravated by the accompanying loss of family property.

Furthermore, if one looks at the beautiful farmsteads and historical buildings of Stellenbosch and environs, the contribution of artisans of Eastern and slave origin is unmistakable. Cape cuisine and the Afrikaans language were likewise significantly influenced by the Eastern and slave heritage.

Hence, despite the structural violence of segregation and apartheid, there are close historical ties of place and blood among different people living in Stellenbosch and these ties may potentially serve as a basis for an 'us' in Stellenbosch. 'We' are different members of one 'us' in Stellenbosch and surroundings; in some cases we are related by blood. We do not argue that the task of building social cohesion in Stellenbosch is as simple as rediscovering some pre-existing 'us'. Given the historical destruction of Stellenbosch communities by segregation and apartheid, we must renegotiate who 'we' are and what it is that binds 'us' together. As Moïsi aptly affirms, "in the age of globalisation, the relationship with the Other has become more fundamental than ever" (2009:20).

Cosmopolitan justice: the legal, moral and philosophical bases

The ties of place and ties of blood may, however, exclude some, especially the African residents of Stellenbosch and others of all races and classes who have migrated from other parts of South Africa, as well as foreign nationals who have made Stellenbosch their home.

The work of Seyla Benhabib may help us to find a wider basis for social cohesion than ties of place and blood. Benhabib notes that since the adoption of the Universal Declaration of Human Rights in 1948, an international human rights regime has emerged. She describes this regime as "a set of interrelated and overlapping global and regional regimes that encompass human rights treaties as well as customary international law or international soft law" (2006:27).

In her first lecture, Benhabib draws on Kant's three levels of right and presents the duty of hospitality in terms of cosmopolitan right, not as a virtue of sociability, but as a right that belongs to all human beings by virtue of their status as potential participants in a world republic:

> Cosmopolitan norms of justice, whatever the condition of their legal origination, accrue to individuals as moral and legal persons in a worldwide civil society.
>
> [*Ibid.*:16]

This is an extremely important statement. Although nation states may be party to bringing into being international human rights regimes, it is important to note that cosmopolitan norms of justice accrue to individuals, not to states. The implication is that the manner in which nation states treat both the citizens and residents within their borders cannot be regarded as a private affair anymore (which serves as justification for the intervention of the North Atlantic Treaty Organisation (NATO) countries in Libya in 2011 amid the perpetration of gross human rights violations by the state organs of that country).

Benhabib rightly points to the applicability of the right to universal hospitality to citizens and residents alike. A nation state may decide to extend or to refuse to extend citizenship rights and state protection to strangers who reside within its boundaries. And in the case of citizens, it is conceivable that it might, for some or no reason, want at some stage to denaturalise sectors of it citizens on account of religion, race, ethnicity, language or culture (for example, South Africa under apartheid, Bosnia and Ruanda) and withdraw from those sectors citizenship rights and state protection. These are typical cases where cosmopolitan norms of justice must overrule the republican sovereign's democratic rule.

Benhabib's concept of cosmopolitan norms of justice is appealing, given the global age that we live in, where livelihoods are sought by all, both inside and across national boundaries. We find it attractive also because of its aptness for the diverse (multicultural, multifaith, etc.) South African society, as well as the fact that many foreign nationals (especially nationals of other African countries) flee their home countries because of war and famine to seek a better future in South Africa. Benhabib's concept of cosmopolitan norms of justice thus provides us with an important legal, moral and philosophical basis for social cohesion.

Sustainability: the existential basis and environmental, economic and social elements
Stellenbosch is known for its natural beauty, including the mountains, rivers, vineyards and oak trees – a proud heritage shared by all inhabitants of the town. The continued expansion of the town and increase in population put tremendous pressure on infrastructure that has not expanded at the same rate. It is exactly the beautiful natural resources of Stellenbosch that are under threat at present. Homeless people with nowhere else to go end up in squatter camps that are often located in environmentally sensitive areas, while the lifestyle of the affluent demands more and more resources (energy, water, imported foods) and contribute disproportionately high volumes of waste to the municipal landfill. The fight to save the natural resources of Stellenbosch and the prospect of a common sustainable future may serve as another potential basis for social cohesion by galvanising the inhabitants of the town and providing them with an opportunity to transcend racially linked or individualist thinking and to start thinking of an interconnected 'us' in Stellenbosch.

Ties of place and blood, cosmopolitan justice and sustainability are good starting points for our quest for social cohesion in Stellenbosch. We now turn to what may practically be done to advance it.

What to do?

A common vision and values
Earlier in this chapter, we asked the question "What binds us together?" and resolved that it should be something more profound and enduring than personalities or symbols. The authors are of the opinion that social cohesion may be built around a common vision and values that provide hope that the status quo can change. Ruben Alves (quoted in Boesak 2005:239) reminds us that "what drives us is not the belief in the possibility of a perfect society, but rather the belief in the non-necessity of this imperfect order". This implies visioning exercises that are inclusive of all the sectors of the Stellenbosch community.

Building a movement around a common vision and values
A real movement, which comprises conflict-sensitive development- and cohesion-minded people from various disciplines, must lead an authentic and organic (dialogue, problem solving and action) process that will inspire people to see a prototype of a preferred common future that responds constructively to ongoing – but hopefully diminishing – threats to the well-being and sustainable development of all who live in Stellenbosch. Otto Scharmer talks about "leading from the future as it emerges" and "exploring the future by doing and experimenting" (2007:203). This process must have the unconditional mandate of government, business, academia and civil society as a whole.

The authors are part of an emergent Stellenbosch social cohesion movement, the participants of which have links to eighty-three networks within the greater Stellenbosch area. At a visioning exercise in February 2010 at the Kayamandi Corridor, participants agreed that values are of supreme importance, placing special emphasis on connectedness, caring, compassion and *ubuntu*, as opposed to unfettered personal interest, greed, opulence and conspicuous consumption.

Once a firm vision and values are in place, inspirational leaders who live by them are required. A specific type of leadership is needed in this movement, one with the ability to connect the missions of various role players in such a way as to promote cooperation and dialogue in the interest of the town. Endorsement should be sought for the work of the movement, and it would be ideal if the Mayor and Rector of Stellenbosch University could act as its patrons.

An infrastructure for conflict transformation
A local infrastructure for conflict transformation (peace for short) and development has to be grown organically. An institutional mechanism with a clear mandate to serve as a safe space for generative dialogue and problem solving is needed as a counterbalance to the normal political spaces of contestation and competition. If the only spaces for interaction in a community are confrontational, party political spaces, the community is in trouble. There has to be safe spaces where leaders meet "to resolve the tensions and mistrust [...] and to build

(or rebuild) their capacity to work effectively together across all of the country's lines of ethnic and political division", because "democracy depends as much upon cooperation as upon competition" (Wolpe & McDonald 2008:138-139). Local peace building structures across the world have proven to be invaluable, proactive transformers of potentially destructive conflict into generative win-win solutions (Odendaal forthcoming).

Pruitt (2007:4) argues that generative dialogue shifts people from unconstructive attitudes and thinking towards constructive ones, as illustrated below.

Building a coalition – an alternative network of giving and receiving

From	To
• Seeing others as separate and different, defined by their roles, their positions on the issues, or their place in a hierarchy	• Seeing others as fellow human beings: 'we are in this together' and all have something important to contribute
• Seeing oneself as separate from the problem situation and looking to others to change in order to resolve it	• Seeing oneself as part of the system that sustains the situation and accepting responsibility for changing oneself
• Disconnecting relationships within stuck problem situations	• Creative relationships energised by mutually owned ideas for addressing problems
• Acceptance of dysfunctional societal structures and systems	• Commitment to promoting change toward healthy societal structures and systems

An infrastructure for peace should become an official cornerstone of governance and socio-economic processes, so that no one in the greater Stellenbosch area should have any doubt that this community is developing a habit of engaging in dialogue and problem solving as a first response to any challenging situation.

A social network as a safety net is the essence of community mobilisation for sustainable growth in relationships and development. What is needed is a civil society coalition that may put in place alternative networks of giving and receiving to augment traditional distribution networks (for example, the tax system). What is needed is not coercion, but cohesion – the more difficult route.

Building capabilities

Social cohesion initiatives will have to prioritise investment in building the human capital of those presently poor in order to reduce inequality. The inability of the current education system to prepare the youth for responsible citizenship and skilled work constitutes a violation of their fundamental human rights, because they are condemned to receive inferior income (if any) in a whirlpool of a series of poverties. Van der Berg (2010:3) points out that "reducing inequality substantially is currently unlikely without a massive increase in the human capital of those presently poor", but adds that "prospects in this regard are inauspicious".

New towns and corridors

Segregated residential areas – the physical legacy of separation – should be addressed through a spatial development framework for cohesion. New towns that reflect social cohesion thinking must be built and serious consideration should be given to establishing corridors between segregated residential areas. The municipality clearly has a crucial role to play in this respect.

Linking local and national efforts

Any effort from civil society to build cohesion must have linkages to efforts to foster cohesion at the national level. In 2009, the Human Sciences Research Council (HSRC) and the Department of Arts and Culture (DAC) held a Social Cohesion Colloquium around the theme of 'Building a Caring Nation'. The conference was meant to be the start of a national dialogue on social cohesion and offered a number of recommendations, including the following:

- the promotion of social cohesion should be integrated into the work of all government departments;
- the concept of *ubuntu* must be promoted among all South African citizens;
- all structures of civil society must be involved in the promotion of social cohesion; and
- more dialogue should be facilitated around issues that hinder social cohesion, such as xenophobia, racism and social inequities.

The call has been made at national level. Sadly, the DAC appears to have dropped the ball and (as of January 2011) the recommendations of the colloquium have not been implemented.[4] It is now civil society's patriotic duty to answer that call and make calls of their own as they partner with all stakeholders at various levels to heal society.

Imagine ...

Imagine a Stellenbosch where every year, thousands of citizens gather on Die Braak in a massive festive celebration of the successes of the local municipality, the university, the economic sector, and civil society networks. Citizens honour successive municipal councils with an award for twenty years of excellence, justice and development. The Stellenbosch Gross Wellness Index, developed by Stellenbosch University, shows that citizens are extremely satisfied with the levels of sustainable development, environmental protection, cultural preservation, and good governance achieved.

Citizens are proud of the 'new' heritage and legacy that add fame to Stellenbosch: the birthplace of the apartheid system has become the model of 'conflictability', the ability to transform conflict into healthy opportunities for growth and cohesion. The local Cohesion Forum, which has become an official local governance institution for problem solving and dialogue, has inspired a wave of other local peace building structures across South Africa and Africa. Stellenbosch is known as a place where people's default reaction to challenges is to talk to one another and then to act as one.

Unemployment is at its lowest levels. A predicted housing deficit and backlog of twenty-five years has been wiped out in less than ten. Local industries and economic actors led a turnaround strategy to reduce inequality, because they realised that it was in their interest to do so. Vibrant local civil society networks and forums – aided by competent, locally trained facilitators – form the backbone of constructive partnerships between citizens and the government and business sectors. Not a single contentious issue is left unaddressed. Conflict is seen as healthy; debates are vibrant; dialogue is of a high standard; and the ability to respond constructively to conflict is continuously developed.

No problem is able to divide its citizens, because everyone – yes, even the local politicians – has shifted from destructive competition to constructive collaboration. They understand the concept of 'we-ness', and talk about 'cowesion'. No one is left behind and everyone, including foreigners from across the borders, has a sense of belonging. Stellenbosch is now hailed as the prototype of nation-building. Language barriers broke down as people learned to speak one another's language. There is a culture of sharing and injustice and socio-economic inequality are seen as a toxic threat to Stellenbosch as a whole. Citizens are proud of the way they take responsibility for their own actions and how they defend a common human value system.

Those skeptics who would argue that the above is a fantasy should take the words of St Francis of Asissi to heart: "Start by doing what is necessary; then do what is possible; and suddenly you are doing the impossible." As we practice the art and skill of creating a new legacy, we find ourselves among ...

> *[... the] ordinary men and women of our country who are daily weaving a memory, beading a legacy, cutting a spoor, telling a story, and loading into these bowls of history a future for all our people driven by the humane vision of ubuntu, the deep-seated sentiment that makes us who we are.*
>
> [Mbeki 2004:n.p.]

---

4   The Department of Arts and Culture's Social Cohesion Directorate organised the much anticipated Social Cohesion Summit in July 2012.

REFERENCES

AHURI. 2005. Housing, housing assistance and social cohesion in Australia. [Accessed 10 October 2012] http://www.ahuri.edu.au/publications/projects/p50300

Biscombe H (ed). 2006. *In ons bloed*. Stellenbosch: Sun Press.

Boesak A. 2005. *The Tenderness of Conscience: African Renaissance and the Spirituality of Politics*. Stellenbosch: Sun Press.

Cabrera M. 2002. Living and surviving in a multiply wounded country. [Accessed 10 October 2012] http://www.envio.org.ni/articulo/1629

Club of Madrid. 2011. The Shared Societies Project. [Accessed 9 September 2012] http://www.clubmadrid.org/en/programa/the_shared_societies_project

Council of Europe. 2004. Revised strategy for Social Cohesion. [Accessed 9 September 2012; European Committee for Social Cohesion (CDCS)] http://www.coe.int/t/dg3/socialpolicies/socialcohesiondev/source/RevisedStrategy_en.pdf

CSVR (Centre for the Study of Violence and Reconciliation). 2008. *Adding Injury to Insult. How exclusion and inequality drive South Africa's problem of violence.* [Accessed 9 October 2012] http://www.csvr.org.za/index.php?option=com_content&view=article&id=2454:adding-injury-to-insult-how-exclusion-and-inequality-drive-south-africas-problem-of-violence

Easterly W. 2006. *Social Cohesion, Institutions, and Growth.* Working Paper No 94. Washington, DC: Centre for Global Development.

Fairbairn B. 2006. *Cohesion, Adhesion, and Identities in Co-operatives.* Centre for the Study of Cooperatives. Saskatoon, Canada: University of Saskatchewan.

Gade C. 2009. New perspectives on the *ubuntu* philosophy and the South African Truth and Reconciliation Commission. Paper delivered at the Beyond Reconciliation Conference. University of Cape Town, Cape Town: 2-6 December.

Giliomee H. 2007. *Nog altyd hier gewees: Die storie van 'n Stellenbosse gemeenskap.* Cape Town: Tafelberg Publishers.

Gobodo-Madikizela P. 2010. Working through a past of shared pain. *Sunday Times*, 14 November. [Accessed 9 September 2012] http://www.timeslive.co.za/sundaytimes/article760799.ece/Working-through-a-past-of-shared-pain

Gould C. 1998. *Rethinking democracy*. Cambridge: Cambridge University Press.

Heese HF. 1984. *Groep sonder grense: Die rol en status van die gemengde bevolking in die Kaap, 1652-1795.* Bellville: University of the Western Cape.

Hendler P. 2010. Housing policy: Let's talk about an alternative to what there is. *Cape Times: Insight*, 14 September:9.

iCoCo (Institute of Community Cohesion, University of Coventry). 2005. Community cohesion: Seven steps. A Practitioner's Toolkit. [Accessed 9 September 2012] http://resources.cohesioninstitute.org.uk/Publications/Documents/Document/Default.aspx?recordId=82

IJR (Institute of Justice and Reconciliation). 2010. Annual Report. [Accessed 9 September 2012] http://www.ijr.org.za/publications/pdfs/IJR%20AR%202010.pdf

Jenson J. 1998. Mapping Social Cohesion: The State of Canadian Research. CPRN Study F/03. [Accessed 9 September 2012] http://www.rwbsocialplanners.com.au/spt2006/Social%20Cohesion/Mapping%20social%20cohesion.pdf

Koelble T. 2003. Building a new nation: solidarity, democracy and nationhood in the age of circulatory capitalism. In: D Chidester, P Dexter & W James (eds). *What holds us together: Social cohesion in South Africa.* Cape Town: Human Sciences Research Council (HSRC) Press.

Mandela NR. 2011. *Nelson Mandela by Himself: The Authorised Book of Quotations.* Johannesburg: Pan Macmillan.

Max-Neef M. 1991. *Human scale development: Conception, application and further reflections.* London: Apex Press.

Mbeki T. 2004. Remarks of the President of South Africa, Thabo Mbeki, on the occasion of the celebration of National Heritage Day. Kimberley: 24 September. [Accessed 9 September 2012] http://www.dfa.gov.za/docs/speeches/2004/mbek0927.htm

Moïsi D. 2009. *The Geopolitics of emotion: How cultures of fear, humiliation and hope are reshaping the world.* London: Bodley Head.

Odendaal AA. Forthcoming publication on Local Peace Committees. Washington, DC: United States Institute of Peace (USIP).

Pruitt B. 2007. The generative change community: cases about the meaning of 'generative dialogic change processes'. *Reflections*, 8(2). Society for Organizational Learning (SOL). [Accessed 9 September 2012] http://connection.ebscohost.com/c/articles/25946929/generative-change-community-cases-about-meaning-of-generative-dialogic-change-processes-research

RSA DAC (Republic of South Africa. Department of Arts and Culture). 2010. Presentation on social cohesion to Portfolio Committee: May. [Accessed 9 September 2012] http://www.pmg.org.za/files/docs/1000526dac.ppt

Scharmer O. 2007. *Theory U: Leading from the future as it emerges.* Cambridge: Society for Organisational Learning.

South African History Online. n.d. Simon van der Stel, commander and governor of the Cape, is born. [Accessed 12 October 2012] http://www.sahistory.org.za/dated-event/simon-van-der-stel-commander-and-governor-cape-born

Statistics South Africa. 2010. *Labour Force Surveys: 2005-2010*. Pretoria: Statistics South Africa.

Stellenbosch Municipality. 2009. *Integrated development plan: May 2009*. Stellenbosch: Stellenbosch Municipality. [Accessed 11 October 2012] http://www.stellenbosch.gov.za/jsp/e-lib/list.jsp?catid=173

Stellenbosch Municipality. 2011. *Integrated Development Plan for the municipal area of Stellenbosch*. Stellenbosch: Stellenbosch Municipality.

Sustainability Institute. n.d. *Stellenbosch sustainable human settlement strategy*. [Accessed 12 October 2012] http://www.sustainabilityinstitute.net/newsdocs/document-downloads/cat_view/52-talks-presentations-press?start=10

Turzi M. 2008. Social cohesion in China: Lessons from the Latin American experience. *The Whitehead Journal of Diplomacy and International Relations*, 9(1):129-144. [Accessed 9 September 2012] http://www.blogs.shu.edu/diplomacy/files/archives/12%20Turzi.pdf

Tutu D. 1999. *No future without forgiveness*. New York: Double Day.

Van der Berg S. 2010. Current poverty and income distribution in the context of South African history. Stellenbosch Economic Working Papers: 22/10 October. Stellenbosch: Stellenbosch University.

Wilkinson R & Pickett K. 2009. *The spirit level: why greater equality makes societies stronger*. London: Penguin Books.

Wolpe H & McDonald S. 2008. Democracy and peace-building: rethinking the conventional wisdom. *The Round Table*, 97(394):137-145.

Chapter 20

BUSINESS
Transformation towards sustainability

ALAN BRENT, PIETER VAN HEYNINGEN & SUMETEE PAHWA-GAJJAR

Introduction

The fact that the conceptual divide between the Global North and Global South is not as clear in South Africa as in other parts of the world presents the South African business sector with unique challenges. For example, South African businesses enjoy first-world infrastructure, (reasonable) energy reliability and stable market structures; but at the same time, they face a number of developmental demands, such as a high degree of inequality in income, welfare and education levels. Nowhere is this more prevalent than in Stellenbosch, which may be one of the most unequal localities in South Africa. On the positive side, this very fact implies that there is enormous potential for transformation here. The business community has a vital role to play in unlocking this potential by providing leadership for the creation of a common vision for the future of Stellenbosch – hardly a trivial exercise in itself.

This chapter provides an overview of the Stellenbosch business sector and the dynamic relationship between the pressures and opportunities faced by this sector in a global context, before exploring ways in which the business sector may mobilise itself in order to foster the town's transformation towards sustainability.

Overview of the Stellenbosch economy

The local economy has been described elsewhere in this book, but its salient features may be summarised again here as follows (Stellenbosch Municipality 2008):

- economic activities within the municipal boundaries are diverse and often highly specialised, providing goods and services well beyond regional and even national boundaries;
- the total value of economic activities approaches R4 billion per year, at a growth rate of over 4%, representing a quarter of the District's value addition;
- sector-specific figures indicate that financial and business services are the main contributors to the economy, followed by manufacturing; collectively, these sectors contribute over half of the total economic value;
- trade and government services together make up another quarter of the economy (with a fairly even share);
- agriculture, transport and communication, community, social and personal services, and construction represent the remaining approximately 20% of the economy; and
- sectors experiencing strong growth include finance and business, trade, transport and communication, as well as community, social and personal services.

Stellenbosch Municipality's *Local Economic Development Strategy* (*ibid.*) also highlights the following key or niche sectors whose competitive advantages should be supported in order to stimulate local economic growth:

- services: education, financial, business, administrative and government;
- tourism: wine, food, wellness, historic homesteads, adventure and sport;

- agro-processing: wine, grapes, fruit and flowers;
- wood processing: wood, paper and furniture manufacturing; and
- construction: residential, office and commercial.

Business profile of the growing local economy

Although precise information about the profile of the local business community has yet to be established, it is clear that it is extremely diverse across the socio-economic spectrum. Recent initiatives, such as the Stellenbosch Entrepreneur & Enterprise Development (SEED)[1] organisation, have made significant strides in supporting and strengthening small, micro-, and medium-sized enterprises to enhance Broad-Based Black Economic Empowerment (BBBEE) in the municipality. However, the niche sectors remain dominated by large business conglomerates, such as Remgro, Venfin, Distell and Medi-Clinic Southern Africa, and powerful single role players, such as wine estates, property developers and tourism operators. Historically, these entities all benefited from the previous political dispensation and an unregulated (local) marketplace, which translated into high profits for a minority upper class. Smaller commercial interests (which are of increasing importance, especially in the manufacturing, services, and technology development sectors) are represented by groupings such as the Chamber of Commerce. Those businesses that have benefited most from being located in the Stellenbosch Municipality typically also capitalise on opportunities offered by the marketplace outside the regional and national borders. In consequence, they have also been the first to come to grips with the notion of sustainability and its implications for conducting business (Haywood, Brent, Trotter & Wise 2010).

Changes in the responsibilities and roles of the global business community

Since the formation of bodies such as the World Business Council for Sustainable Development (WBCSD) and the International Institute of Sustainable Development (IISD) in the early 1990s, the business world has seen a plethora of definitions for 'sustainable development' and 'sustainability'. Most, if not all, of these definitions are based on interpretations that ensure that the goals and needs of corporations themselves are not compromised (Ketola 2007). In this way, the concept of sustainable development has been misappropriated to offer a 'business-as-usual' management solution, as long as businesses "consider and report" (Haywood *et al*. 2010) the implications of their actions from economic, environmental and social perspectives. It appears that the implicit assumption underlying these definitions is that sustainability principles will ultimately lead to sustainable growth and development within society, regardless of whether business takes a proactive role in implementing them or not.

By providing the goods and services demanded by the public, businesses meet vital societal needs. However, in doing so – whether as a result of their resource use, the processes they apply, or the products they manufacture – they have frequently been viewed as major contributors to environmental degradation and at least partially guilty of compromising human rights and dignity through their activities (Haywood *et al*. 2010). The need for businesses to strike a balance between financial viability on the one hand, and environmental sensitivity and social responsiveness on the other, is therefore quite apparent. The escalating consumption of natural resources has advanced human development at a growing environmental cost. Furthermore, human development has not been consistent across the whole of society, resulting in increasing levels of poverty and inequality, which are currently addressed by means of political interventions such as the Millennium Development Goals (MDGs). Thus, the implicit assumptions underlying the (business) definitions of sustainable development appear invalid (Hamann 2006).

Present-day businesses face unexpected risks and considerable uncertainty pertaining to the societies and ecosystems upon which their activities depend. This has resulted in a call to move beyond the traditional perspective of the basic and most fundamental purpose of business, namely to continually increase the shareholder value of the company in a responsible and ethical manner, to one that includes the imperative to also continually improve goods and services for a growing population at affordable prices in environmentally

---

1 SEED was established in 2010 and is a partner organisation of the Greater Stellenbosch Development Trust (GSDT), which was established in 2002.

sustainable ways (World Business Council for Sustainable Development (WBCSD) 2006). This shift is being driven by numerous forces and factors, some of which are listed in Table 20.1.

Table 20.1 Forces and factors that drive the increasing responsibilities of businesses
[Adapted from Haywood et al. 2010]

Force / Factor	Comment
The lack of will and capacity within government to protect and provide social and environmental goods and services	The 2002 World Summit on Sustainable Development emphasised the requirement for private sector action to supersede that of the public sector to meet the global goals of sustainable development.
Rapidly deteriorating natural and social environments that are reaching critical thresholds beyond which it is not known how business operations will be forced to operate	Businesses now have to make decisions in the context of extreme scarcities of the natural resources (stocks and flows) upon which businesses depend, forcing them to reconsider the way they define themselves and operate within their environment.
	Businesses can no longer operate under the neoclassical economic model, where profitability is the ultimate goal, natural resource scarcities are not acknowledged or accounted for, and the environment is considered merely as the provider of inputs to production and sinks for wastes.
	They must also recognise the fundamental uncertainty of the emergent properties of complex social-ecological systems in which businesses are embedded: not only does it become impossible to determine the probabilities of potential outcomes, but also to predict the actual outcomes/impacts. This basic recognition requires an entirely different approach and suite of decision-making tools.
Increasing public awareness of environmental and social problems and demand for 'green' and socially responsible products and processes	Examples of increased awareness include attentiveness to the carbon footprint of purchased goods and social misconduct in the supply chain, such as the use of child labour.
Stricter environmental and social standards and controls imposed on business processes (internationally sanctioned and enforced)	The compliance of companies and their supply chains to the requirements of 'eco-labels' in order to obtain access to international markets is one example.
Availability of technologies to improve the effectiveness and quality of products at lower cost	All of the above forces imply both a greater demand for appropriate technologies to buffer business and a (potential) inability to supply such technologies to meet those demands.

The business community has traditionally been good at devising plans and strategies to deal with the associated risks of their operational environments, especially where the impacts and probabilities of these risks are of a localised nature. However, the characteristics of these risks, as shaped by some of the forces and factors listed in Table 20.1, are rapidly changing in terms of:

- frequency (for example, the increasing incidence of crop failure, decreasing rainfall events, increasing floods, increasing child labour in value chains, and others);
- magnitude (for example, substantially heavier downpours of rain, longer droughts, 'blood' minerals, and others);
- predictability (past trends and relationships can no longer be used to predict or make projections of the future and the complex (non-linear) and interrelated nature of social and ecological systems is increasingly being understood and shown to be impossible to model and predict with any degree of accuracy and confidence); and
- threshold effects (many threats or impacts are likely to cause systems to switch from one state to another as social and ecological systems are pushed towards critical thresholds (extreme scarcities and variability)).

As part of the planning and strategising process, such risks (and their concomitant opportunities) must be converted into corporate action. The immediate task for business in dealing with the characteristics of risk is to understand how to adapt to and mitigate present-day social-ecological challenges in order to expand their business activities and achieve economic growth (Haywood et al. 2010). The complexity of these challenges threaten the very existence of businesses, especially in terms of the fundamental uncertainty arising from global change and unpredictability, and necessitates a fundamental shift in the way they conduct business and understand and implement sustainability (Azapagic 2003). Instead of reducing those practices that are perceived to be unsustainable, businesses should rather be working towards strengthening the sustainability of the systemic underpinnings of the economy and society (Ehrenfeld 2005). Large corporations and business groupings are starting to understand that, in order to evaluate the risks to their business operations posed by the changing global climate, a more systematic approach to sustainability is required – an approach that entails understanding the social-ecological systems in which business activities take place (Haywood *et al.* 2010).

Innovation as a means to collective transformation

Stellenbosch represents a rare (if not unique) micro-environment, one that is ideally suited to innovation – especially because the communities within the municipality are supported by local or nearby learning and research institutions. New knowledge and information flows are of utmost importance for the rapid transformation of micro-environments toward sustainability (Van Heyningen & Brent 2011). Business communities are usually most able to make rapid transitions, since they are well-versed in organisational mechanisms that are designed to operate in changing economic climates. In fact, according to Kipp (2001), innovation and adaptability are the primary coping mechanisms and strategies of fast moving firms. However, to do this they need new knowledge (and technology) at increasing rates of learning and dissemination within the respective firms (Argote & Ingram 2000).

These micro-environments normally adopt changes more readily, because they can be protected and artificially isolated from the macro-economic dynamics at various levels until they become more robust (Van Heyningen & Brent 2011). The level of isolation depends on numerous factors, such as legislation and political will, as well as financial independence. Hence, the ability to innovate, learn and adopt new technologies is largely dependent on numerous external factors associated with the characteristics of the micro-environment within which it operates and are often beyond the control of the firm. For this reason, a collective and inclusive sustainability-oriented innovation system approach, which allows, supports, or enables rapid transitions at the micro-environment level, is necessary (Van Heyningen & Brent 2011). This means that government and society must provide support and enable environments within which organisations, clusters and firms are able to change.

Firms often perceive transformation towards sustainability as adopting new technologies, which are erroneously viewed as synonymous with innovation. However, this addresses neither the organisational functioning of the firm, nor its core business strategy. It is therefore imperative that a more comprehensive understanding of innovation be relayed to the firms and organisations operating within these micro-environments. Businesses are becoming increasingly aware of their obligation to make a difference in society (Haywood *et al.* 2010). The sustainability-oriented innovation system approach (Stamm, Dantas, Fischer, Ganguly & Rennkamp 2009) specifically addresses the question of how business functioning and regional development may be made inclusive of individuals who find themselves at the margins of the formal economy.

Transformation towards sustainability

It has been argued that an innovation system approach, which demands a change from unsustainable systems toward more sustainable ones, is required to achieve a transition towards sustainability (Stamm *et al.* 2009). However, transitions also require direction, and the use of strategic visioning for regional development is gaining prominence (Holbrook & Wolfe 2005). Transitions toward sustainability require that numerous factors work together to weave a complex web of change. The weight and importance of each of these factors and components are varied and very much dependent on their immediate environment or landscape. This is the reason why the innovation system approach is useful in developing appropriate policy and institutional structures to support rapid change through innovation.

Innovation networks within the micro-environment enable transitions towards sustainability through socio-technical and socio-cognitive innovation processes (Geels & Schot 2007). This method represents a greater commitment to the social elements necessary within the sustainability-oriented innovation system. This kind of transformation should place greater emphasis on the role of the firm or the organisation – precisely because the benefits of innovation and learning for sustainability translate into greater competitiveness for the firm or organisation, and eventually also the region.

Stellenbosch as a micro-innovation environment for sustainability: The case of Technopark

Collaboration between businesses has been proven to enhance the innovative culture within micro-environments. Technopark is one example in the greater Stellenbosch area where the potential to become an innovation environment for sustainability may be realised.

Technopark was originally conceived as a science park in the late 1970s, when Prof Christo Viljoen, then vice-dean of the Engineering Faculty at Stellenbosch University, went to Taiwan. There he visited Shinshu Science Park, one of the world's revered successes. He brought the concept over to Stellenbosch and managed to obtain buy-in from local and national government. The park was established in 1985, with the financial support of the Industrial Development Corporation (IDC). It housed an incubator and innovation lab and was jointly managed by a committee and the local municipality. The entrance of firms was very slow for numerous reasons, including a poor understanding of the concept and benefits of a science park within the community; in consequence, the management was forced to allow it to become market-oriented and to relax the strict entrance criteria.

Today the park is well developed (albeit not as a traditional science park) and accommodates a variety of businesses, ranging from satellite manufacturers, banks and financial institutions to the headquarters of a liquor company. However, there are also numerous firms that fit the profile of traditional science park businesses; more importantly, they fit the profile of the modern concept of 'innovation hubs', including a number of information and communication technology (ICT) businesses, engineering firms, financing firms, and others. The more modern concept of 'innovation hubs' or 'innovation environments' allows for mixed-use spaces, living spaces, and interactive and attractive spaces for citizens (Yigitcanlar, Velibeyoglu & Martinez-Fernandez 2008). What all these 'innovation environments' have in common is a strong link to knowledge centres or universities, which provides incentives for collaboration and mutual learning (Youtie & Shapira 2008). The concept also makes provision for looser boundaries, which attracts a variety of visitors – and thus also new knowledge and diversity – to the 'innovation hub'. The vibrancy of modern 'science cities' are known to stimulate innovation and mutual learning in all sectors. This vibrancy is often created through incorporating art and cultural elements into the environment; examples are Newcastle Science City in north-east England and the Gold Coast Pacific Innovation Corridor in Queensland, Australia (Couchman, McLoughlin & Charles 2008). Universities benefit because it allows them to align certain research projects with real-world ventures, while firms benefit because they receive a flow of new knowledge, information and expertise, which increases their competitiveness. And all parties benefit from various forms of joint training and education programmes or internships for students.

Technopark is represented by the Technopark Owners Association (TPOA), which was established as a representative body to resolve some of the service delivery issues within the park. A preliminary study by Van Heyningen & Brent (2011), including interviews and contact sessions with the Technopark community, revealed that the mandate of the park was still in a state of flux, which led to the establishment of the Sustainable Innovation Stellenbosch Network (SISN). This initiative aims to engage business owners, directors, and managers to form an interest group or forum around the issue of a future mandate of Technopark. More specifically, it is designed to test the viability of business interests by creating a network for collaboration and innovation for sustainability. The project will require collaboration with Stellenbosch University and other institutions of learning and research and the local and regional governments, as well as inter-firm collaborations. Both the successes and failures of the SISN initiative will provide valuable insights into the business culture of Stellenbosch, which will allow for the identification of obstacles and opportunities in order to formulate collective strategies for change toward more sustainable trajectories in the business community.

The potential to change at the organisational level: the case of Spier

Spier's ongoing journey towards sustainability is characterised by three distinct phases. The first phase represents the pioneering years, from the purchase of the estate in 1993 to about 2002. This exciting phase of its evolution laid the foundation for future endeavours and created the 'Spier' brand, which is now well-established both in South Africa and abroad. It also included increased efforts to achieve commercial success and organisational stability within the wine and leisure sector from 2002 to mid-2003. The second phase began when Adrian Enthoven, a shareholder, entered the business and appointed a dedicated Director for Sustainable Development. Under the Director's leadership, various disparate efforts under the banner of sustainable development were consolidated, as was the reporting thereof in an annual sustainability report.

Alongside the organisational processes aimed at embedding sustainability into the core of the business, are three key ongoing endeavours aimed at reducing the estate's ecological footprint: the search for effective renewable energy solutions for its existing infrastructure, experiments with waste water treatment, and solid waste recycling.

The adoption of 10-year macro-goals, such as carbon neutrality and poverty alleviation, signifies the current phase in the evolution of the business, where sustainability is meant to guide each and every business decision. With the aim of meeting its macro-goals, Spier maintains a healthy portfolio of ecologically sensitive activities, such as regular river cleansing, clearing alien vegetation, maintaining a compost site, wetland conservation, and biodynamic and organic farming. As of the end of the 2010 financial year, Spier management has been focussing on tracking progress on a predefined sustainability trajectory through consolidated environmental reporting and footprint analyses.

The current focus on using the Global Reporting Initiative (GRI) format stands in stark contrast to the ecological concepts that underpinned Spier's strategy in the development framework of 1999. The vision more than a decade ago was to create a sustainable 'micro-ecology' in the Cape Winelands through a combination of attention to heritage and culture, wealth generation, meeting social and economic needs, and providing infrastructure to encourage new community lifestyles compatible with an equitable vision for South Africa. At present, Spier management aims to achieve its sustainability objective through the identification of innovation opportunities at the individual employee level, exemplified by, among others, lighter weight bottles, recyclable wine cases, humidifiers and solar bulbs in the cellar, the harvesting of rain water from cellar buildings, and the installation of water meters and water saving devices. Senior management is now tasked with devising an effective strategy to further increase levels of organisation-wide innovation and enable vision-led investments in sustainability against a background of ongoing economic ups and downs.

Spier's continual striving for a sustainable future provides key lessons to aid other business in transforming towards sustainability. Firstly, business managers and shareholders must understand that their organisations are complex systems made up of sub-systems, which may exist in multiple stable regimes at any given time. Each regime is a manifestation of management or shareholder response to shocks and pressures, thus exhibiting system adaptability. The norms, beliefs and value systems of agent-actors, including senior managers and intermediaries, determine the human response and, therefore, system adaptability. Social learning and changing values towards sustainability-oriented behaviour thus become critical in enabling the adoption of resilience-based strategies, policies and implementation tools. Collaborative sustainability-oriented innovation is paramount to the success of this process. For example, the Spier Group interacts with the Sustainability Institute and the Centre for Renewable and Sustainable Energy Studies (CRSES) of Stellenbosch University, both of which contain experts with distinct mental models and knowledge areas, to provide a platform for social learning within the state and the business.

Social-ecological systems as a framework for developing sustainability approaches for business management

Approaches to sustainability and resource management have traditionally been based on a presumed ability to predict probabilistic responses to external drivers (Walker, Carpenter, Anderies, Abel, Cumming, Janssen, Lebel, Norberg, Peterson & Pritchard 2002). However, as such predictions and projections have become less viable due to the complexity of external global drivers, these methods for analysing risk have been overwhelmed and the goal of sustainable development has come to be seen as non-achievable (Fiksel 2006). One of the main reasons why efforts at improving sustainability are failing is that scientists, decision makers and implementers are attempting to find solutions from within the same mental paradigm, using the same tools and adopting the same worldviews that threaten sustainability in the first place (Du Plessis 2008). Businesses must therefore change the paradigm within which they operate from a mechanistic worldview (as embodied in neoclassical economics theory) to a systemic worldview (Haywood et al. 2010).

Responding to the challenges of sustainability requires insight into the characteristics of sustainable systems and a fundamental rethinking of the ways in which all business activities are designed, built, operated and evaluated (Bakshi & Fiksel 2003). Sustainability is not an end in itself, but rather one of the characteristics of a dynamic, evolving system (Fiksel 2006). Businesses must understand that they are embedded in the cyclical processes of the social-ecological system in which they operate (see Figure 20.1) and are dependent on this system for the resources they require and the surety of supply, as well as the wastes it absorbs. They must stop thinking of themselves as separate from the wider ecological context and stop competing with the social-ecological systems in which they operate. Instead, businesses should accept that they are part of these systems and therefore co-evolving within them.

If businesses recognise the dynamic interactions between nature and society and operate as part of their social-ecological systems, they will better appreciate how dependent they are upon those systems and be better equipped to address the particular risk and uncertainty elements pertaining to them in order to establish resilience perspectives and adaption mechanisms – which in turn will bring them closer to achieving long-term sustainability (Haywood et al. 2010). However, dealing with the complexity of social-ecological systems will require fundamental changes in decision-making processes.

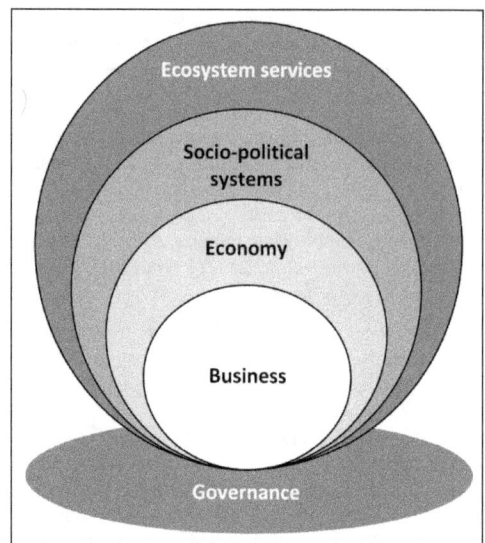

Figure 20.1 Businesses are embedded sub-systems of larger social-ecological systems [Adapted from the Department of Environmental Affairs (RSA DEA) 2008]

Resilience and adaptibility

The key principles of social-ecological systems are that they are complex and adaptive, with properties of self-organisation and emergence (Walker et al. 2002). There is a shift underway in the manner in which systems are considered: from prediction and control to understanding the resilience of a given system in order to provide a basis for adaptive systems management (Korhonen & Seager 2008).

The concept of resilience has emerged as an essential characteristic of complex systems. Social-ecological system resilience is defined as the capacity of a system to absorb disturbance and adapt to change in a manner that allows it to retain the same function, structure and identity (Walker, Gunderson, Kinzig, Folke, Carpenter & Schultz 2006). More specifically in a business context, resilience is the capacity of a business to survive, adapt and grow in the face of turbulent change. Business management can destroy or build resilience, depending on how the social-ecological system organises itself in response to management actions. Faced with a dynamic and unpredictable business environment, management theorists are increasingly identifying the need for resilience (Hamel & Välikangas 2003).

Resilience is therefore the potential of a social-ecological system (including business) to retain a particular configuration and maintain its feedbacks and functions, as well as its capacity to reorganise itself following disturbance-driven change (Walker et al. 2006). Resilient social-ecological systems are able to cope, adapt, or reorganise themselves without sacrificing the provision of ecosystem services.

Adaptive capacity, which may lead to a new equilibrium, is an important element of resilience. It comprises the learning, flexibility, problem-solving capacity and knowledge storage (Folke, Carpenter, Elmqvist, Gunderson, Holling & Walker 2002) that enable resilient business systems to grow in the face of uncertainty and unforeseen disruptions. The concept of adaptability is closely related to that of adaptive capacity and describes an organisation's ability to change its practices, resource allocations, designs, relationships, or other aspects of the business in response to changing conditions (Korhonen & Seager 2008). Adaptability strategies for a business operation may involve new technology or re-engineering but on a resilience trajectory, so that the basic social-ecological system remains recognisable as it comprises all the elements of the original, acceptable ecological state. Adaptability also implies the ability of the businesses in the social-ecological system to learn and by so doing enable decision makers within the organisation to better understand the interaction between the organisation, its operations, its products or services, and the social-ecological system.

In short, managing risk is about understanding the system in which one operates, understanding the resilience of that system, and taking appropriate action to adapt to any changes in order to ensure continued survival and economic viability within the system. The key element in managing risk in relation to social-ecological system resilience is understanding where resilience resides in the system, and when and how it may be lost or gained.

Conclusion

There is certainly no magic (or 'green' for that matter) formula for transforming economies or regions toward sustainable trajectories. It is something that takes time and requires changes in socio-cognitive behaviour through learning and building competence through a network of societal actors, as well as various support mechanisms and concentrations of efforts. While democratic processes may be just, they are also slow. This in itself may be seen as a hindrance to more rapid transformations. Community mindsets are often resistant to change if there is no immediate benefit or visible threat. In Stellenbosch, some citizens are keen for change and renewal; others are content with the *status quo* – especially in the business community, where there is limited evidence of change.

Change ultimately occurs through collective effort and happens most rapidly when societies are faced with dire challenges or shocks. Although it may not yet be apparent, a continuation of unsustainable business practices and resource consumption in Stellenbosch (and elsewhere) will inevitably result in such challenges or shocks. The business community and corporate citizens therefore have a responsibility to accommodate potential future crises in their operations and models in order to avert them. This will require substantial ongoing marketing, collaboration and lobbying within the larger society that constitutes Stellenbosch. To this end, many envisage Stellenbosch as a golden opportunity for crafting, through a collective effort, a sustainability-oriented innovation strategy that may drive transformation towards a sustainable trajectory. The adaptive nature of business means that it has a crucial role to play in crafting such a strategy and, in so doing, to capitalise on emerging markets created by, for example, the green economy, while uplifting Stellenbosch society as a whole.

REFERENCES

Argote L & Ingram P. 2000. Knowledge transfer: A basis for competitive advantage in firms. *Organizational Behavior and Human Decision Processes*, 82(1):150-169.

Azapagic A. 2003. Systems approach to corporate sustainability. A general management framework. *TransIChemE*, 81(B):303-316.

Bakshi BR & Fiksel J. 2003. The quest for sustainability: challenges for process systems engineering. *AIChE Journal*, 49(6):1350-1358.

Couchman PK, McLoughlin I & Charles DR. 2008. Lost in translation? Building science and innovation city strategies in Australia and the UK. *Innovation: Management, Policy & Practice*, 10(2-3):211-223.

Du Plessis C. 2008. A conceptual framework for understanding social-ecological systems. In: M Burns & A Weaver (eds). *Exploring Sustainability Science: A Southern African Perspective*. Stellenbosch: Sun Press.

Ehrenfeld J. 2005. The roots of sustainability. *Sloan Management Review*, 46(2):23-25.

Fiksel J. 2006. Sustainability and resilience: Towards a systems approach. *Sustainability: Science, Practice and Policy*, 2(2):14-21.

Folke C, Carpenter S, Elmqvist T, Gunderson L, Holling CS & Walker B. 2002. Resilience and sustainable development: Building adaptive capacity in a world of transformations. *Ambio*, 31(5):437-440.

Geels FW & Schot J. 2007. Typology of sociotechnical transition pathways. *Research Policy*, 36(3):399-417.

Hamann R. 2006. Can business make decisive contributions to development? Towards a research agenda on corporate citizenship and beyond. *Development Southern Africa*, 23(2):1-21.

Hamel G & Välikangas L. 2003. The quest for resilience. *Harvard Business Review*, 81(9):52-63.

Haywood LK, Brent AC, Trotter DH & Wise R. 2010. Corporate sustainability: A social-ecological research agenda for South African business. *Journal of Contemporary Management*, 7:326-346.

Holbrook JA & Wolfe DA. 2005. The innovation systems research network: A Canadian experiment in knowledge management. *Science and Public Policy*, 32(2):109-118.

IISD (International Institute for Sustainable Development). 1992. *Business strategies for sustainable development: Leadership and accountability for the '90s*. Winnipeg: IISD.

Ketola T. 2007. Ten years later: Where is our common future now? *Business Strategy and Environment*, 16: 171-189.

Kipp M. 2001. Mapping the business innovation process. *Strategy & Leadership*, 29(4):37-39.

Korhonen J & Seager TP. 2008. Beyond eco-efficiency: A resilience perspective. *Business Strategy and the Environment*, 17:411-419.

RSA DEA (Republic of South Africa. Department of Environmental Affairs). 2008. *The National Framework For Sustainable Development*. [Accessed 13 April 2012] http://www.environment.gov.za/?q=content/documents/strategic_docs/national_framework_sustainable_development

Stamm A, Dantas E, Fischer D, Ganguly S & Rennkamp B. 2009. *Sustainability-oriented innovation systems: Towards decoupling economic growth from environmental pressures?* Bonn: German Development Institute (DIE).

Stellenbosch Municipality. 2008. *Local Economic Development Strategy*. 3rd Draft. Stellenbosch: Stellenbosch Municipality.

Van Heyningen P & Brent AC. 2011. Towards developing a theoretical paradigm for sustainability-oriented innovation systems. *Proceedings of the 17th annual International Sustainable Development Research Conference*. New York: The Earth Institute, Columbia University.

Walker B, Carpenter S, Anderies J, Abel N, Cumming G, Janssen M, Lebel L, Norberg J, Peterson GD & Pritchard R. 2002. Resilience management in social-ecological systems: A working hypothesis for a participatory approach. *Conservation Ecology*, 6(1):14.

Walker B, Gunderson L, Kinzig A, Folke C, Carpenter S & Schultz L. 2006. A handful of heuristics and some propositions for understanding resilience in social-ecological systems. *Ecology and Society*, 11(1):13.

WBCSD (World Business Council for Sustainable Development). 2006. *From Challenge to Opportunity. The role of business in tomorrow's society*. Geneva: WBCSD.

Yigitcanlar T, Velibeyoglu K & Martinez-Fernandez C. 2008. Rising knowledge cities: The role of urban knowledge precincts. *Journal of Knowledge Management*, 12(5):8-20.

Youtie J & Shapira P. 2008. Building an innovation hub: A case study of the transformation of university roles in regional technological and economic development. *Research Policy*, 37(8):1188-1204.

Chapter 21

HEALTH AND WELLNESS
The burden of disease in Stellenbosch

BOB MASH

Introduction

This chapter sets out to describe the burden of disease in the Stellenbosch area[1] and explore the underlying risk factors that erode the potential for health and wellness, before considering possible approaches to promote improved health and wellness during the next decade and beyond.

Health services

In terms of the public sector, the Stellenbosch sub-district is served by the Stellenbosch District Hospital and three satellite Community Day Centres (CDCs) in Kayamandi, Franschhoek/Groendal and Cloetesville. The Kayamandi CDC is fed by five smaller clinics; the Franschhoek/Groendal CDC by two mobile clinics; and the Cloetesville CDC by three smaller clinics and four mobile ones. The clinics and day centres are operated by nurses with support from visiting doctors. Although the district hospital has various medical officers (and more recently a specialist family physician), their numbers limit the support that the hospital is able to extend to the day centres and clinics.

As for the private sector, there is one private hospital in Stellenbosch and a large network of general practitioners in private practice. Stellenbosch University also runs its own student health services on campus.

In addition to these services, there are a number of agencies in the non-government sector that are active in rendering support and counselling people (and their families) who struggle with alcohol and substance abuse. Likewise, there are several HIV-related organisations that promote prevention of HIV/AIDS and provide counselling, testing, and support for HIV/AIDS patients. Yet, other projects focus on the care of street children and feeding schemes for children from poor communities.

The burden of disease

Table 21.1 lists the top ten causes of premature death in Stellenbosch.

---

1 Stellenbosch is a sub-district of the Winelands District of the Western Cape. Much of the available information relating to health is at the district and not sub-district level.

Table 21.1   Top ten causes of premature death in Stellenbosch [Groenewald, Bradshaw, Krige, Van Niekerk, Champion, Karstens, Verwey & Van der Merwe n.d.]

Rank	Cause	Percentage
1	HIV/AIDS	13.5
2	Homicide	11.1
3	Road traffic accidents	9.3
4	Tuberculosis	7.3
5	Stroke	4.7
6	Pneumonia	3.9
7	Diarrhoea	3.8
8	Ischaemic heart disease	3.8
9	Diabetes mellitus	3.4
10	Drowning	3.2

The picture presented in Table 21.1 is typical of most communities in South Africa. HIV/AIDS (in combination with tuberculosis) is by far the leading cause of premature death, followed by homicide (that is, the result of interpersonal violence) and road traffic accidents – hence injuries and violence together constitute the second most significant burden of 'disease'. Interestingly for Stellenbosch, drowning (presumably related to swimming in farm dams) is also a major cause of premature death. The third significant disease burden is related to maternal and child health, as may be seen from the high incidence of mortality resulting from pneumonia and diarrhoea. Non-communicable chronic disease constitutes the fourth major cause of death, which is presented here in the number of deaths resulting from stroke, ischaemic heart disease and diabetes.[2]

Figures 21.1a and 21.1b show the distribution of deaths in the Winelands District by age and gender, as well as the contribution of the four main causes of death for each age category.

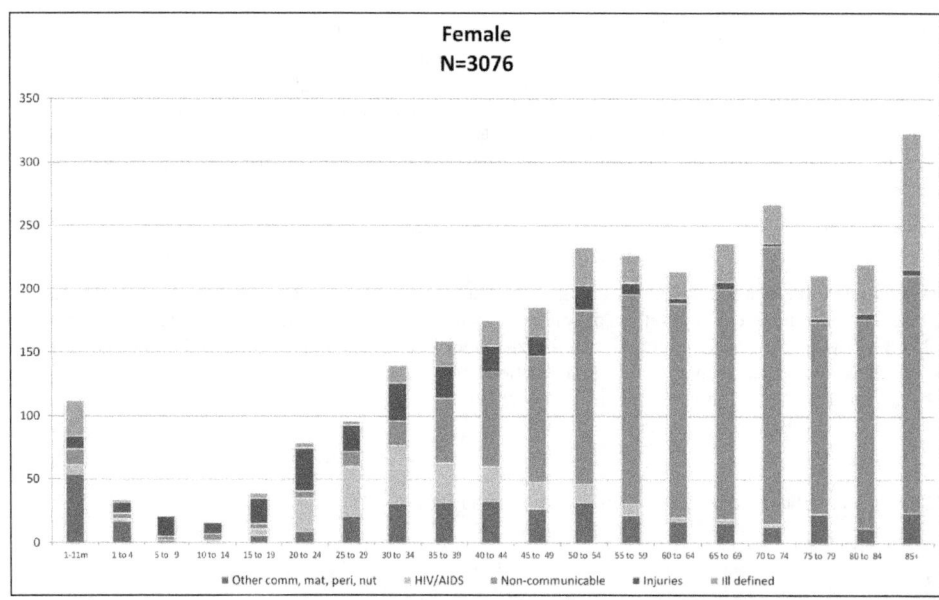

Figure 21.1a   Age distribution of female deaths in the Winelands District (comm = communicable diseases; mat = maternal causes; peri = perinatal conditions; nut = nutritional deficiencies) [Groenewald et al. n.d.]

---

2   For the Winelands District as a whole, chronic obstructive pulmonary disease (COPD) and low birth weight with respiratory distress syndrome also make the list of the ten most significant causes of premature death.

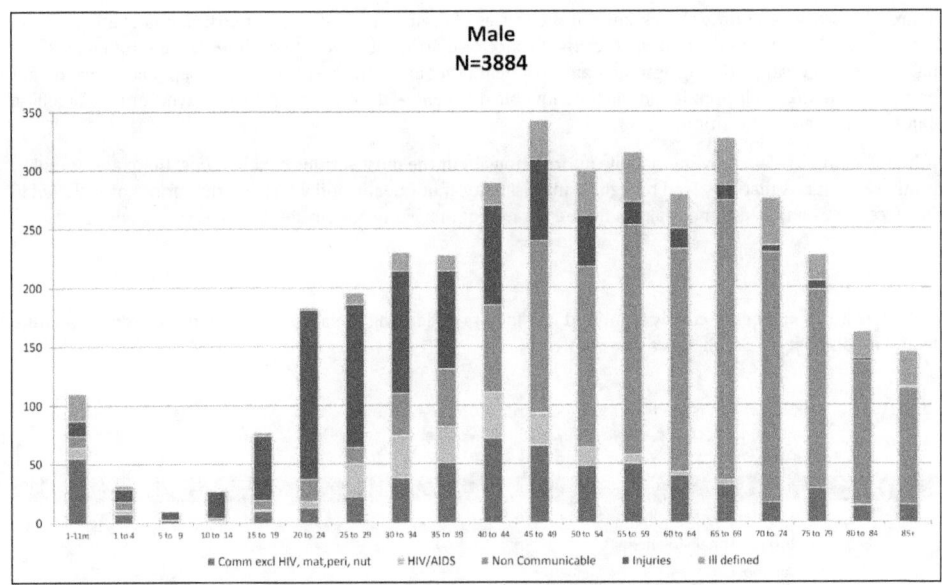

Figure 21.1b   Age distribution of male deaths in the Winelands District [Groenewald et al. n.d.]

It is clear that the infant mortality is substantial among both males and females. During preschool and school age years, the mortality rates drop, but subsequently rise again in the late teens and early twenties. A comparison of Figures 21.1a and 21.1b shows that young women suffer a disproportionate number of deaths from HIV/AIDS,[3] while the number of deaths attributable to injuries and violence is higher among young men. Among both groups, deaths from non-communicable chronic disease start increasing from middle age and dominate the picture among the elderly.

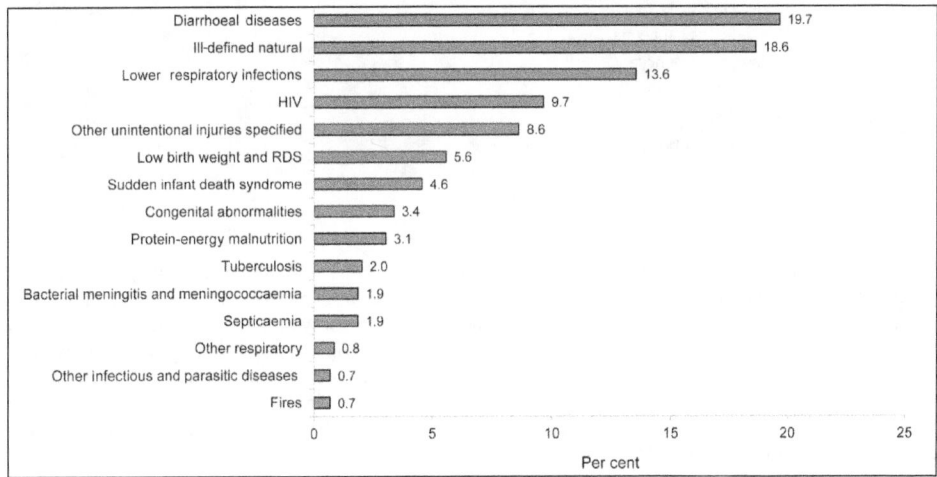

Figure 21.2   Causes of post-neonatal infant deaths in the Cape Winelands and Overberg, 2004-2006 [Groenewald et al. n.d.]

---

3   It should be noted that HIV also increases vulnerability to infections such as tuberculosis, gastroenteritis (diarrhoea), and respiratory tract infections (pneumonia and tuberculosis).

Figure 21.2 provides a further breakdown of the causes of premature death among post-neonatal infants (from 1 month to 12 months) and confirms the significance of diarrhoea, pneumonia, HIV/AIDS and tuberculosis as major causes of death in this age group. Causes of death that are exclusive to this age group (and hence do not feature among the adult population) include low birth weight with respiratory distress syndrome, congenital abnormalities and malnutrition.

Although mortality data provide valuable information about the most serious health issues, there are a number of other conditions that may lead to significant disability, if not death, and which are therefore not included in this data; for example, asthma is also a frequent cause of morbidity among children and young adults.

Risk factors

Table 21.2 lists a number of clearly identified risk factors underlying the abovementioned causes of premature death and disability in South Africa as a whole.

Table 21.2   Contribution of risk factors to South African disability adjusted life years (DALYs) [Norman, Bradshaw, Schneider, Joubert, Groenewald, Lewin, Steyn, Vos, Laubscher, Nannan, Nojilana, Pieterse & the South African Comparative Risk Assessment Collaborating Group 2007b]

Rank	Risk factor	% DALYs
1	Unsafe sex / sexually transmitted infections	31.5
2	Interpersonal violence	8.4
3	Alcohol harm	7.0
4	Tobacco smoking	4.0
5	High body mass index (excess bodyweight)	2.9
6	Childhood and maternal underweight	2.7
7	Unsafe water sanitation and hygiene	2.6
8	High blood pressure	2.4
9	Diabetes	1.6
10	High cholesterol	1.4
11	Low fruit and vegetable intake	1.1
12	Physical inactivity	1.1
13	Iron deficiency anaemia	1.1
14	Vitamin A deficiency	0.7
15	Indoor air pollution	0.4
16	Lead exposure	0.4
17	Urban air pollution	0.3

Unsafe sex

Unsafe sex is undoubtedly the immediate cause of the majority of cases of HIV/AIDS, the transmission rate of which is increased by other sexually transmitted infections. Sex becomes more unsafe in the case of multiple partners, especially if they are concurrent, when teenage girls have sex with older men, and when coercion and violence are involved. This risk factor is most significant among teenagers and young adults, with women particularly at risk. In 2008, the prevalence of HIV in the Cape Winelands was estimated at 12%, based on the infection rate among women tested in the course of antenatal care during that year; and during the period 2009 to 2010, there were 1 215 registered cases of tuberculosis in Stellenbosch, with a successful treatment rate of 81.7% (Cape Winelands District Office 2011).

Interpersonal violence

Interpersonal violence is made up of intimate partner violence (62%), unspecified community and family violence (29%), and child sexual abuse (9%) (Norman, Bradshaw, Schneider, Jewkes, Mathews, Abrahams, Matzopoulos, Vos & the South African Comparative Risk Assessment Collaborating Group 2007a). South Africa has the world's highest reported rate of intimate femicide (9 per 100 000); and in cases where the perpetrator is known, one in every two murdered women is killed by their intimate partner (Mathews, Abrahams, Martin, Vetten, Van der Merwe & Jewkes 2004). Moreover, interpersonal violence is not only related to homicide, but also to increased HIV transmission (Jewkes, Dunkle, Nduna & Shai 2010).

Alcohol harm

Alcohol abuse is rampant in South African society – and as Stellenbosch lies at the very heart of the wine industry, it is no exception. The Winelands District has one of the highest rates of foetal alcohol syndrome (FAS) in the world and extraordinarily high rates (as much as 8.8%) have been recorded among children of school entry age (Nutrition Information Centre of the University of Stellenbosch (NICUS) 2010). FAS causes growth retardation and a range of neurological impairments, which ultimately impact on educational ability and development. Moreover, in a general sense, alcohol abuse is linked to vulnerability to unsafe sex – and therefore HIV – and is also a common co-factor in interpersonal violence, homicide and road traffic accidents (Norman 2007b).

Tobacco smoking

Tobacco smoking is an important risk factor for a host of diseases, including tuberculosis, stroke, ischaemic heart disease, COPD, and lung cancer. Combined with hypercholesterolaemia, diabetes and hypertension, it also significantly increases the risk of cardiovascular disease.

High body mass index

South Africa is an obese society, particularly among women. Overall, 56% of adult women and 29% of men may be classified as overweight/obese, with a mean body mass index (BMI) of 27.3 kg/m^2 for women and 23.4 kg/m^2 for men; and 17% of one- to nine-year-olds and 7% of male and 25% of female school children are overweight or obese (Joubert, Norman, Bradshaw, Goedecke, Steyn & Puoane 2007a).

Obesity is related to hypertension, stroke, diabetes, ischaemic heart disease, osteoarthritis and various cancers. In addition, risk factors associated with diet and physical activity are closely related to obesity. Unhealthy diets have been characterised as:

> [... high in] saturated fat, particularly of animal origin, and [with] an imbalance between the different polyunsaturated fatty acids. This diet is also very high in salt, cholesterol, alcohol, sugar and energy intake, and very low in fibre, vitamin and trace element intake.
>
> [Steyn, Fourie & Temple 2006:1]

Like obesity, unhealthy diets are clearly linked to high blood pressure, diabetes, ischaemic heart disease and stroke. At the same time, South Africans are a sedentary people, with 63.2% of men and 75.3% of women inactive or not sufficiently active (<150 minutes of physical activity per week). In terms of inactivity, South African levels of 43.4% to 48.5% compared unfavourably with the 2007 global average of 17% and the African average of 10% (Joubert, Norman, Lambert, Groenewald, Schneider, Bull & Bradshaw 2007b).

While overweight is an important risk factor, maternal and childhood underweight is equally significant – and the two problems may even coexist in the same household.

Unsafe water and poor sanitation

Unsafe water and poor sanitation are obviously risk factors for diarrhoea, which is the leading cause of premature deaths among infants.

Approaches to improving health and wellness

The above discussion demonstrates the complex interconnections between the risk factors and diseases that underpin the burden of disease in Stellenbosch. In the present section, possible approaches to improving health and wellness at a local level are suggested, with specific reference to some of the most significant contributors to premature deaths in the Cape Winelands area. The importance of an intersectoral approach – to 'think health' in all policies – is also discussed. It is not possible to give a comprehensive overview of all preventive strategies; the intention here is rather to demonstrate the breadth of thinking required to tackle these issues.

Reducing unsafe sex

The factors that lead young people to engage in unsafe sex may be thought of as a triad of influences that are intrapersonal, proximal, or distal to the person. From an intrapersonal perspective, important factors include self-esteem, stage of physical and emotional development, age of sexual debut, and biological factors that increase vulnerability in women. Proximal factors include social and relational aspects, such as the number of sexual partners, gender norms (for example, the notion that women should concede to the desires of their male partners), intergenerational and transactional sex, and attitudes to pregnancy and condom usage. Distal factors are more contextual and include poverty, violence and the media.

Up till now, health services have largely focused on technical approaches to reducing the risk of HIV transmission through the provision of condoms and encouraging HIV testing, as well as providing access to anti-retroviral drugs where indicated; more recently, male circumcision has also been promoted.

While a technical approach to reducing transmission may be useful, it will not impact on many of the important abovementioned factors underlying unsafe sexual practices. Increasing social capital and recreational opportunities for youth may reduce casual unsafe sex and strengthen self-esteem; and educational programmes that not only provide information, but also encourage a critical consciousness about gender issues, may also reduce vulnerability. A tough approach to underage drinking and programmes to address substance abuse and sexual coercion may also help. Transactional and intergenerational sex may be reduced by poverty reduction, educational success, and improved job opportunities.

Reducing injuries and violence

Health services usually adopt a reactive stance to injuries and violence, in the sense that they treat the consequences and not the underlying causes. Intimate partner violence is a hidden scourge in South African society implicating several governmental sectors, including the health services, the police, the justice system, and social services. Apart from their obvious physical injuries, many abused women who approach health services present with psychological symptoms. Usually the underlying issue goes unrecognised; yet, even when the abuse is noted, management remains superficial and not comprehensive (Joyner & Mash 2010). Attempts are being made to improve recognition and management in primary care, but management requires the cooperation and support of other sectors. The police must be committed to implementing the Domestic Violence Act (No 116 of 1998) and to assisting women in accessing the courts, while social services should enable support groups and shelters and provide counselling services. Engagement with male perpetrators may also be needed, as well as policy to reduce access to firearms.

Violence of all kinds, and road traffic accidents in particular, is often accompanied by alcohol or other substance abuse. Serious attention should be given to reducing substance abuse and treating people with harmful use and dependency. Norman (2007b) suggest the following strategies to diminish alcohol abuse and reduce the number of road accidents resulting from it:

- strict enforcement of the minimum purchasing age;
- restricting the sale of alcohol in terms of trading hours or days;
- increased taxation of alcohol;
- strict enforcement of existing laws pertaining to drunk driving and increased sobriety checkpoints;
- administrative suspension of driving licences;

- graduated licensing for novice drivers;
- (brief) intervention for dangerous drivers;
- improved access to counselling and treatment services; and
- improved traffic management in urban environments (which may also reduce exposure to lead and outdoor pollution).

Improving maternal and child health

In terms of maternal and child health, a number of recommendations relating to health services have been identified. Broadly speaking, they include the following (Pattison 2009):

- improving programmes aimed at preventing mother to child transmission of HIV to ensure that mothers are not overlooked, whether in terms of testing or of accessing anti-retrovirals – keeping HIV transmission to an absolute minimum is not only critical to reducing infant mortality rates; it is also entirely achievable. Preventing mothers from acquiring HIV during pregnancy and breastfeeding is also important, because the risk of transmission is much higher in this group;
- improving clinical skills in terms of obstetric and neonatal emergencies, routine midwife care and postnatal care;
- implementing existing maternal and neonatal guidelines;
- improving coordination and communication in postnatal care;
- improving clinical governance, auditing and monitoring;
- ensuring that constant, coherent and consistent health messages are given to communities, patients and health care workers;
- improving transport and referral systems for women in labour and/or ill neonates; and
- ensuring sufficient staffing and equipment at all facilities.

Again, considering the main causes of premature death among children, it is obvious that access to safe water and a hygienic environment are major issues in diarrhoeal disease, as is knowledge about the use of a sugar-salt solution in preventing dehydration. Therefore, it is important that issues around sanitation, municipal water supplies, polluted rivers and community-based education be addressed in preventing these conditions.

Reducing non-communicable chronic disease

As far as these diseases are concerned, the creation of non-obesogenic environments, supported by policies that encourage and enable physical activity, is important. This may involve improving community safety for people exercising outdoors, improving access to green spaces, encouraging the use of bicycles and public transport, providing physical activities at school, and encouraging physical activity in the workplace. Dietary factors are clearly important as well; here policy should focus on enabling access to affordable fruit and vegetables and reducing access to unhealthy processed and fast foods, especially at schools. The salt content of food is of particular concern in causing high blood pressure. Education, urban design, transportation and food policy are therefore all important.

Reducing COPD

Tobacco smoking is an important risk factor[4] and more effort should be expended in trying to reduce it. A number of preventative measures may be useful, including continued increased taxation of tobacco products, community-based education about health risks associated with smoking, restricting smoking in public places and workplaces, banning cigarette advertising, and improving support for people trying to stop smoking, such as counselling, support groups, or nicotine replacement therapy (currently little structured assistance is offered through the health services).

---

4   The burning of biomass or kerosene is also an important risk factor for COPD. It may be significantly reduced by the electrification of households.

Reducing drowning

As mentioned before, drowning is (somewhat unexpectedly) one of the leading causes of premature death in the Stellenbosch area, presumably as a result of swimming in farm dams. This should be further investigated, but possible interventions may include educating people about the risks of drowning, limiting access to bodies of water and providing learn-to-swim programmes.

General health and wellness

Health services have traditionally focused on treating disease and disability, on the assumption that this will help people move towards health and greater wellness. While the absence of disease may be an important factor in wellness, it is plainly not the only one. Travis and Ryan (2004) describe the wellness paradigm as a continuum, from premature death at the one end to health and wellness at the other (Figure 21.3). Health services tend to attempt to bring people back to the neutral midpoint, but do not usually assist them in making the full journey to health and wellness.

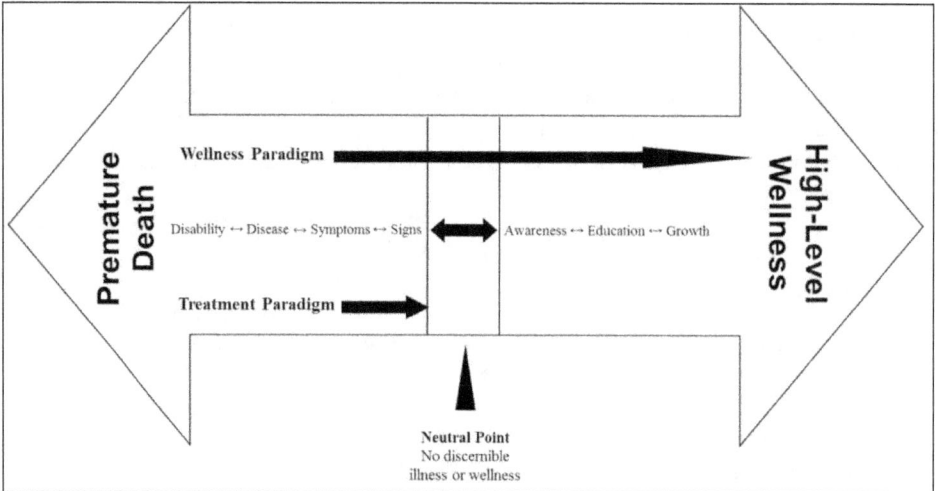

Figure 21.3 The wellness paradigm [Travis & Ryan 2004]

Benedict defines high levels of health and wellness as:

> *[...] giving good care to your physical self, using your mind constructively, expressing your emotions effectively, being creatively involved with those around you, and being concerned about your physical, psychological and spiritual environments.*
>
> [Quoted in Christodoulou 2011:435]

In similar vein, Ardell describes the wellness thusly:

> *The eventual goal of wellness is the actualisation of one's true psychophysical/spiritual potential. Wellness is Maslow's notion of self-actualisation carried to its natural extension as growth towards the full integration of mind, body, spirit and environment. This end point is not really an end. It is not perfection. Better than a continuum in this case is a spiral model, continually cycling through the ebb and flow of life towards higher levels of wellness.*
>
> [Quoted in Christodoulou 2011:436]

Conclusion

The causes of premature death and their underlying risk factors have been identified for the Stellenbosch area. Clearly, we all have some control and choice over our behaviour and personal risk factors. Nevertheless, interventions to prevent premature death and disability by reducing these risk factors require that policy makers think broadly across all sectors and that they realise that, in isolation, health services have only limited potential impact. Moving beyond a focus on death and disability to a more integrated understanding of health and wellness may also be important. In conjunction with policy at a national and provincial level, policy makers at local government level should plan cost-effective interventions that may contribute to improving health and wellness in their communities.

REFERENCES

Cape Winelands District Office. 2011. Personal communication. Stellenbosch: 31 January.

Christodoulou M. 2011. Chapter 14: Integrative Medicine. In: B Mash (ed). *Handbook of Family Medicine*. 3rd Edition. Cape Town: Oxford University Press.

Groenewald P, Bradshaw D, Krige F, Van Niekerk M, Champion F, Karstens T, Verwey G & Van der Merwe W. n.d. *Causes of death and premature mortality in the Cape Winelands and Overberg Districts 2004-2006*. Cape Town: Medical Research Council and Department of Health.

Jewkes R, Dunkle K, Nduna M & Shai N. 2010. Intimate partner violence, relationship power inequity, and incidence of HIV infection in young women in South Africa: a cohort study. *Lancet*, 376(9734):41-48.

Joubert J, Norman R, Bradshaw D, Goedecke J, Steyn N & Puoane T. 2007a. Estimating the burden of disease attributable to excess body weight in South Africa in 2000. *South African Medical Journal*, 97(8):683-690.

Joubert J, Norman R, Lambert EV, Groenewald P, Schneider M, Bull F & Bradshaw D. 2007b. Estimating the burden of disease attributable to physical inactivity in South Africa in 2000. *South African Medical Journal*, 97(8):725-731.

Joyner K & Mash B. 2010. How to provide comprehensive, appropriate care for survivors of intimate partner violence. In: K Joyner (ed). *Aspects of Forensic Medicine: An Introduction for Healthcare Professionals*. Cape Town: Juta.

Mathews S, Abrahams N, Martin L, Vetten L, Van der Merwe L, & Jewkes R. 2004. *"Every six hours a woman is killed by her intimate partner": A National Study of Female Homicide in South Africa*. Medical Research Council of South Africa (MRC) Policy Brief (5). Pretoria: MRC & Centre for the Study of Violence and Reconciliation (CSVR).

Norman R, Bradshaw D, Schneider M, Jewkes R, Mathews S, Abrahams N, Matzopoulos R, Vos T & the South African Comparative Risk Assessment Collaborating Group. 2007a. Estimating the burden of disease attributable to interpersonal violence in South Africa in 2000. *South African Medical Journal*, 97(8):653-656.

Norman R, Bradshaw D, Schneider M, Joubert J, Groenewald P, Lewin S, Steyn K, Vos T, Laubscher R, Nannan N, Nojilana B, Pieterse D & the South African Comparative Risk Assessment Collaborating Group. 2007b. A comparative risk assessment for South Africa in 2000: Towards promoting health and preventing disease. *South African Medical Journal*, 97(8):637-341.

Nutrition Information Centre of the University of Stellenbosch (NICUS). *Foetal alcohol syndrome*. [Accessed 31 January 2011] http://www.sun.ac.za/nicus

Pattison RC (ed). 2009. *Saving babies 2006-2007: Sixth perinatal care survey of South Africa*. Pretoria: Tshepesa Press.

Steyn K, Fourie J & Temple N (eds). 2006. *Chronic diseases of lifestyle in South Africa: 1995-2005*. Technical Report. Cape Town: MRC.

Travis JW & Ryan RS. 2004. *Wellness Workbook*. 3rd Edition. Berkeley: Celestial Arts.

Chapter 22

WELL-BEING
Changing human behaviour

LESLIE SWARTZ & KEES VAN DER WAAL

Introduction

Readers of this book and many others take it as read that environmental sustainability is an important goal globally and important for the quality of life of all people. Surveys of public opinion tend to show that despite some obfuscation and mistrust in science, people say that they are concerned about environmental issues (Uzzell 2010). South Africa is not an outlier when overall attitudes to the environment are compared across a range of countries with varying income levels (Struwig 2010), but attitudes in this country vary across dimensions of class, race, location and education. Part of the problem is not that there are insufficient people who believe that environmental issues are important, but that there are many competing issues for people to face and think about, which may have an effect on the immediacy with which we take action (or whether we take action at all) on environmental issues.

It is a difficult task to understand and to know how to intervene in terms of changing practices and behaviours pertaining to environmental sustainability. Consider, for example, the argument developed by Struwig (2010) to explain the finding of the South African Social Attitudes Survey that people living in urban informal environments scored significantly lower on an environmental concern index than did people living in urban formal or rural formal or informal environments:

> Respondents living in urban informal areas cared much less about the environment than respondents residing in urban formal, rural informal and rural formal areas. The relative deprivation theory suggests that people living in polluted and degraded areas get used to the situation and that the outcry is likely to come from people living in cleaner areas who become exposed to the polluted side [...] However, there is also a danger that constant exposure to pollution and environmentally hazardous situations would start to take precedence over less concrete issues, such as civil rights issues. However, once the issue of civil rights and human dignity is connected to environmental deprivation, it can become a major issue, capable of influencing opinion and creating unrest and dissatisfaction. The 2007 riots related to inadequate service delivery in several areas in South Africa would appear to support this.
>
> [Struwig 2010:215]

This rather confusing paragraph embodies one of the core dilemmas faced by behavioural scientists with an interest in environmental and sustainability issues. It is not possible to divorce environmental concerns from other concerns (nor should we try to do so), but when we try to think about these concerns in a broader context, we run the risk of becoming lost in a mire of speculation about a wide range of factors affecting human behaviour. In this regard, the challenges concerning developing a coherent and systematic approach to understanding and intervening in behavioural issues affecting environmental sustainability are not dissimilar to the challenges associated with attempts to prevent HIV infection (Tomlinson, Rohleder, Swartz, Drimie & Kagee

2010). HIV issues, like environmental ones, are closely linked to broader social concerns, including issues of social exclusion. There is ample evidence that many people who are well informed about HIV prevention and who have been counselled on the issues at stake, are, for a range of reasons, unable to take action to protect themselves. This is not a question of whether people 'care' about HIV infection (to use Struwig's term when he talks about environmental concerns in the quotation above), but rather a question of what is and is not structurally possible given the realities of people's lives.

These reflections on the confusion around behaviour change with respect to environmental issues and HIV should be familiar to anyone who has tried to instigate behaviour change in Stellenbosch. Every Integrated Development Plan formulated and adopted by Stellenbosch Municipality since 1994 has identified social divisions as a key problem, and much emphasis has been given to integration. A wide range of cultural, educational, environmental and religious initiatives have been taken to bridge the divides in an attempt to build an idealised image of a harmonious, integrated Stellenbosch community. Almost all appeal in some way to the rationality of shared norms and behaviours, coupled to a romantic depiction of the beauty of the environment that all who live in Stellenbosch share. But anyone who has thought about these challenges or been involved in such actions will know how difficult all this is in practice. They all express intense frustration and many eventually give up because of the limits to wider impacts. Even councillors with wards that include constituencies drawn from richer and poorer areas or from areas dominated by different racial groups seem ill-equipped in terms of techniques to communicate with all in ways that allow their messages to be heard; and officials know only too well what it means to call public participation meetings on key policy documents in the City Hall that are not attended by all constituencies, but usually only influential (generally white-dominated) elite stakeholder groups. And when people from poorer communities do manage to get to meetings to discuss development proposals, consultants do PowerPoint presentations and engage in professional discourse that is unintelligible even for those with a university education. All this is reinforced by developers who manage to get proposals that reinforce separation approved by the municipality. Klapmuts is probably one of the few emerging towns in the world that has no centre and is simply a collection of totally disconnected walled security villages with no economic base or means of social interaction.

This difficulty in thinking about and effecting changes in people's behaviour concerning environmental issues, though at this stage by no means resolvable, suggests that it may be helpful to look in two directions. Firstly, we need to be as clear as we can about what we know about how behaviour change has been shown to work in a range of contexts; and secondly – and this is crucial if we wish to address the social embeddedness of environmental issues – we need to think more carefully about the limits of these behavioural models and what we may do to address them. We shall consider each of these areas in turn and provide case material from the Stellenbosch context.

Understanding behavioural issues in creating sustainable environments

In 2002, the psychologist Daniel Kahneman was awarded the Nobel Prize for economics (see Kahneman 2003). With his colleague Amos Tversky and others, he had developed the field known as Prospect Theory, and his work is credited with having been key to the establishment of the field known as behavioural economics. The details of this work are not relevant to this chapter; however, it is important to note that a key contribution of this field was the insight (supported by empirical research) that, even where the outcomes of decisions are apparently amenable to rational thinking, where risk and uncertainty is involved a range of non-rational factors inform the decisions that people make. For example, though it had been recognised for a long time that emotion and sentiment play an important part in the behaviour of currency markets and those who trade on them, behavioural economics brought into sharp focus as never before the extent, range, and relative contribution of these non-rational factors.

Given Kahneman's Nobel Prize and the attendant publicity, and the appeal of his work and that of others as a new approach to understanding economic behaviour, it is not surprising that work informed explicitly or implicitly by this tradition has become attractive to people interested in a range of policy issues. It is especially attractive to people interested in changing behaviour in fields where there are clearly non-rational elements at play, including health behaviour and behaviour related to environmental sustainability. The Institute for Government in the United Kingdom, an independent think tank that claims to "provide evidence based [*sic*] advice that draws on best practice from around the world" (2012:n.p.), has launched a project it terms 'Mindspace'. The project is conceived to lie at the interface between behavioural economics and social psychology, and focuses on what it terms "influencing behaviour through public policy" (Dolan, Hallsworth, Halpern, King & Vlaev 2009). It was commissioned by the British Cabinet Office and derives its name from what Paul Dolan and the co-authors of the *Mindset: Influencing behaviour through public policy* report,[1] which outlines the founding principles of the project, view as "nine of the most robust (non-coercive) influences on our behaviour" (2009:8), as summarised in Table 22.1.

Table 22.1 The Mindspace checklist of influences on behaviour [Dolan *et al.* 2009:8]

Messenger	We are heavily influenced by who communicates information
Incentives	Our responses to incentives are shaped by predictable mental shortcuts, such as strongly avoiding losses
Norms	We are strongly influenced by what others do
Defaults	We 'go with the flow' of preset options
Salience	Our attention is drawn to what is novel and seems relevant to us
Priming	Our acts are often influenced by subconscious cues
Affect	Our emotional associations can powerfully shape our actions
Commitments	We seek to be consistent with our public promises, and reciprocate acts
Ego	We act in ways that make us feel better about ourselves

It is worth providing a slightly more comprehensive description of each of the Mindspace terms (following Dolan and colleagues). What follows is a summary of the central Mindspace argument with reference to the South African and Stellenbosch situation, with commentary and criticism of the approach provided further on in the chapter.

Messenger

As has clearly been shown in a number of contexts, a crucial factor in how we weigh information relates to our views on the person or organisation supplying that information. For example, Dolan *et al.* (2009) note that information is more likely to be believed when it is perceived to come from experts in the field. In addition, they point out that demographic similarities between the person giving or receiving the message tend to increase the extent to which information results in behaviour change. They further argue that peer acceptance is also an important factor.

A key issue which is as yet under-researched in the South African and Stellenbosch social context is the question of the extent to which demographics and identity may play a role in influencing behaviour, and may even trump the perception of expertise. As community projects in the mental health field have shown, there is a long history of mistrust of authorities, superimposed on past histories of racial and class exploitation, in South Africa (Swartz, Gibson & Gelman 2002). In contemporary South Africa, perceptions of cronyism and corruption may also play a part. If an institution such as Stellenbosch University, for example, wishes to play a key part in behaviour change related to environmental issues, a central question that needs to be addressed (as is widely known) is the image of the university in the public imagination. It is by no means a given that it would be automatically trusted as a messenger in terms of behaviour change. For some, Stellenbosch University may be a trusted figure of authority; for others it may be a symbol of oppression and exclusion, not to be trusted.

In fact, Stellenbosch University is partnering with Stellenbosch Municipality to address the challenges of social cohesion and development in the town, but the internal local government politics have undermined this effort. The two contending large political parties in this region have used every opportunity to undermine each other, and have ultimately created a dysfunctional environment due to their own strategic interests for narrow victories. The capacity of the municipality to be an agent of development and change by partnering with the university has thus been severely curtailed.

---

1 See http://www.instituteforgovernment.org.uk/publications/mindspace and Dolan *et al.* 2009.

Incentives

The question of incentives to change behaviour is an issue that has concerned behavioural scientists for decades. Dolan *et al.* (2009) mention a number of principles on which to develop incentives to support behaviour change. A key principle is that immediate incentives tend to be more effective in changing behaviour than delayed ones.

In resource-poor contexts the question of incentivising behaviour becomes complicated. Many incentives are costly in the short term and may be difficult to implement with short-term budgeting, even if they promise important long-term benefits. In the South African and Stellenbosch context, which is characterised by a high degree of inequality, this needs careful consideration. Relative deprivation (the perception that one is given less than others for no good reason) may be a factor impeding pro-social behaviour. If one's behaviour is incentivised in a context of relative deprivation, it is possible that the incentive may lose its perceived value.

Norms

Social norms affect behaviour and should be borne in mind when planning programmes to change behaviour. According to Dolan and his co-authors, programme planners must create conditions appropriate to the development of social norms that may support behaviour change explicitly and use existing social networks to reinforce these norms.

In a complex social environment where different groups may subscribe to differing norms, the question of 'social norming' raises the challenge of how to make use of overarching norms that are acceptable to a range of people. Although the concept of *ubuntu* is not without its critics (Mji, Gcaza, Swartz, McLachlan & Hutton 2011), it may prove a useful basis for establishing explicit norming. The Social Cohesion movement described elsewhere[2] is an example of a group that has mounted what may be described in this context as an attempt at social norming that bridges the divides that characterise the way social life is conducted and structured in Stellenbosch. Ultimately, norms and incentives in Stellenbosch are embedded in a social structure that is still highly fragmented and unequal. Progress in the field of sustainable development and behaviour change, based on mutual trust, can only be made when underlying power issues are addressed.

Defaults

People will tend to maintain previous behaviour by default and 'making change' is difficult (Anderson 2003; Ritov & Baron 2004). Moreover, when offered choices, people are more likely to accept a default option than to opt to actively go against it. One example of how this fact may affect work on environmental sustainability is that if the more sustainable option is presented as the default option when people are asked to make choices about their options regarding waste disposal, they are, for example, more likely to accept that option than if they are required to actively consider and choose the sustainable option.

Salience

As a result of information overload, people filter out stimuli all the time. For this reason, behaviour change messages need to be simple and presented in a context of minimal distraction (Dolan *et al.* 2009). In contexts where people live very different lives and may be subject to different messages and stimuli, it may be helpful to think about slightly different salient stimuli for different groups. On the other hand, the assumption that people who live in what appear to be very separate communities will necessarily differ radically in their reaction to interventions may not be borne out in practice, especially where there are cross-cutting media usage patterns (such as social media).

Priming

People respond to subliminal cues and respond to 'priming' without being aware of being primed (Dolan *et al.* 2009). As advertisers who use 'teasers' know, people may respond more strongly to a message if they have been prepared for it previously, whether subliminally or overtly. Priming may therefore be a useful

---

2   See Section 4: Social Cohesion by Jerome Slamat, Thumakele Gosa and Chris Spies (p. 287).

technique in campaigns to change behaviour. However, there are a range of ethical questions associated with priming, some of which we shall touch on later in this chapter.

Affect

Some would regard the heart of a behavioural economics to be the recognition that emotions matter and that they affect behaviour (Dolan *et al*. 2009). Dolan and colleagues cite a study[3] in which the researchers substantially increased rates of hand-washing in Ghana, but not by basing a campaign around the rational reasons to increase hand-washing or by promoting soap use. Instead, as the researchers had found that hand-washing in this population was associated with feelings of disgust at having dirty hands and worries about contamination, they developed a media campaign showing the protagonist spreading contamination after not washing her hands with soap. The evoked emotions of worry about contamination and disgust were sufficient to significantly increase hand-washing.

Commitment

As practitioners involved in the exercise industry know, publicly committing to long-term behaviour change increases the likelihood that it will indeed change. The immediate act of writing down or signing a document declaring the intent to change one's behaviour assists with behaviour change (Dolan *et al*. 2009).

Ego

People tend to behave in ways that enhance their view of themselves (Dolan *et al*. 2009). If behaviours that are consistent with environmental sustainability are presented in such a way as to add to a person's self-esteem or status in a group, this is likely to enhance the chances that he or she will engage in such behaviour. If programmes are able to link sustainability behaviours to other valued personal norms, it may increase their success.

The Mindspace approach (though at times somewhat superficial and selective in its treatment of data from research in social and cognitive psychology and behavioural economics) provides a very useful summary of key lessons learned from research. The authors of the document provide examples illustrating the usefulness of the Mindspace approach in various projects, as well as a template of processes to follow in developing programmes.[4]

Of necessity, the *Mindspace* report simplifies some rather more complex arguments (see, for example, Gowdy 2008 & Kazdin 2009). Kazdin correctly characterises climate change as what he and others have termed a "wicked problem" (a description that may very well apply to other sustainability issues as well). He (2009:343) neatly summarises the key characteristics of a 'wicked problem' as follows:

- There is no single, definitive, or simple formulation of the problem.
- The problem is not likely to be the result of an event (for example, a tsunami or hurricane), but rather of a set of intersecting trends that co-occur and co-influence each other.
- The problem has embedded in it other problems – including other 'wicked problems'. There is no one solution – no single, one-shot effort – that will eliminate the problem.
- The problem is never likely to be solved.
- Multiple stakeholders are likely to be involved. This fact leads to multiple formulations of what 'really' is the problem, and therefore also to multiple possible legitimate or appropriate solutions.
- Values, culture, politics and economics are likely to be involved in the problem, as well as in possible strategies to address it.

---

3 See: Scott BE, Lawson DW & Curtis V. 2007. *Hard to handle: understanding mothers' hand-washing behaviour in Ghana*. London: The Hygiene Centre, Department of Infectious and Tropical Diseases, London School of Hygiene and Tropical Medicine.

4 The details of these processes are too lengthy to elaborate here. However, the complete document is available as a free download in pdf format at http://www.instituteforgovernment.org.uk/publications/mindspace. See also Dolan *et al*. 2009.

- Information as a basis for action will be incomplete, because of the uniqueness of the problem and the complexities of its interrelations with other problems.
- The problem is likely to be unique and therefore does not easily lend itself to previously tried strategies.

Kazdin's recognition of the complexity of the climate change problem is to some extent a corrective to what at first blush may seem an overly optimistic and even somewhat evangelistic tone set by the excitement associated with new developments in behavioural economics and related fields. Toolkits of the Mindspace-type are very helpful, but it is also important to question whether technologies of this type are sufficient for us to make the best difference we can in behavioural and social processes, especially in the South African context. It is to this question that we turn in the following section.

Are there limits to behavioural models?

Reading behavioural economics work is exciting and interesting, but the field does position both the reader and the objects of public policy in particular ways. Social marketing[5] commonly has excellent goals and looks towards the greater good, but the field is exemplified by a rather instrumentalist view of people and behaviour change. The optimism of a new field may lead researchers to ignore the extent to which there are substantial barriers and difficulties in implementing behaviour change on a large scale (see, for example, Gifford 2011 and Swim, Stern, Doherty, Clayton, Reser, Weber, Gifford & Howard 2011).

The issues are, however, not simply empirical: we need to think about what methods and assumptions are most helpful as we address complex issues. David Uzzell, a professor of environmental psychology at the University of Surrey, presented the annual joint British Academy/British Psychological Society lecture in 2010. In this lecture, Uzzell makes the telling point that programmes such as Mindspace promise what he terms "the quick-fix solution, the language of management" (2010:881) and notes that they focus on changing automatic behaviour not readily available to consciousness. He then raises a question that is crucial for the field as a whole – and probably especially relevant for interventions in countries like our own, where there are high levels of intergroup and interpersonal suspicion and mistrust. Interventions that emphasise behaviour change may be very useful; but, Uzzell asks: "Do we really want to change behaviours without changing minds? Is this the kind of society we want?" (*ibid.*). Clearly (and other chapters in this book bear testimony to this), it is not. Technicist solutions to complex social problems have their place, but they are both ethically problematic and incomplete. As noted at the outset of this chapter, part of the challenge of making changes for sustainability is that sustainability issues are so entwined with others relating to how society is organised. In South Africa, more clearly than in many other contexts, sustainability issues interface very strongly with questions about how we work to create a society of true partnerships and sociality.

A study addressing questions of mental health promotion and the prevention of mental disorders in poorly-resourced contexts suggested some basic principles that should be adhered to in the development of innovation programmes (see Swartz 2010). We have adapted this list slightly in Table 22.2.

---

5 The definition of social marketing is contested, but an oft-quoted definition is that by Andreasen (1994:110): "Social marketing is the adaptation of commercial marketing technologies to programs designed to influence the voluntary behaviour of target audiences to improve their personal welfare and that of the society of which they are a part."

Table 22.2 Basic building blocks for innovation [Adapted from Swartz 2010]

Requirements for successful implementation of mental health promotion and mental disorder prevention efforts	What is it?	What may happen when it is not in place?
Good governance	A broad political system that is supportive of human rights and the development of human potential	Political systems that are oppressive or neglectful of their citizens are not conducive to sustainability.
Infrastructure	A system in place that enables the proper delivery of interventions	It is difficult to provide interventions on anything but a small and local scale in the absence of a properly functioning local government system.
Record-keeping	A system that enables interventions to be noted down and monitored.	If a programme does not keep records of what it does and the impacts of its actions, it is not possible to keep track of what is being done and what needs to be done.
Data-driven decision-making	A system in which decisions are made as far as possible on the basis of knowledge of what is happening in the world in which the organisation operates, and of what works and what does not	Where systems operate on the basis of impressions only, or on the basis of what seems politically or interpersonally the best thing to do, it becomes almost impossible to track whether prevention and promotion activities have an impact.
Sufficient staff	Sufficient staff, and sufficiently skilled staff, to implement and monitor interventions	In the absence of sufficient staff it is sometimes possible to use other resources creatively, but it is not reasonable to expect very few people to do many jobs; and asking for too much may lead to poor motivation and burnout.
Motivated staff	Staff who are keen to implement interventions	If staff do not believe in or are opposed to interventions, they may undermine or interfere with the success of the interventions.
Cultural acceptability	Interventions are acceptable to people in the community	If interventions are not culturally acceptable, they will not be adopted and may be rejected as useless.

There is nothing especially original about the above list of basic requirements; nor, at first glance, do they appear to have much in particular to do with behaviour change or well-being with regard to sustainability. The fact is, though, that these kinds of infrastructural arrangements, which are sometimes taken for granted as being in place in other countries, are not always present. Without some degree of predictability and sense of control over the social world and the social infrastructure, it is very difficult for people to change their behaviour – and especially difficult to maintain changes.

Catherine Campbell's (2003) incisive analysis of the failure of a very well thought out AIDS prevention programme in South Africa demonstrates that, though behavioural techniques are important in making changes in behaviour and social relationships, they are not enough. In addition, we must understand the importance of political will for change, and we must support and develop social cohesion and accountability. In subsequent work, Campbell and co-authors Nair, Maimane and Gibbs refer to "AIDS competent communities" (2009:221), where members collaborate and support one another in bringing about change in the social factors that create HIV risks; for example, reducing the stigma attached to HIV in the community, promoting sexual behaviour change, and supporting people in the community who are living with HIV. None of this can be achieved without strong and sustained community engagement – an engagement that requires the building of trust and the promise of the long-term involvement of innovation-minded people.

There is much to learn from Campbell's model in terms of understanding the difficulties and possibilities for developing programmes that promote sustainability, in at least two senses of the word: sustainability in terms of environmental issues, as narrowly understood; and sustainability in terms of programmes that have a future. Bracher (2009) suggests that many social problems (and environmental degradation may be one of them) may be exacerbated or maintained by a failure on the part of those with power to explore underlying identity needs and issues. In South Africa, more than elsewhere, identity issues are writ large, and the politics of exclusion continues to be a fine art.

Social cohesion and the sustainability of interventions in the Dwars River Valley

To illustrate the importance of social relations, trust and identity in sustainable development, one may look at two recent interventions in the Dwars River Valley between Stellenbosch and Franschhoek.

At Boschendal, the new owner of a group of farms and Anglo American – the previous owner – agreed on a plan for the economic and social development of the area. The *Boschendal Sustainable Development Initiative* (BSDI) document outlined the way in which local communities would be involved in the planning of development interventions, and how they would benefit from the sale of the subdivisions of the farms to wealthy new residents (for what was termed "gentlemen's wine estates") (Dennis Moss Partnership 2008; Gary Player 2004:n.p.). Secure lifestyle land units would be created on the land that was formerly worked by the farm workers, who in turn were removed from the farms and relocated in a new town, Lanquedoc, created with the help of government housing subsidies and substantial contributions from the multinational corporation. Many of the farm workers lost their jobs in the shift from wine production as a result of the combination of a more mechanised production process and the realisation of the real estate value of the land through its sale to the new owners.

While the BSDI contained an extensive discussion of sustainable development and referred in detail to contemporary international agreements and policies to justify the approach used in the plan, even the preliminary stages of its actual execution ran into several difficulties, indicating problems around its sustainability. The planners emphasised public participation in the BSDI planning process and consulted with local communities of farm workers and the residents of the neighbouring settlements. Nevertheless, scepticism about the eventual realisation of the promised benefits was rife and many formal objections, as well as court cases, public protest marches and forced evictions, indicated strong resistance to the plan from a part of the population.

The objections and resistance had diverse origins, including heritage and landscape interests (represented by the South African Heritage Resources Agency, who was responsible for assessing and ruling on them); farm workers whose dependents were not given access to the new housing; and people in the neighbouring settlements, who objected on the grounds of the impact the plan would have on the character of their town and the difficulties for future generations in terms of accessing land in the area. The various objections also pointed out that the creation of a wealthy class of landowners living in exclusive and secure estate conditions would reinforce the existing deep social and economic divisions in the valley. Much of the contestation was also related to a lack of trust in the large companies who had successively owned the land in the past and, despite relatively progressive labour relations, many questioned the capitalist nature of the development. Moreover, the neighbouring settlements claimed that parts of the land rightfully belonged to them.

In the event, as a result of the global economic crisis and the loss in the land value due to the retardation of its realisation during the years of trying to resolve the many objections and conflicts, the new owners were forced to sell off some of the land after a few years and to find new financial partners.

The BDSI was undermined by a historical lack of trust, divisions in the population in the valley, identity issues, and high levels of inequality, and serves as a clear illustration of how difficult it is to reach sustainable development, including community participation, where economic interests seem to dominate other considerations.

At Solms-Delta, a smaller sustainable development initiative has taken off (Solms-Delta 2011) and at present appears to be a comparatively much more successful project than the BSDI, due not only to its smaller scale, but also its more inclusive and radical approach. The land owner, Mark Solms, realised that the nature and extent of the negative cumulative impact of slavery, apartheid and commercial farming on farm workers had to be addressed. In order to address this, he involved the workers in a collaborative search for material, personal and emotional legacies from the past, which resulted in a process of reconciliation and the establishment of a museum on the farm that documented and interpreted that past by facing it in all its human intensity. In her inaugural address as Professor of Psychology at the University of Cape Town, Pumla Gobodo-Madikizela (2010) referred to the way in which the transformative dialogue of the people at Solms-Delta, working through the trauma of the past and its complicated legacies, contributed to true transformation and the creation of new meaning, and noted that the representation of the narratives of the families on the farm in the museum validated them and assisted everyone involved in reflecting on the past in order to find new understandings and trust.

The underlying approach of the Solms-Delta initiative was based on a radical understanding of reconciliation and partnership, resulting in a participative management style that attends to the basic needs of the workers, with a special focus on housing, education and skills development. The documentation and celebration of local popular music and an annual harvest festival (among many other initiatives) indicated the radical and inclusive nature of social, production and management relations in this initiative. Workers became co-owners and co-managers of the land, which was extended by mortgaging the existing property as security to buy neighbouring land.

The small scale of the Solms-Delta initiative permits the building of personal relations of trust, the provision of immediate benefits and an openness to the complexity inherent in any change process, thus enhancing the social, and ultimately also the economic, sustainability of this farming enterprise. While no initiative is perfect, in this instance, sustainability and inclusivity are practised as much as possible, thus preventing the obstacles faced by the much larger and more rigorously planned project at Boschendal. The BSDI was, in the first instance, fundamentally dependent on its real estate-based financial success as devised by expert planners – and therefore so much more vulnerable on other fronts – with sustainability serving as the ideology underpinning the planning blueprint. In contrast, the Solms-Delta initiative is fundamentally dependent on the success of its combination of inclusive sociality and economic production, driven by all the people on the farms and including needs pertaining to identity and *homo ludens* ('human at play'). Sustainability, in this instance, was much more a practice than an ideology.

Conclusion: Reflections on behaviour change in Stellenbosch

What does this mean for the role of Stellenbosch University as a key player in contributing to sustainability? At the centre of much of what we know about behaviour and behaviour change and sustainability is the recognition that issues are interrelated. If Stellenbosch University is to assume the key role it can play, issues of identity and exclusion need to be addressed by the organisation as a whole. Sustainability requires ongoing trust and relationships of mutuality rather than of manipulation. For this to occur, all parties must be clear about their *bona fides* as places of inclusion and engagement at all levels, not just at the environmental one as narrowly understood.

While much of the process of sustainable development in Stellenbosch will happen outside of Stellenbosch University, the staff and students there may play an important role as brokers of sustainable development. It would be both arrogant and quite simply wrong to assume that marginalised groups with fewer resources than the university should by definition be relegated to a minor role. Recognising this fact requires university staff and students to be cognisant of their own position and status and issues of power. They may become involved as citizens and researchers in local situations, where their knowledge and research experience may contribute to sustainable development planning and practice. Students from a wide range of disciplines may become more actively involved in community interaction in the area through the process of service learning. By using these experiences as a valuable knowledge resource, the university may be able to become more involved in and contribute to sustainable development through a critical and compassionate engagement in local initiatives. The in-depth knowledge to be gained by involvement in the social process may be turned into research knowledge, as well as public knowledge. Through their engagement with the complexities of small-scale local contexts, academics and students become 'translators' of sustainable development (see Latour 1993; Lewis & Mosse 2006) by scaling up from there to make their knowledge and analysis useful for wider application. In this process, the possibility emerges that they will be transformed themselves as citizens to overcome the divisions that are so deeply ingrained in the backgrounds of divided South African communities. In turn, this should help them to contribute to the commons of the social and natural sphere that is Stellenbosch.

REFERENCES

Anderson C. 2003. The psychology of doing nothing: Forms of decision avoidance result from reason and emotion. *Psychological Bulletin*, 129(1):139-166.

Andreasen AR. 1994. Social marketing: Ist definition and domain. *Journal of Public Policy and Marketing*, 13(1): 108-114.

Bracher M. 2009. *Social symptoms of identity needs: Why we have failed to solve our social problems and what to do about it.* London: Karnac.

Campbell C. 2003. *Letting them die: Why HIV prevention programmes fail.* Oxford: James Currey.

Campbell C, Nair Y, Maimane S & Gibbs A. 2009. Strengthening community responses to AIDS: Possibilities and challenges. In: P Rohleder, L Swartz, S Kalichman & L Simbayi (eds). *HIV/AIDS in South Africa 25 years on: Psychosocial perspectives.* New York: Springer NY.

Curtis VA, Garbraah-Adoo N & Scott B. 2007. Masters of marketing: Bringing private sector skills to public health partnerships. *American Journal of Public Health*, 97(4): 634-641.

Dennis Moss Partnership 2008. *Boschendal Sustainable Development Initiative – Overview & Summary.* [Accessed 5 July 2011] http://www.dmp.co.za/downloads/initiatives/boschendal-sdi-case-study-section-1.pdf

Dolan P, Hallsworth M, Halpern D, King D & Vlaev I. 2009. *Mindspace: Influencing behaviour through public policy.* London: Institute for Government.

Gary Player. 2004. *New Owners of Boschendal Wine Estate: Gary Player to Act as Spokesman.* [Accessed 3 October 2012] http://www.garyplayer.com/news/news_detail/new_owners_of_boschendal_wine_estate_gary_player_to_act_as_spokesman/

Gifford R. 2011. The dragons of inaction: Psychological barriers that limit climate change mitigation and adaptation. *American Psychologist*, 66(4):290-302.

Gobodo-Madikizela P. 2010. The Face of the Other: Human dialogue at Solms-Delta and the meaning of moral imagination. Inaugural lecture. Cape Town: University of Cape Town.

Gowdy JM. 2008. Behavioral economics and climate change policy. *Journal of Economic Behavior and Organization*, 68(3-4):632-644.

Kahneman D. 2003. Maps of bounded reality: Psychology for behavioral economics. *The American Economic Review*, 93(5):1449-1475.

Kazdin AE. 2009. Psychological science's contributions to a sustainable environment. *American Psychologist*, 64(5):339-356.

Latour B. 1993. *We have never been modern.* Cambridge, MA: Harvard University Press.

Lewis D & Mosse D (eds). 2006. *Development brokers and translators: The ethnography of aid and agencies.* Bloomfield: Kumarian.

Mji G, Gcaza S, Swartz L, McLachlan M & Hutton B. 2011. An African way of networking around disability. *Disability and Society*, 26(3):365-368.

Ritov J & Baron I. 2004. Omission bias, individual differences, and normality. *Organizational Behavior and Human Decision Processes*, 94(2):74-85.

Solms-Delta. 2011. *Solms-Delta Wine Estate.* [Accessed 5 July 2011] http://www.solms-delta.co.za

Struwig J. 2010. South Africans' attitudes towards the environment. In: B Roberts, M wa Kivulu & YD Davids (eds). *South African Social Attitudes. 2nd Report. Reflections on the Age of Hope.* Cape Town: Human Sciences Research Council (HSRC) Press.

Swartz L. 2010. Contextual issues. In: I Petersen, A Bhana, AJ Flisher, L Swartz & L Richter (eds). *Promoting mental health in scarce-resource contexts: Emerging evidence and practice.* Cape Town: HSRC Press.

Swartz L, Gibson K & Gelman T (eds). 2002. *Reflective practice: Psychodynamic ideas in the community.* Cape Town: HSRC Press.

Swim JK, Stern PC, Doherty TJ, Clayton S, Reser JP, Weber EU, Gifford R & Howard GS. 2011. Psychology's contributions to understanding and addressing global climate change. *American Psychologist*, 66(4):241-250.

Tomlinson M, Rohleder P, Swartz L, Drimie S & Kagee S. 2010. Broadening psychology's contribution to addressing issues of HIV/AIDS, poverty and nutrition: Structural issues as constraints and opportunities. *Journal of Health Psychology*, 15(7):972-981.

Uzzell D. 2010. Collective solutions to a global problem. Extracts from the joint British Academy/British Psychological Society Annual Lecture. *The Psychologist*, 23(11):880-883.

Chapter 23

EDUCATION

Socially critical education for a sustainable Stellenbosch

LESLEY LE GRANGE, CHRIS REDDY & PETER BEETS

Introduction

In this chapter we discuss some challenges for education in Stellenbosch in view of achieving a more sustainable future for the town. Our usage of the term 'sustainability' includes both ecological sustainability and social justice principles. Ecological sustainability concerns people/nature relationships and includes the following elements: interdependence (people are part of nature and dependent on it); biodiversity (all life forms should be respected by people); living lightly on the earth (biophysical resources should be used carefully and degraded ecosystems should be restored); and interspecies equity (all life forms have value independent of their perceived importance to humans). The social justice principle concerns people/people relationships and includes the following elements: basic human needs (the needs of all individuals and societies should be met within the constraints of the planet's resources); inter-generational equity (future generations should be left with a planet that has similar benefits to those enjoyed by present generations); human rights (all persons should enjoy fundamental freedoms); and participation (all persons in communities should be empowered to exercise responsibility for their own lives) (Fien 1993).

Our discussion of education challenges for Stellenbosch over the next two decades will be informed by the understanding of sustainability described above, but before beginning this discussion, we shall first provide some background information. The chapter is therefore divided into the following main sections: environment and environmental problems; environment and education; environment in South Africa's national curriculum frameworks; the education context in Stellenbosch; cases of partnerships between Stellenbosch University and schools; challenges for education with regard to sustainability in Stellenbosch; and socially critical education for a sustainable Stellenbosch.

Environment and environmental problems

Environmental problems have reached unprecedented levels and few would disagree that our planet is on the brink of ecological disaster. Many environmental problems, such as climate change, transcend national borders, but their effects are felt by local communities. As the temperature of our oceans (including ocean currents) increase annually, floods and droughts will become more prevalent – and Stellenbosch might not be spared these. Residential areas and developments on the floodplain of the Eerste River might therefore come under threat. History reminds us that, when Europeans such as Simon van der Stel settled in Stellenbosch, the Eerste River was so large that it had islands on which Europeans settled. There are of course also several localised environmental problems, such as the pollution of the Eerste River and its tributaries as a consequence (of among others things) economic activity such as commercial farming. As the population of the town increases, greater demands are placed on water and energy resources and there are challenges around waste disposal, and so on. These challenges are discussed in greater detail elsewhere in this book. What we wish to show is that environmental problems have multiple and interacting dimensions, and that any educational response to environmental problems requires an appreciation of this fact. Figure 23.1 shows that the biophysical dimension forms the base that supports all life and all human activity – the latter having produced the interacting social, economic and political dimensions.

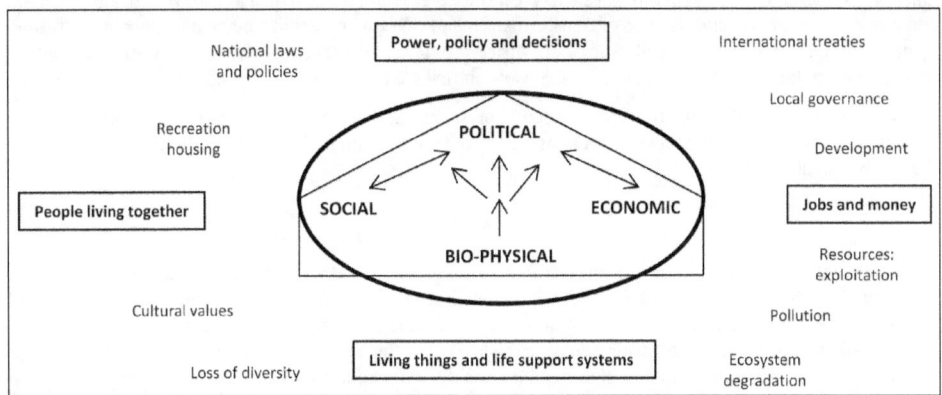

Figure 23.1   The four dimensions of environment [Adapted from Janse van Rensburg 1994]

By way of illustration: a polluted river is not simply a problem of the biophysical dimension – which may be solved through technical measures – but might have multiple causes, such as economic activity in the area, poor living conditions, and poor policy, decisions and regulation by local government. The effects of an environmental problem will also traverse various environmental dimensions.

Environment and education

The relationship between environment and education is not a simple one. There is a common belief around the world that education may solve (or largely contribute to solving) societal ills, including environmental problems. However, education *per se* might not be the answer; in fact, it may be part of the problem. Many of those who contribute most to environmental degradation hold university degrees, and many would have taken environmental courses as part of their degree programmes. We know that teaching students about environmental problems and risks is not enough, because they learn what Orr (1992) refers to as the 'lesson of hypocrisy' – that it is sufficient to learn about environmental deterioration without actually having to do anything about it.

Yet, there is a deeper issue at play here, which we may only explore briefly here. Postmodern writers have argued that the crisis in education and the environmental crisis have a common cause: the unquestioning belief in the modern progress story. In line with this thinking, Schreuder (1995) argues that environmental crises are inextricably linked to education crises. He argues that, in South Africa, the complex interplay between social, environmental and educational dilemmas has reached a crisis, which he attributes to either poor education or 'miseducation', or a combination of both. In terms of Schreuder's argument, poor education is the result of a political dispensation in which education resources were inequitably distributed, benefiting a minority of the population and leaving the vast majority of people with little or no access to quality education; while 'miseducation', which characterises the education of the privileged minority of the South African population, may be attributed to hegemonic influences of modernist economic and political ideologies on education philosophy and practice, which are often referred to as typical of a dominant positivist social paradigm. One consequence of miseducation is a lifestyle characterised by environmental illiteracy, overt consumerism, a lack of environmental sensitivity, and overexploitation of natural resources (*ibid.*). Lotz (1996) argues that education systems in which both poor education and miseducation thrive have pedagogical climates characterised by (among others) highly structured curricula, rote learning, inappropriate teacher-learner ratios, authoritarianism, teacher-dominated pedagogy, and poor-quality resource materials or a lack of resource materials.

On the social justice side, in educational theory there is what is referred to as 'reproduction' or 'correspondence theory'. Simply put, it holds that schools function to reproduce inequalities in society (or that schools mirror societal inequalities). Proponents of this theory have produced a body of literature critiquing the role of schools in society. For example, neo-Marxist writers (such as Bowles and Gintis (1976)) argue that schools function to reproduce class structures in society. On the other hand, others (such as Giroux and Simon (1984)) argue that schools may serve as sites of transformation and that one may move beyond languages of critique by developing languages of possibility. In South Africa, we have certainly seen schools being sites of political transformation, for example, during the 1976 Soweto uprising and the 1980 school boycotts.

As stated in the introduction, the above discussion serves as background information for analysing the current education context in Stellenbosch in order to identify education challenges to the goal of achieving a sustainable Stellenbosch.

Environment and the national curriculum

Environmental education was first formally introduced into the South African school curriculum in 1997, when a new curriculum framework (*Curriculum 2005*) was instituted. *Curriculum 2005* has since been revised, and in 2002 the *Revised National Curriculum Statement* (RNCS) for General Education and Training (GET)[1] was introduced. The RNCS has eight learning areas and is based on five principles, one of which pertains to social justice, a healthy environment, human rights and inclusivity. Two important points are made in the elaboration of this theme: the South African curriculum should play a role in creating awareness of the relationship between the different elements of this principle; and the principle should not be advanced in a single learning area only, but instead be integrated into the discursive terrains of all learning areas. In 2006, a new curriculum was phased in for Further Education and Training (FET), with – as in the case of GET – environmental concerns being infused into the learning outcomes and assessment criteria of all subjects.

While the inclusion of environmental concerns in national curriculum frameworks might have been a necessary and important step, there were particular challenges associated with this development in the South African context. The first challenge relates to the fact that post-apartheid South Africa's curriculum frameworks were outcomes based.[2] Although Outcomes-Based Education (OBE) might provide opportunities for innovative teachers to do creative work, Le Grange and Reddy (1997) argue that OBE and environmental education are incompatible. They point out that the perceived mechanistic, reductionist and instrumentalist epistemology of OBE might be antithetical to the holistic understanding of knowledge in environmental education – a widely accepted principle in environmental education circles worldwide. They also point out that OBE tends to favour a narrow conceptualisation of environmental education, which entails moulding learners through behaviour modification.

In addition, some government officials have suggested that environmental education in South Africa has come to an end – since environment is included in the curriculum, all that is required is environmental learning. Le Grange (2004) cautions that this view is uncritical of what has been included in curriculum statements and argues against the notion of environmental learning because, among other things, it makes education tantamount to an economic transaction.

A third challenge pertains to the growth of the technology of 'performativity' in South African education under the influence of the ascendancy of neoliberal politics globally (see Le Grange 2006), resulting in a focus on measurable outputs (such as student achievement) rather than on what is taught or when and how teaching occurs.

---

1 'General education' refers to the first ten years of compulsory schooling and the first four levels of Adult Basic Education and Training (ABET) in South Africa.

2 OBE has been in force in South African schools for the past 14 years. Although the national curriculum has been revised and OBE will be discarded, it will remain in effect in certain grades until the new curriculum is fully phased in.

Finally, a distinction is made in curriculum theory between 'the curriculum intended' and the 'curriculum-in-use'. This distinction helps us to understand that intentions are not necessarily mirrored by classroom practices – much depends on educational contexts, the capital learners bring to classrooms, and on what teachers do there. With regard to the latter, Fullan (1991:117) writes: "Educational change depends on what teachers do and think – it's as simple and complex as that."

All of the above may thwart efforts at advancing environmental education, and once again brings to the fore the question so neatly formulated by Robottom (1996:24):

> *Given its critical orientation, is environmental education better served by remaining permanently peripheral – a form of border pedagogy – rather than an institutionalised subject within the curriculum?*

The educational context in Stellenbosch

Back in 1994, Penderis and Van der Merwe noted that, despite the abolition of the Group Areas Act in 1991, very little residential integration had occurred in Stellenbosch and that the apartheid pattern of racially segregated residential areas still remained – and it is reasonable to claim that, at present, the situation remains largely the same, with the spatial morphology of Stellenbosch still conforming to that of the apartheid era. Our interest here is to assess the extent to which this fact has shaped the current educational landscape of the town, and the associated possibilities for – or threats to – achieving a more sustainable Stellenbosch in the foreseeable future.

When formal education was first introduced in Stellenbosch, it was aimed at serving the needs of European settlers. Indigenous peoples and slaves were denied access to education. The first education institution established in the town was a Dutch Reformed Church theological seminary, in 1859. A year later, the first school – for the daughters of Rhenish missionaries (the Rhenish Institute) – was opened, followed by a school for boys in 1866 (Stellenbosch Gymnasium, the present-day Paul Roos Gymnasium) and Bloemhof Girls High School in 1874 (for Afrikaans-speaking, white learners). The massification of schooling saw many more schools being founded by churches and successive governments, with schooling extended to coloured and African learners through the promulgation of various laws, including the establishment of the Department of Coloured Affairs in 1958 and Bantu Education Act of 1953. Prior to and during the apartheid era, education for different population groups was disparately funded. During the years 1976 to 1977, for example, the *per capita* expenditure in education was as follows: R551 per white learner; R236.13 per Indian learner, R185.16 per coloured learner; and R54.08 per African learner (Blignaut 1981). At the start of South Africa's democracy in 1994, the per capita expenditure per learner stood at the ratio of 4:3:2,6:1 for white, Indian, coloured and African learners respectively (Sedibe 2009). So although the situation had improved by 1994, gross inequalities remained.

Today there are twenty-eight public schools in Stellenbosch, of which nineteen are primary and nine secondary schools. Eight of these schools are situated on the periphery of the town. There are also three independently owned pre-primary schools. Eight of the public schools are former Cape Education Department (CED) (for white education) schools; sixteen are former House of Representative (HOR) (administering coloured education) schools; and two are former Department of Training (DET) (for African schooling) schools. There are no former Indian schools in Stellenbosch. Two public schools were established after 1994 and are administered by the Western Cape Education Department (WCED).

In order to redress some of the inequalities of the past, the National Norms and Standards for School Funding was amended in 2009, with schools being divided into National Quintiles (NQ) (based on three poverty indicators – income, unemployment rates and the level of education of the community in which the school is located) and the NQ category determining the per capita amount that a school receives for operational costs (see Table 23.1).

Table 23.1   National table of targets for school allocations, 2009-2011 [Adapted from the Department of Education (RSA DoE) 2008]

National Quintile	Funding allocation per learner		
	2009	2010	2011
NQ 1	R807.00	R855.00	R916.00
NQ 2	R740.00	R784.00	R840.00
NQ 3	R605.00	R641.00	R687.00
NQ 4	R404.00	R428.00	R459.00
NQ 5	R134.00	R147.00	R158.00
No fee threshold	R605.00	R641.00	R687.00

With the exception of one school, all former white schools in Stellenbosch fall in the NQ5 category and all former DET schools fall in the NQ1 one. Former HOR farms schools and the two township schools founded after 1994 also fall in the NQ1 category. All other schools fall in the NQ2, 3 and 4 categories (see Appendix 1). However, despite efforts to redress the past inequalities, educational disparities still persist. Schools allocated in NQ1 may charge exorbitant school fees, which enable their governing bodies to employ additional teachers, thereby keeping their teacher-learner ratios favourable. In fact, teacher-learner ratios reflect apartheid legacies. For example, the teacher-learner ratio of Rhenish Girls High (a former CED school) is 1:16; that of Lückhoff High School (a former HOR school) 1:29; and that of Kayamandi High School (a former DET school) 1:33.

Very little racial integration has occurred in Stellenbosch schools. Where it has happened is in former CED schools where middle- and upper-class coloured and African parents have enrolled their children. In short, schools in Stellenbosch continue to mirror the apartheid spatial morphology, reflecting the deep racial, class, and urban-rural divides still evident in Stellenbosch society. All of this does not augur well for a sustainable Stellenbosch. We shall return to this in a later section, but first, we will turn to a discussion of two instances of collaborative environmental education work between Stellenbosch University and schools in the town.

Partnerships between Stellenbosch University and schools in Stellenbosch

The partnerships described below involve collaborative projects between the Environmental Education Programme University of Stellenbosch (EEPUS), schools in Stellenbosch, and other role players.

The School Water Analysis Programme (SWAP)

The Schools Water Analysis Programme (SWAP) kits[3] and materials were developed as part of a process of research and development geared towards a research-based approach towards professional development in environmental education. For this project (1992-1998) the Department of Water Affairs and Forestry (DWAF)[4] partnered with EEPUS to provide a research-based professional development programme for teachers and learners in schools, with DWAF and the former World Wildlife Fund (now the World Wide Fund for Nature (WWF)) providing information and funding to make professional development opportunities for teachers in schools possible. DWAF and the WWF also provided information about important catchment areas that they considered to be at risk and in need of investigation. The partnership essentially advanced some of the educational aims of environmental education as developed at the Tbilisi Conference of 1977: knowledge acquisition, awareness creation, skills development, attitude changes, participation, learning through experience, and empowerment/capacity enhancement.

---

3   SWAP kits are low-cost water monitoring resource materials that were developed from the original Global Rivers Environmental Education Network (GREEN) kits in America. They include basic scientific tests and other data collecting techniques that may be used for rivers and catchment areas in general. SWAP was initially developed as an environmental education resource for teachers and learners at schools, but has also found application in community-based projects and adult education programmes in various contexts.

4   In 2009, the former DWAF was split, with the forestry responsibility being transferred to the new Department of Agriculture, Forestry and Fisheries and the old department being renamed to the Department of Water Affairs.

The role of Stellenbosch University staff was to provide expertise related to the SWAP resource, as well as professional development programmes for teachers. This involved the implementation of an in-service workshop programme for teachers developed by Stellenbosch University staff. Interactive workshops were held at four sites (close to schools) selected by EEPUS staff. During these workshops, a facilitator from Stellenbosch University introduced teachers from the selected schools to the SWAP resource, followed by on-site SWAP activities at the water source in the vicinity of the school to familiarise teachers with the resource and its functionality. Further follow-up work included assisting teachers to identify curriculum possibilities for water testing and accompanying teachers and learners on field trips in order to familiarise them with the SWAP field tests and facilitate learner activities. Several monitoring sites were set up along the Eerste and Plankenbrug rivers, and learners visited these sites regularly to develop data based on the tests included in the SWAP kit. Many schools in Stellenbosch were involved in the project during the early to middle 1990s, but this involvement has not been sustained. However, the programme has been extended to other areas in Greater Cape Town over the years, and pre-service teachers at the Faculty of Education of Stellenbosch University still get an introduction to the SWAP kits and visit the sites along the Eerste and Plankenbrug rivers.

In the context of this project, Stellenbosch University staff supported applied and collaborative research. The research was based on active learning, was investigative, and focused on action to improve teaching practice. The process of introducing teachers to the SWAP resource was interactive, provided opportunities for discussion, and made possible the development of skills that would enable teachers to approach their work differently when applying these skills. It was believed that such an approach to research partnerships might lead to the involvement of the broader school community or civil society, promote mutually beneficial partnerships, and enable innovative teaching practices.

The Rietenbosch Wetland Rehabilitation Programme (RWRP)

The RWRP (2005-present) has as its aims the rehabilitation and development of the wetland on the Rietenbosch Primary School premises in Cloetesville, Stellenbosch. The school is situated in a relatively socio-economically poor community and is, for historical reasons, not very well resourced. The overall environment of the area is rather drab and offers its residents little by way of aesthetic beauty or recreational resources.

The RWRP is a multi-stakeholder project involving EEPUS, the Department of Environmental Affairs (DEA), the Rupert Trust, South African Breweries and Rietenbosch Primary school. Funding provided thus far has been divided between the rehabilitation of the wetland and the development of wetland-focused educational resources for use by teachers from various schools. The wetland area will then serve as an educational resource for teachers, learners at this and neighbouring schools, and the broader community of Cloetesville.

The project effectively aims to restore and develop the Rietenbosch wetland as a multi-purpose resource in the hope of realising various benefits:

- an ecological benefit (maximising the quality and diversity of the wetland habitat);
- an educational benefit (the ecological improvement will allow the wetland to serve as a resource for professional teacher development in Environmental Education in the formal school curriculum);
- community involvement (the school community will maintain the wetland through a volunteer parent system working in tandem with wetland experts);
- recreational benefits (a picnic spot will provide further opportunities for the community and visitors to interact with care towards the environment);
- the creation of a youth club to provide both schoolgoing youth and those who have left school with diversion (especially as a counter to crime and violence); and
- the development of learning support material (a package for environmental education aligned to the national curriculum).

Reflections on SWAP and the RWRP

As mentioned earlier, the SWAP project had multiple benefits, which are also envisioned for the RWRP. However, partnership projects of this kind are not without their attendant challenges. In the case of SWAP, responses from participants from all the schools involved reflected both positive and negative experiences.

Teachers indicated that learners enjoyed working with the resource and that they found it novel and innovative. However, they also mentioned that school timetables and organisation proved too rigid in some cases, which complicated implementation of the programmes and constrained the use of the SWAP kits in their teaching.

The broader social context of the schools involved appeared to be an important factor influencing the outcomes of the project and the responses of communities of parents and learners also reflected this. At two of the schools, where the surrounding community seemed to be economically depressed and unemployment was rife, there was minimal community assistance or participation, limited cooperation from parents,[5] and learners could not manage the resource; while other schools in the community enjoying much better socio-economic conditions recorded strong cooperation from parents in terms of training and the provision of transport, even going so far as to commit to action to improve water quality in the local environment.

For Stellenbosch University, the partnership provided an opportunity to develop new knowledge about the possible use of SWAP resource kits in school settings and professional teacher development. In fact, the partnership enabled the professional development of teachers in many ways. One teacher mentioned that working with SWAP had made her aware of the location of the river and water quality problems at the site, issues she had never thought of before. Teacher development was also evident in teachers' comments that they had learned new skills and alternative ways of teaching and developing relationships with colleagues. Teachers also mentioned that working with the resource had provided them with ideas for their teaching practice and had inspired them to consider ways of improving their existing practice. It had also increased awareness of the local environment and stimulated interest in applying new resources and approaches to their work.

Educational challenges and threats to achieving sustainability in Stellenbosch

As mentioned earlier (and throughout this book), Stellenbosch is beleaguered by gross social and economic inequalities, and these inequalities are reflected and reproduced in the schooling system. It is the authors' contention that the social scars of racial, class and urban-rural division, which overlay the town's natural and aesthetic beauty, might thwart efforts to achieve a more sustainable Stellenbosch within the foreseeable future. Social and economic injustice poses a threat to both the social justice and ecological sustainability principles mentioned in the opening to this chapter. Inequalities in any broader community will serve as barriers to collective action aimed at improving environmental risk positions. Moreover, when affluence and poverty are present in the same community (and this is certainly the case in Stellenbosch), the potential is there for both miseducation and poor education to thrive. Pedagogies are likely to be teacher-dominated, curricula highly structured, and teacher-learner ratios inappropriate (among other things). In terms of the latter, the difference in teacher-learner ratios between privileged, former white schools and poor black schools located in townships are stark (see Appendix 1). Such a context will not provide opportunities for schools to engage critically with local environmental problems. In the case of SWAP, we saw that although teachers found the kits and materials useful and innovative, implementation was in some instances constrained by rigid school timetables and organisation (which may also offer a reason for why SWAP has not been sustainable, in the sense that schools have not taken ownership of the programme; it only functioned when Stellenbosch University staff were leading and driving the process). The introduction of the new Curriculum and Assessment Policy Statement (CAPS) in 2012, which involves even more structured curricula and a stronger emphasis on school textbooks, only exacerbates the problem. Thus, the educational challenge here is how best to ameliorate this situation.

As mentioned earlier, environmental concerns were included in the learning outcomes and assessments standards of all learning areas of the GET, as well as all FET subjects. This was a necessary and positive step. However, the inclusion of environmental concerns in national curriculum frameworks does not necessarily translate into the implementation of environmental education programmes that are meaningful (programmes that empower learners to engage local environmental problems). There are several potential barriers to meaningful implementation of environmental education in Stellenbosch schools, of which we shall mention only a few here:

---

5   When parents in one area were asked to send plastic bottles and ice cream dishes to school, nothing was forthcoming. Further investigation revealed that these items were sought-after resources in local households, because they served as water containers and food storage containers.

- Although environmental concerns have been incorporated in national curriculum frameworks, in many instances this inclusion involves cosmetically tagging the word 'environment' onto learning outcomes and assessment standards (see Le Grange 2010 for a detailed discussion).
- This cosmetic 'tagging-on' of environment might leave some teachers confused, while others may opt not to incorporate environmental concerns in their programmes. We know that teachers engage policy rhizomatically – some may comply, while others may resist or even subvert policy (see Honan 2004).
- Because environmental education has not been authentically and coherently conceptualised in South Africa's new CAPS teachers who do not have a strong background in environmental science (and/or a related field) may not have the confidence to implement environmental education programmes (bearing in mind that they have not received in-service education programmes specifically focused on environmental education).

The challenge lies in overcoming these barriers to ensure that programmes that may contribute to a more sustainable Stellenbosch are implemented in Stellenbosch schools.

Education for a sustainable Stellenbosch

Given that environmental concerns have been included in successive national curriculum frameworks, one might assume that some form of education about environmental problems occurs in Stellenbosch schools. And, with the opportunities afforded by the town of Stellenbosch and its surrounds, one might also assume that at least some teachers take learners into natural settings where they may engage directly with the 'more than human' world (non-human nature). However, we wish to argue here that, although knowledge about the environment and experiences in natural settings are important, it is not sufficient if schools are to play a meaningful role in helping Stellenbosch to become more sustainable. In order to defend this position, it is useful to look at three approaches towards (environmental) education: education *about*, *in* and/or *for* the environment.

Distinguishing between education *in*, *about* and/or *for* the environment – as first suggested by Lucas in his PhD thesis of 1972 (in Gough 1997) – has become an established way of thinking in environmental education. According to Fien (1993), education *about* the environment emphasises knowledge about natural systems and processes; education *in* the environment emphasises learners' experience in the environment as a means of developing their competencies and values clarification capacities; and education *for* the environment has an overtly critical agenda of values education, social change and transformation through action-based exploration and involvement in resolving environmental problems. Fien notes that these three approaches have often been associated with ideological positions: neo-classical/vocational, liberal progressive, and socially critical theory, respectively. The neo-classical/vocational orientation sees education as preparation for work, which Fien characterises as education that uncritically accepts existing social structures and hierarchies, which may thus perpetuate elitism, injustice, and class and gender inequalities, and favour economic growth at the cost of environmental protection. The liberal progressive orientation views education as preparation for life and emphasises individual excellence and achievement. The socially critical orientation towards education contends that education should not merely prepare learners for the world of work, but that it should engage society and social structures directly. A socially critical education develops constructively critical thinking, not just in individuals, but also in group processes (Kemmis, Cole & Sugget 1983).

Table 23.2 provides an overview of the above three approaches to environmental education.

Table 23.2   Approaches to environmental education [Adapted from Kemmis, Cole & Suggett 1983 and Fien 1993]

	About	In/through	For
Aims	Vocational	Liberal/progressive	Socially critical
View of knowledge	Pre-existing in the community; developed by experts	Sensory awareness through contact with surroundings	Emerging from inquiry into contextual issues
Role of teachers	Dispensers of knowledge and authority figures	Providers of experience	Co-investigators and knowing adults; resourceful individuals organising critical and collaborative projects with learners and the community
Role of learners	Passive receivers	Active learners	Active learners/investigators
Role/nature of texts	Source of authority	Provides guidelines and often involves fieldwork	Emerges from the inquiry

Education *about* and *in* the environment does not have an interest in addressing the underlying causes of environmental degradation in local communities. It has no interest in transforming society in order to improve environmental risk positions. For this reason, we argue that any contribution to a more sustainable Stellenbosch through education may only be possible through socially critical education. What might this entail?

Firstly, it requires an investigation of local environmental problems and their underlying causes – learners must understand the interacting biophysical, social, political and economic dimensions of environmental problems in Stellenbosch. Associated with this is a need for learners to understand why Stellenbosch remains a segregated society almost two decades after South Africa became a democracy. Investigating local environmental problems will enable learners to understand that these are produced both by poor living conditions and affluent lifestyles. Such an investigation will help learners, teachers and community members to realise that local environmental problems (and their solutions) are complex, and that addressing these problems requires collaboration on the part of all sectors of the community – and may thus serve as a basis for crossing boundaries within the community. Secondly, a socially critical education for a sustainable Stellenbosch requires teachers to change their roles and to understand that knowledge and texts are to be viewed differently.

A socially critical education for a sustainable Stellenbosch will also require less structured curricula, reorganisation of schools in terms of timetables, etc., and weaker boundaries between schools, classrooms and communities. It might require doing things against the grain. A model similar to that of the Wildlife and Environment Society of South Africa (WESSA) Eco-Schools Programme (see WESSA Eco-Schools 2010) could be implemented as part of a sustainable education process. Essentially, this would involve developing a policy for environmental education at school level, which could in turn involve a variety of aspects related to school life and activities, including recycling programmes, energy and water audits, heritage interests, and learner community initiatives promoting better environmental practices.

Setting up such a programme must be a collective and collaborative effort involving various schools. The process could be similar to Paulo Freire's (1972; 1993) suggestions for developing a 'critical consciousness' by way of generative themes collaboratively identified and developed by the various role players (teachers, learners and parents) in school communities in the town of Stellenbosch. In this way, instead of perpetuating a set of individual policies essentially representing the insular interests of particular schools, themes important to all may form the basis of a town-wide school policy. While the policy should be general in nature, so as to be appropriate for all the schools in the town, it should also be flexible enough to accommodate and address the particular local issues of each school context. These are not easy challenges and overcoming them will "inevitably involve sweat, personal risk, intellectual daring and inconvenience" (Orr 1992:166) – a way of doing things that may well run counter to the modes that currently dominate curriculum and classroom practice in Stellenbosch schools.

Conclusion

In writing this chapter we decided not to focus on how education (and its institutions) in Stellenbosch might be sustained, as such a position may place education in Stellenbosch on an unsustainable path. Put differently, if education in Stellenbosch does not undergo radical transformation, then education institutions will continue to contribute to the currently unsustainable trajectory of the town. The path to a sustainable Stellenbosch is a socially critical one, which may be kick-started by tapping into existing partnerships and assimilating the lessons learned from the two partnership projects described in this chapter.

The experience of both the SWAP and RWRP projects teaches us that partnerships between schools, universities and other organisations are complex ones that are difficult to establish and to sustain. If success is to be achieved, total commitment is required from all partners. It would be prudent to be mindful of the experiences of Watson and Fullan (1992) which suggest that, in order to establish meaningful and beneficial relationships, school-university partnerships must develop a shared vision that transcends self-interest. We also support the argument of Sandholtz and Finnen (1998) that school-university partnerships should focus on developing relationships rather than on programme specifics, as this cultivates trust and facilitates informal discussions and interactions between researchers and teachers that often lead to the attainment of the other stated outcomes. In line with both these arguments, we suggest that, if an Eco-Schools model is to be implemented in the Stellenbosch area, it should be one that focuses on collaborative efforts between schools rather than insular eco-policies developed by individual schools.

In closing this chapter and our defence of a socially critical educational practice for a sustainable Stellenbosch, we would like to part with these words by Susanne Kappeler (1986:212):

> [We] do not really wish to conclude and sum up, rounding off the argument so as to dump it in a nutshell for the reader. A lot more could be said about any of the topics [we] have touched upon ... [we] have meant to ask the questions, to break out of the frame ... The point is not a set of answers, but making possible a different [educational] practice ...

REFERENCES

Blignaut S. 1981. *Statistics in education in South Africa, 1968-79*. Braamfontein: South African Institution of Race Relations.

Bowles S & Gintis H. 1976. *Schooling in capitalist America: Educational reform and the contradictions of economic life*. London: Routledge and Paul.

Fien J. 1993. *Education for the environment: Critical curriculum theorising and environmental education*. Geelong, Victoria: Deakin University Press.

Freire P. 1972. *Pedagogy of the oppressed*. New York: Seabury.

Freire P. 1993. *Pedagogy of the city*. New York: Continuum.

Fullan M. 1991. *The new meaning of educational change*. London: Cassell.

Giroux HA & Simon R. 1984. Curriculum study and cultural politics. *Journal of Education*, 166:226-238.

Gough A. 1997. *Education and the environment: Policy trends and the problem of marginalisation*. Melbourne: The Australian Council for Educational Research.

Honan E. 2004. (Im)plausibilities: A rhizo-textual analysis of policy texts and teachers' work. *Educational Philosophy and Theory*, 36(3):267-281.

Janse van Rensburg E. 1994. Social transformation in response to the environment crisis: The role of educational research. *Australian Journal of Environmental Education*, 10:1-20.

Kappeler S. 1986. *The pornography of representation*. Minnesota: University of Minnesota Press.

Kemmis S, Cole P & Suggett D. 1983. *Orientations to curriculum and transition: Towards the socially-critical school*. Melbourne: Victorian Institute for Secondary Education.

Le Grange L. 2004. Against environmental learning: why we need a language of environmental education. *Southern African Journal of Environmental Education*, 21:234-240.

Le Grange L. 2006. Quality assurance in South Africa: A reply to John Mammen. *South African Journal of Higher Education*, 20(6):903-909.

Le Grange L. 2010. Environment in the Mathematics, Natural Sciences and Technology learning areas for General Education and Training in South Africa. *Canadian Journal of Mathematics, Science and Technology Education*, 10(1):13-26.

Le Grange L & Reddy C. 1997. Environmental education and outcomes-based education in South Africa: A marriage made in heaven? *Southern African Journal of Environmental Education*, 17:12-18.

Lotz H. 1996. The development of environmental education resources materials for junior primary education through teacher participation: The case of the We Care primary project. PhD thesis. Stellenbosch: Stellenbosch University.

Orr D. 1992. *Ecological literacy: Education and the transition to a postmodern world.* Albany, NY: State University of New York Press.

Penderis SP & Van der Merwe IJ. 1994. Kayamandi hotels, Stellenbosch: Place people and policies. *South African Geographical Journal*, 76(1):33-38.

Randolph C. 2002. Partnerships: Promoting equity and excellence in science. *The Science Teacher*, October:10-10.

Robottom I. 1996. Permanently peripheral? Opportunities and constraints in Australian environmental education. *Southern African Journal of Environmental Education*, 14:20-26.

RSA DoE (Republic of South Africa. Department of Education). 2008. *South African Schools Act 1996 (Act No 84, 1996): National Norms and Standards for School Funding.* Government Notice 17 October 2008, No 1089. Pretoria: Department of Education.

Sandholtz J & Finan E. 1998. Blurring the boundaries to promote school-university partnerships. *Journal of Teacher Education*, 49(1):13-25.

Schreuder D. 1995. Delusions of progress: A case of reconceptualising environmental education. *Southern African Journal of Environmental Education*, 15:18-25.

Sedibe G. 2009. The achievement gap between learners who are assessed in a primary language and those assessed in a non-primary language in the Natural Sciences learning area. MA thesis. Stellenbosch: Stellenbosch University.

Watson N & Fullan M. 1992. Beyond school district-university partnerships. In: M Fullan & A Hargreaves (eds). *Teacher development and educational change.* London: The Falmer Press.

WCED (Western Cape Education Department). 2011. *Central Education Management Information System (CEMIS).* [Accessed 15 February 2011] http://www.cemis.wcape.gov.za/pls/cemis

WESSA (Wildlife and Environment Society of South Africa) Eco-Schools. 2010. *What we do.* [Accessed 24 March 2011] http://www.wessa.org.za/what-we-do/eco-schools.htm

APPENDIX 1

Teacher-learner ratios in Stellenbosch schools (compiled from WCED 2011)

SCHOOL NAME	SITUATION	TYPE	EXDEPT	LOLT	NAT. QUINTILE	Per LEARNER allocation	Teacher/ learner ratio	PRE GRR	GRR	GR1	GR2	GR3	GR4	GR5	GR6	GR7	GR8	GR9	GR10	GR11	GR12	LSEN	TOTAL # learner	TOTAL # teacher
Devonvallei Primary	Periphery	Primary School	HOR	Afr	NQ1	R 855.00	1:29	0	29	38	33	20	28	27	28	30	0	0	0	0	0	0	233	8
Ikaya Primary	Centre	Primary School	DET	Par:Xho/Eng	NQ1	R 855.00	1:32	0	211	196	166	174	123	157	204	180	0	0	0	0	0	0	1411	44
J.J. Rhode Primary	Periphery	Primary School	HOR	Afr	NQ1	R 855.00	1:26	8	27	26	34	30	30	32	32	45	0	0	0	0	0	0	264	10
Kayamandi Primary	Centre	Primary School	WCE	Par:Xho/Eng	NQ1	R 855.00	1:33	0	88	92	90	62	59	65	94	100	0	0	0	0	0	0	650	20
Kayamandi Secondary	Centre	Secondary School	DET	Eng	NQ1	R 855.00	1:32	0	0	0	0	0	0	0	0	0	190	233	168	175	0	931	29
Koelenhof RK Primary	Periphery	Primary School	HOR	Par:Afr/Eng	NQ1	R 855.00	1:30	0	70	128	122	113	105	53	51	45	0	0	0	0	0	0	687	23
Makupula Secondary	Centre	Secondary School	WCE	Eng	NQ1	R 855.00	1:32	0	0	0	0	0	0	1	0	2	171	272	214	87	46	0	793	25
Vlottenburg Primary	Periphery	Primary School	HOR	Afr	NQ1	R 855.00	1:28	0	51	44	54	37	44	43	41	50	0	0	0	0	0	0	364	13
Lynedoch Primary	Periphery	Intermediate School	HOR	Afr	NQ2	R 784.00	1:31	0	34	39	24	42	44	29	41	26	26	0	0	0	0	0	305	10
Stellenzicht Secondary	Periphery	Secondary School	HOR	Afr	NQ2	R 784.00	1:25	0	0	0	0	0	0	0	0	0	142	195	126	73	72	0	608	24
Weber Gedenk NGK Primary	Periphery	Primary School	HOR	Afr	NQ2	R 784.00	1:29	0	64	104	86	72	86	63	78	110	0	0	0	0	0	0	669	23
Bruckner de Villiers Primary	Centre	Primary School	HOR	Afr	NQ3	R 784.00	1:33	0	30	36	43	43	38	37	39	35	0	0	0	0	0	0	301	9
Cloetesville Primary	Centre	Primary School	HOR	Afr	NQ3	R 784.00	1:29	0	76	81	55	74	73	77	82	66	0	0	0	0	0	0	584	20
P.C. Petersen Primary	Periphery	Primary School	HOR	Afr	NQ3	R 784.00	1:25	0	50	68	58	61	44	59	66	88	0	0	0	0	0	0	494	20
St. Idas RK Primary	Centre	Primary School	HOR	Afr	NQ3	R 784.00	1:25	0	27	32	41	32	32	29	32	29	0	0	0	0	0	0	254	10
A.F. Louw Primary	Centre	Primary School	CED	Par:Afr/Eng	NQ4	R 428.00	1:26	0	48	87	83	83	74	52	57	60	0	0	0	0	0	0	544	21
Cloetesville High	Centre	Secondary School	HOR	Afr	NQ4	R 428.00	1:30	0	0	0	0	0	0	0	0	0	268	375	275	240	121	0	1299	43
Idasvallei Primary	Centre	Primary School	HOR	Afr	NQ4	R 428.00	1:32	0	76	118	113	95	127	104	112	112	0	0	0	0	0	0	857	27
Luckhoff Secondary	Centre	Secondary School	HOR	Afr	NQ4	R 428.00	1:29	0	0	0	0	0	0	0	0	0	224	208	166	123	92	0	813	28
Rietenbosch Primary	Centre	Primary School	HOR	Afr	NQ4	R 428.00	1:28	50	71	116	93	83	105	63	78	0	0	0	0	0	0	13	636	23
Pieter Langeveldt Primary	Centre	Primary School	HOR	Afr	NQ4	R 428.00	1:29	0	60	102	87	80	97	96	96	117	0	0	0	0	0	0	735	25
Bloemhof High	Centre	Secondary School	CED	Afr	NQ5	R 161.00	1:19	0	0	0	0	0	0	0	1	0	1	139	141	148	134	142	0	706	38
Eikestad Primary	Centre	Primary School	CED	Afr	NQ5	R 147.00	1:21	0	105	91	107	80	107	107	88	100	0	0	0	0	0	0	785	37
Paul Roos Gimnasium	Centre	Secondary School	CED	Par:Afr/Eng	NQ5	R 147.00	1:18	0	0	0	0	0	0	0	0	0	238	232	247	221	222	0	1160	66
Rhenish High	Centre	Secondary School	CED	Eng	NQ5	R 192.00	1:16	0	0	0	0	0	0	0	0	0	156	150	145	130	115	0	696	44
Rhenish Primary	Centre	Primary School	CED	Eng	NQ5	R 147.00	N/A	37	72	75	85	61	78	66	91	69	0	0	0	0	0	0	634	N/A
Stellenbosch High	Centre	Secondary School	CED	Afr	NQ5	R 150.00	1:19	0	0	0	0	0	1	0	0	0	121	122	120	123	126	0	613	33
Stellenbosch Primary	Centre	Primary School	CED	Afr	NQ5	R 147.00	1:24	0	68	75	83	77	91	82	94	68	0	0	0	0	0	0	638	27
TOTAL								95	1257	1548	1457	1319	1386	1243	1411	1411	1695	1860	1674	1299	1111	19	18,664	

LOLT – language of learning and teaching

Chapter 24

SPORT

Devising a game plan for a sustainable Stellenbosch

JULIAN F. SMITH

Introduction

The history and prerequisites with regard to sport in Stellenbosch are key co-determinants in the pursuance of a sustainable town.

Sport as an ideologically significant social phenomenon or activity has been widely recognised for its impact on individuals, groups, local communities, regions, and the world in general. Black and Nauright (1998:2) state that sport does not operate in a "separate sphere". Consequently, its impact may be observed locally, regionally, nationally, continentally and globally. Sport has been shown to construct identities, but also to destroy them; it has been used to create cohesion, but also to fragment; it has united, but it has also divided; it has been the cause of pride, but also of shame; it has been about the creation of opportunities, but also about the abuse of opportunities; it has been used to advance broad, common interests, but also narrow self-interest; it has been a terrain of harmony, but also of struggle; it has contributed to the creation of hope, but has also led to despair.

While sport has grown into a mega industry globally, at a local level it is bound to be greatly impacted by political forces and social and economic conditions. Thus, in order for sport to contribute significantly to a sustainable Stellenbosch, some prerequisites pertaining to these forces and conditions may be accepted to apply. These include: political stability in local government; continuity in local administration; a local sports plan that articulates with a provincial and a national sport framework or plan and receives due prominence and sufficient resources in (municipal) integrated development plans; adequate support from provincial and national sport federations; strategic partnerships between the municipality, the university and other local institutions and structures; the availability of and access to sports facilities and open spaces that may be used for practice; and duly recognised exceptional sport achievements.

This chapter will present a brief overview of the sport landscape in general and its implications for sport in (a sustainable) Stellenbosch, with some reference to international practice that has proved successful and that might find useful application in the Stellenbosch context, before moving on to a brief stocktake of Stellenbosch sport past and present; the assets it has at its disposal to contribute to a sustainable town; and the hurdles it will have to clear in order to do so.

TEXT BOX 1

Professional sport, which has become an industry with global reach and power, is described as "sports in which athletes receive payment for their performance" (Wikipedia 2011a). The amounts involved are truly astronomical; for example, the Indian Premier League pays its cricketers salaries of about USD1.5 million for 45 days' worth of 'work' each year. With top players and athletes gaining superstar status, it is only to be expected that sport administrators would want to capitalise on fans' devotion by a variety of means, including "the sale, display, or use of sport or some aspect of sport so as to produce income" (Balisunset 2008:n.p.). The money to be made does not stop there, however. With fans literally willing to go to the ends of the earth to support their teams or see their favourite sport stars in action, sport tourism has grown into a multi-billion dollar industry (in 2011, sport tourism contributed more than R6 billion to the South African economy (The Crest Online 2012)). Sport tourism is not, of course, limited to fans; major sport events (or so-called 'mega events') often spawn corollary events that have their own economic spin-offs in terms of local hospitality and tourism industries, such as the Sport Mega Events and Their Legacies Conference, which was held in Stellenbosch in 2009 during the build-up to the 2010 FIFA World Cup. Chan (2010) notes eleven variables that determine the economic value of sport, namely: manufacturing/ merchandising, consumer spending, corporate activity/sponsorship, event hosting, public sector grants/funding, broadcasting/television rights, employment/jobs/salaries, sports betting, sport tourism, international funding and tax. A sustainable Stellenbosch would arguably have taken these factors into account and put in a much greater effort to capitalise on sport tourism than it did for the 2010 FIFA World Cup.]

Match preview

The prize

Generally speaking, a sustainable future Stellenbosch may be expected to have the following features: environment-friendly development; environmentally sensitive service delivery (including the delivery of water and electricity and the handling of sewage); socially responsible and equitable spatial use; good governance (including sound, uncorrupted and incorruptible financial control and robust procurement systems); responsible, town-serving – as opposed to self-serving and party-serving – local politicians; a redress of past injustices and inequalities in a fair and balanced manner that promotes reconciliation and social cohesion; and the promotion of a healthy lifestyle.

What may we reasonably expect from the game?

Reflecting on the role that sport may play in ensuring a sustainable Stellenbosch, it is necessary to determine how and what sport may reasonably be expected to contribute.

▣ General expectations

The following definition of sport is representative of how this social activity is generally described:

> *A sport is an organised, competitive, entertaining, and skilful activity requiring commitment, strategy, and fair play, in which a winner can be defined by objective means (and) it is governed by a set of rules or customs.*
>
> [Wikipedia 2011b]

While sport, as entertainment, has become increasingly professionalised and commercialised, it is also viewed as a tool to drive political and social change. Millward (2007:n.p.) describes the interaction between sport and ideology as follows:

> *[...] the way in which the former, as a distinct form of leisure activity, impacts on the body of ideas which reflect the beliefs of a social group or political system. Indeed, the ideological capacity of sport, rather than religion, might sensibly be considered to be the new opiate of the people.*

The history of sport in South Africa serves as a prime example of how sport was used both to support an unjust and untenable system, and to oppose and undermine it. Local and foreign examples of anti-apartheid sport actions and strategies abound, for example the 'No normal sport in an abnormal society' approach of the South

African Council on Sport (SACOS) era; the support of sport boycotts and vehement opposition to 'rebel' tours; the disruption of sport events overseas; and the active pursuit of non-racialism in sport.[1]

There is currently great emphasis on the potential contribution of sport to general social improvement in terms of human development and health, gender equality and peace-building, with concepts such as 'education through sport' and 'peace through sport' gaining some prominence. Ban Ki-Moon has actively propagated the notion that sport is a valuable tool for achieving the Millennium Development Goals (MDGs), stating that it "exemplifies that very spirit [required for the realisation of the MDGs]: teamwork, fair play and people collaborating for a common goal" (People's Daily Online 2010:n.p.).[2] The United Nations Inter-Agency Task Force on Sport for Development and Peace report entitled *Sport as a Tool for Development and Peace: Towards achieving the United Nations Millennium Development Goals* makes various notable recommendations in this regard, some of which Stellenbosch would do well to bear in mind if it is to become a sustainable town:

- *sport should be better integrated into the development agenda as a useful tool;*
- *programmes promoting sport for development need greater attention and resources by (local) government;*
- *communications-based activities using sport should focus on well-targeted advocacy and social mobilisation, particularly at the local level; and*
- *forming partnerships is the most effective way to implement programmes using sport for development.*

[United Nations Inter-Agency Task Force on Sport for Development and Peace 2005:4]

Expectations of sport in a rural context

Historically, sport in non-metropolitan areas has not enjoyed the same exposure and prominence as sport in urban settings. It inevitably also does not have the profile to attract major sponsorship and funding. This does not, however, detract from the potential of sport as a force for social betterment. On the contrary, it could be used to support and enhance various local objectives.

Sport in rural areas in South Africa typically face daunting challenges, which contribute to lower levels of participation and often result in the discontinuation of some sport codes and the closure of clubs. Lack of vision and leadership at a national and provincial level, misguided priorities, and poor local governance may all be regarded as contributing factors to this situation.

With its high-quality amenities, proximity to Cape Town, and international profile as a university town and tourist destination – as well as the increasing presence of the chairmen and directors of some of South Africa's largest corporations (see Formby 2007) – Stellenbosch is hardly a typical rural town. Yet, in terms of sport and development, it may glean some valuable lessons from international practice and initiatives on a local (or non-metropolitan) level. One example is the Local Sports Councils (LSCs) established by the Sussex Association of Local Councils (SALC) "to promote and coordinate sports activities at club and neighbourhood level" in local authority areas (SportScotland 2011:n.p.). These LSCs are usually assisted by a sport development officer (or similar) to help clubs develop their facilities and activity programmes and advise them on coaches and training schemes, as well as to assist them in accessing financial support.

On a rather more expansive scale, the International Olympic Committee (IOC) launched an Olympic Youth Development Centre in Zambia, in partnership with government and international and national sports federations and donors, as part of its Sports for HOPE project (Olympic.org 2010). The aim of the project is to combine the pleasure associated with sport with learning and development opportunities, and it serves as a model that a town such as Stellenbosch, with the considerable resources at its disposal, should be able to replicate. Coincidentally, in 2010, Stellenbosch University launched its own HOPE Project, with the objective of

---

1 The role of sport in both upholding the apartheid regime and supporting the anti-apartheid 'struggle' has been extensively documented, interpreted and contextualised from various points of views and in multiple forms. See, for example, Black and Nauright (1998) and Grundlingh, Odendaal and Spies (1995) for a revealing analysis of the link between Afrikaner nationalism and rugby; and Reddy (n.d.) about the role of sports boycotts in the liberation struggle.

2 See also David (2010) in this regard.

creating opportunities and solutions to societal problems through the application of knowledge, science and technology. What needs to be added are national, provincial and local government endorsement and support, as well as the support of donors (of which Stellenbosch has no shortage). Additional support may be enlisted from Stellenbosch Tourism, which thus far does not seem to have been included in a town-wide strategy.

Expectations of sport in higher education

Stellenbosch is, of course, a university town, and sport is often a very prominent extra-curricular activity on university or college campuses. While it has not escaped the forces of professionalisation and commercialisation, the sport activities of institutions of higher education (HE) may be used to advance the vision of these institutions and are often used as a vehicle and barometer of change; and they may play a key role on a national level in supporting and complementing the national sport agenda in tandem with the appropriate national sport federations.[3] Appropriate strategic partnerships between the HE sector and government (through its respective departments) are of great importance in this regard.

A comprehensive audit of Scottish and British university and college sport by Sport England represents a very good example of a fundamental and serious approach to the role of higher education in sport. The exercise provided excellent information, which may be used to define and plan the role of HE sport to enhance national objectives. The foreword to the subsequent report of the audit (Universities UK 2004) contains two contributions that are particularly relevant to the topic of a sustainable Stellenbosch:

> *The findings [...] demonstrate that higher education institutions across the country make a very important contribution to sport. This contribution is not just through the fantastic opportunities they offer students and staff but also through those they extend to a wide cross section of people in their local communities and top-level competitors in this country. More exciting, however, is the prospect that this review raises of higher education institutions playing an even greater role – one that may see them becoming co-ordinating hubs of provision in their local communities – maximising the synergy they can bring between education, volunteering, participation and competitive success. Sport England looks forward to building even stronger working relationships with higher education to realise the potential they offer for people to start, stay and succeed in sport.*
>
> [Patrick Carter (Chair: Sport England) in Universities UK 2004:3]

> *In realising the ambitious agenda for sport set out in Game Plan in 2002 the Government recognised the role that higher education institutions could play. I am pleased to see in this publication that higher education institutions are already working to meet the agenda and in many cases reaching out further to engage with their communities through sport in imaginative ways and to provide top-class facilities to enable elite athletes to develop and perform at the highest levels. I look forward to higher education institutions and sports bodies building on the firm foundations of partnership detailed in this publication so that all parts of society are able to access sport at appropriate levels with all the health, societal and recreational benefits that sport brings.*
>
> [Rt Hon Richard Caborn MP (Minister for Sport) in Universities UK 2004:3]

In general, sport in the HE context presents numerous possible benefits. These include the promotion of participation, health and positive values (such as those associated with Olympism); skills development; assisting with the development of appropriate policies at local, regional and national level; and the generation of scientific knowledge through sport science, sport medicine and coaching research and expertise, and the dissemination of this knowledge through publications, colloquia, conferences, etc. Once sport is viewed as an educational instrument, it may contribute to the effective management of sport as a strategic asset, encourage community engagement, create opportunities for access and development, and contribute to the attainment of national performance goals (for example, a high number of Olympic gold medallists).

---

3   See Shehu (2000) in this regard.

▣ *Expectations of sport in a university town*

A key feature of a university town is that it is dominated by its university population. This population is generally described as highly educated and largely transient, with numerous subcultures and a high number of people living non-traditional lifestyles; therefore, university towns must have a high tolerance for non-conformist music, culture and politics. The university provides employment and business opportunities, and the economy of the town is closely linked to it and supported by the entire university structure. Moreover, the history of such towns is often intertwined with the history of the universities located in them, so that, apart from being important places of scientific and academic endeavour, university towns are often also centres of political, cultural and social influence.

The dominance of a university in a town makes it advisable (if not obligatory) for it to have a formalised relationship with the local authority. This may be structured in different ways, such as regular meetings between the vice-chancellor and the mayor (as in the case of the Rector-Mayor Forum in Stellenbosch), which may take place in the context of a formalised memorandum of understanding that empowers the entities to undertake joint projects and to share important information.

Match review

Stellenbosch past

Sport has always been a very prominent feature of life in Stellenbosch. However, in predictably South African mode, the document history of sport in the town is skewed, focusing just about exclusively on white sport. Typically, black sport participation and achievements have been underreported and generally ignored. While records for white sport in Stellenbosch are readily available, it is only recently that black sport history has become visible – and only through the deliberate efforts of those who recognise that this non-recognition reproduces inequality. This record reflects a fascinating wealth of talent and achievement on the part of players and administrators of colour. Two examples will serve to illustrate this point: Hilton Biscombe's fascinating *Sokker op Stellenbosch* (2010), jointly commissioned by Stellenbosch Municipality and Stellenbosch University during the run-up to the 2010 FIFA World Cup, which explores the history of soccer participation in the coloured and black communities of the region; and local cricketer and administrator Matt Segers, who was honoured with a Lifetime Sports Achiever Award in 2003 but has yet to be recognised by the town for his role as educator and sport administrator.

Thompson (1978:81) remarks on the "interdependent relationship between the social institution of sport and the dominant ideology in contemporary society". In pre-1994 Stellenbosch, as everywhere else in South Africa, sport not only contributed to the establishment and perpetuation of the apartheid system of separate, racially-based identities, thus working against social cohesion; it actively contributed to racial segregation: separate sport facilities, separate teams, unequal funding, etc. The result was racially segregated, inferior and superior facilities, teams, recognition and opportunities. Stellenbosch University sport did not stand outside of this *status quo*: the Matie Rugby Club was established in 1875, but its first team never included a black player until 1999 (with the inclusion of scrumhalf Sunu Gonera), while the first athletics meet at Coetzenburg to include athletes of all races took place in 1976, almost a century after very first athletics competition on Die Braak, in 1885 (Van der Merwe 1984). One may assume that the spectators at Coetzenburg in 1976 were exclusively (or at least predominantly) white. Whites dominated the university, the local authority, and predictably also sport in town, while black sportsmen, -women and administrators were relegated to the periphery or made their contribution to sport in anonymity.

Stellenbosch present

Deliberate efforts have been made in recent years to create a more equitable sport setup in Stellenbosch. The municipality has tried to improve sport facilities in previously disadvantaged communities, albeit not always effectively or within the framework of a long-term plan. Similarly, Stellenbosch University has actively tried to identify and recognise talent and create opportunities through improved access to facilities and expertise and by providing financial support. The pursuance of representative university teams and encouragement of participation in black communities have become accepted norms for university sport codes. Through the efforts of its four sport-related units (the Department of Sport Science, Centre for Human Performance Science (CHPS), Stellenbosch Sport Performance Institute (SUSPI), and Matie Sport) the university has made a significant impact locally, regionally and nationally, and has earned some international (continental and global) recognition.

At this stage of the development of the town, it would therefore appear fair to say that two major institutions in Stellenbosch, the municipality and the university, have the shared objective of harnessing their resources and expertise to create equal opportunities for all its respective stakeholders – mindful of its unjust and discriminatory past and the need for a shared future – in the interest of the town and a better future for its residents.

TEXT BOX 2

The four sport-related units of Stellenbosch University have been charged since 2004 to work together within the framework of the institutional Sport Plan, which aims to establish the university as the most respected destination in the world for sport education and sport experiences. The objectives of each unit may be summarised as follows:

- The academic **Sport Science Department:** to provide formal training to students that enables them to become leaders in their field; to perform relevant research and community service; and to use their material and human resources to the advantage of the department, the university and the community.

- The post-graduate **Centre for Human Performance Sciences (CHPS):** to design and deliver interdisciplinary learning programmes, research projects, and community interaction activities in the field of human performance (sport, exercise and wellness).

- The commercially-oriented **Stellenbosch University Sport Performance Institute (SUSPI):** to commercialise the sport assets of the university and to provide a range of services to the sport, fitness and wellness industries.

- The academic support service section, **Matie Sport:** to build and market the excellent image of the university through the creation of an outstanding centre for competitive, residence-based and recreational sport through the provision of appropriate services.

The significance and value of these four units have over time been proven and has the potential to increase substantially. Collectively, they contribute to (among other things) research, skills development and training, events management, the provision of facilities, community engagement projects and initiatives, and volunteerism. Through the work of these units, the university has been recognised for its leadership role with regard to sportspeople with disabilities, its sport science research outputs, its training and rehabilitation expertise, its events management capabilities, and the quality of its facilities. Academic collaboration with local and international partners is a prominent feature, and the image and reputation of the town is promoted through the provision of services to state departments, educational institutions, provincial and national sports teams, and international sportsmen and -women.]

Stellenbosch future?

Realistic expectations with regard to the potential of sport to contribute towards the sustainability of the town include significant contributions through sport-related services; research, training and community engagement; and facility renewal, redevelopment and management in partnership with the business, leisure, hospitality and tourism sectors. In addition, one may reasonably expect it to promote increased participation; pursue robust procurement policies; systemically introduce the recognition of achievements; and pay special attention to youth, women and persons with disabilities.

'Hope' has become a quite prominent concept in thinking and planning for the future. At the international level, the concept has been linked to the achievement of the MDGs, with specific emphasis on 'Sport for Development and Peace'. The *Sport as a Tool for Development and Peace* report of the United Nations Inter-Agency Task Force, which was commissioned to determine (among other things) what concrete steps should be taken to use sport as an instrument to promote development and peace, found "that well-designed sport-based initiatives are practical and cost-effective tools to achieve objectives in development and peace" (Sportanddev. org 2008:n.p.).

Elements of hope through sport include facilities and resources (such as sport equipment), structures and systems, talent identification and management, commitment and application, technical expertise and the link to science and education, and the creation of opportunities. Significantly, Stellenbosch University has approved a Sport Youth Initiative (SYI) as one of several HOPE Project programmes aimed at applying its expertise and scientific knowledge in support of the attainment of the MDGs. The SYI focuses specifically on the identification of talent and the provision of learning support to create opportunities for learners selected from local schools. Taking into account all the pluses of the university (and the town), it may not be far-fetched to approach the United Nations Office for Partnerships to pursue new collaborations and alliances and to try to get access to the United Nations Fund for International Partnerships (UNFIP).

Apart from the MDGs and the HOPE Project, the following relevant points of reference should also be kept in mind:

- The five **national priorities**, namely: creating decent work and sustainable livelihoods, education, health, rural development and agrarian reform, and the fight against crime and corruption (Zuma 2011).
- The **'A United Province'** approach of the Western Cape Provincial Government (WCPG): Although this programme remains to be translated into credible, coherent, systemic and sustainable action, its goals may be summarised as poverty reduction, the prevention of anti-social behaviour, and the creation of economic employment opportunities (Meyer 2011).
- The **municipal priorities**, as expressed in the Integrated Development Plan (IDP) (which critically also informs budget allocations). While trying to articulate with national and provincial priorities and major issues, such as planning for sustainability and the delivery of efficient services, the Stellenbosch Municipality Integrated Development Plan 2007-2011 makes scant mention of the potential of and the need for sport to be integrated into this planning (Stellenbosch Municipality 2007).

Team resources

Facilities

As far as sport facilities are concerned, Stellenbosch has a wonderful base to build on. In addition to those of the various schools and Boland College, there are eighteen municipal sport facilities (all bar one are referred to as 'sports grounds'), while Stellenbosch University has an extensive array of sport fields and facilities, with several improvements, extensions and new facilities in the pipeline.[4] The most recent addition (January 2012) to this already very impressive range, is the Stellenbosch Academy of Sport (SAS), which offers a wide variety of on-site facilities and services, including "indoor and outdoor training fields, a state-of-the-art gym, access to leading sports scientists, doctors, coaches, physiotherapists and biokineticists, cutting-edge conference facilities, comfortable athlete accommodation and performance-nutrition catering" (SAS 2012:n.p.).

---

4   Existing facilities include an athletics stadium, two artificial surface hockey fields and nine grass hockey fields, a rugby stadium and fourteen rugby fields, nine squash courts, eight netball courts, five soccer fields, three swimming pools, twenty-eight tennis courts, and an indoor sport centre (Stellenbosch University Facilities Management 2011).

TEXT BOX 3

In terms of creating new facilities, care should be taken to ensure that they are not only functional, but will also contribute to the beauty, general appeal and physical sustainability of the town, as well as social cohesion and active, healthy lifestyles. 'Short-termism' should be abandoned in favour of proper (long-term) planning. In his study of the cost-benefit relationship of sports facilities, Tim Chapin (2002) highlights several important factors to be taken into account, especially where sport facilities are viewed as tools for economic development. While huge public spending is lavished on stadiums and arenas, decision makers often lack sufficient information or understanding about the real costs and benefits of such facilities. Chapin argues strongly for an informed assessment of both the economic and the non-economic impacts of sports facilities – noting that the former is often overstated and the latter underrated – and asserts that, even if the construction of sport facilities generally fail as economic development tools, the non-economic benefits, especially in terms of "social/psychic impacts, image impacts, political impacts and development impacts" (*ibid.*:9), may nevertheless provide ample justification for building them.

One way of gaining sufficient information with a view to planning and providing facilities is through comprehensive audits, such as the Sport England HE sport audit mentioned earlier. The organisation has developed a 'Sports Facilities Calculator', which it describes as a "planning tool which helps to estimate the amount of demand for key community sports facilities that is created by a given population" (Sport England 2011). The Department for Communities and Local Government (United Kingdom) also provides some useful approaches and mentions several elements which should be taken into account. Apart from also stressing "robust assessments of the existing and future needs of their communities for open space, sports and recreational facilities" (2002:5), it also advises on matters such as the setting of standards and multiple use.]

Sport science and services

In terms of sport-related science and services, the town is equally uniquely endowed. As indicated earlier, the university may contribute through its (academic) Sport Science Department, the CHPS, and SUSPI, while the SAS also offers access to a range of sport science and medicine experts. In addition, the private Stellenbosch MediClinic hospital also offers specialised sport-related services, rehabilitation and care.

Planning and management

The Integrated Planning Committee, a sub-committee of the Rector-Mayor Forum, offers an ideal platform for discussions pertaining to sustainability issues and could be instrumental in formulating a long-term sport strategy that articulates with the appropriate regional, provincial and national frameworks and plans of the respective political governing structures and sport federations.

Game plan

Sport has the potential to be (and should be) a key feature of development in Stellenbosch. Ensuring that sport makes an appropriate and systemic contribution to a sustainable Stellenbosch is, however, not a simple task, with many interrelated factors having to be taken into account. It is against this background that a blueprint for matching sport and sustainability in Stellenbosch may now be considered.

A sustainable Stellenbosch would have functional, attractive, ecologically sensitive, and accessible facilities catering for multiple sport codes and recreational activities. Inevitably, there would be a historically determined hierarchy of sport codes, which would be respected and the sport assets at its pinnacle adequately resourced to ensure the quality and attractiveness of facilities and to maximise their commercial value (for example, through commercial sponsorships and the selling of associated services). This scenario should see Stellenbosch being world-renowned for specific sport codes, both in terms of performance and the quality of related services, as well as its sports infrastructure. It would, however, also accommodate other, less prominent, sport codes and give them due recognition, for example, through greater exposure (such as promotional activities sponsored by the town), appropriate municipal resource allocations, and a formalised award system. In the same vein, sport in Greater Stellenbosch (which has a huge historical disadvantage) would not be neglected and would in fact benefit from the progress made in the town proper. In this regard, the Stellenbosch Sport and Recreation Council (SSRC) – which consists of two representatives each from the seventeen sport codes practiced in the greater Stellenbosch area – would play an important role in transcending parochial interests in favour of serving the common town interest.

TEXT BOX 4

> Rugby is undisputedly Stellenbosch's greatest sport asset. The town is home to the biggest and one of the most successful rugby clubs in the world (Stellenbosch Rugby Football Club, commonly known as Maties Rugby), and both the increasingly successful Springbok Sevens side and the Western Province Rugby Institute are based there. Stellenbosch has hosted numerous national competitions and was also co-host of the International Rugby Board (IRB) Junior Rugby World Championships in 2012. Maties Rugby has served as 'laboratory' for IRB rule experimentation and, through the Maties Rugby Academy: Development, the club is involved with rugby development in Mitchell's Plain, Khayelitsha, Stellenbosch, Kylemore, Kayamandi, Genadendal and Malmesbury, with more than ninety schools involved in the programme in Mitchell's Plain alone (Maties Rugby 2012). Moreover, the sport has the potential to grow into an even greater asset through the realisation of as yet unrealised possibilities and opportunities. However, the dominance of rugby poses the risk of preventing other (competing) sport codes with great potential, such as soccer, from growing. A careful balance should be pursued.

But does Stellenbosch have what it takes?

The best possible outcome will only be achieved if several objectives are met. The value of sport for the development of social capital must be realised. It is also important that a multiplicity of relationships, synergies and planning objectives are negotiated and facilitated between stakeholders and role players who are willing to commit appropriate resources to a sustainable town. Proper audit-generated information should be gathered as a basis for planning. Realistic targets should be set, pursued and monitored, while sport should be mainstreamed and its potential benefits maximised. In order to achieve this, there must clearly be stability in the local government structures and administration, and the various stakeholders should work together to create (developmental) opportunities (especially for youth and women) and foster a sense of communal identity based on shared values, so that individuals would relate their identity to a 'town' identity, which should in turn contribute to the optimisation of volunteerism. Finally, the potentially negative impact on the environment should be limited, and health and well-being systemically promoted.

Game, set and match!

When the sustainability of Stellenbosch is reflected on, it does not imply that the town might cease to exist at some point in the future. Rather, it implies a concern for its continued existence at a level and in a way that is an improvement on how the town is currently governed and managed and how social relationships and dynamics unfold.

Stellenbosch is privileged to have major sport assets, which, if used judiciously, may make a significant contribution to the sustainability of the town. As a university town, its sport-related scientific and intellectual assets offer a major advantage, complementing the many sport facilities available. The environmentally sensitive management of these facilities (and appropriate design of new ones) creates further real possibilities for sport to contribute to sustainability in the town. In addition, the town is home to several major business interests, which could be harnessed in a proper town plan. Another advantage is the fact that Stellenbosch has already established itself as world-class destination in terms of sport-related facilities and services and also enjoys international renown as a leisure tourism destination; promoting and optimising sport tourism is therefore a logical next step and something that should be addressed in the town plan.

In a sense, sport is synonymous with Stellenbosch; yet, thus far it has not featured prominently or in a systemic way in contributing towards the sustainability of the town. This may be ascribed to the fact that it has long been a contested terrain where power relations are reproduced. The historically unequal and divided nature of sport, as well as the absence of a town strategy to counter these historical challenges, have undermined the full use of the available possibilities. In consequence, sport has mostly reflected and reproduced inequalities and stereotypes, and attempts to fully use the potential of sport to the advantage of the town have been relatively insignificant.

Yet sport, with its intrinsic levelling and unifying influence, offers real opportunities to bring people together. This fact should be judiciously exploited; for example, by establishing a coalition of interests to advance the positive impact of sport. Such a coalition may – subsequent to a process of comprehensive consultation with all involved, and with due cognisance of the planning priorities of national and provincial government and sport federations – take deliberate and carefully considered steps to ensure the systemic integration of sport into the pursuance of local objectives. Building on successful examples from elsewhere in the world, and dealing seriously with issues that represent local barriers to the optimisation of sport, a sustainable Stellenbosch should ensure redress and a shared vision for the future. Stellenbosch University's HOPE Project and its links to the Millennium Development Goals may create a focal point for a town-wide approach to sport and attract significant international interest. The formal Memorandum of Understanding between the university and Stellenbosch Municipality and the increased cooperation between them concerning matters of strategic importance constitute a good foundation for the town to benefit from its vast sport assets.

Stellenbosch has the potential to be a winning town as far as sustainability over the next few decades is concerned. However, as Tiger Woods has said, "there is an art to winning, a skill, an approach, a belief that winners have" (quoted in Spander 2007:n.p.). Much depend on the political dynamics in general, and the political stability in particular, of the town. It is incumbent on the local authorities to honour not only their constitutional mandate, but also to have the moral integrity to move the town from an inequitable past to a more equitable and sustainable future.

REFERENCES

Balisunset. 2008. *The Commercialization of Sport*. [Accessed 7 July 2011] http://www.hubpages.com/hub/The-Commercialization-of-Sport

Biscombe H. 2010. *Sokker op Stellenbosch*. Stellenbosch: Rapid Access Publishers.

Black DR & Nauright J. 1998. *Rugby and the South African nation: sport, culture, politics and power in the old and new South Africas*. Manchester: Manchester University Press.

Chan J. 2010. Economic value of sport. MBA research proposal. Pretoria: University of Pretoria.

Chapin T. 2002. *Identifying the real costs and benefits of sports facilities*. [Accessed 7 July 2011] http://www.arroyoseco.org/671_chapin-web.pdf

David R. 2010. *The impact and effectiveness of sport on the Millennium Development Goals*. [Accessed 7 July 2011] http://assets.sportanddev.org/downloads/research_paper_on_sport_and_the_mdgs.pdf

Department for Communities and Local Government (United Kingdom). 2002. *Planning Policy Guidance 17: Planning for Open space, Sport and Recreation*. [Accessed 20 September 2012] http://www.communities.gov.uk/publications/planningandbuilding/planningpolicyguidance17

Formby H. 2007. South Africa's Elite Spawned at Paul Roos. *Financial Mail*, 30 March. Edited version by blog admin team. [Accessed 7 July 2011] http://www.prgees.blogspot.com

Grundlingh A, Odendaal A & Spies B. 1995. *Beyond the try line: Rugby and South African society*. Johannesburg: Ravan.

Maties Rugby. 2012. *About us*. [Accessed 21 September 2012] http://www.matiesrugby.co.za/about-us

Meyer I. 2011. *Western Cape Department of Cultural Affairs, Sport and Recreation: 2011/2012 Budget Speech*. [Accessed 7 July 2011] http://www.info.gov.za/speech/DynamicAction?pageid=461&sid=17366&tid=30994

Millward P. 2007. Sport and Ideology. In: G Ritzer (ed). *Blackwell Encyclopaedia of Sociology*. [Accessed 7 July 2011] http://www.sociologyencyclopedia.com/public/tocnode?id=g9781405124331_yr2011_chunk_g978140512433115_ss1-12

Olympic.org. 2010. *Zambia's youngsters thrilled about new sports complex*. [Accessed 7 July 2011] http://www.olympic.org/development-through-sport/zambias-youngsters-thrilled-about-new-sports-complex

People's Daily Online. 2010. *Sports valuable tool for achieving MDGs: UN chief*. [Accessed 7 July 2011] http://www.english.peopledaily.com.cn/90001/90779/90867/7147217.html

Reddy ES. n.d. *Sports and the liberation struggle: a tribute to Sam Ramsamy and others who fought apartheid sport*. [Accessed 7 July 2011] http://scnc.ukzn.ac.za/doc/SPORT/SPORTRAM.htm

SAS (Stellenbosch Academy of Sport). 2012. *About*. [Accessed 20 September2012] http://www.sastraining.co.za/

Shehu J. 2000. Sport in higher education: An assessment of the national sports development policy in Nigerian universities. *Assessment & Evaluation in Higher Education*, 25(1):39-50.

Spander A. 2007. Tiger Woods perfects art of winning. *The Telegraph*, 14 August. [Accessed 7 July 2011] http://www.telegraph.co.uk/sport/golf/2318969/Tiger-Woods-perfects-art-of-winning.html

Sportanddev.org. 2008. *Sport for development and peace towards achieving the Millennium Development Goals*. [Accessed 7 July 2011] http://www.sportanddev.org/learnmore/sport_education_and_child_youth_development2/index.cfm?uNewsID=23

Sport England. 2011. *Sports Facility Calculator*. [Accessed 7 July 2011] http://www.sportengland.org/facilities__planning/planning_tools_and_guidance/sports_facility_calculator.aspx

SportScotland. 2011. *Local Sports Councils and Local Authorities*. [Accessed 7 July 2011] http://www.helpforclubs.org.uk/TopicNavigation/Funding/SALSC+and+Local+Authorities.htm

Stellenbosch Municipality. 2007. *Stellenbosch Municipality Integrated Development Plan 2007-2011: Executive Summary*. [Accessed 7 July 2011] http://www.stellenbosch.gov.za/jsp/util/document.jsp?id=774

Stellenbosch University Facilities Management. 2011. Personal communication. Stellenbosch: 15 June.

The Crest Online. 2012. *Spotlight on sport tourism*. [Accessed 21 September 2012] http://www.thecrestonline.co.za/spotlight-on-sports-tourism

Thompson RW. 1978. Sport and ideology in contemporary society. *International Review for the Sociology of Sport*, 13(2):81-94.

United Nations Inter-Agency Task Force on Sport for Development and Peace. 2005. *Sport as a tool for development and peace: Towards achieving the United Nations Millennium Development Goals*. [Accessed 7 July 2011] http://www.un.org/sport2005/resources/task_force.pdf

Universities UK. 2004. Participating and performing: sport and higher education in the UK. [Accessed 7 July 2011] http://www.universitiesuk.ac.uk/Publications/Documents/sport.pdf

Van der Merwe FJG. 1984. *Honderd jaar Matie-atletiek*. Stellenbosch: Stellenbosch-Atletiekklub.

Wikipedia. 2011a. *Professional sports*. [Accessed 7 July 2011] http://en.wikipedia.org/wiki/Professional_sports

Wikipedia. 2011b. *Sport*. [Accessed 7 July 2011] http://en.wikipedia.org/wiki/Sport

Zuma JG. 2011. *2011 Local Government Manifesto: Together we can build better communities – Message from the President*. [Accessed 7 July 2011] http://www.anc.org.za/docs/manifesto/2011/lge_manifestotxtl.pdf

Chapter 25

CHILDREN

Imagining a Stellenbosch where children come first

EVE ANNECKE, NALEDI MABEBA,
MAGDELIEN DELPORT & JESS SCHULSCHENK

Introduction

A multitude of challenges face children today, including an entirely different reality from that into which their parents were born. Global challenges of climate change, devastating environmental degradation and growing poverty will shape the ways in which they live out their lives, regardless of whether they are conscious of these challenges or not. Refreshed views of the role of ecological learning – and the centrality of sustainability within this – are needed in all our communities if we are to offer young people hope of not only surviving, but flourishing as they grow today and lead tomorrow.

This chapter will present a brief overview of the current situation of youth in Stellenbosch, as well as offer some suggestions of how to realise a vision of positive development in Stellenbosch that puts children first.

The context

Stellenbosch is considered to be one of the most unequal places in the world. The deep reaches of inequality are reflected in the state of the education system, high levels of unemployment, and the overall economic well-being of Stellenbosch residents. An estimated 64.4% of the population in the Winelands District Municipality are under the age of 35. Education in Stellenbosch mirrors national realities, such as the fact that of the less than 50% of learners who enrol in Grade 1 and stay in school up to Grade 12, only 67.8% manage to matriculate. While certain schools produce outstanding results, others have pass rates as low as 37% (Department of Basic Education (RSA DBE) 2010). Our national education system is failing our youth by not providing them with the qualifications or the potential to develop beyond (mostly) unskilled labour.

Education, however, is not the only key to success. Children grow up in complex environments constituted by a multiplicity of factors that include people, friends, places and institutions. Inequality in Stellenbosch is perhaps most visible in the marked differences between children's realities according to their area of residence, income grouping, language and race. These factors largely determine which schools they attend, the opportunities to which they have access, the social support they receive, and the quality of their recreational time. Currently there is little movement across these 'borders' and the chances of different children's realities interconnecting and transforming one another are slim. Considering the fact that today's children are the future of Stellenbosch and South Africa, it would appear essential to ensure that all children are granted the opportunity to develop to their full potential, regardless of the circumstances into which they are born.

Envisioning a children-centred Stellenbosch

It would be Stellenbosch's far-sighted vision to create a town where children are prioritised and included in all planning, development and financial decisions. A children-centred Stellenbosch could be envisaged as a place where children are cherished and supported from birth up to the point where they are ready to step into adulthood. This view values children as the very fabric of communities and suggests that the prosperity of communities depend on children's opportunities to develop, grow and learn to their full potential during their early years. This clearly requires appropriate institutions, such as pre- and post-natal care, high quality and affordable Early Childhood Development (ECD) centres, good primary and secondary schools, social and psychological support, and a variety of channels that children may pursue to attain further education or develop market-related skills.

However, we will argue here that, beyond this obvious institutional support, a 'child-friendly' town is required – one that prioritises the well-being of children as integral to the well-being of the town. What features might exemplify such a town? There would be many public spaces that are easily accessible, clean, safe, beautiful and natural. Free access and affordable transport to nature reserves and places of arts and culture will promote a holistic development of children's interests. All places of education will have green campuses, where food gardens, sustainably built classrooms and biodiversity corridors on school grounds demonstrate sustainability-in-action to students and the broader communities. Opportunities will be constantly created for children and youth to contribute meaningfully to the sustainable development of their town, and will be designed in such a way as to facilitate interaction between children from different socio-economic and racial groups, which will contribute significantly towards social cohesion in this 'model' town. In short: in a children-centred town, all efforts will made to ensure that children feel welcome and are active, stimulated and participate fully in town life.

The remainder of this chapter will look at three phases of children's development, namely early childhood development (age 0-9), children of school-going age (10-16), and youth (16-21). For each phase, case studies are provided of ways in which Stellenbosch is already supporting its youth and suggestions are offered to improve the children-centred focus of Stellenbosch. In each case, collaborative leadership from within the relevant local authorities, educational institutions, businesses and non-profit organisations is vital in order to turn this 'wish list' into reality.

Early childhood development (ECD)

Early childhood development (age 0-9) is widely recognised as fundamental to children's development in terms of cognitive, social and emotional abilities (Barnett & Ackerman 2006). It is therefore vitally important to ensure that all young children of Stellenbosch have the support, stimuli and security required to ensure optimum ECD.

Beyond basic learning, ECD is also a good time to introduce children to the wonders of our universe, the diversity of nature, the beauty of our planet, and the importance of each person's story in the universal timeline (Herbert 2008). Rather than separating ECD from parents and communities, as is often the case with ECD centres, ECD facilities have the possibility of being community development hubs where children are central to multiple interrelated projects (Ball 2005). Following an integrated approach, ecological, social and poverty-related challenges are simultaneously addressed, so that the ECD space becomes one where ecology and equity meet.

The elements discussed below may prove useful in thinking about an integrated approach to ECD and the establishment of ECD hubs.

Nature first

It is undeniable that current lifestyle and consumption patterns are unsustainable on a both a local and global level. If children are expected to lead the way towards alternative futures, then early childhood education should emphasise the interconnectedness of humans and nature and create opportunities for children to develop a love of and sense of stewardship towards nature. In early childhood education, the prioritisation of nature in ECD curricula builds upon children's natural curiosity and fascination with the natural world. Equally important is to provide children with opportunities to spend time exploring, playing, testing limits, and trying to understand nature. This may be enhanced by 'green' ECD buildings and playgrounds, outings, greenery in the classroom, or food gardening.

Community development

The process of nurturing children to become agents of change starts at an early age. This implies that crèches, schools, aftercare facilities and community centres are firmly rooted in the local context, and that these institutions actively contribute to the creation of resilient and sustainable communities. High-quality ECD, therefore, is characterised by its contribution to the establishment of individual and collective capabilities for sustainable community development. These capabilities include a sense of stewardship towards nature; the ability to find local and cost-effective ways to work towards sustainable lifestyles; being able to relate to and work together with people from different races and language groups; breaking negative cycles (such as violence or addiction); and formulating new ways to walk into the future. In this way, educational institutions may serve as hubs for sustainable development in a way that coordinates developmental efforts, encourages synergies, promotes inter-generational learning, facilitates access to services, and brings together families, schools and communities.

Nutrition and well-being

Although there are many factors that are critical to children's development and well-being, the importance of consistent access to nutritious food cannot be underestimated. Nutritious food provides the building blocks for physical and brain development, which is strongly linked to further educational attainment, linguistic and social development, and psychosocial disorders (Chilton, Chyatte & Breaux 2007). Food is also one of the most important connections between humans and nature. One way of emphasising this connection is through school or residential food gardens, which provide numerous opportunities for environmental education, outdoor learning, physical exercise and healthy eating habits. The connectedness of the food system (local and global) to resource use, land ownership, energy, pollution, water and soil conservation, hunger, trade and climate change is an ideal entry point for sustainability discussions and the development of a systemic understanding of the food that we consume each day. Food gardens also offer a sustainable and cost-effective way of circumventing rising food prices and supermarket domination.

School age (10-18)

There are many factors beyond school and academics that have profound influences on children. In Stellenbosch, some children have access to a plethora of activities in and after school to keep them occupied, stimulated and inspired. Other children, however, face afternoons of boredom because their schools offer no extramural activities such as sports, drama or art. It is nothing new to say that boredom often leads to crime, teenage pregnancy, vandalism and gangsterism, which correlates with higher school-dropout rates and poor academic performance. A children-centred Stellenbosch may seek to ameliorate this situation in the ways described below.

Constructive activity

One way to counteract the negative effects of boredom is to find ways to keep children active in meaningful ways. In Stellenbosch, some initiatives already exist that seek not only to keep children busy, but to help them develop their passions, strengths and talents. One of these programmes is the 'Keep Them Active' programme, which emerged from the four-week 'Keep Them Safe' initiative that ran during the 2010 FIFA World Cup. 'Keep Them Safe' was aimed at harnessing the energy generated by a major occasion such as the soccer world championship and channelling it towards the creation of holistic spaces where youth are equipped with tools and experiences that may help them to reach their full potential and to live sustainable and balanced lives based on positive values. The initiative was made possible with the financial and physical support of many local and international organisations. The momentum gained during this four-week programme was so strong that it was subsequently transformed into a longer term programme called 'Keep Them Active'.

Natural spaces

It is impossible to overemphasise the importance of beautiful natural spaces. In an age of computers, Internet and television, the concern is that children do not spend adequate time immersed in the great, natural outdoors.

There are many reasons why it is imperative for Stellenbosch to create spaces where children may play and explore and discover nature freely. The positive and healing effects of nature have been shown repeatedly. On a physical level, uninhibited play in natural settings is critical for the development of fine and gross motor skills, muscle development, balance and the exploration of physical limits. Natural settings have also been shown to greatly increase the creativity, diversity and inclusiveness of children's play (Bell & Dyment 2008). Two of the greatest potential barriers between children and the outdoors are the attitudes, fears and perceptions of adults (often related to safety concerns or biophobia) and a lack of access to natural spaces. There are many wonderful natural spaces in and around Stellenbosch, as well as many opportunities to transform barren school grounds into living green campuses. The challenge for a sustainable Stellenbosch is to improve children's access to these spaces by means of walkways, affordable transport and entry fees, and adequate safety precautions. Ecological learning – connecting with nature through being in nature – cannot be left to schools alone. Creating space for this to occur in all childhoods may be the greatest gift the town may bestow on its own future.

Active youth participation

The children of today are the leaders, community members and citizens of tomorrow. In a children-centred sustainable Stellenbosch, children would be used to their role as active participants in community life and community decision-making. This requires the creation of spaces where children and youth may contribute meaningfully to society and where adults develop the capacity to respect and listen to younger voices. A possible vehicle towards achieving this end may be the provision of a platform allowing youth to discuss, plan and take action around pressing issues (such as water, waste, electricity and food) in Stellenbosch. Another may be collaborative art projects – for example, closing the central streets on a Saturday to create an exhibition for the town hall. Such a joint purpose between children and youth from different parts of Stellenbosch may be a uniting factor allowing them to bridge previous divides and work together towards common futures.

Young adults (18-35)

Fifty-one per cent (51%) of South Africans between the ages of 15 and 24 are unemployed, which is more than double the national unemployment rate of 25% (South African Institute of Race Relations (SAIRR) 2011) (see also Appendix 1, Figure C). Clearly, a Grade 12 certificate is not enough to guarantee a decent job after high school. Furthermore, many high school learners drop out of school before their matric year due to many reasons, including poverty, teenage pregnancy or familial pressure (see Appendix 1, Figure B for the level of education in Stellenbosch). Other learners complete Grade 12, but do not perform well enough to be admitted to a tertiary institution. It is therefore important to find ways of empowering these learners with practical skills that will enable them to earn a decent living. The ideas presented below may prove useful in this regard.

Crafts for skills

There is tremendous value in learning craft-based skills for the discipline and creativity it encourages as values to be carried through into other aspects of life. A return to craft that encourages purpose and durability, as well as beauty, would redefine the way we engage with material objects and perhaps has a role to play in challenging the modern conception of progress as the accumulation of material wealth and consumerism, to be valued above purpose, quality and beauty. Craft can and should be integrated into learning environments, with special focus given to craft-based technical colleges that may provide young people with practical skills towards meaningful employment.

Barefoot engineers: Vocational skills for meaningful work

Craft-based skills may be integrated with an emerging sector of 'green' jobs in agro-ecological food production, land-derived materials, ecological construction, and resource-related fields (such as water efficiency, renewable energy, alternative sanitation systems, etc.) and a vision of empowering 'barefoot engineers' to design and implement simple and affordable solutions to sustainable living for all. Barefoot engineering courses are very practical in nature and consist mostly of learning through mentorship and experience. The purpose of these courses is to equip students with concrete knowledge and skills that not only contribute to their own survival, but also serve to realise the vision of a sustainable Stellenbosch. Versions of these programmes have already begun to be offered by the Sustainability Institute, Learning for Sustainability FET College and others in the Stellenbosch region.

Table 25.1 Examples of potential 'barefoot engineering' courses

Theme	Description	Practical skills
Organic vegetable gardening	Learning the basics of growing vegetables, including soil preparation, seed sowing and times, companion planting, harvesting and natural pest control.	• Prepare a bed in the gardens • Plant seeds into sowing trays • Design a small garden (based on available space in learner's home environment) • Container gardening: plant seedlings out into recycled containers
Composting & worm farming	An introduction into the fundamentals of making compost (beds and DIY systems) with an overview of worm farming.	• Prepare a simple compost preparation in the gardens • Build and prepare your own worm farm container
Nutritious cooking	A fun programme to learn new skills in the kitchen for preparing, cooking and enjoying harvested vegetables. Overview of various vegetables, nutritious properties and how best to prepare meal plans on a budget.	• Prepare six common and nutritious vegetables from the region • Divide into teams to prepare various dishes for a shared meal with participants • Practical recipe book
Indigenous planting	An introduction to planting indigenous and water-wise gardens.	• Plant identification in the gardens • Soil preparation in the gardens • Designing your own small garden • Planting from cuttings and seed
Biogas digesters	Learning the basics of how biogas digesters operate and installation of a domestic unit.	• Experience of a biogas system • Construct a micro biogas project • Go through the construction of a larger scale biogas project
Solar cooking and clay ovens	Practical experience of building two solar cookers (parabolic and box) and a simple clay oven from local and recyclable materials.	• Build a parabolic solar cooker from old mirrors • Build a box solar cooker • Build a clay oven from sand and mud • Enjoy meals prepared by the various new systems (weather dependent for solar cooker)
Energy efficiency	Learning the basics of a simple energy audit and measures to improve domestic energy efficiency.	• Audit a house's energy consumption • Identify and install interventions: timers, lighting, sensors, insulation, thermostats • Take home: starter energy efficiency kit
Water systems	Basic do-it-yourself knowledge about how to save and reuse water domestically, including basic plumbing, rainwater harvesting and grey water recycling.	• Understanding the domestic water system • Redirecting grey water for irrigation • Installing a rainwater harvesting tank • Retrofitting toilets, taps and showers
Recycling systems	Learn how to implement a simple recycling system for the home, office or school.	• Visit to the local landfill site • Separate waste streams: waste identification and sorting • Set up a system that works in a particular area (including bin identification and signage) • Education and awareness-raising techniques

Conclusion

While the importance of formal education, social development and psychological support structures cannot be underestimated, we make the case too for a town that takes all children seriously. Their vibrancy and energy will stretch our leadership to reconsider what might be precious and what is expendable. A Stellenbosch that establishes norms and values celebrating children as the lifeblood of the town will surely see futures that have built participative citizenship, alive to the ethos and values of sustainability. The (implicit and explicit) involvement of children at every level to lead the quiet development of cultures that integrate around art and music, transport, food, nature, water, waste and energy – as well as living without ostentation and within limits – may create untold possibilities for a sustainable Stellenbosch.

REFERENCES

Ball J. 2005. Early childhood care and development programs as 'hook' and 'hub' for inter-sectoral service delivery in First Nations communities. *Journal of Aboriginal Health*, 2:1-43.

Barnett WS & Ackerman DJ. 2006. Costs, benefits and long-term effects of early care and education programs: Recommendations and cautions for community developers. *Journal of the Community Development Society*, 37(2):86-100.

Bell AC & Dyment JE. 2008. Grounds for health: the intersection of green school grounds and health-promoting schools. *Environmental Education Research*, 14(1):77-90.

Chilton M, Chyatte M & Breaux J. 2007. The negative effects of poverty and food insecurity on child development. *Indian Journal of Medical Research*, 126:262-272.

Herbert T. 2008. Eco-intelligent education for a sustainable future life. In: I Pramling-Samuelsson & Y Kaga (eds). *The contribution of early childhood development to sustainable societies*. Paris: United Nations Educational, Scientific and Cultural Organisation (UNESCO).

Montezuma R. 2005. Facing the environmental challenge: the transformation of Bogota, Colombia, 1995-2000: Investing in citizenship and urban mobility. *Global Urban Development*, 1(1):1-10.

Peñalosa E. 2004. Social and Environmental Sustainability in Cities. Proceedings of the International Mayors Forum on Sustainable Urban Energy Development. Kunming, PR, China: 10-11 November.

Peñalosa E & Ives S. 2004. *The politics of happiness*. [Accessed 4 March 2011] http://www.yesmagazine.org/issues/finding-courage/the-politics-of-happiness

RSA DBE (Republic of South Africa. Department of Basic Education). 2010. *School Realities*. Pretoria: Government Printer.

SAIRR (South African Institute of Race Relations). 2010. *2009/2010 South Africa Survey*. [Accessed 4 March 2011] http://www.sairr.org.za

Stellenbosch Municipality. 2011. *Integrated Development Plan for the municipal area of Stellenbosch*. Stellenbosch: Stellenbosch Municipality.

Zietsman HL. 2007. *Recent changes in the population structure of Stellenbosch Municipality*. [Accessed 21 June 2012] http://www.stellenbosch.gov.za/jsp/util/document.jsp?id=778

APPENDIX 1

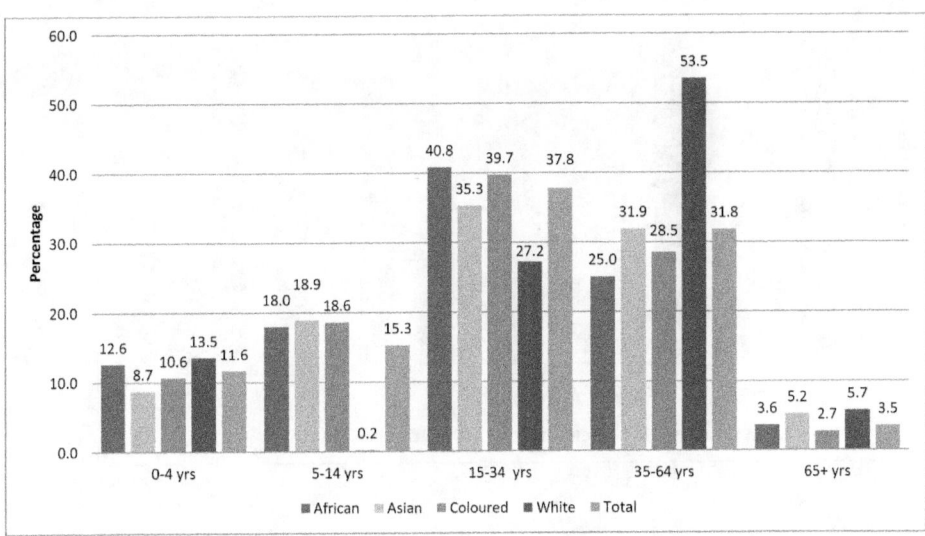

Figure A: Estimated age distribution in the Stellenbosch population, 2010 [Adapted from Stellenbosch Municipality 2011]

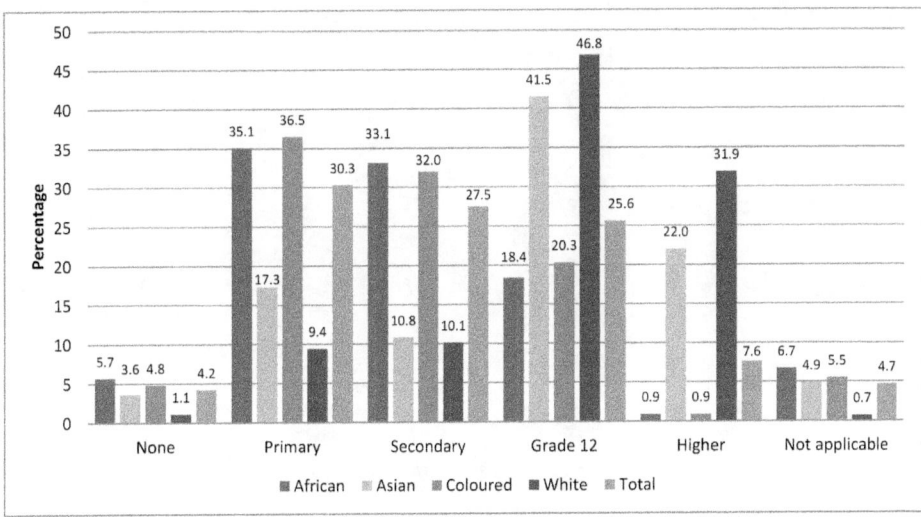

Figure B: Estimated education levels in Stellenbosch Municipality by population group, 2006 [Adapted from Zietsman 2007]

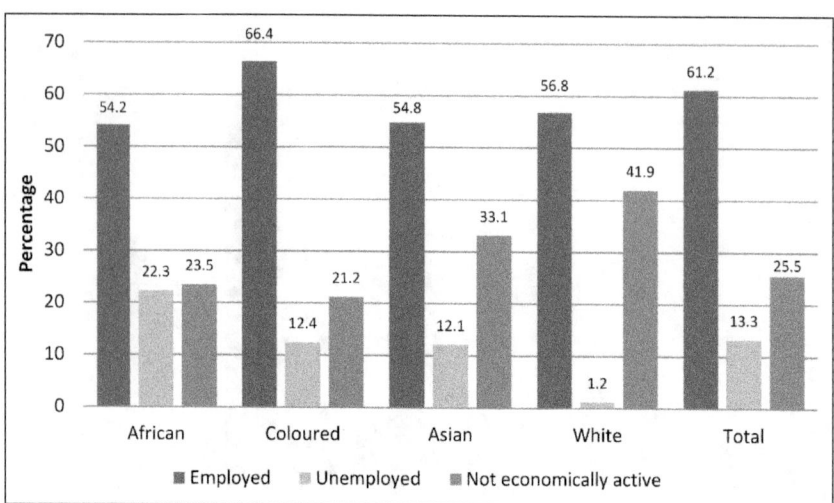

Figure C: Estimated employment status in Stellenbosch by population group, 2010 [Adapted from Stellenbosch Municipality 2011]

Note: 'Not economically active' refers to learners or students; homemakers or housewives; pensioners or retired persons, or persons who are too old to work or unable to work due to illness or disability; seasonal workers not presently working; persons who choose not to work; and persons younger than 15 and older than 65.

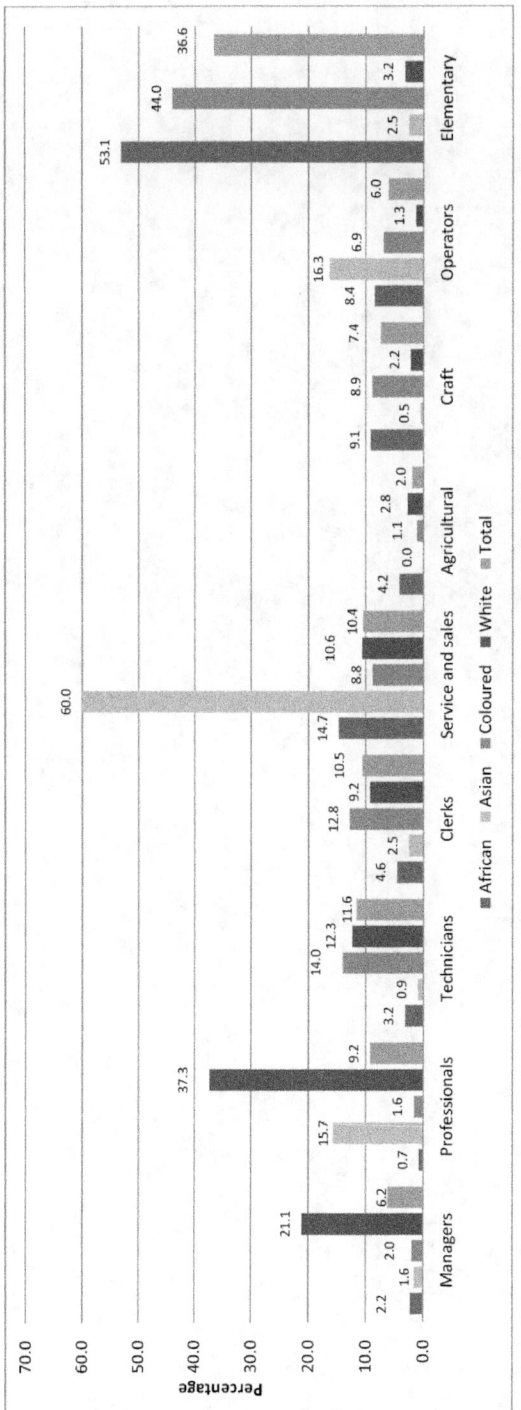

Figure D: Estimated occupational composition in Stellenbosch by population group, 2006 [Adapted from Zietsman 2007]

Conclusion

Conclusion

MARK SWILLING

At one level, this book is *about* the town of Stellenbosch. Written largely by academics and researchers employed by (or studying at) Stellenbosch University, the various chapters aim to deliver on the intentions articulated in the introduction, namely to provide a volume that will assist citizens, policy-makers, municipal officials, university administrators, business people, academics, students, development workers, and visitors to better understand what Stellenbosch is about.

The primary conclusion that stems from the section introductions and respective chapters is that the wide range of unresolved problems facing Stellenbosch will require responses that go way beyond what Stellenbosch Municipality can deal with on its own. Unless a meaningful partnership emerges between Stellenbosch Municipality, the Western Cape Provincial Government, Stellenbosch University, major business interests and the key civil society formations, a compelling vision of a future Stellenbosch and related concrete actions to achieve it will not be forthcoming. In this regard, the formation of Stellenbosch 360 as a voluntary forum that includes most of these stakeholders is clearly a step in the right direction.[1] While the majority of participants fully understand the business and social dynamics of the town, very few appreciate the fact that, unless approximately R1 billion is found to overcome the infrastructure backlogs created by over a decade of underinvestment by the municipality, their best efforts will not save the town from deteriorating. All of them stand to lose if future investments go elsewhere because the municipality cannot approve major new developments. Nor may Stellenbosch University hope to fully implement its master plan for expanding and integrating the campus.

That said, the scope of this book goes beyond the economics of municipal finance. It also points to the significance of a wide range of interventions pertaining to art, health, housing, biodiversity, food, and other issues that could make a difference. The gradual evolution of organised civil society groupings – such as the Social Cohesion movement, Stellenbosch 360, and the various local initiatives linked to the Informal Settlements Network (ISN) – suggests that spaces are opening up for individual and collective action on many of these issues.

At another – possibly deeper – level, this is also a book about Stellenbosch University's relationship with its home town and how it may foster and support a longer-term vision for the town. All the conclusions drawn by the authors point to the need for a new long-term vision for Stellenbosch, one that takes the positive from the past and builds on it, while charting a new course for the future. Stellenbosch is commonly referred to as a 'university town'. In fact, the Memorandum of Understanding (MOU) between Stellenbosch Municipality and Stellenbosch University specifically refers to the long-term vision of a 'sustainable university town' as the purpose of the collaboration between the two institutions. Signed on 31 July 2007, the MOU[2] marks a decisive shift in the relationship between the university and the town. A key sentence of the document states:

> *Stellenbosch University and Stellenbosch Municipality express herewith their Understanding: That they have a responsibility, individually and jointly, to address the challenges facing Stellenbosch and to work towards a better future for its inhabitants, temporary or permanent. [...] That they will pursue international interaction and initiatives to their mutual benefit.*

---

1   For more information, see http://www.stellenboschtourism.co.za/stellenbosch-360/about-stellenbosch-360
2   This document was framed and now hangs in the office of the Mayor.

The discussion about the relationship between universities and their urban contexts has a long and distinguished history (see Bender 1988). Bender distinguishes between the Anglo-American 'pastoral tradition' typified most clearly by Oxford and Cambridge, and the 'City University' tradition that "associates the university at its best with the city at its best" (Bender 1988:4). While the former was influenced by a profound anti-urbanism that took pride in consciously and deliberately disconnecting the polite, elite and 'worldly' culture of the university from the vulgarities of the town, the latter tradition – as in the case of London University and New York University – took the needs and context of the city (and, inevitably, also the funding provided by the great builders of these cities) as its point of departure.

So what is Stellenbosch University? What was it in the past? What is it becoming? Does it aspire to the Anglo-American pastoral tradition or the more embedded 'City University' tradition? Can it be both? What does the MOU with the town represent?

In addressing such questions Bender reflects on three universities, all of which were established within a decade of the year 1200, in Bologna, Paris and Oxford, respectively (Bender 1988). These institutions not only founded the modern global university movement, but also represented three distinct archetypal traditions that manifest themselves in various combinations in most modern universities. The University of Bologna effectively established the 'City University' tradition: organised in the same way as the guilds of the emerging commercial class, it became the university *in and of* the city of Bologna, its prestige rising and falling with the prosperity of the city. The University of Paris "was both in and not of the city *and* in and of the city" (Bender 1988:5). While it aspired to be intellectually independent of its context in order to engage in the disinterested pursuit of lofty ideas, its concerns with the material realities of everyday life were profoundly urban and not related to matters of court, the monastery, or conquest. In this sense, one may say that the Paris tradition is about being both local and global at the same time. On the other hand, Oxford University represents the archetypal pastoral campus: disdainful of the town and its ways, explicitly elitist and (in later centuries) obsessed with empire – the ultimate 'elsewhere'. This is of course unsurprising: unlike Paris and Bologna, Oxford was a small town that existed largely to serve the university, and not the other way round. The last thing its founders had in mind was that graduates should remain in Oxford when there was a whole world 'out there' to go and conquer.

When the then Rector of Stellenbosch University, Professor Chris Brink, initiated what he called his 'Town and Gown' strategy in 2005, he repeatedly opined that, as Rector, he did not have a relationship with the Mayor of the town.[3] As the town was legally deracialised and amalgamated after 1994, local political and managerial elites – who were not nearly as thoroughly networked into the university as their apartheid-era predecessors – came to power within the municipality. Indeed, there was a great deal of scepticism in these circles about the *bona fides* of the university, as well as genuine fears about its capacity to overwhelm the municipality's internal intellectual capacity to formulate plans and policies for a non-racial future. Unsurprisingly, the natural response was to view Stellenbosch University as the 'Oxford of Africa'. After all, like the town of Oxford, Stellenbosch (as town) is often seen from Stellenbosch University's perspective as defined by the very presence of the university.

After visits to other university towns in the world, Prof Brink came to the conclusion that Stellenbosch University should help found a new global movement of university towns. Underpinning this conviction was a realisation that there were places in the world where universities had become the economic drivers of local economies that had positioned themselves within a globalised knowledge economy. This represented two closely connected dynamics: the universities in these places had become intensely embedded in their respective locales, which, at the same time, positioned them within a globalised knowledge economy, where ideas had replaced labour and capital as the key drivers of economic development. The lines from the MOU between Stellenbosch University and Stellenbosch Municipality quoted above capture this twin dynamic: 'challenges facing Stellenbosch' and 'international interaction'. Is this local-global dynamic the contemporary manifestation of the Paris tradition in the Stellenbosch context? It is certainly not the Oxford tradition, and it is more than the Bologna tradition.

In fact, it may be worth considering whether there is an alternative – an 'African tradition'. As part of its initiative to establish a 'Pan-African University', the African Union commissioned a review of higher education on the African continent. The opening statement of the subsequent report defines the problem as follows:

---

3   During the course of 2012, Prof Andreas Van Wyk (who preceded Prof Brink as Rector) made a statement in a sub-committee meeting of the Rector-Mayor Forum to the effect that Prof Brink was not the first Rector to initiate working relations with the town, insisting that these relations had existed for some time already.

> *There is a growing concern about the quality and standard of higher education in Africa, particularly with respect to contribution to research and research capacity. Current pedagogies and incentive structures in African universities and higher institutions are discipline-based, focused on literacy and publications, organized into faculties with rigid boundaries and hence discourage collaboration, trans-disciplinary research and responsible innovation. Curriculum has been seldom reviewed in light of emerging development challenges in many countries. For Africa's quest to achieve socio-economic growth and for sustainable development to become a reality, it is imperative not only to produce high quality output from highly qualified professionals, but also to produce and adapt knowledge and technologies relevant to its development.*
>
> [Urama, Swilling & Acheampong 2012:5]

The statement suggests that the aspiration to replicate the Oxford tradition in Africa is – and always has been – inappropriate, not least because it is unaffordable. Using this central idea as its point of departure, the report calls for a shift that draws on the 'Paris' and 'Bologna' traditions and makes recommendations that are quite closely aligned to the ideals of Stellenbosch University's HOPE Project.

To come back now to this book and the suggestion that it should be read as more than just a book *about Stellenbosch*: it should also be read as an inflection point in the history of the relationship between the town and the university. The arguments and conclusions developed here mark a decisive intervention, showing how researchers from within the academy may collaborate to reflect on the challenges facing the town in a way that builds Stellenbosch University's contribution to the growing *corpus* of writing on the transdisciplinary co-production of solutions-oriented knowledge. Admittedly, this book is not in and of itself the product of a transdisciplinary methodology, but it does emerge from the construction of a new set of 'town and gown' relationships in the various sub-structures of the Rector-Mayor Forum that do provide a basis for ongoing transdisciplinary co-production of knowledge. In this sense, this book is therefore also *about Stellenbosch University*. It marks a particular moment in the evolution of a body of knowledge that is both about the material realities of the town within which the university is based and about the forging of a research practice that contributes to our global understanding of the connections between innovation, knowledge production and sustainability.

It is against this background that the conclusions contained in this book should be judged. They should be read as an opening dialogue that enriches an engaged 'glocal' tradition with a normative commitment to sustainability, social justice and sufficiency. As the arguments and conclusions of this book suggest, in practice this means invoking the local challenges that Stellenbosch University should address by virtue of it being both *in and of* the town, but also recognising that the university is part of a much wider global discourse about alternatives that are inseparable from the solutions that may be forged at the local level. The link is not just about looking after the university's 'backyard', while its real work lies beyond the town – as these chapters reveal (and following on Prof Johan Hattingh's introduction), it is within the local that we discover the enormous complexities that are elided from the abstractions of global solutions. Engaging the local is where we encounter the dangers that Hattingh warns of; yet, it is also the source of much of the inspiration that will be required to translate hope into the practical realities expected by future generations.

REFERENCES

Bender T. 1988. *The University and the City: from medieval origins to the present*. New York: Oxford University Press.

Urama K, Swilling M & Acheampong EN. 2012. *Role and contribution of research and postgraduate training in strengthening and sustaining the African Higher Education and Research Space* (AHERS). Commissioned by the Working Group on Higher Education (WGHE) of the Association for the Development of Education in Africa (ADEA). Nairobi: African Technology Policy Studies Network.

INDEX

A

abandoned spaces 65
active youth participation 336
Adopt-a-River programme 136
aerated waste water treatment plants 144
age distribution of population, 2010 (estimated) 339
agriculture
    challenges 195-196
    farmer case studies 196-200
    South African agricultural landscape 193
    Stellenbosch agricultural system 193-195
    sustainability and the polycrisis 192
    water use 207
        see also soil health
alien vegetation 133
alternative solar thermal applications 154-157
alternative waste water management systems 143-145
appropriate student living space, designing 64
automobility 173-174

B

'barefoot engineers' 336-337
behavioural models, limits to 305-306
behaviour change *see* human behaviour change
Biodiversity and Wine Initiative 223-224
biodynamic and organic farming practices 224
biogas digesters 156
biolytic systems 144
Black Economic Empowerment and the reduction of inequalities 88-89
Bloch, Ernst, principle of hope 15-17
Blue Drop Certification Programme 130-131
Breaking New Ground (BNG) plan 73
'builders' rubble, banned from landfills 119
business
    economy of Stellenbosch, overview 282-283
    global business community, changes in roles and responsibilities 283-285
    innovation system approach 285
    local economy, business profile 283
    micro-innovation (case study) 286
    potential to change at organisational level (case study) 286-287
    resilience and adaptability 288-289
    social-ecological systems framework 287-289
    strategies 234
    transformation towards sustainability 285-287
        see also economy and employment; Local Economic Development (LED); tourism

C

Cape Action for People and the Environment (C.A.P.E.) 221-223
Cape Winelands Biosphere Reserve 221, *222*
case studies
    Biodiversity and Wine Initiative (BWI) 199-200
    Boschendal 307
    cluster waste water management system 145-146
    Dwars River Valley 307-308
    Integrated Production of Wine (IPW) 199-200
    Lynedoch EcoVillage waste water management 145-146
    micro-innovation 286
    potential to change at organisational level 286-287
    Solms-Delta 307-308
    Spier 286-287
    Stellenbosch 199-200
    Swartland 196-199
    Technopark 286
    tourism 96
    transport 170-171
    'underperforming' schools 265-266
children and youth
    active youth participation 336
    age distribution of population, 2010 (estimated) *339*
    'barefoot engineers' 336-337
    community development 335
    constructive activity 335
    crafts for skills 336
    Early Childhood Development (ECD) 334-335
    education levels by population group, 2006 (estimated) *339*
    employment status by population group, 2011 (estimated) 340
    maternal and child health, improving 297
    natural spaces 335-336
    'nature first' in Early Childhood Development (ECD) 334
    nutrition and well-being 335
    occupational composition by population group, 2006 (estimated) 341
    of school age (10-18) 335-336
    section introduction 235
    vision for Stellenbosch 333-334
    vocational skills for meaningful work 336-337
    young adults (18-35) 336-337
      see also education
climate change, effect on ecological assets, 220
cluster waste water management system (case study) 145-146
community development and early childhood development 335
community involvement
    ecological services 228
    school governance 267
community participation in sustainable food systems 111-112
composting toilets 144
concentrated solar power (CSP) 155

congestion   161-163
congestion charging   170-171
Constitution of the Republic of South Africa, 1996, land tenure security   49
consumer participation in recycling project, encouraging   121
crafts for skills   336
cultural tourism   95
cycle lanes   *171*

D

domestic water use   206
double street refuse bins   119
dwelling unit density   *26*

E

ecological assets of Stellenbosch
   Cape Floral Kingdom   216-217
   described   216-218
   Fynbos   216-217
   interventions to protect   220-224, 228
   planning and ecology, reconciling   224-225, 228-229
   threats to   218-220
ecological services
   community involvement   228
   individual responsibility   228
   knowledge transfer   228
     see also ecological assets of Stellenbosch; river health; water quality
economy and employment   258-259
   see also business
economy of Stellenbosch, overview   282-283
ecosan systems   144
ecosystem degradation   7
education
   approaches to environmental education   318
   challenges and threats to achieving sustainability   316-317
   context in Stellenbosch   313-314
   crafts for skills   336
   Early Childhood Development (ECD)   334-335
   education levels by population group, 2006 (estimated)   *339*
   and environment   311-312
   environment and environmental problems   301-311
   environment and national curriculum   312-313
   partnerships between university and schools   314-316
   Rietenbosch Wetland Rehabilitation Programme (RWRP)   315-316
   schools, governance and community involvement   267
   School Water Analysis Programme (SWAP)   314-315, 316
   survey   263-267
   for a sustainable Stellenbosch   234, 317-318
   teacher-learner ratios   321
   'underperforming' schools (case study)   265-266
   vocational skills for meaningful work   336-337
     see also children and youth

education levels by population group, 2006 (estimated) *339*
employment status by population group, 2011 (estimated) 340
energy
    alternative solar thermal applications 154-157
    biogas digesters 156
    concentrated solar power (CSP) 155
    current system 151-152
    energy efficiency 152-153
    energy neutrality 157-158
    parabolic troughs and linear Fresnel systems 154-155
    photovoltaic panels 155-156, *157*
    solar water heaters 153-154
    strategies and technologies to transform system 152-157
    wind turbines, small 156, *157*
energy sector, lessons for waste water management 146-147
Enkanini (informal settlement) 69-71
environment
    education 311-312
    four dimensions of *311*
environmental (habitat) water needs 207-208
ethics of hope
    Christian ethics of hope, of Harvie, J 14-15
    HOPE Project 13
    ideology critique 16-17
    latency and tendency 16
    pedagogy of hope, of Paolo Freire 17-18
    principle of hope, of Ernst Bloch 15-17
    theology of hope, of Jürgen Moltman 14-15
Extension of Security of Tenure Act (ESTA), 1997 49-50, 53

F

farm workers and occupiers, their land tenure security 50-51
fire patterns in fynbos, unusual 219
food insecurity 7
food security
    possible interventions to increase 109-112
    status 104-106
food systems
    analysis of the Stellenbosch food system 106-108
    community participation 111-112
    conclusions about the current food system in Stellenbosch 108-109
    consumption *105*, 107-108, 109, 111
    distribution 106-107, 109, 111
    macro trends influencing 103-104
    production 101, 106, 108, 110
formalisation of informal settlements 72-73
Franschhoek, spatial planning 27
freight and congestion 161, 167-168
Freire, Paolo, pedagogy of hope 17-18

G

gated developments 62
Ghoema Route 96
global economic recession 6
global polycrisis 8
global warming 7
Green Drop Certification Programme 140
green waste, banned from landfills 119
groundwater 206

H

Harvie, J., and ethics of hope 14-15
health and wellness
    alcohol harm 295, 296
    approaches to improvement 296-298
    burden of disease 291-294
    drowning, reducing 298
    general health and wellness 298
    health services 291
    high body mass index 295
    injuries and violence, reducing 296-297
    interpersonal violence 294-295, 296
    maternal and child health, improving 297
    non-communicable chronic disease, reducing 297
    obesity 295
    premature death, causes 291-294
    risk factors 234, 294-295
    tobacco smoking 295, 297
    unsafe sex 294, 296
    unsafe water and poor sanitation 295
    water quality, unsafe 295
healthcare 262-263
heritage
    'African' experience in Stellenbosch 238
    Afrikaner middle class, rise of 238
    brown people of Stellenbosch 241
    conservationist lobby 239
    and culture 233
    Group Areas Act, forced removals 240-241
    historical touristic area 237-238
    inclusive, 'new' heritage 240-241
    material and discursive underpinnings 238-240
    Rupert, Anton 239-240
    self-representation 236-238
hope, philosophies 9-10
HOPE Project 13
housing
    inadequate 273
    low-income housing policies 24-25

housing
- RDP houses 72-73
- section introduction 23
- student living space, appropriate 64
- *see also* informal settlements

human behaviour change
- basic building blocks for innovation 305-306
- behavioural issues in creating sustainable environments, understanding 301-305
- Boschendal (case study) 307
- Dwars River Valley (case study) 307-308
- limits to behavioural models 305-306
- MINDSPACE 302-305
- social cohesion in the Dwars River Valley 307-308
- in Stellenbosch 308
- sustainability of interventions in the Dwars River Valley 307-308

I

ideology critique 16-17
inadequate housing, as risk to social cohesion 273
income, poverty and inequality 259-261
incrementalism *see* upgrading
industrial and technological water use 207
inequality 7, 272
informal settlements
- Breaking New Ground (BNG) plan 73
- Enkanini 69-71
- formalisation 72-73
- funding 77-79
- infrastructure alternatives, workable 79-81
- iShacks 80-81
- policy changes towards 73-75
- population in Stellenbosch 68-69
- RDP houses 72-73
- Slum Dwellers International (SDI) 75-77
- upgrading 73-80

Integrated Development Plans 37
Integrated Waste Management Plan, implementation 117-119
invasive alien plants 219
iShacks 80-81

K

Klapmuts, spatial planning 27

L

landfill
- management 119
- non-compliance 117, *118*
- redevelopment 122, *123*

landowners, and land tenure security issues 50
land reform 23

land tenure security on farmland
    constitutional and legislative context 49
    Constitution of the Republic of South Africa, 1996, 49
    Extension of Security of Tenure Act (ESTA), 1997 49-50, 53
    farm workers and occupiers 50-51
    historical background 49
    key challenges 53-54
    landowners 50
    local authorities 51
    municipal commonage 51-53
    role players 50-51
law enforcement and transport operations 166
linear Fresnel systems and parabolic troughs 154-155
Local Economic Development (LED)
    Black Economic Empowerment 88-89
    employment, income and social development 88
    enterprise mix 86
    factors and forces shaping 84-89
    future-orientated strategy, critical steps 90-92
    future priorities 93
    future unfolding of LED process 92-93
    gross domestic product 87-88
    holistic strategy, driving 89-91
    infrastructure 85
    introduction 23
    local business development 88
    location 85
    population growth 87
    progress with LED in Stellenbosch 86-89
    resources 85
    socio-political stability 86
    sustainability of development 89
    tourism 95-96
    viable economic sectors 85-86
loss and fragmentation of natural habitats 218-219
low-income housing policies 24-25
Lynedoch EcoVillage waste water management (case study) 145-146

M

material flows 7
microbial pathogens in water 136
microbial risk, sources *128*
Midlands Meander 96
Millennium Development Goals (MDGs) 13
Moltman, Jürgen, theology of hope 14-15
multi-barrier treatment of raw water 129
multimodality transport systems 170
municipal commonage 51-53

N

natural habitats, loss and fragmentation 218-219
non-motorised transport (NMT) 165-166

O

occupational composition by population group, 2006 (estimated) 341
oil peak 7
organic and biodynamic farming practices 224
overarching policy frameworks 37-38

P

parabolic troughs and linear Fresnel systems 154-155
past-productivist countryside 62-63
'Pay as you throw' system 119
photovoltaic panels 155-156, *157*
planetary boundaries 8
planning and ecology, reconciling 224-225
political leadership, unstable 273
political power in Stellenbosch 9
pollution from human activity 135-136
polycrisis
    agriculture and sustainability 192
    and food systems 103-104
population and demographic profile 257, *258*
population growth as factor in Local Economic Development (LED) 87
poverty and inequality
    economy and employment 258-259
    education 263-267
    governance and community involvement 267
    healthcare 262-263
    income, poverty and inequality 259-261
    measurements and trends of poverty in South Africa 256-257
    population and demographic profile 257, *258*
    poverty in South Africa, measurements and trends 256-257
    section introduction 233-234
    'underperforming' schools (case study) 265-266
principle of hope, of Ernst Bloch 15-17
public space and art
    Cosy for a Rhino project 250-253
    fragile common 244-245
    heritage practice of Solms-Delta 248-250
    Predators and Prey Urban Sculpture Tour of Dylan Lewis 245-248
    Safe house knitting project 250-253
    section introduction 233
public transport 166

R

RDP houses 72-73
recreational water use 208

Recycling at Source Project, implementation 119-121
Rietenbosch Wetland Rehabilitation Programme (RWRP) 315-316
river health 131-133, 136
road safety 168-169
Route 360 96
rural areas within Stellenbosch Municipality 27-28
rural gating, protecting the rural landscape against 65

S

sanitation system of Stellenbosch, current performance 141-142
septic tanks 143
shared space *see* public space and art
Slum Dwellers International (SDI) 75-77
small wind turbines 156, *157*
social cohesion
    action to achieve 277-279
    defined 269-270
    Dwars River Valley 307-308
    identity and unity 274-277
    inadequate housing 273
    inequality 272
    political leadership, unstable 273
    section introduction 234
    social cowesion 271
    ties of blood and place 275-276
    two worlds, keeping alive 271-273
    ubuntu 270-271
    unemployment 272
    unequal societies, unsustainability of 273-274
    vision 279
soil health
    agro-ecology 186-188
    complexity thinking 186-188
    concept 184-185
    local context 185-186
    measurement, difficulties 185
    soils in Stellenbosch 180-184
    Swartland 195-196
solar water heaters 153-154
South Africa, green economy policies 8
Spatial Development Framework (SDF) 2010 220-221
spatial planning
    Apartheid era 32-34
    democracy era, to 2000 34-36
    Development Facilitation Act (DFA) (Act 67 of 1995), summary of principles 45
    in first 250 years 28-32
    Franschhoek 27
    growth of town *29*
    introduction 23
    Klapmuts 27

spatial planning
    lessons for the future 43
    National Environmental Management Act (NEMA), 1998 46-47
    scenarios 38-42
    Stellenbosch Municipality, in 2012 24-28
    Stellenbosch Municipality (WC024) Spatial Development Framework, 2005 38-41
    Stellenbosch Spatial Development Framework, 2006 41-42
    transformed local government era, from 2000 37-42
sport
    Centre for Human Performance Sciences (CHPS) 327
    economic benefits 322-323
    expectations 323-326, 327-328
    in higher education 325
    history 234, 326
    hopes 328
    Matie Sport 327
    planning 329-330
    present situation 234, 326-327
    resources 328-329
    in a rural context 324-325
    Sports Science Department of Stellenbosch University 327
    Stellenbosch University Institutional Sports Plan 327
    Stellenbosch University Sport Performance Institute (SUSPI) 327
    in a university town 326
stability of political power 9
Stellenbosch and Environs Sub-Regional Plan, 1995 35-36
Stellenbosch Divisional Council Regional Development Scheme, 1967 32-34
Stellenbosch Guide Plan, 1988 34
Stellenbosch Spatial Development Framework (SDF) 60
studentification of urban spaces 61-62
student living space, appropriate 64
suburbs 25
surface water 205-206
sustainability
    agriculture and the polycrisis 192
    challenges and threats to achieving 316-317
sustainable environments, behavioural issues in creating 301-305
sustainable tourism 95-96

T
tariff charges 133-134
teacher-learner ratios in Stellenbocsh schools 321
technoburbia 66
theology of hope, of Jürgen Moltman 14-15
ties of blood and place 275-276
tourism
    case studies 96
    Local Economic Development (LED) 95-96
      see also heritage
townships
    Franschhoek 27
    Stellenbosch 25

transport
    automobility 173-174
    case studies 170-171
    congestion 161-163
    congestion charging 170-171
    current situation 160-166
    cycle lanes *171*
    freight and congestion 161, 167-168
    issues contributing to congestion and poor road safety 165-166
    law enforcement and transport operations 166
    long term solutions 168
    minibuses 171
    multimodality 170
    non-motorised transport (NMT) 165-166
    public 166
    rethinking, with a view to the future 167-169
    road safety 163-166, 168-169
    working hours 167

U
ubuntu 270-271
unemployment, as threat to social cohesion 272
unequal societies, unsustainability of 273-274
unstable political leadership 273
unusual fire patterns 219
upgrading of informal settlements 73-80
urban majority 7
urban spaces
    abandoned spaces 63, 65
    appropriate student living space, designing 64
    current urban sociospatial relations 61-63
    gated developments 62
    introduction 23
    meaningful urban space 60-61
    partitioning of space 63
    past-productivist countryside 62-63
    rural gating 65
    Stellenbosch context 57-60
    studentification 61-62
    sustainable urban space 64-66
    technoburbia 66

V
vocational skills for meaningful work 336-337

W
waste management (solid waste)
    Integrated Waste Management Plan, implementation 117-119
    landfill, non-compliance 117

waste management (solid waste)
	planning for the future 121-124
	Recycling at Source Project, implementation 119-121
	waste loop, closing 124
	waste minimisation 119-121
waste water management
	alternative systems 143-145
	central systems, background 139-141
	cluster system (case study) 145-146
	decentralised systems 143-145
	energy sector, lessons from 146-147
	Lynedoch EcoVillage (case study) 145-146
	projects in other developing counties, lessons from 147-148
	sanitation system of Stellenbosch, current performance 141-142
water quality
	deterioration, effect on ecosystems 219-220
	measurement 208-210
	pollution 210-211
	strategies for a sustainable future 211-213
	threats to 127-129
	unsafe water and poor sanitation, consequences for health and wellness 295
water resources, limited 131-134
water scarcity 101
water services
	Adopt-a-River programme 136
	alien vegetation 133
	Blue Drop Certification Programme 130-131
	challenges 131-136
	infrastructure, ageing and inadequate 134-135
	microbial monitoring 130
	microbial pathogens 136
	microbial risk *128*
	monitoring and management 127-131
	multi-barrier treatment of raw water 129
	operational monitoring 130
	pollution from human activity 135-136
	supplied by Stellenbosch Municipality 127
water supply
	resources 131-134, 204-206
	strategies for a sustainable future 211-213
	use 206-208
well-being *see* human behaviour change
wind turbines, small 156, *157*
Winelands Integrated Development Framework (WIDF), 2000 36
workable infrastructure alternatives for informal settlements 79-81
Working for Water (WfW) 223
working hours and traffic congestion 167

Y
young adults (18-35) 336-337
youth *see* children and youth

Acronyms

A

AADT	Average Annual Daily Traffic
ABET	Adult Basic Education and Training
AIDS	Acquired Immune Deficiency Syndrome
AMCHUD	African Ministerial Conference on Housing and Urban Development
ANC	African National Congress
ARC	Agricultural Research Centre
ARMCANZ	Agricultural and Resource Management Council of Australia and New Zealand
ASPO	Association for the Study of Peak Oil and Gas
AU	African Union
AWTS	Aerated Waste Water Treatment System

B

BBBEE	Broad-Based Black Economic Empowerment
BDAASA	Biodynamic Agriculture Association of South Africa
BEE	Black Economic Empowerment
BESTUFS	Best Urban Freight Solutions
BMI	Body mass index
BNG	Breaking New Ground
BRICS	Brazil, Russia, India, China and South Africa
BSDI	Boschendal Sustainable Development Initiative
BWI	Biodiversity and Wine Initiative

C

CAPS	Curriculum Assessment and Policy Statement
C.A.P.E	Cape Action for People and the Environment
CBD	Central business district
CDC	Community Day Centre
CED	Cape Education Department
CEO	Chief executive officer
CHPS	Centre for Human Performance Sciences
COPD	Chronic obstructive pulmonary disease
CORC	Community Organisation Resource Centre
CRSES	Centre for Renewable and Sustainable Energy Studies
CSA	Community-supported agriculture (initiative)
CSIR	Council for Scientific and Industrial Research
CSP	Concentrating solar power
CSVR	Centre for the Study of Violence and Reconciliation

D

DA	Democratic Alliance
DAC	Department of Arts and Culture
DAFF	Department of Agriculture, Forestry and Fisheries
DALYs	Disability-Adjusted Life Years
DANIDA	Danish International Development Agency
DAWC	Department of Agriculture, Western Cape
DBE	Department of Basic Education
DC	Direct current
DEA&DP	Department of Environmental Affairs and Development Planning
DEAT	Department of Environmental Affairs and Tourism
DET	Department of Education and Training
DFID	Department for International Development (United Kingdom)
DHS	Department of Human Settlements
DLA	Department of Land Affairs
DME	Department of Minerals and Energy
DMR	Department of Mineral Resources
DoE	Department of Education
DORA	Division of Revenue Act
DoT	Department of Tourism
DWA	Department of Water Affairs
DWAF	Department of Water Affairs and Forestry
DWQ	Drinking water quality

E

ECD	Early childhood development
EDC	Endocrine disrupting chemical
EEDSM	Energy efficiency and demand-site management
EEPUS	Environmental Education Programme University of Stellenbosch
EGSs	Ecosystem goods and services
EMF	Environmental Management Framework
EPA	Environmental Protection Agency
ESSP	Earth System Science Partnership
ESTA	Extension of Security of Tenure Act
ESTIF	European Solar Thermal Industry Federation
ETS	Evapotranspiration seepage
EU	European Union

F

FAO	Food and Agricultura Organisation
FAS	Foetal alcohol syndrome
FEDUP	Federation of the Urban Poor
FET	Further Education and Training
FMCG	Fast Moving Consumer Goods

G

GDP	Gross Domestic Product
GET	General Education and Training
GIS	Geographic information system
GR	Green Revolution
GREEN	Global Rivers Environmental Education Network
GRI	Global Reporting Initiative
GSDT	Greater Stellenbosch Development Trust
GVA	Gross Value Added
GVA-R	Regional Gross Value Added

H

HAWT	Horizontal axis wind turbines
HDI	Human Development Index
HE	Higher education
HIV	Human immunodeficiency virus
HOR	House of Representatives
HSL	Household subsistence level
HSP	Human settlement plan
HSRC	Human Sciences Research Council

I

IAASTD	International Assessment of Agricultural Knowledge, Science and Technology
iCoCo	Institute of Community Cohesion
ICT	Information and communication technology
IDC	Industrial Development Corporation
IDP	Integrated Development Plan
IEA	International Energy Agency
IFPRI	International Food Policy Research Institute
IIED	Internationl Institute for Environment and Development
IISD	International Institute for Sustainable Development
IJR	Institute for Justice and Reconciliation
IOC	International Olympic Committee
IPCC	Intergovernmental Panel on Climate Change
IPW	Integrated Production of Wine
IRB	International Rugby Board
IRP	Integrated Resource Plan
ISN	Informal Settlement Network

J

JSE	Johannesburg Stock Exchange

L

LAB	Local Action for Biodiversity
LED	Local economic development
LEI	Low external input
LP	Liquefied petroleum
LOLT	Language of learning and teaching
LSC	Local Sport Council
LTMS	Long-Term Mitigation Strategy

M

MA	Master of Arts
MAB	Man and the Biosphere (programme)
MDG	Millennium Development Goal
MEA	Millennium Ecosystem Assessment
MEC	Member of the Executive Council
MIG	Municipal Infrastructure Grant
MLL	Minimum living level
MOU	Memorandum of Understanding
MRC	Medical Research Council of South Africa
MRF	Material Recovery Facility
MSFM	Municipal Services and Financial Model

N

NAMC	National Agriculture Marketing Council
NATO	North Atlantic Treaty Organisation
NBRI	National Building Research Institute
NGO	Non-governmental organisation
NHMRC	National Health and Medical Research Council
NICUS	Nutrition Information Centre of the University of Stellenbosch
NMT	Non-motorised transport
NNP	New National Party
NPO	Non-Profit Organisation
NQ	National Quintiles
NRF	National Research Foundation
NSBA	National Spatial Biodiversity Assessment

O

OBE	Outcomes-Based Education

P

PCUs	Positive Chill Units
PDL	Poverty datum line
PES	Payments for Ecosystem Services
PPP	Public-Private Partnership

PSDF	Provincial Spatial Development Framework
PV	Photovoltaic

R

RADP	Recapitalisation and Development Programme
RDP	Reconstruction and Development Programme
RENAC	Renewables Academy
RNCS	Revised National Curriculum Statement
RTMC	Road Traffic Management Corporation
RWRP	Rietenbosch Wetland Rehabilitation Programme

S

SACOS	South African Council on Sport
SALC	Sussex Association of Local Councils
SANBI	South African National Biodiversity Institute
SAS	Stellenbosch Academy of Sport
SDC	Stellenbosch Divisional Council
SDF	Spatial Development Framework
SDI	Slum Dwellers International
SEED	Stellenbosch Entrepreneur and Enterprise Development
SI	Sustainability Institute
SIDA	Swedish International Development Agency
SISN	Sustainable Innovation Stellenbosch Network
SITT	Stellenbosch Infrastructure Task Team
SM	Stellenbosch Municipality
SMZ	Special Management Zone
SSRC	Stellenbosch Sport and Recreation Council
STTMP	Stellenbosch Town Transport Master Plan
SU	Stellenbosch University
SUSPI	Stellenbosch University Performance Institute
SWAP	Schools Water Analysis Programme
SWH	Solar water heater
SYI	Sport Youth Initiative

T

TCTA	Trans Caledon Tunnel Authority
TfL	Transport for London
TPOA	Technopark Owners Association

U

UISP	Upgrading of Informal Settlements Programme
UN	United Nations
UN DESA	United Nations Department of Economic and Social Affairs
UNDP	United Nations Development Programme
UNEP	United Nations Environmental Programme
UNESCO	United Nations Educational, Scientific and Cultural Organisation
UNFIP	United Nations Fund for International Partnerships
UN-HABITAT	United Nations Human Settlements Programme
UNIDO	United Nations Industrial Development Organisation
UNISA	University of South Africa
USB	University of Stellenbosch Business School
UV	Ultraviolet

V

VIW	Vertically Integrated Wetland (system)

W

WBCSD	World Business Council for Sustainable Development
WCED	Western Cape Education Department
WCPG	Western Cape Provincial Government
WCPT	Western Cape Provincial Treasury
WCRSC	Western Cape Regional Services Council
WDC	Winelands District Council
WESSA	Wildlife and Environment Society of South Africa
WfW	Working for Water
WHO	World Health Organisation
WI	Worldwatch Institute
WIDF	Winelands Integrated Development Framework
WSDP	Water Service Delivery Plan
WWF	World Wide Fund for Nature
WWQ	Waste Water Quality
WWTWs	Waste Water Treatment Works

www.ingramcontent.com/pod-product-compliance
Lightning Source LLC
Chambersburg PA
CBHW081327230426
43667CB00018B/2857